A Course in
Discrete Mathematical Structures

A Course in
Discrete Mathematical
Structures

L R Vermani
Kurukshetra University, India

Shalini Vermani
Apeejay School of Management, India

Imperial College Press

ICP

Published by

Imperial College Press
57 Shelton Street
Covent Garden
London WC2H 9HE

Distributed by

World Scientific Publishing Co. Pte. Ltd.
5 Toh Tuck Link, Singapore 596224
USA office: 27 Warren Street, Suite 401-402, Hackensack, NJ 07601
UK office: 57 Shelton Street, Covent Garden, London WC2H 9HE

British Library Cataloguing-in-Publication Data
A catalogue record for this book is available from the British Library.

A COURSE IN DISCRETE MATHEMATICAL STRUCTURES

ISBN-13 978-1-84816-696-7
ISBN-10 1-84816-696-6
ISBN-13 978-1-84816-707-0 (pbk)
ISBN-10 1-84816-707-5 (pbk)

Typeset by Stallion Press
Email: enquiries@stallionpress.com

Printed in Singapore.

To
Juhi, Parth & Siddharth

CONTENTS

PREFACE

This book presents a selection of topics from set theory, logic, combinatorics, graph theory and algebra (both linear and abstract), which we found to be essential and useful for students of mathematics, computer science and computer engineering. It is intended as a textbook for Bachelor and Master of Technology (computer engineering and information technology) and Master of Science (both in computer science and mathematics) levels. As a prerequisite the material presented requires a background in mathematics up to the end of secondary/high school level and a bit of college algebra. The book can be used for a two-semester course in discrete structures where, as for a one-semester course, some topics may be either omitted or not covered in full. The material included has been so chosen as it may be needed for any of the above-mentioned courses of many universities. Our aim in this book has been to present the material in a precise and readable manner, but giving complete logical arguments.

The authors, while teaching a course in discrete mathematics to the students of computer science and computer engineering at various institutes found that most books available on the subject either give too many topics and too much material or sometimes do not cover all the topics needed. The authors then prepared their own notes (separately) for this purpose, joined together to combine these into one set of notes and expanded the same to give the notes the shape of the present book. Care has been taken to include a large number of solved examples for the benefit of students not very sound in the background mathematics. Several unsolved exercises are included in each chapter (at the end of the chapter and also in the body text of the chapter). The presentation, throughout, is very simple and straightforward, but lucid. We expect that the book will be well received by students as well as teachers teaching the course and the ones using such materials in their study.

The term discrete structures/mathematics does not convey a fixed meaning and more topics could have been included here. But due to limitations of time and space, and other reasons, they have not made it into this title. We do intend to add (in a future edition) a chapter each on (a) fuzzy sets and fuzzy logic (necessary for artificial intelligence) and (b) finite automata. Although there are books available on these topics, an introductory chapter on each may be of help.

While teaching the course and preparing our notes, the books by (i) Kenneth H. Rosen (2003/2005), (ii) B. Kolman, R. C. Busby and S. C. Ross (2005), (iii) C. L. Liu (2000), (iv) L. L. Dornhoff and F. E. Hohn (1978), (v) R. Lidl and G. Pilz (1998), (vi) N. Jacobson (1964) were freely used. We would like to place on record our appreciation for these authors.

The authors would like to thank Avneesh Bhasin for typesetting the four chapters containing figures and diagrams. We would like to place on record our appreciation for our colleague Harish Chugh for helping us in converting the material to the format required. Without his help the project may perhaps have taken much longer to complete. The authors would like to thank the authorities of their respective institutes (i) Panipat Institute of Engineering and Technology, Samalkha and (ii) Apeejay Institute of Management, Dwarka, New Delhi, for providing the necessary facilities during the last stages of the project.

L. R. Vermani
Shalini Duggal

Chapter 1

SETS

One concept that pervades almost the whole of mathematics and many other disciplines is the notion of sets. George Cantor is considered the founder of set theory. In this chapter we study set theory and follow Cantor's original intuitive approach but also build it axiomatically at times. So our approach is a fine blend of intuitive ideas and the theory built on axioms.

1.1. Preliminaries

Intuitively speaking, a set is a well-defined collection of objects. For example, the collection of all the students of B.Tech/M.Tech computer engineering in the college is a set. There is one thing inherent in the statement, 'well-defined collection of objects' in the definition of a set. By this we mean that, given a set and some object, we should be able to decide whether the object is a member of the set under consideration, or not. Objects which form the set are all distinct (some authors do not insist on this but we shall do so). For example, all the chairs lying in this room do not form a set, for if we take chairs out and place them among other chairs, we cannot identify the chairs lying in this room. Objects in the set are called **members** or **elements** of the set. Now we will give a more formal definition of a set. Let P be a given property. A **set** is a collection of objects having the given property P. Sets are normally denoted by capital letters such as S, T, U, X, Y, A, B, C etc. and elements or objects or members of a set are denoted by lower case letters. If S is a set and x is an object in S, we write $x \in S$ while if x is not an element of S we write it as $x \notin S$. We may describe a set S by writing all the elements of S, if it were possible, and to enclose these within brackets. For example, (a) {Ram, Rahim, Krishan, Uma} is a set consisting of Ram, Rahim, Krishan and Uma; (b) {1, 3, 5, 7, 9} is the set consisting of the odd integers 1, 3, 5, 7 and 9; (c) {1, 2, 3, ..., 99, 100} is the set consisting of the integers from 1 to 100. Here the ellipsis after 3, and before 99 indicates the integers beyond 3 and up to 98; (d) {1, 3, 5, ..., 2n + 1, ...} is the set consisting of all odd positive integers; (e) {2, 4, 6, ..., 2n,...} is the set of all even positive integers. The sets in examples (b) to (e) above may also be expressed in the set builder notation or in a more formal manner as

(b)   {$n|n$ is an odd positive integer less than 10}
      = {$n|n$ is an integer, it is odd and $1 \leq n \leq 10$}
(c)   {$n|n$ is a positive integer less than 101}
      = {$n|n$ is an integer and $1 \leq n \leq 100$}
(d)   {$n|n$ is an odd positive integer}
      = {$2n + 1|n$ is a non-negative integer}
      = {$m|m = 2n + 1$, n a nonnegative integer}
(e)   {$n|n$ is an even positive integer}
      = {$2n|n$ is a positive integer}
      = {$m|m = 2n$, n a positive integer}.

A set which has no elements is called an **empty set** or a **null set** and is denoted by { } or φ. Empty sets may be obtained more formally by using a property P which is self-contradictory. For example, the set of all positive integers less than 0 or the set of all odd integers which are divisible by 2 or the set of all real numbers which are the roots of the polynomial $x^2 + 1$. In the set builder notation we may describe these sets as

$$\{n|n \text{ is a positive integer and } n < 0\}$$
$$\{n|n \text{ is an odd integer and 2 divides } n\}$$
$$\{a|a \text{ is a real number and } a^2 + 1 = 0\}.$$

We have not placed any restriction on the elements of a set. For example, we may have sets like

{Ram, Rahim, London, New Delhi, New York, 1, 2, 3}; or

{{a, b, c}, 1, 2, 3, 4, x, y, z} in which one of the elements is {a, b, c},

which itself is a set; or {$\varphi, \{\varphi\}$, {1, 2, 3}, 1, 2, 3}. Observe that {φ} is not an empty set but is a set having one element which is φ the empty set. Or {$a, \{a\}, \{\{a\}\}, \{\{\{a\}\}\}, b, \{b\}, \{\{b\}\}$}.

Recall that while defining a set we have not placed any order in which the elements of a set are listed. Thus {1, 2, 3}, {1, 3, 2} and {2, 1, 3} all represent the same set, the elements of which are 1, 2 and 3. Sets in which elements are written in a particular order are called ordered sets and will be studied later. We next establish notations for some well-known sets. Throughout, we write

N = The set of all natural numbers = $\{1, 2, 3, \ldots, n, \ldots\}$;
  which is also called the set of counting numbers;
Z^+= The set of all non-negative integers =$\{0, 1, 2, 3, \ldots, n, \ldots\}$;
  which is the same as the set N of natural numbers together with the number 0;
Z = The set of all integers = $\{0, \pm 1, \pm 2, \pm 3, \ldots\}$;
Q = The set of all rational numbers = $\{p/q \mid p, q \text{ integers}, q \neq 0\}$;
Q^*= The set of non-zero rational numbers
  = $\{p/q \mid p, q \text{ integers}, p \cdot q \neq 0\}$.

There are infinitely many numbers which are not rational numbers. Examples of such numbers are $\sqrt{2}, \sqrt{3}, \sqrt{5}, 3\sqrt{2}, \pi$ etc. These are real numbers, none of which is a rational number. Numbers which can be represented as points on the real line are called **real numbers**. There is a deep theorem in mathematics which says that between every two rational numbers there are infinitely many real numbers. We do not go into details about these numbers but we introduce the notations:

R = The set of all real numbers;
R^* = The set of all non-zero real numbers;
C = The set of all complex numbers = $\{a + ib \mid a, b \in R \text{ and } i = \sqrt{-1}\}$;
C^* = The set of all non-zero complex numbers
  = $\{a + ib \mid a, b \in R, i = \sqrt{-1} \text{ and } a \neq 0 \text{ or } b \neq 0\}$.

Corresponding to every complex number there is a point in the plane and corresponding to every point in the plane there is a complex number.

The number of elements in a set S is called the **order of the set** and is denoted by $o(S)$ or $|S|$. If the order of S is finite, we call S a **finite set** and otherwise we call it an **infinite set**. Also in this case we say that S is an infinite set. Observe that all sets N, Z, Q, R and C are infinite sets. The set of all the roots of the cubic $x^3 + 3x^2 + 8$ is a finite set of order 3. This is so because every polynomial of degree n has n roots. Given two sets A and B, we say that A is a **subset** of B if every element of A is in B and we express it by writing $A \subseteq B$. If A is a subset of B and there is an element $b \in B$ which is not in A, then A is called a **proper subset** of B and this situation is expressed by $A \subset B$. Two sets A and B are said to be equal and written as $A = B$ if A is a subset of B and B is a subset of A. It is clear that if A is a subset of B and B is a subset of C, then A is a subset of C. Also, every set is a subset of itself. Following are some examples.

1. The set of B.Tech computer engineering second-year students in this college is a subset of the set of all B.Tech second-year students in the college.
2. The set of all second-year engineering students in this college is a subset of the set of all engineering students in the college.
3. The set of all people in Delhi is a subset of the set of all people in India.
4. The set $\{1, 3, 5, 7, 9\}$ is a subset of the set $\{1, 2, 3, 4, 5, 6, 7, 8, 9\}$.
5. The set of all odd integers is a subset of the set of all integers.
6. The set of all even integers is a subset of the set of all integers.
7. The set of all integers which are multiples of 3 is not a subset of the set of all odd integers as 6 is a multiple of 3 but is not an odd integer.
8. The set N of all natural numbers is a subset of the set of all integers, the set Z of all integers is a subset of the set of all rational numbers, the set Q of all rational numbers is a subset of the set of all real numbers, the set of all real numbers is a subset of the set of all complex numbers.
9. Let n be a positive integer. The set $S = \{e^{2\pi ik/n} | 0 \leq k < n\}$ of all nth roots of unity is a subset of the set C of all complex numbers. Observe that, if for $k < l < n, e^{2\pi ik/n} = e^{2\pi il/n}$ then $e^{2\pi i(l-k)/n} = 1$ or $\cos 2\pi(l - k)/n + i \sin 2\pi(l - k)/n = 1$, which is possible only if $2\pi(l - k)/n$ is a multiple of 2π. However, this is not possible as $1 \leq l - k < n$. This proves that the set S has order n. We shall consider this set again when we discuss groups. Observe that S is a subset of the set R of all real numbers only when $n = 1$ or 2.
10. The set $\{1, -1\}$ is a subset of the set $\{1, -1, i, -i\}$ where $i = \sqrt{-1}$.
11. Let $M_2(R)$ denote the set of all square matrices of order 2 with real entries, $M_2(R)^*$ be the set of all nonsingular square matrices of order 2 with real entries and $M_2(R)^{**}$ be the set of all square matrices of order 2 with real entries which have determinant 1. Then $M_2(R)^{**}$ is a subset of $M_2(R)^*$ and $M_2(R)^*$ is a subset of $M_2(R)$.
12. The set of vowels in the English alphabet is a subset of the set of all English alphabets.

1.2. Algebra of Sets

Sets that we study are subsets of a certain larger set. This larger set of which every set under study is a subset is called the **universal set**. The universal set may change with context. For example, if we talk of sets of numbers, we may take the set C of complex numbers as the universal set. It may sometimes be enough to consider the set R of real numbers as the universal set. When we talked of students of B.Tech computer engineering second year in this college we take the set of all engineering students in the college as the universal set. For the set of all the people in Bombay, we may take the set of all people in India as the universal set. For the sets under consideration we will write X or E or U for the universal set. Let A, B be two sets. We define $A \backslash B$ or $A - B$ by $A \backslash B = \{x \in A | x \notin B\}$ which clearly is a subset of A. The set $X \backslash A = \{x \in X | x \notin A\}$ where X is the universal set is called the **complement** of A and is denoted by A' or A^c or $\bar{A}$.

We define **union** $A \cup B$ and **intersection** $A \cap B$ of two sets A, B by $A \cup B = \{x \in X | x \in A \text{ or } x \in B\}, A \cap B = \{x \in X | x \in A \text{ and } x \in B\}$.

It is clear from the definitions that A, B are subsets of $A \cup B$ while $A \cap B$ is a subset of A as well as B. The following are then clear:

(a) $\varphi' = X$, $X' = \varphi$ (b) $(A')' = A$
(c) If $A \subseteq B$, then $B' \subseteq A'$ (d) $A \cap A' = \phi$, $A \cup A' = X$.

Moreover,

$A \cup B = B \cup A$, $A \cap B = B \cap A$, $A \cup A = A$, $A \cap A = A$,
$A \cup X = X$ and $A \cap X = A$.

Also, if A is a subset of B, then $A \cup B = B$ and $A \cap B = A$. The converse is also true and we can have:

Theorem 1.1. For sets A and B, (a) $A \cup B = B$ if and only if $A \subseteq B$; (b) $A \cap B = A$ if and only if $A \subseteq B$.

Proof. Suppose that $A \subseteq B$. Let $x \in A \cup B$. Then $x \in A$ or $x \in B$. As $A \subseteq B, x \in A$ implies that $x \in B$. Then $x \in B$ or $x \in B$ i.e. $x \in B$. Hence $A \cup B \subseteq B$. Also $B \subseteq A \cup B$. Therefore $A \cup B = B$. On the other hand, if $A \cup B = B$, as $A \subseteq A \cup B$, we have $A \subseteq B$. Proof of (b) follows on similar lines.

Union and intersection of sets are related through laws known as distributive laws.

Theorem 1.2. For any sets A, B, C,

(a) $A \cap (B \cup C) = (A \cap B) \cup (A \cap C)$;
(b) $A \cup (B \cap C) = (A \cup B) \cap (A \cup C)$.

Proof. (a) Let $x \in A \cap (B \cup C)$. Then $x \in A$ and $x \in B \cup C$ or $x \in A$ and $(x \in B$ or $x \in C)$. This means that $(x \in A$ and $x \in B)$ or $(x \in A$ and

$x \in C$) i.e. $x \in A \cap B$ or $x \in A \cap C$. Thus $x \in (A \cap B) \cup (A \cap C)$. This proves that

$$A \cap (B \cup C) \subseteq (A \cap B) \cup (A \cap C). \qquad (1.1)$$

Next, let $x \in (A \cap B) \cup (A \cap C)$. Then $x \in A \cap B$ or $x \in A \cap C$ i.e. ($x \in A$ and $x \in B$) or ($x \in A$ and $x \in C$).

This is equivalent to saying that $x \in A$ and ($x \in B$ or $x \in C$). i.e. $x \in A$ and $x \in B \cup C$. Therefore $x \in A \cap (B \cup C)$. This proves that

$$(A \cap B) \cup (A \cap C) \subseteq A \cap (B \cup C). \qquad (1.2)$$

Combining (1.1) and (1.2) gives

$$A \cap (B \cup C) = (A \cap B) \cup (A \cap C).$$

(b) Since $B \cap C \subseteq B$ and $B \cap C \subseteq C$,

$$A \cup (B \cap C) \subseteq A \cup B \quad \text{and} \quad A \cup (B \cap C) \subseteq A \cup C.$$

Combining these we get

$$A \cup (B \cap C) \subseteq (A \cup B) \cap (A \cup C).$$

For the reverse inclusion, let $x \in (A \cup B) \cap (A \cup C)$. Then

$$x \in A \cup B \quad \text{and} \quad x \in A \cup C$$

or

$$(x \in A \text{ or } x \in B) \quad \text{and} \quad (x \in A \text{ or } x \in C).$$

This implies that $x \in A$ or $\{x \in B$ and $x \in C\}$ i.e. $x \in A$ or $x \in B \cap C$ which together imply that $x \in A \cup (B \cap C)$. This proves that

$$(A \cup B) \cap (A \cup C) \subseteq A \cup (B \cap C).$$

Hence $A \cup (B \cap C) = (A \cup B) \cap (A \cup C)$.

Theorem 1.3. For any sets A and B,
   (a) $A \backslash B = A \cap B'$,   (b) $(A \cup B)' = A' \cap B'$,   (c) $(A \cap B)' = A' \cup B'$.

Proof. (a) $A \backslash B = \{x \in A \mid x \notin B\} = \{x \in A \mid x \in B'\} = A \cap B'$.

(b) Let $x \in (A \cup B)'$. Then $x \notin A \cup B$ and so $x \notin A$ and $x \notin B$ which is the same thing as saying that $x \in A'$ and $x \in B'$. This shows that $x \in A' \cap B'$. Thus $(A \cup B)' \subseteq A' \cap B'$.

Next, let $x \in A' \cap B'$. Then $x \in A'$ and $x \in B'$ which is equivalent to saying that $x \notin A$ and $x \notin B$. Therefore $x \notin A \cup B$ or that $x \in (A \cup B)'$. Hence $A' \cap B' \subseteq (A \cup B)'$.

Combining the two inclusions, we get $(A \cup B)' = A' \cap B'$.

(c) Let $x \in (A \cap B)'$. Then $x \notin A \cap B$. Therefore either $x \notin A$ or $x \notin B$ which implies that $x \in A'$ or $x \in B'$ i.e. $x \in A' \cup B'$. Thus $(A \cap B)' \subseteq A' \cup B'$. On the other hand, if $x \in A' \cup B'$, then $x \in A'$ or $x \in B'$ i.e. $x \notin A$ or $x \notin B$. This shows that $x \notin A \cap B$ or that $x \in (A \cap B)'$. This proves that $A' \cup B' \subseteq (A \cap B)'$. Combining it with the reverse inclusion already proved, we get $(A \cap B)' = A' \cup B'$.

The two results as at (b) and (c) above are called **De Morgan's Laws.**

Example 1.1. For any sets A, B, C prove that

(a) $A \cap (B \backslash C) = (A \cap B) \backslash (A \cap C)$,
(b) $A \cup (B \backslash C) = (A \cup B) \backslash (C \backslash A)$,
(c) $A \backslash B \subseteq B'$,
(d) $(A \backslash B) \backslash C = A \backslash (B \cup C)$.

Solution. Since $A \backslash B = A \cap B'$ (c) follows.

(a)
$$
\begin{aligned}
(A \cap B) \backslash (A \cap C) &= (A \cap B) \cap (A \cap C)' \\
&= (A \cap B) \cap (A' \cup C') \\
&= (A \cap B \cap A') \cup (A \cap B \cap C') \\
&= A \cap B \cap C' \\
&= A \cap (B \cap C') \\
&= A \cap (B \backslash C).
\end{aligned}
$$

(b)
$$
\begin{aligned}
(A \cup B) \backslash (C \backslash A) &= (A \cup B) \backslash (C \cap A') \\
&= (A \cup B) \cap (C \cap A')' \\
&= (A \cup B) \cap (C' \cup (A')') \\
&= (A \cup B) \cap (C' \cup A) \\
&= (A \cap (C' \cup A)) \cup (B \cap (C' \cup A)) \\
&= A \cup ((B \cap C') \cup (B \cap A)) \\
&= (A \cup (B \cap A)) \cup (B \cap C') \\
&= A \cup (B \cap C') = A \cup (B \backslash C)
\end{aligned}
$$

which completes the proof of (b).

(d) $(A \backslash B) \backslash C = (A \backslash B) \cap C' = (A \cap B') \cap C' = A \cap (B' \cap C')$

$$= A \cap (B \cup C)' = A \backslash (B \cup C).$$

Let A, B be two sets. By the **symmetric difference** of the sets A, B we mean the set $(A \backslash B) \cup (B \backslash A)$ and it is denoted by $A \sim B$ or $A \oplus B$. Now

$$A \oplus B = (A \backslash B) \cup (B \backslash A) = (A \cap B') \cup (B \cap A')$$

$$= ((A \cap B') \cup B) \cap ((A \cap B') \cup A')$$

$$= ((A \cup B) \cap (B' \cup B)) \cap ((A \cup A') \cap (B' \cup A'))$$

$$= ((A \cup B) \cap X) \cap (X \cap (A \cap B)')$$

$$= (A \cup B) \cap (A \cap B)' = (A \cup B) \backslash (A \cap B).$$

Thus the symmetric difference $A \oplus B$ consists of all elements which are in A or in B but not in both.

Two sets A and B are said to be **mutually disjoint** or **mutually exclusive** if they do not have any element in common, i.e. if $A \cap B = \varphi$. Observe that if A, B are mutually exclusive sets, then the symmetric difference $A \oplus B$ is just $A \cup B$. On the other hand if $A \subseteq B$, then $A \backslash B = \phi$ and $A \oplus B = (A \cup B) \backslash (A \cap B) = B \backslash A = B \cap A'$.

Example 1.2. For any sets $A, B, C, A \cap (B \oplus C) = (A \cap B) \oplus (A \cap C)$.

Solution.

$(A \cap B) \oplus (A \cap C)$

$$= ((A \cap B) \backslash (A \cap C)) \cup ((A \cap C) \backslash (A \cap B))$$

$$= ((A \cap B) \cap (A \cap C)') \cup ((A \cap C) \cap (A \cap B)')$$

$$= ((A \cap B) \cap (A' \cup C')) \cup ((A \cap C) \cap (A' \cup B'))$$

$$= ((A \cap B \cap A') \cup (A \cap B \cap C')) \cup ((A \cap C \cap A') \cup (A \cap C \cap B'))$$

$$= (A \cap B \cap C') \cup (A \cap C \cap B')$$

$$= (A \cap (B \cap C')) \cup (A \cap (C \cap B'))$$

$$= A \cap ((B \cap C') \cup (C \cap B'))$$

$$= A \cap ((B \backslash C) \cup (C \backslash B)) = A \cap (B \oplus C).$$

Example 1.3. Let A, B, C be any sets. Given that $A \cap C \subseteq B \cap C$ and $A \cap C' \subseteq B \cap C'$, prove that $A \subseteq B$.

Solution. Let X be the universal set. It follows from the definition of the complement of a set that $X = C \cup C'$. Now

$$A = A \cap X = A \cap (C \cup C') = (A \cap C) \cup (A \cap C') \subseteq (B \cap C) \cup (B \cap C')$$

$$= B \cap (C \cup C') = B \cap X = B.$$

Thus, $A \subseteq B$.

Using the above we can deduce that if A, B, C are sets with $A \cap B = A \cap C$ and $A' \cap B = A' \cap C$, then $B = C$.

Exercise 1.1. We have used in some of the proofs earlier that if A, B, C, D are sets such that $A \subseteq B$ and $C \subseteq D$, then $A \cap C \subseteq B \cap D$. Give a formal proof of this observation.

In view of the above, if A, B, C and D are sets such that $A \subsetneq B$ and $C \subsetneq D$, then $A \cap C \subseteq B \cap D$ and $A \cup C \subseteq B \cup D$. It is not essential that $A \cap C \subsetneq B \cap D$ and $A \cup C \subsetneq B \cup D$.

For example, let $A = \{1, 2, 3\}, B = \{1, 2, 3, 5, 7\}, C = \{1, 2, 4\}$ and $D = \{1, 2, 4, 6\}$. Then $A \subsetneq B, C \subsetneq D$ but $A \cap C = \{1, 2\} = B \cap D$.

This example also shows that $A \cap C = B \cap C$ but $A \neq B$.

For the observation about union, we take $A = \{1, 2, 3\}, B = \{1, 2, 3, 4\}$, $C = \{1, 2, 4\}$ and $D = B$. Then $A \subsetneq B, C \subsetneq D$ and $A \cup C = \{1, 2, 3, 4\} = B \cup D$. This example also shows that $A \cup B = A \cup C$ while $B \neq C$.

1.3. Venn Diagrams

Sometimes it is convenient to visualize results about unions, intersections, complementation and their combinations using two-dimensional figures or diagrams. In this pictorial representation, universal set is represented by a rectangle and sets are represented by circular, elliptic or any curved regions. If A, B are any two sets then their intersection and union are represented by the shaded region in Figs. 1.1 and 1.2 below while the complement A' of A is given by the shaded region in Fig. 1.3, $A \backslash B$ is given by Fig. 1.4 and the symmetric difference $A \sim B = A \oplus B$ is given by Fig. 1.5.

Exercise 1.2. If A, B, C are any sets indicate using Venn diagrams the proofs of the distributive laws

(a) $A \cap (B \cup C) = (A \cap B) \cup (A \cap C)$;
(b) $A \cup (B \cap C) = (A \cup B) \cap (A \cup C)$.

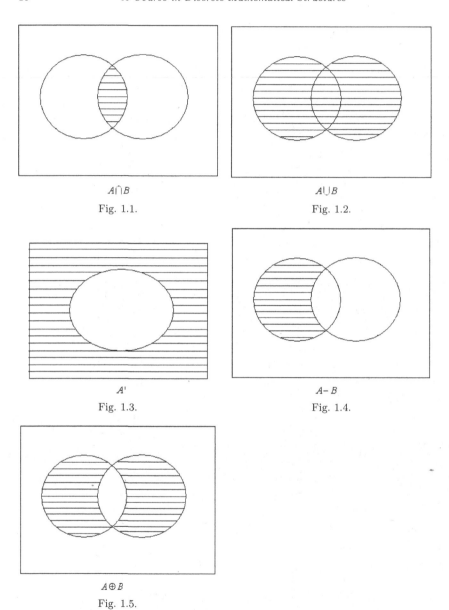

$A \cap B$

Fig. 1.1.

$A \cup B$

Fig. 1.2.

A'

Fig. 1.3.

$A - B$

Fig. 1.4.

$A \oplus B$

Fig. 1.5.

1.4. Power Set

We will now consider power set of a set and binary relations on sets. Let S be a non-empty set. The collection of all subsets of S is called the **power set** of S and is denoted by $\mathfrak{P}(S)$. For example, if

(a) $S = \{1\}$, then $\mathfrak{P}(S) = \{\phi, S\}$;
(b) $S = \{1, 2\}$, then $\mathfrak{P}(S) = \{\phi, \{1\}, \{2\}, S\}$;
(c) $S = \{1, 2, 3\}$, then

$$\mathfrak{P}(S) = \{\phi, \{1\}, \{2\}, \{3\}, \{1,2\}, \{1,3\}, \{2,3\}, S\}.$$

In case (a) the number of elements in $\mathfrak{P}(S)$ is $2 = 2^1$; in case (b) the number of elements in $\mathfrak{P}(S)$ is $4 = 2^2$; and in case (c), the number of elements in $\mathfrak{P}(S)$ is $8 = 2^3$.

We shall prove later that if S is a finite set of order n, then $o(\mathfrak{P}(S)) = 2^n$. Binary relations on sets shall also be considered later.

1.5. Countable Sets

Let A, B be two non-empty sets. By a **map** f from A to B written as f: $A \to B$ we mean a rule or a law which associates to every element of A a unique element of B. The unique element of B which is associated with an element a of A is denoted by $f(a)$ and is called the f-image of a or the image of a under f. For example, if $A = B = R$, the set of real numbers, we may define $f: R \to R$ by $f(a) = |a|$, $a \in R$, where $|a|$ denotes the magnitude of a. Recall that $|a| = a$ if $a \geq 0$ and $|a| = -a$ if $a < 0$.

We may consider some other maps $f : R \to R$ as

(a) $f(a) = -a, a \in R$,
(b) $f(a) = a^2, a \in R$,
(c) $f(a) = e^a, a \in R$,
(d) $f(a) = 2a + 3, a \in R$.

Observe that we have not defined any map like $f(a) = 1/a + 2$ or $1/a - 1$ etc. because in the first case the right-hand side is not defined for $a = -2$ while it is not defined for $a = 1$ in the second case. We have also not considered things like $f(a) = \sqrt{a}$ for if $a < 0$, then $\sqrt{a}$ is not a real number. However, (e) if A is the set of all positive real numbers and B is the set of all real numbers in the interval $(0, 1)$, we may define $f(a) = 1/a + 2$, $a \in A$. (f) Let $A = \{1, 2, 3\}$, $B = \{a, b\}$. Then the assignments in Figs. 1.6–1.8 below are maps while those in Figs. 1.9 and 1.10 below are not maps.

(g) Let N be the set of all natural numbers and Z be the set of all integers. Then

   (i) $f : N \to N$ given by $f(n) = 2n$, $n \in N$;
   (ii) $f : Z \to Z$ given by $f(n) = 2n$, $n \in Z$

are maps.

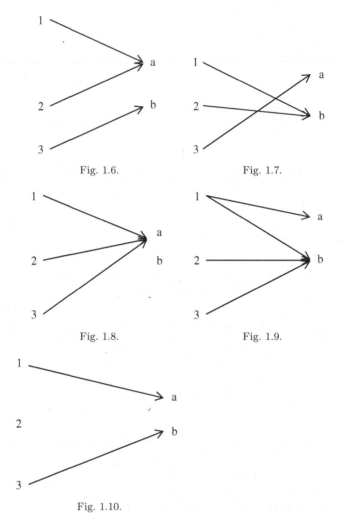

Fig. 1.6.

Fig. 1.7.

Fig. 1.8.

Fig. 1.9.

Fig. 1.10.

(h) $f : C \to R$ given by $f(a + ib) = a^2 + b^2$, $a, b \in R$, is a map.

(i) Let $M_2(R)$ be the set of all square matrices of order 2 over R.
   Then $f : M_2(R) \to R$ defined by $f(A) = det A$, $A \in M_2(R)$ is a map.

(j) $f : R \to C$ defined by $f(\alpha) = e^{i\alpha} = \cos\alpha + i\sin\alpha$, $\alpha \in R$, is also a map.

Let A, B be non-empty sets. A map $f : A \to B$ is called

(1) **injective** or **one–one** or **monomorphic** if for $a_1, a_2 \in A, a_1 \neq a_2$ implies $f(a_1) \neq f(a_2)$. Equivalently, if $f(a_1) = f(a_2), a_1, a_2 \in A$, then $a_1 = a_2$;

(2) **surjective** or **onto** or **epimorphic** if for every $b \in B$, there exists an $a \in A$ such that $f(a) = b$ i.e. every element of B is the image of an element of A;

(3) **bijective** or **isomorphic** if f is both one–one and onto.

Example 1.4. The maps in examples (a), (d), (e) above are bijective, the maps in examples (c), (g)((i) and (ii)) are injective but not surjective while the maps in (b), (h), (j) are neither injective nor surjective. The map in (i) is surjective but not injective.

Example 1.5. Define a map $f : N \to Z$ by $f(2k+1) = k+1, k \geq 0$ and $f(2k) = -k, k \geq 0$. Since $f(2k+1) > 0$ while $f(2k) \leq 0$, for no k and l, $f(2k+1) = f(2l)$. Also $f(2k+1) = f(2l+1)$ implies that $k+1 = l+1$ which gives $k = l$ and $f(2k) = f(2l)$ shows that $-k = -l$ or that $k = l$. Thus the map f defined is one–one. For any positive integer $k, k = f(2k-1)$ and $-k = f(2k)$.) Hence the map f is onto as well. Thus the map f is bijective.

Definition. Given two sets P and Q, we say that there is a **one-to-one correspondence** between the elements of P and the elements of Q if it is possible to pair off the elements in P and Q such that every element in P is paired off with a distinct element of Q and vice versa, i.e. every element of P is associated with a unique element of Q and every element of Q is associated with exactly one element of P. This is equivalent to saying (in precise formal language) that there exists a map $f : P \to Q$ which is bijective.

Definition. A set S is said to be **countable** if it is finite or the elements of S are in one-to-one correspondence with the elements of N — the set of natural numbers. Sometimes a set S, the elements of which are in one-to-one correspondence with the elements of N, is called **denumerable**. Then a set S is countable if either it is finite or denumerable. Thus for infinite sets there is no difference between denumerable and countable sets. A set which is not countable is called **uncountable**.

Theorem 1.4. Every infinite subset of a denumerable set is denumerable.

Proof. Let A be a denumerable set and B be an infinite subset of A. Let $f : N \to A$ be a map which is one–one and onto. Then the elements of A are $f(1), f(2), f(3), \ldots$ and, therefore, the elements of B look like $f(i_1), f(i_2), \ldots$ for $i_1 < i_2 < i_3 < \ldots$. Define a map $g : N \to B$ by $g(n) = f(i_n), n \in N$. By the very definition g is onto. Suppose that $g(n) = g(m)$.

Then $f(i_m) = f(i_n)$ which implies that $i_m = i_n$, the map f being one–one. However, for $m \neq n, i_m \neq i_n$, by our choice. Therefore $i_m = i_n$ implies that $m = n$. Hence the map g is one–one as well. This proves that B is denumerable.

Theorem 1.5. The set $Z^+ \times Z^+ = \{(m, n) | m, n \in Z^+\}$ of all ordered pairs $(m, n), m, n \in Z^+$ is denumerable.

Proof. Observe that a set is denumerable if the elements of the set can be listed in a row so that the elements can be called the first, the second, the third, etc. Keeping this in mind, we list the elements of $Z^+ \times Z^+$ as follows:

$$(0, 0), (0, 1), (1, 0), (0, 2), (1, 1), (2, 0), (0, 3), (1, 2),$$

$$(2, 1), (3, 0), (0, 4), (1, 3), (2, 2), (3, 1), (4, 0), \ldots$$

i.e. we first list the pair $(0, 0)$, then all pairs the sum of the entries of which is 1 starting with the pair having first entry 0, then all pairs (a, b) with $a + b = 2$, starting with $(0, 2)$, then all pairs (a, b) with $a + b = 3$, then $a + b = 4$, and so on every time the first pair being $(0, a + b)$. The above fashion lists all the elements of $Z^+ \times Z^+$ showing that the set $Z^+ \times Z^+$ is denumerable.

A careful examination of the first few terms of the above listing of elements of $Z^+ \times Z^+$ suggests that there may be defined a map $f : Z^+ \times Z^+ \to N$ by $f(m, n) = (1/2)(m + n)(m + n + 1) + m + 1, m, n \in Z^+$ which gives the above listing.

Exercise 1.3. Prove that the map $f : Z^+ \times Z^+ \to N$ defined above is injective. (Hint: Let $(a, b), (c, d) \in Z^+ \times Z^+$ and suppose that $f(a, b) = f(c, d)$. Consider four possible cases, namely:

(a) $a \neq c, b \neq d, a + b \neq c + d$.
(b) $a \neq c, b \neq d$ but $a + b = c + d$.
(c) $a = c, b \neq d$.
(d) $a \neq c, b = d$.

Now $f(a, b) = f(c, d)$ implies that $(a + b)^2 + 3a + b = (c + d)^2 + 3c + d$ which ultimately leads to $(a + b + c + d + 1)(a - c + b - d) + 2(a - c) = 0$.

This implies that $a - c = km$ and $a - c + b - d = -k$ or $-2k$ for some positive integer k. The second alternative leads to $a + b + c + d + 1 = m < a - c + 1$ which is obviously not possible. In the case of first alternative, $a + b + c + d + 1 = 2(a - c)/k$. For $k > 1$, this again leads to an obvious contradiction, while for $k = 1$, this leads to $a = b+3c+d+1$ which again can be shown to lead to a contradiction. In case $(b-d)/(a-c)$ is an integer, we must have $1 + (b-d)/(a-c) = -1$ and $a + b + c + d + 1 = 2$. This implies that three of a, b, c, d are zero and one of these is 1. This is a contradiction to case (i). Therefore, in case (i) it follows that $f(a,b) = f(c,d)$ cannot happen (Cases (ii), (iii) and (iv) are much simpler than this case).

As a consequence of this result it follows that $Z^+ \times Z^+$ may be identified with a subset of N and so is denumerable. Also the set $N \times N$ being an infinite subset of the denumerable set $Z^+ \times Z^+$ is denumerable The set Z^+ is also denumerable in view of the map $f : Z^+ \to \mathbb{N}$ given by $f(a) = a + 1$, $a \in \mathbb{Z}^+$ being bijective.

Theorem 1.6. The set Q of all rational numbers is denumerable.

Proof. We can prove this result in a few steps.

1. Consider the subset $A = \{(m,n) \mid m, n \in N$ and $g.c.d\ (m,n) = 1\}$. This is an infinite subset of the set $N \times N$. Since $N \times N$ is denumerable, A is denumerable. Let Q^+ be the set of all positive rational numbers. Define a map $f : A \to Q^+$ by $f(m,n) = m/n$, $(m,n) \in N \times N$. Since every positive rational number is of the form p/q, where p, q are positive integers and $g.c.d(p,q) = 1$, therefore $p/q = f(p,q)$. Thus f is onto. In fact, every positive rational number can be uniquely expressed in the p/q, p, q positive integers with $g.c.d.(p,q) = 1$, the map f is one–one as well. Hence Q^+ is denumerable.

2. We have proved in Example 1.5 that the map $f : N \to Z$ given by $f(m) = -m/2$ if m is even and $(m + 1)/2$ if m is odd is bijective. Therefore, the set Z of integers is denumerable.

   Define a map $g : N \times N \to Z \times Z$ by $g(m,n) = (f(m), f(n))$, m, $n \in N$. Let m, n, m', $n' \in N$ and suppose that $g(m,n) = g(m',n')$. Then $(f(m), f(n)) = (f(m'), f(n'))$ which implies that $f(m) = f(m')$ and $f(n) = f(n')$. The map f being injective, it follows that $m = m'$, $n = n'$ or that $(m,n) = (m',n')$. Hence g is one–one. Let a, $b \in Z$. The map f being onto, there exist m, $n \in N$ such that $a = f(m)$ and $b = f(n)$. Then $(a,b) = ((f(m), f(n)) = g(m,n)$. Hence g is onto as well. Thus g is a bijective map. The set $N \times N$ being denumerable, it follows that $Z \times Z$ is denumerable.

3. Let $B = \{(m,n) \mid m, n \in Z, n \neq 0$ and $g.c.d(m,n) = 1\}$. Then B is an infinite subset of the denumerable set $Z \times Z$ and, so B is denumerable. Now define a map $h : B \to Q$, where Q is the set of all

rational numbers, by $h(m, n) = m/n$, $(m, n) \in B$. Since every rational number can be uniquely written in the form m/n, where $m, n \in Z$, $n \neq 0$ and $g.c.d.(m, n) = 1$, the map h is one–one and onto. The set B being denumerable, it follows that the set Q is denumerable. We next describe a practical approach to enlist all the positive rational numbers.

List all the positive rational numbers with denominator 1 in the first row, the numbers with denominator 2 in the second row, the numbers with denominator 3 in the third row and so on. The elements in each row being in one-to-one correspondence with positive integers are denumerable. Similarly the elements in every column are denumerable. Thus the table is:

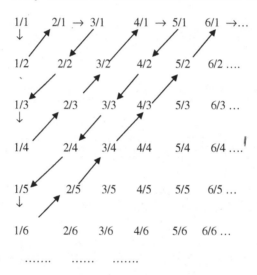

Then start listing the positive rational numbers as indicated. This procedure lists all the positive rational numbers. Observe that in the listing there will appear numbers which have been listed previously. All such numbers will be removed from the listing or a number which has occurred earlier again appears is omitted (i.e. the duplication is removed by omitting a number which occurs the second time). Observe that in the above procedure we first list the number 1, then all numbers p/q with $p + q = 3$, then all numbers p/q with $p + q = 4$, then numbers p/q with $p + q = 5$ and so on omitting from the listing a number which has already appeared in the listing.

Theorem 1.7. The set of all real numbers is uncountable.

Proof. If the set R of real numbers is denumerable, then every infinite subset of R is denumerable. However, we prove that the set of all real numbers in the interval $(1, 2)$ is not denumerable. From this it will then follow that the set of real numbers is uncountable. We prove by contradiction that the set of real numbers in the interval $(1, 2)$ is uncountable. So, suppose that the set of all real numbers in $(1, 2)$ is denumerable. Let these numbers be listed as $a_1, a_2, a_3, \ldots, a_i, \ldots$.

Let the decimal representations of these numbers be

$$a_1 = 1.a_{11} \quad a_{12} \quad a_{13} \quad a_{14} \quad \cdots$$

$$a_2 = 1.a_{21} \quad a_{22} \quad a_{23} \quad a_{24} \quad \cdots$$

$$a_3 = 1.a_{31} \quad a_{32} \quad a_{33} \quad a_{34} \quad \cdots$$

$$\cdots \qquad \cdots \qquad \cdots$$

$$a_i = 1.a_{i1} \quad a_{i2} \quad a_{i3} \quad a_{i4} \quad \cdots$$

$$\cdots \qquad \cdots \qquad \cdots$$

Then a_{ij} all belong to the set $\{0, 1, 2, \ldots, 9\}$. Consider a real number b with decimal representation $b = 1.b_1 \, b_2 \, b_3 \ldots$ where the decimal digits $b_1, b_2, b_3, \ldots$ are defined by $b_i = 3$, if $a_{ii} \neq 3$ i.e.

$$a_{ii} \in \{0, 1, 2, 4, 5, 6, 7, 8, 9\} \quad \text{and} \quad b_i = 4 \text{ if } a_{ii} = 3.$$

Thus $b_i \neq a_{ii}$ for any i or b_i is not the ith decimal digit of a_i for any i. Two numbers are equal if and only if the corresponding entries in the decimal representations of the two are equal. It follows that the number b does not appear in the above listing of the real numbers in the interval $(1, 2)$. However, $1 < b < 2$ so that $b \in (1, 2)$. This is a contradiction and it follows that the set of real numbers in the interval $(1, 2)$ is not denumerable and, hence, R is uncountable.

Remark: The procedure given for proving that the set of positive rational numbers is denumerable can also be used to prove that *a denumerable union of denumerable sets is denumerable.*

1.6. Some Special Maps (Functions)

In this section we give a brief description of some functions defined on real numbers which are quite useful and may appear frequently. These include, among others, polynomial functions, logarithmic functions and exponential functions. In a polynomial function every real number x is associated with a polynomial in x such as $f(x) = 2x + 1$, $f(x) = x^2 + x + 1$, $f(x) =$

$1/2x^2 - x + 1$ etc. In general, a polynomial function may neither be one–one, nor onto. However, a linear polynomial function such as $f(x) = ax + b$, where a, b are fixed real numbers, is always one–one as well as onto. Observe that in a linear polynomial function $f(x) = ax + b$, $x \in R$, a is non-zero. In case a is zero, this function no longer remains a linear function and becomes a constant function which associates the fixed real number b to every real number x. The function $f(x) = x^2$, $x \in R$ is neither one–one, nor onto as there is no real number x for which $x^2 = -1$ and $f(1) = 1^2 = (-1)^2 = f(-1)$.

In **logarithmic function** $f(x) = log^x$ to every real number x is associated the unique real number a such that $x = 2^a$. Observe that for every real number a, $2^a > 0$. Thus the logarithmic function is defined on the set of positive real numbers. We have used the notation log^x for the logarithm of x to the base 2. Logarithmic functions may be defined to any base b. In that case we write log_b^x to mean the number a such that $x = b^a$ and call log_b^x to be the logarithm of x to the base b. When the base of logarithm is the number 10, we call such a logarithm as natural logarithm and we write log_{10}^x as *In x*.

Another function that is defined on the set N of natural numbers and called the **factorial function** is defined for every natural number n as $f(n) = n! = n(n-1)\ldots 1$ (i.e. the product of the first n natural numbers) and is called factorial n. We define $0! = 1$. Observe that $3! = 3 \times 2 \times 1 = 6$, $5! = 5 \times 4 \times 3 \times 2 \times 1 = 120$ and $7! = 7 \times 6 \times 5 \times 4 \times 3 \times 2 \times 1 = 5040$.

We next define two other functions from the set of all real numbers to the set of integers.

Definition. The **floor function** denoted by $\lfloor \ \rfloor$ is the function: $R \to Z$ which assigns to the real number x the largest integer which is less than or equal to x. Thus $\lfloor x \rfloor$ denotes the largest integer $\leq x$. On the other hand we define the **ceiling function** denoted by $\lceil \ \rceil : R \to Z$ to be the function which associates to the real number x the smallest integer which is greater than or equal to the real number x. Thus $\lceil x \rceil$ is the smallest integer $\geq x$. Observe that

(a) $\lfloor x \rfloor = \lceil x \rceil$ for every integer x;
(b) both the floor function $\lfloor \ \rfloor$ and the ceiling function $\lceil \ \rceil$ are onto;
(c) neither of the functions $\lfloor \ \rfloor$ and $\lceil \ \rceil$ is one–one.

The floor function is also sometimes called **the greatest integer function** and analogously the ceiling function may be called **the least integer function.**

In analysis both floor function and the ceiling function are cited as functions which have infinitely many discontinuities. These functions are in fact discontinuous for every $x \in Z$. To fix up our ideas about these functions, note that

$\lfloor 1/3 \rfloor = 0$; $\lceil 1/3 \rceil = 1$; $\lfloor -1/3 \rfloor = -1$; $\lceil -1/3 \rceil = 0$; $\lfloor 2.3 \rfloor = 2$;

$\lceil 2.3 \rceil = 3$; $\lfloor \sqrt{5} \rfloor = 2$; $\lceil \sqrt{5} \rceil = 3$.

The following properties of the floor and the ceiling function are clear from their definitions. If n is an integer, then

(a) $\lfloor x \rfloor = n$ if and only if $n \leq x < n+1$;
(b) $\lfloor x \rfloor = n$ if and only if $x - 1 < n \leq x$;
(c) $\lceil x \rceil = n$ if and only if $n - 1 < x \leq n$;
(d) $\lceil x \rceil = n$ if and only if $x \leq n < x + 1$.

Combining (b) and (d) above, we also have

$$x - 1 < \lfloor x \rfloor \leq x \leq \lceil x \rceil < x + 1.$$

The floor and the ceiling functions for x and $-x$ are related by

(e) $\lfloor -x \rfloor = -\lceil x \rceil$; (f) $\lceil -x \rceil = -\lfloor x \rfloor$.
For (e), let $\lceil x \rceil = n$. Then by (d) above $x \leq n < x + 1$. This leads to $-x \geq -n > -x - 1$ or $-x - 1 < -n \leq -x$ which by (b) implies that $-n = \lfloor -x \rfloor$. Hence $\lfloor -x \rfloor = -\lceil x \rceil$. The result (f) may be proved on similar lines by using (a) and (c).

Theorem 1.8. For any integer n and real number x,

(a) $\lfloor x + n \rfloor = \lfloor x \rfloor + n$; (b) $\lceil x + n \rceil = \lceil x \rceil + n$.

Proof. Suppose that $\lceil x \rceil = m$, where m is an integer. Then by (c) above $m - 1 < x \leq m$. Adding n to the numbers in this inequality, we have $m + n - 1 < n + x \leq m + n$ in which $m + n$ is again an integer. By (c) again, $\lceil n + x \rceil = m + n = \lceil x \rceil + n$. This proves (b) above. The result (a) may be proved on similar lines.

Theorem 1.9. For any real number x, prove that

$$\lfloor 2x \rfloor = \lfloor x \rfloor + \lfloor x + 1/2 \rfloor.$$

Proof. Let $x = n + \epsilon$, where n is an integer and ϵ is a real number with $0 \leq \epsilon < 1$. Then by the definition of the floor function $\lfloor x \rfloor = n$. We consider two cases: $\epsilon < 1/2$ and $\epsilon \geq 1/2$.

Case (a) $0 \leq \epsilon < 1/2$. Then $0 \leq 2\epsilon < 1$. Therefore $2x = 2n + 2\epsilon$ where $2n$ is an integer and $0 \leq 2\epsilon < 1$. Hence $\lfloor 2x \rfloor = 2n$. Also $x + 1/2 = n + \epsilon + 1/2$ and $0 \leq \epsilon + 1/2 < 1$. Therefore $\lfloor x + 1/2 \rfloor = n$. Hence

$$\lfloor 2x \rfloor = 2n = n + n = \lfloor x \rfloor + \lfloor x + 1/2 \rfloor.$$

Case (b) $1/2 \leq \epsilon < 1$. Then $1 \leq 2\epsilon < 2$ or $0 \leq 2\epsilon - 1 < 1$. Now $2x = 2n + 2\epsilon = 2n + 1 + (2\epsilon - 1)$, with $0 \leq 2\epsilon - 1 < 1$, so that $\lfloor 2x \rfloor = 2n + 1$. Also $x + 1/2 = n + \epsilon + 1/2 = n + 1 + (\epsilon - 1/2)$, with $0 \leq \epsilon - 1/2 < 1/2$. Therefore $\lfloor x + 1/2 \rfloor = n + 1$. Thus $\lfloor 2x \rfloor = 2n + 1 = n + n + 1 = \lfloor x \rfloor + \lfloor x + 1/2 \rfloor$.

Example 1.6. Data stored on a computer disc or transmitted over a data network are usually expressed as strings of bytes. Each byte is made up of 8 bits. How many bytes are required to encode 125 bits of data?

Solution. To determine the number of bytes needed, we need to divide 125 by 8. The quotient which is 15 is such that 15 bytes will be able to store $15 \times 8 = 120$ bits of data. We are still left with 5 bits of data for which another byte will be used. Then the number of bytes needed is 16 which equals $\lceil 125/8 \rceil$ i.e. the smallest integer which is $\geq 125/8$.

Example 1.7. In asynchronous transfer mode (ATM) (which is a communications protocol used on backbone networks), data are organized into cells of 53 bytes. How many ATM cells can be transmitted in one minute over a connection that transmits data at the rate of 128 kilobits per second?

Solution. In one minute, this connection can transmit $128000 \times 60 = 7680000$ bits. Each ATM cell is 53 bytes long i.e. it is $53 \times 8 = 424$ bits long. Then the number of cells that can be transmitted in 1 minute is $\leq 7680000/424$ and it has to be a whole number i.e. an integer. This means that $\lfloor 7680000/424 \rfloor = \lfloor 18113 + 11/53 \rfloor = 18113$ ATM cells can be transmitted in 1 minute over a 128 kilobits per second connection.

Exercise 1.4. How many ATM cells as described in the above example can be transmitted in 10 seconds over a link operating at the rate of

(a) 300 kilobits per second?
(b) 500 kilobits per second?
(c) 1 megabit per second (1 megabit = 1,000,000 bits)?

1.6.1. *The characteristic function*

Let U be a universal set. For any subset A of U, we can define a map $f_A : U \to B = \{0,1\}$ by $f_A(x) = 0$ if $x \notin A$ and 1 if $x \in A$.

The map f_A is called the **characteristic function of** A. The characteristic functions of the subsets of the universal set U have the following properties:

Theorem 1.10. Let A, B be subsets of U. Then, for every $x \in U$,

(a) $f_{A \cap B}(x) = f_A(x) f_B(x)$;
(b) $f_{A \cup B}(x) = f_A(x) + f_B(x) - f_A(x) f_B(x)$;
(c) $f_{A'}(x) = 1 - f_A(x)$;
(d) $f_{A \oplus B}(x) = f_A(x) + f_B(x) - 2 f_A(x) f_B(x)$.

Proof. (a) If $x \in A \cap B$, then $x \in A$ and $x \in B$ so that $f_{A \cap B}(x) = 1 = f_A(x) f_B(x)$. If $x \notin A \cap B$, then either $x \notin A$ or $x \notin B$ and we have $f_A(x) = 0$ or $f_B(x) = 0$ and $f_{A \cap B}(x) = f_A(x) \, f_B(x)$.

(b) If $x \in A \cup B$, then either x belongs to both A and B or x belongs to one of A, B and not the other. In the first case

$$1 = f_{A \cup B}(x) = f_A(x) + f_B(x) - f_A(x) f_B(x)$$

while in the other case one of $f_A(x), f_B(x)$ is zero and the other is 1.

Therefore, $f_A(x) + f_B(x) - f_A(x) f_B(x) = 1 = f_{A \cup B}(x)$.

If $x \notin A \cup B$, then $x \notin A$ and $x \notin B$ so all three of $f_{A \cup B}(x), f_A(x)$ and $f_B(x)$ are 0 and, so, $f_{A \cup B}(x) = f_A(x) + f_B(x) - f_A(x) f_B(x)$.

(c) $x \in A'$ if and only if $x \notin A$. Therefore one of $f_A(x), f_{A'}(x)$ is 0 and the other is 1. It then follows that $f_{A'}(x) = 1 - f_A(x)$.

(d) Recall that $A \oplus B = (A - B) \cup (B - A) = A \cup B - A \cap B$. If $x \in A \oplus B$, then $x \in A \cup B$ but $x \notin A \cap B$ and one of $f_A(x), f_B(x)$ is 1 and the other is 0 but $f_{A \cap B}(x) = 0$. Therefore

$$1 = f_{A \oplus B}(x) = f_A(x) + f_B(x) - 2 f_A(x) f_B(x).$$

On the other hand, if $x \notin A \oplus B$, then $x \notin A \cup B - A \cap B$. This means that either $x \in A \cap B$ or $x \notin A$ and $x \notin B$ i.e. either x is in both A and B or x is in neither of A, B. In the first case $f_A(x) + f_B(x) - 2 f_A(x) f_B(x) = 1 + 1 - 2.1 = 0$ while in the second case $f_A(x) + f_B(x) - 2 f_A(x) f_B(x) = 0$. Thus in both the cases $f_{A \oplus B}(x) = 0 = f_A(x) + f_B(x) - 2 f_A(x) f_B(x)$.

Consider the universal set U and the power set $\wp(U)$ of U i.e. the set of all subsets of U. To every subset A of U corresponds a function $f_A : U \to \mathbb{B}$. Thus we obtain a collection of maps (**characteristic maps**) the elements of which are in one-to-one correspondence with the subsets of U. This set or collection of maps is denoted by B^U i.e.

$$B^U = \{f_A : U \to B \mid A \in \wp(U)\}.$$

1.7. Partitions of Sets

Definition. Let A be non-empty set. A collection or set $\{A_i\}, i \in K$, of subsets A_i of A, where K is an index set is called a partition of A if
(a) $A_i \neq \emptyset$ for every $i \in K$;
(b) $A_i \cap A_j = \emptyset$ for $i, j \in K$ and $i \neq j$,
(c) $A = \cup A_i$, where i runs over all the elements of K.

Each subset A_i of A appearing in the partition is called a **block** of the partition.

Partitions of sets appear in a natural way in the study of groups and rings. In fact the collection of all left cosets of a subgroup H of a group G provides a partition of the set G. We shall consider this problem later while studying groups. Another situation where partitions appear naturally arises while studying equivalence relations on sets. Indeed, equivalence classes determined by an equivalence relation on a set provide a partition of the set. We study these also later in the book. A set can have many partitions.

Example 1.8. If $A = \{1, 2, 3\}$, then $P_1 = \{\{1\}, \{2\}, \{3\}\}, P_2 = \{\{1, 2\}, \{3\}\}, P_3 = \{\{1\}, \{2, 3\}\}, P_4 = \{\{2\}, \{1, 3\}\}$ are partitions of A.

Example 1.9. If $A = \{1, 2, 3, 4\}$, then

$$P_1 = \{\{1\}, \{2\}, \{3\}, \{4\}\}, \quad P_2 = \{\{1\}, \{2\}, \{3, 4\}\},$$
$$P_3 = \{\{1\}, \{3\}, \{2, 4\}\}, \quad P_4 = \{\{1\}, \{4\}, \{2, 3\}\},$$
$$P_5 = \{\{2\}, \{3\}, \{1, 4\}\}, \quad P_6 = \{\{2\}, \{4\}, \{1, 3\}\},$$
$$P_7 = \{\{3\}, \{4\}, \{1, 2\}\}, \quad P_8 = \{\{1, 2\}, \{3, 4\}\},$$
$$P_9 = \{\{1, 3\}, \{2, 4\}\}, \quad P_{10} = \{\{1, 4\}, \{2, 3\}\},$$
$$P_{11} = \{\{1\}, \{2, 3, 4\}\}, \quad P_{12} = \{\{2\}, \{1, 3, 4\}\},$$
$$P_{13} = \{\{3\}, \{1, 2, 4\}\}, \quad P_{14} = \{\{4\}, \{1, 2, 3\}\}$$

are some partitions of A. However, $\{\{1\}, \{2, 3\}, \{1, 2\}\}$ is not a partition of the set $\{1, 2, 3\}$ nor is $\{\{1, 2\}, \{2, 3\}\}$ a partition of this set. Similarly $\{\{1, 2\}, \{2, 3, 4\}\}, \{\{1, 2\}, \{2, 3\}, \{4\}\}, \{\{1, 4\}, \{2, 3\}, \{4\}\}$ are not partitions of the set $\{1, 2, 3, 4\}$.

Example 1.10. The collection $\{2Z, 2Z + 1\}$, where $2Z$ is the set of all even integers and $2Z + 1$ is the set of all odd integers, is a partition of the set of all integers. Another partition of Z is $\{3Z, 3Z + 1, 3Z + 2\}$, where $3Z + i, 0 \leq i \leq 2$ is the set of integers which on division by 3 leave the remainder i.

Example 1.11. Recall that a complex number $a + ib$, where a, b are real numbers, is called pure imaginary if $a = 0$ but $b \neq 0$. Let iR^* denote the set of all pure imaginary numbers, with $i = \sqrt{-1}$. Then the collection $\{R, iR^*\}$ is a partition of the set C of complex numbers.

Example 1.12. Consider the set A of the English alphabet, V the set of vowels and C the set of consonants in A. Then $\{V, C\}$ is a partition of the set A.

For a set A, the collection $\{A\}$ has only one element, namely the set A itself also provides a partition of the set A. Such a partition is called **trivial partition**. It is only non-trivial partitions which are of interest.

For obtaining partitions of a subset B of a set A from a given partition of the set A, we have

Theorem 1.11. Let A be a non-empty set, B be a non-empty subset of A and $\{A_i\}$, $i \in K$ be a partition of A. Then $\{A_i \cap B | i \in K \text{ with } A_i \cap B \neq \varphi\}$ is a partition of B.

Proof. Let J be a subset of K consisting of all $j \in K$ for which $A_j \cap B \neq \varphi$. Let $b \in B$. Then $b \in A = \cup A_i, i \in K$. Thus, there exists a $j \in K$ such that $b \in A_j$. Then $b \in A_j \cap B$ and $A_j \cap B \neq \varphi$. This proves that $j \in J$ and, so, $J \neq \varphi$. The above also proves that $B \subseteq \cup(A_j \cap B)$ where j runs over all the elements of J. As $A_j \cap B \subseteq B$ for every $j \in J$, $\cup(A_j \cap B) \subseteq B$. Hence $B = \cup(A_j \cap B)$.

Consider $A_i \cap B$ and $A_j \cap B$, where $i, j \in J$ and $i \neq j$. Then $A_i \cap A_j = \varphi$ and $(A_i \cap B) \cap (A_j \cap B) = (A_i \cap A_j) \cap B = \varphi \cap B = \varphi$. This completes the proof that $\{A_j \cap B \,|\, j \in J\}$ is a partition of B.

Definition. A partition $\wp = \{B_i | i \in L\}$ of a set A is called a **refinement of a partition** $\Pi = \{A_i | i \in K\}$ of A if for every $i \in L, B_i$ is a subset of A_j for some $j \in K$. If for some $i \in L, B_i$ is a proper subset of some $A_j, j \in K$, then $\wp$ is called a **proper refinement** of Π. Every partition Π of A is a refinement of itself.

Consider a partition $\Pi = \{\{1\}, \{2,3\}, \{4,5\}\}$ of the set $A = \{1, 2, 3, 4, 5\}$. Then $\Pi_1 = \{\{1\}, \{2\}, \{3\}, \{4,5\}\}$, $\Pi_2 = \{\{1\}, \{2,3\}, \{4\}, \{5\}\}$ and

$\Pi_3 = \{\{1\}, \{2\}, \{3\}, \{4\}, \{5\}\}$ are refinements of Π but $\{\{1, 2\}, \{3\}, \{4, 5\}\}$ is not a refinement of Π. Observe that Π, Π_1, Π_2, Π_3 are the only possible refinements of Π.

Exercise 1.5.

1. Find all possible nontrivial partitions of the set $A = \{1, 2, 3, 4, 5\}$.
2. If $A = \{1, 2\}$ there is one nontrivial partition of A, namely $\{\{1\}, \{2\}\}$. For $A = \{1, 2, 3\}$, we have calculated all the possible partitions of A and these are 4 in number. For a set having 4 elements, we have obtained 14 nontrivial partitions which are all the possible ones. For a set having 5 elements, prove or disprove that the number of nontrivial partitions is 66. If this number is not 66, find the correct number.
3. For a set having 5 elements, find the number of partitions having two blocks each.
4. For a set having 6 elements, find the number of partitions having (i) two blocks each; (ii) three blocks each.
5. Among the partitions P_1, P_2, P_3, P_4 of $A = \{1, 2, 3\}$ obtained in Example 1.8, decide the partitions which are refinements of some other partitions.
6. Among the partitions P_1 to P_{14} of $A = \{1, 2, 3, 4\}$ obtained in Example 1.9, determine the partitions which are refinements of some other partitions.
7. Prove that every refinement P_1 of a partition P of a non-empty set $A, P_1 \neq P$, is a proper refinement of P.

1.8. The Minset and Maxset Normal Forms

Let X be a universal set and $\{A_i\}, 1 \leq i \leq r$ be a collection of subsets of X. A set obtained from φ, X and $A_1, A_2, \ldots, A_r$ by the use of set operations (complementation), $\cup$ and $\cap$ is called a set generated by $A_1, A_2, \ldots, A_r$. Consider, for example, two subsets A, B of X. Then

$$\varphi, X, A, B, A', B', A \cap B, A \cap B', A' \cap B, A' \cap B', A \cup B, A \cup B',$$
$$A' \cup B \quad \text{and} \quad A' \cup B' \tag{1.3}$$

are some of the sets generated by A, B.

Example 1.13. Consider sets generated by A, B which involve three of the sets A, A', B, B'. Such sets are of the form

$$(A^* \cup B^*) \cup C, \quad (A^* \cup B^*) \cap C, \quad (A^* \cap B^*) \cup C, \quad (A^* \cap B^*) \cap C \tag{1.4}$$

where A^* is A or A', B^* is B or B' and C is either A or A' or B or B'.

We consider the case when C is either A or A'. Observe that

$$(A^* \cup B^*) \cup C = A^* \cup B^* \qquad \text{if } C = A^*$$
$$= X \qquad \text{if } C \neq A^*;$$
$$(A^* \cup B^*) \cap C = (A^* \cap C) \cup (B^* \cap C)$$
$$= A^* \qquad \text{if } C = A^*$$
$$= B^* \cap C \qquad \text{if } C \neq A^*;$$
$$(A^* \cap B^*) \cup C = (A^* \cup C) \cap (B^* \cup C)$$
$$= A^* \qquad \text{if } C = A^*$$
$$= B^* \cup C \qquad \text{if } C \neq A^*;$$
$$(A^* \cap B^*) \cap C = A^* \cap B^* \qquad \text{if } C = A^*$$
$$= \varphi \qquad \text{if } C \neq A^*.$$

Thus the sets in (1.4) are sets generated by A, B involving two of A, B, A', B'. Interchanging A and B in the above argument, we can prove that the sets in (1.4) when C is either B or B' are also sets generated by A, B involving at most two of A, B, A', B'. By using iteration or a simple induction argument, we can prove that the sets in (1.3) are all the possible sets generated by A, B. An argument similar to the above also gives

Theorem 1.12. Every set generated by a collection $\{A_i\}$, $1 \leq i \leq r$ of subsets of X involves at most r of the sets $A_1, \ldots, A_r, A'_1 \ldots, A'_r$.

Let $\{A_i\}$, $1 \leq i \leq r$, be a collection (or set) of subsets of a universal set X. A set of the form $\cap A_i^*$, $1 \leq i \leq r$, where each A_i^* is either A_i or A'_i is called a **minset** or **minterm generated by** $A_1, A_2, \ldots, A_r$. Since every A_i^* has two choices, there are 2^r such minsets.

For example, if we consider a set

(a) $\{A_1, A_2\}$ of two subsets of X, the minsets generated by A_1, A_2 are

$$A_1 \cap A_2, A_1 \cap A'_2, A'_1 \cap A_2, \quad \text{and} \quad A'_1 \cap A'_2.$$

(b) $\{A_1, A_2, A_3\}$ of three subsets of X, then the minsets generated by A_1, A_2, A_3 are

$$A_1 \cap A_2 \cap A_3, \quad A_1 \cap A_2 \cap A'_3, \quad A_1 \cap A'_2 \cap A_3, \quad A'_1 \cap A_2 \cap A_3,$$
$$A_1 \cap A'_2 \cap A'_3, \quad A'_1 \cap A'_2 \cap A_3, \quad A'_1 \cap A_2 \cap A'_3 \quad \text{and} \quad A'_1 \cap A'_2 \cap A'_3.$$

All minsets need not be non-empty. Consider for example,

$$X = \{1, 2, \ldots, 9\}, A_1 = \{1, 2\} \text{ and } A_2 = \{3, 4\}.$$

One of the minsets in this case is $A_1 \cap A_2$ which is the empty set. In general, if we take r subsets A_i, $1 \leq i \leq r$ of X such that $\cap A_i = \varphi$, then one of the minsets is the empty set. Also, if one of the subsets $A_1, A_2, \ldots, A_r$, say A_r, is the universal set X itself, then all minsets in which A'_r appears are the empty set. This situation may be avoided by taking all the subsets $A_1, A_2, \ldots, A_r$ proper.

The concept of minsets is related to the partitions of a set and this relationship is provided by

Theorem 1.13. The set of all non-empty minsets generated by $A_1, A_2, \ldots, A_r$ constitutes a partition of the set X.

Proof. It is enough to prove that every $x \in X$ belongs to one and only one minset. Let $x \in X$. For any i, $1 \leq i \leq r, x \in A_i$ or $x \in A'_i$. Thus there exists a minset $\cap A_i^*$, $1 \leq i \leq r$, to which x belongs. If possible, suppose that x belongs to two distinct minsets, say, $\cap B_i$ and $\cap D_i$, where for every $i, B_i = A_i$ or A'_i and for every j, $D_j = A_j$ or A'_j. The two minsets being distinct, there exists an i, $1 \leq i \leq r$, such that $B_i \neq D_i$. But B_i being either A_i or A'_i and similarly for D_i, in one of the minsets A_i appears while in the other A'_i appears. But $\cap B_j \subseteq B_i$ and $\cap D_j \subseteq D_i$. Therefore $x \in B_i \cap D_i = A_i \cap A'_i = \varphi$ — which is a contradiction. Thus every $x \in X$ belongs to one and only one minset.

Observe that later half of the above proof amounts to proving that *every two distinct minsets are disjoint.*

For the sake of convenience a minset $\cap A_i^*$, $1 \leq i \leq r$, is denoted by $M_{\delta_1 \delta_2 \ldots \delta_r}$ where each δ_i is either 0 or 1 according to the rule that $\delta_i = 0$ if $A_i^* = A'_i$ and $\delta_i = 1$ if $A_i^* = A_i$. For example, if A, B, C are three subsets of X and we think of A as A_1, B as A_2 and C as A_3, then

$$M_{111} = A \cap B \cap C, \quad M_{110} = A \cap B \cap C', \quad M_{101} = A \cap B' \cap C,$$

$$M_{011} = A' \cap B \cap C, \quad M_{100} = A \cap B' \cap C', \quad M_{010} = A' \cap B \cap C',$$

$$M_{001} = A' \cap B' \cap C, \quad M_{000} = A' \cap B' \cap C'.$$

Example 1.14. Consider subsets A, B of X. As already seen. all the sets generated by A, B are given in (1.3). Moreover, all the minsets generated by A, B are

$$A \cap B, \quad A \cap B', \quad A' \cap B, \quad A' \cap B'.$$

By Theorem 1.13.,

$$X = (A \cap B) \cup (A \cap B') \cup (A' \cap B) \cup (A' \cap B').$$

Then

$$
\begin{aligned}
A &= A \cap X \\
&= A \cap ((A \cap B) \cup (A \cap B') \cup (A' \cap B) \cup (A' \cap B')) \\
&= (A \cap (A \cap B)) \cup (A \cap (A \cap B')) \cup (A \cap (A' \cap B)) \cup (A \cap (A' \cap B')) \\
&= (A \cap B) \cup (A \cap B') \cup \varphi \cup \varphi \\
&= (A \cap B) \cup (A \cap B').
\end{aligned}
$$

Similarly,

$$
\begin{aligned}
B &= (A \cap B) \cup (A' \cap B); \\
A' &= (A' \cap B) \cup (A' \cap B'); \\
B' &= (A \cap B') \cup (A' \cap B').
\end{aligned}
$$

Also then

$$
\begin{aligned}
A \cup B &= (A \cap B) \cup (A \cap B') \cup (A \cap B) \cup (A' \cap B) \\
&= (A \cap B) \cup (A \cap B') \cup (A' \cap B); \\
A' \cup B &= (A' \cap B) \cup (A \cap B) \cup (A' \cap B) \cup (A' \cap B') \\
&= (A \cap B) \cup (A' \cap B) \cup (A' \cap B'); \\
A \cup B' &= (A \cap B) \cup (A \cap B') \cup (A' \cap B'); \\
A' \cup B' &= (A \cap B') \cup (A' \cap B) \cup (A' \cap B').
\end{aligned}
$$

Thus all non-empty sets generated by A, B have been expressed as union of some minsets generated by A, B. This result is true in general and we have

Theorem 1.14. Every set generated by $A_1, A_2, \ldots, A_r$ is either φ or is equal to a union of distinct minsets generated $A_1, A_2, \ldots, A_r$.

When a set is φ or is expressed as a union of distinct minsets, it is said to be in **minset normal form** or **minset canonical form**.

Example 1.15. Consider three subsets A, B, C of a universal set X. The minsets generated by A, B, C are

$$A \cap B \cap C, \quad A \cap B \cap C', \quad A \cap B' \cap C, \quad A' \cap B \cap C, \quad A \cap B' \cap C',$$
$$A' \cap B \cap C', \quad A' \cap B' \cap C \quad \text{and} \quad A' \cap B' \cap C'.$$

Then, proceeding as in Example 1.14, we get

$$A = (A \cap B \cap C) \cup (A \cap B \cap C') \cup (A \cap B' \cap C) \cup (A \cap B' \cap C')$$
$$B = (A \cap B \cap C) \cup (A \cap B \cap C') \cup (A' \cap B \cap C) \cup (A' \cap B \cap C')$$
$$C = (A \cap B \cap C) \cup (A \cap B' \cap C) \cup (A' \cap B \cap C) \cup (A' \cap B' \cap C)$$

and A', B', C' can be obtained from the expressions for A, B, C by changing A to A', B to B' and C to C' respectively. Thus

$$A' = (A' \cap B \cap C) \cup (A' \cap B \cap C') \cup (A' \cap B' \cap C) \cup (A' \cap B' \cap C')$$
$$B' = (A \cap B' \cap C) \cup (A \cap B' \cap C') \cup (A' \cap B' \cap C) \cup (A' \cap B' \cap C')$$
$$C' = (A \cap B \cap C') \cup (A \cap B' \cap C') \cup (A' \cap B \cap C') \cup (A' \cap B' \cap C').$$

Once these are obtained, any non-empty set generated by A, B, C can then be expressed as union of distinct minsets by substituting for A, B, C, A', B', C' from above. For example,

$$(A \cap B') \cup (B \cap (A \cup C'))$$
$$= (A \cap B') \cup ((B \cap A) \cup (B \cap C'))$$
$$= ((A \cap B') \cup (A \cap B)) \cup (B \cap C')$$
$$= (A \cap (B' \cup B)) \cup (B \cap C')$$
$$= (A \cap X) \cup (B \cap C')$$
$$= A \cup (B \cap C')$$
$$= (A \cup B) \cap (A \cup C')$$
$$= (A \cup (A' \cap B \cap C) \cup (A' \cap B \cap C')) \cap ((A \cup (A' \cap B \cap C')$$
$$\quad \cup (A' \cap B' \cap C'))$$
$$= ((A \cup (A' \cap B \cap C')) \cup (A' \cap B \cap C)) \cap ((A \cup (A' \cap B \cap C'))$$
$$\quad \cup (A' \cap B' \cap C'))$$
$$= (A \cup (A' \cap B \cap C')) \cup ((A' \cap B \cap C) \cap (A' \cap B' \cap C'))$$
$$= (A \cup (A' \cap B \cap C)) \cup \varphi$$
$$= A \cup (A' \cap B \cap C')$$
$$= (A \cap B \cap C) \cup (A \cap B \cap C') \cup (A \cap B' \cap C) \cup (A \cap B' \cap C')$$
$$\quad \cup (A' \cap B \cap C').$$

Exercise 1.6. Find the minimal normal form of $((A \cup D') \cap (B' \cup C')) \cup (A \cap B \cap D)$ generated by A, B, C, D.

Definition. Let $\{A_i\}$, $1 \leq i \leq r$, be a collection of subsets of a universal set X. Then a set of the form $\cup A_i^*$, $1 \leq i \leq r$, where for every i, $1 \leq i \leq r$, A_i^* is either A_i or A_i', is called a **maxset** generated by $A_1, A_2, \ldots, A_r$. As in the case of minsets, there are 2^r maxsets generated by $A_1, A_2, \ldots, A_r$.

Observe that $(\cup A_i^*)' = \cap (A_i^*)'$ and for every $i, 1 \leq i \leq r, (A_i^*)'$ is either A_i or A_i'. Thus every maxset generated by $A_1, A_2, \ldots, A_r$ is the complement of a minset generated by $A_1, A_2, \ldots, A_r$ and conversely. Similar to the expression for minsets, a maxset $\cup A_i^*$ generated by $A_1, A_2, \ldots, A_r$ is denoted by $\tilde{M}_{\delta_1 \delta_2 \ldots \delta_r}$, where every δ_i is either 0 or 1 with $\delta_i = 0$ if $A_i^* = A_i$ and $\delta_i = 1$ if $A_i^* = A_i'$. Writing A as A_1, B as A_2 and C as A_3, we observe that the maxset $A' \cup B \cup C'$ is $\tilde{M}_{101}$. However $A' \cup B \cup C' = (A \cap B' \cap C)' = M_{101}'$. In general (it is clear from the definition of minset and maxset) that $\tilde{M}_{\delta_1 \delta_2 \ldots \delta_r} = M_{\delta_1 \delta_2 \ldots \delta_r}'$. In view of this relationship between minsets and maxsets and Theorem 1.14 we have

Theorem 1.15. Every set generated by $A_1, A_2, \ldots, A_r$ is either X or is an intersection of distinct maxsets generated by $A_1, A_2, \ldots, A_r$.

When a set generated by $A_1, A_2, \ldots, A_r$ is expressed as X or as an intersection of distinct maxsets, it is said to be in a **maxset normal form** or **maxset canonical form**.

Example 1.16. Consider subsets A, B of the universal set X. The maxsets generated by A, B are $A \cup B, A \cup B', A' \cup B$ and $A' \cup B'$. Then

$$A = A \cup \varphi = A \cup (B \cap B') = (A \cup B) \cap (A \cup B').$$

Similarly

$$B = (A \cup B) \cap (A' \cup B),$$
$$A' = (A' \cup B) \cap (A' \cup B'),$$
$$B' = (A \cup B') \cap (A' \cup B').$$

Example 1.17. Consider the case of three subsets A, B, C of X. The maxsets generated by A, B, C are

$$A \cup B \cup C, \quad A \cup B' \cup C, \quad A \cup B \cup C', \quad A' \cup B \cup C, \quad A \cup B' \cup C',$$
$$A' \cup B \cup C', \quad A' \cup B' \cup C \quad \text{and} \quad A' \cup B' \cup C'.$$

In this case

$$A = (A \cup B' \cup C') \cap (A \cup B' \cup C) \cap (A \cup B \cup C') \cap (A \cup B \cup C)$$

$$B = (A \cup B \cup C) \cap (A' \cup B \cup C) \cap (A \cup B \cup C') \cap (A' \cup B \cup C')$$

$$C = (A \cup B \cup C) \cap (A' \cup B \cup C) \cap (A \cup B' \cup C) \cap (A' \cup B' \cup C)$$

$$A' = (A' \cup B \cup C) \cap (A' \cup B \cup C') \cap (A' \cup B' \cup C) \cap (A' \cup B' \cup C')$$

$$B' = (A \cup B' \cup C) \cap (A \cup B' \cup C') \cap (A' \cup B' \cup C) \cap (A' \cup B' \cup C')$$

$$C' = (A \cup B \cup C') \cap (A \cup B' \cup C') \cap (A' \cup B \cup C') \cap (A' \cup B' \cup C').$$

We next obtain a maxset normal form of $(A \cap B) \cup (A' \cap C)$.

Using the values of A, A' etc. obtained above

$$A \cap B = \{(A \cup B \cup C) \cap (A \cup B' \cup C) \cap (A \cup B \cup C') \cap (A \cup B' \cup C')$$
$$\cap (A' \cup B \cup C) \cap (A' \cup B \cup C')\}$$

and

$$A' \cap C = \{(A \cup B \cup C) \cap (A' \cup B \cup C) \cap (A \cup B' \cup C)$$
$$\cap (A' \cup B' \cup C) \cap (A' \cup B \cup C') \cap (A' \cup B' \cup C')\}.$$

Set

$$(A \cup B \cup C) \cap (A \cup B' \cup C) \cap (A' \cup B \cup C) \cap (A' \cup B \cup C') = L.$$

Then $A \cap B = L \cap (A \cup B' \cup C') \cap (A \cup B \cup C')$ and

$$A' \cap C = L \cap (A' \cup B' \cup C) \cap (A' \cup B' \cup C').$$

Now

$$(A \cup B' \cup C') \cap (A \cup B \cup C') = (A \cup C') \cup (B \cap B')$$
$$= (A \cup C') \cup \varphi = A \cup C'$$

and

$$(A' \cup B' \cup C) \cap (A' \cup B' \cup C') = (A' \cup B') \cup (C \cap C') = A' \cup B'.$$

Therefore

$$(A \cap B) \cup (A' \cap C) = (L \cap (A \cup C')) \cup (L \cap (A' \cup B'))$$
$$= L \cap (A \cup C' \cup A' \cup B')$$
$$= L \cap (A \cup A' \cup C' \cup B')$$
$$= L \cap X = L.$$

Hence

$$(A \cap B) \cup (A' \cap C) = \{(A \cup B \cup C) \cap (A \cup B' \cup C) \cap (A' \cup B \cup C)$$
$$\cap (A' \cup B \cup C')\}$$

is the required maxset normal form representation of $(A \cap B) \cup (A' \cap C)$.

Unlike the set of all minsets generated by $A_1, A_2, \ldots, A_r$, the set of all maxsets generated by $A_1, A_2, \ldots, A_r$ does not provide a partition of the set X. In fact every two distinct maxsets need not be disjoint. For example, $A_1 \cup A_2 \cup \ldots \cup A_r$ and $A_1' \cup A_2 \cup \ldots \cup A_r$ are distinct maxsets and $(A_1 \cup A_2 \cup \ldots \cup A_r) \cap (\mathbf{A}_1' \cup A_2 \cup \ldots \cup A_r) = (A_1 \cap A_1') \cup (A_2 \cup \ldots \cup A_r) = A_2 \cup A_3 \cup \ldots \cup A_r$ which is not an empty set.

We end this section with the following

Theorem 1.16. Minset normal form (or maxset normal form) of a set S generated by $A_1, A_2, \ldots, A_r$ is unique except for the order of minsets (or maxsets).

As an immediate corollary of this we have

Corollary. Two sets S and T generated by $A_1, A_2, \ldots, A_r$ are equal if and only if S and T have identical minset canonical form (maxset canonical form).

Exercise 1.7.

1. Prove Theorem 1.14.
2. Prove Theorem 1.16.
3. Find the minset and maxset normal forms of the sets
   (i) $B' \cap C$, (ii) $A \cup (B \cap C')$, (iii) $(A' \cap B) \cup (B' \cap (A \cup C))$ generated by A, B, C.
4. Use the minset and maxset normal forms to prove that the complement of $(A \cap B') \cup (A' \cap (B \cup C'))$ is $(A' \cup B) \cap (A \cup (B' \cap C))$.
5. Let $A = \{1, 2, \ldots, 9\}$ and set $A_1 = \{5, 6, 7\}$, $A_2 = \{2, 4, 5, 9\}$, $A_3 = \{3, 4, 5, 6, 8, 9\}$.
   (i) Find all minsets generated by A_1, A_2, A_3.
   (ii) Find all maxsets generayed by A_1, A_2, A_3.
   (iii) Express the sets $A_1', A_1 \cup A_2$ and $A_2' \cup A_3'$ in the minset normal form and maxset normal form.

1.9. Multisets

A sub-branch of a bank applied to be upgraded to a regular branch. For this the sub-branch was asked to supply the monthly lists of all bank account numbers for which there were transactions during the last three months. For a particular account, there could be more than one transaction during a month; some of these may be debit entries and some credit entries. For all these transactions the sub-branch could not list that account only once, for otherwise the branch would lose credit for the remaining transactions of this account number. The lists of these account numbers are not sets as it may contain the same account number more than once. Each of these three lists is called a multiset. Thus a multiset is a collection of objects not necessarily all distinct. A second example of a multiset is the collection of names of all students who appeared in the university examination for discrete mathematics in a particular year. Yet another example of a multiset is the collection of all names in the telephone directory of a town. We may list some other multisets as:

$$A = \{a, a, a, b, b, c, c, d, d, d, e\}$$
$$B = \{a, a, b, b, b, c, c, c, d, e, e, e\}$$
$$C = \{a, a, b, c, d, d\}$$
$$D = \{1, 1, 2, 2, 2, 2, 3, 3, 4, 4, 4, 4, 5, 5\}$$
$$E = \{1, 1, 2, 2, 3, 4, 4, 4, 5, 5\}$$
$$F = \{1, 1, 2, 2, 3, 4, 4, 5, 5, 5, 6\}$$
$$G = \{1, 1, 2, 2, 3, 4, 4, 5, 5, 5\}.$$

The **multiplicity** of an element of a multiset is defined to be the number of times the element occurs in the multiset. For example, in the multiset A, the multiplicity of a is 3, multiplicity of b is 2, multiplicity of c is 2, multiplicity of d is 3 while that of e is 1. In the multiset B, multiplicity of a is 2, of b it is 3, of c it is 3, of d it is 1 and that of e is 3. By ignoring or removing multiplicities of different elements of a multiset, we get a set — the underlying set of the multiset. The set corresponding to the multisets A and B is $\{a, b, c, d, e\}$, that corresponding to the multiset C is $\{a, b, c, d\}$, the set corresponding to the multisets D, E and G is $\{1, 2, 3, 4, 5\}$ while the set corresponding to the multiset F is $\{1, 2, 3, 4, 5, 6\}$. The **order** or the **cardinality** of a multiset is defined to be the order or cardinality of the corresponding (or underlying) set. Thus the order of each of the multisets A, B, D, E, G is 5, the order of C is 4 and that of F is 6.

A multiset X is called a **submultiset** of a multiset Y if the multiplicity of every $x\varepsilon X$ is less than or equal to the multiplicity of x as an element of

Y. For example, C is a submultiset of A, E is a submultiset of D, and G is a submultiset of F.

Observe that every set is always a multiset with the multiplicity of every element being 1. We will now define union, intersection and difference of multisets. These are to be so defined that these are in conformity with union, intersection and difference respectively of sets or that these do not contradict the corresponding definitions for sets.

Let P, Q be multisets. Then

(a) the **intersection** $P \cap Q$ of P and Q is the multiset in which the multiplicity of an element x is the minimum of the multiplicities of x as an element of P and as an element of Q;

(b) the **union** $P \cup Q$ of P and Q is the multiset in which the multiplicity of an element x is equal to the maximum of the multiplicities of x as an element of P and as an element of Q;

(c) the **difference** $P - Q$ of P and Q is the multiset in which the multiplicity of an element x is the multiplicity of x as an element of P minus the multiplicity of x as an element of Q. If this difference of multiplicities of x is ≤ 0, then x does not appear in $P - Q$.

It is fairly easy to see that $P \cap Q$ is a submultiset of P as well as of Q, both P and Q are submultisets of $P \cup Q$ and $P - Q$ is a submultiset of P.

Example 1.18. For the multisets A to G as above

(a) $A \cap B = \{a, a, b, b, c, c, d, e\}$

$A \cup B = \{a, a, a, b, b, b, c, c, c, d, d, d, e, e, e\}$

$A - B = \{a, d, d\}, B - A = \{b, c, e, e\}$

(b) $A \cap C = C$, $A \cup C = A$, $A - C = \{a, b, c, d, e\}$, $C - A = \varphi$

(c) $B \cap C = \{a, a, b, c, d\}$

$B \cup C = \{a, a, b, b, b, c, c, c, d, d, e, e, e\}$

$B - C = \{b, b, c, c, e, e, e\}$, $C - B = \{d\}$

(d) $D \cap E = E, D \cup E = D, D - E = \{2, 2, 3, 4\}$, $E - D = \varphi$

(e) $D \cap F = \{1, 1, 2, 2, 3, 4, 4, 5, 5\}$

$D \cup F = \{1, 1, 2, 2, 2, 2, 3, 3, 4, 4, 4, 5, 5, 5, 6\}$

$D - F = \{2, 2, 3, 4, 4\}, F - D = \{5, 6\}$

(f) $D \cap G = \{1, 1, 2, 2, 3, 4, 4, 5, 5\}$

$$D \cup G = \{1, 1, 2, 2, 2, 2, 3, 3, 4, 4, 4, 4, 5, 5, 5\}$$

$$D - G = \{2, 2, 3, 4, 4\},\ G - D = \{5\}$$

(g) $E \cap F = \{1, 1, 2, 2, 3, 4, 4, 5, 5\}$

$$E \cup F = \{1, 1, 2, 2, 3, 4, 4, 4, 5, 5, 5, 6\}$$

$$E - F = \{4\},\ F - E = \{5, 6\}$$

(h) $E \cap G = \{1, 1, 2, 2, 3, 4, 4, 5, 5\}$

$$E \cup G = (1, 1, 2, 2, 3, 4, 4, 4, 5, 5, 5\}$$

$$E - G = \{4\},\ G - E = \{5\}$$

(i) $F \cap G = G,\ F \cup G = F,\ F - G = \{6\},\ G - F = \varphi.$

We next define the symmetric difference of two multisets. This again needs to be consistent with the definition of symmetric difference of two sets. Recall that the symmetric difference of two sets P and Q is defined to be $(P - Q) \cup (Q - P)$ and it has been shown to be equal to $(P \cup Q) - (P \cap Q)$. Let P, Q be two multisets. As in the case of sets, we define the **symmetric difference** $P \oplus Q$ of P and Q by

$$P \oplus Q = (P - Q) \cup (Q - P).$$

Theorem 1.17. For multisets P and $Q, P \oplus Q = (P \cup Q) - (P \cap Q)$.

Proof. Let $x \in P \oplus Q$ be an element of multiplicity ≥ 1. Then multiplicity of $x = \max\{$multiplicity of x in $P - Q$, multiplicity of x in $Q - P\}$

$$= \max\{((\text{multiplicity of } x \text{ in } P) - (\text{multiplicity of } x \text{ in } Q)),$$
$$((\text{multiplicity of } x \text{ in } Q) - (\text{multiplicity of } x \text{ in } P))\}.$$

Observe that one of

(multiplicity of x in P) − (multiplicity of x in Q) and

(multiplicity of x in Q) − (multiplicity of x in P) is positive and the other is negative. Therefore multiplicity of x equals

[(multiplicity of x in P) − (multiplicity of x in Q)] or

[(multiplicity of x in Q) − (multiplicity of x in P)],

whichever is positive. If multiplicity of x in $P >$ multiplicity of x in Q, then (multiplicity of x in P) – (multiplicity of x in Q) =

(multiplicity of x in $P \cup Q$) – (multiplicity of x in $P \cap Q$).

Similarly, if multiplicity of x in Q > multiplicity of x in P, then

(multiplicity of x in Q) – (multiplicity of x in P) =

(multiplicity of x in $P \cup Q$) – (multiplicity of x in $P \cap Q$).

Thus in either case multiplicity of x equals

(multiplicity of x in $P \cup Q$) – (multiplicity of x in $P \cap Q$).

Thus $P \oplus Q$ is a submultiset of $P \cup Q - P \cap Q$.

On the other hand, let x be in $P \cup Q - P \cap Q$. Then, multiplicity of

x = (multiplicity of x in $P \cup Q$) – (multiplicity of x in $P \cap Q$) =

(multiplicity of x in P) – (multiplicity of x in Q)

or

= (multiplicity of x in Q) – (multiplicity of x in P) whichever is positive. Therefore,

multiplicity of x = multiplicity of x in $P - Q$ or multiplicity of x in $Q - P$

= multiplicity of x in $(P - Q) \cup (Q - P) = P \oplus Q$.

Hence, $(P \cup Q) - (Q \cap P)$ is a submultiset of $P \oplus Q$.

Therefore, $P \oplus Q = (P \cup Q) - (Q \cap P)$.

Example 1.19. We have computed $P - Q$ and $Q - P$ as also $P \cup Q$ and $P \cap Q$ for the multisets A to G as in Example 1.18. We now compute the symmetric difference for pairs of mutisets A to G for which $P - Q$ and $Q - P$ have already been obtained.

1. $A \oplus B = \{a, d, d\} \cup \{b, c, e, e\} = \{a, b, c, d, d, e, e\}$.
2. $A \oplus C = (A - C) \cup (C - A) = (A - C) \cup \varphi = A - C = \{a, b, c, d, e\}$ and

   $(A \cup C) - (A \cap C) = A - C = A \oplus C$.

3. $B \oplus C = (B - C) \cup (C - B) = \{b, b, c, c, e, e, e\} \cup \{d\} = \{b, b, c, c, d, e, e, e\}$. Also

   $$B \cup C - B \cap C = \{a, a, b, b, b, c, c, c, d, d, e, e, e\} - \{a, a, b, c, d\}$$
   $$= \{b, b, c, c, d, e, e, e\} = B \oplus C.$$

4. $D \oplus E = (D - E) \cup (E - D) = (D - E) \cup \phi = D - E = \{2, 2, 3, 4\}$

and

$$(D \cup E) - (D \cap E) = D - E = \{2, 2, 3, 4\} = D \oplus E.$$

5. $D \oplus F = (D - F) \cup (F - D) = \{2, 2, 3, 4, 4\} \cup \{5, 6\} = \{2, 2, 3, 4, 4, 5, 6\}$
and

$$\begin{aligned}
(D \cup F) - (D \cap F) &= \{1, 1, 2, 2, 2, 2, 2, 3, 3, 4, 4, 4, 4, 5, 5, 5, 6\} \\
&\quad - \{1, 1, 2, 2, 3, 4, 4, 5, 5\} \\
&= \{2, 2, 3, 4, 4, 5, 6\} = D \oplus F.
\end{aligned}$$

6. $D \oplus G = (D - G) \cup (G - D) = \{2, 2, 3, 4, 4\} \cup \{5\} = \{2, 2, 3, 4, 4, 5\}$
and

$$\begin{aligned}
D \cup G - D \cap G &= \{1, 1, 2, 2, 2, 2, 2, 3, 3, 4, 4, 4, 4, 5, 5, 5, 5\} \\
&\quad - \{1, 1, 2, 2, 3, 4, 4, 5, 5, \} \\
&= \{2, 2, 3, 4, 4, 5\} = D \oplus G.
\end{aligned}$$

7. $E \oplus F = (E - F) \cup (F - E) = \{4\} \cup \{5, 6\} = \{4, 5, 6\}$
and

$$\begin{aligned}
(E \cup F) - (E \cap F) &= \{1, 1, 2, 2, 3, 4, 4, 4, 5, 5, 5, 6\} \\
&\quad - \{1, 1, 2, 2, 3, 4, 4, 5, 5\} \\
&= \{4, 5, 6\} = E \oplus F.
\end{aligned}$$

8. $E \oplus G = (E - G) \cup (G - E) = \{4\} \cup \{5\} = \{4, 5\}$
and

$$\begin{aligned}
(E \cup G) - (E \cap G) &= \{1, 1, 2, 2, 3, 4, 4, 4, 5, 5, 5\} \\
&\quad - \{1, 1, 2, 2, 3, 4, 4, 5, 5\} \\
&= \{4, 5\} = E \oplus G.
\end{aligned}$$

9. $F \oplus G = (F - G) \cup (G - F) = \{6\} \cup \varphi = \{6\}$
and

$$\begin{aligned}
(F \cup G) - (F \cap G) &= F - G = \{1, 1, 2, 2, 3, 4, 4, 5, 5, 5, 6\} \\
&\quad - \{1, 1, 2, 2, 3, 4, 4, 5, 5\} \\
&= \{6\} = F \oplus G.
\end{aligned}$$

Observe that the result of Theorem 1.17 stands verified in all cases in the above example. Also observe that if Q is a submultiset of P, then $P \oplus Q = P - Q$. In particular, if Q is the empty set($=$ multiset), then $P \oplus Q = P$.

Exercise 1.8.

1. For sets A, B and C, show that

$$(A \cup (B \cap C))' = A' \cap (B' \cup C'),$$

where for a set K, K' denotes the complement of K.

2. Let $A = \{0, 2, 4, 6, 8\}$, $B = \{0, 1, 2, 3, 4\}$ and $C = \{0, 3, 6, 9\}$, find

$$A \cup B \cup C \quad \text{and} \quad A \cap B \cap C.$$

3. Let $A = \{1, 2, 3, 4, 5\}$, $B = \{0, 3, 6\}$ and $C = \{0, 3, 6, 9\}$. Find
   (a) $A \cup B$; (b) $A \cap B$; (c) $A \cap C$; (d) $A \cup C$; (e) $A - B$; (f) $B - A$;
   (g) $A - C$; (h) $B - C$; (i) $C - A$; (j) $C - B$.

4. For any sets A and B show that
   (a) $A \cup (A \cap B) = A$; (b) $A \cap (A \cup B) = A$.

5. Find sets A and B if $A - B = \{1, 5, 7, 8\}$ and $B - A = \{2, 10\}$.

6. List the elements of the set
   (a) $\{n \in Z \mid n^2 - 10n - 24 < 0 \text{ and } 4 < n < 16\}$;
   (b) of prime numbers less than 25;
   (c) $\{q \in Q \mid q^2 - 1 = 15 \text{ and } q^3 = 60\}$.

7. Find the power set of A when
   (a) $A = \{\varphi\}$; (b) $A = \{\varphi, a, \{a\}\}$; (c) $A = \wp(\{a\})$.

8. For $A = \{a, b, \{a, c\}, \phi\}$, determine the following sets:
   (a) $A \backslash \{a\}$; (b) $A \backslash \phi$; (c) $A \backslash \{\phi\}$; (d) $A \backslash \{a, b\}$; (e) $\{a\} \backslash A$.

9. If A is a subset of the universal set X, prove that
   (a) $A \oplus A' = X$; (b) $A \oplus \varphi = A$; (c) $A \oplus X = A'$.

10. For any sets A, B show that $(A \oplus B) \oplus B = A$.

11. For sets A, B, C, if $A \oplus C = B \oplus C$, does it necessarily imply that $B = C$?

12. Give an example to show that intersection of two countable infinite sets is (a) finite; (b) again countable and infinite.

13. If A is a finite set having n elements, prove that A has exactly 2^n distinct subsets.
    (Refer to the section on mathematical induction in Chapter 3.)

14. Let S be a non-empty set and $\wp(S)$ be the power set of S. In $\wp(S)$ define addition and multiplication as follows: For A, $B \in \wp(S)$, define

    (a) $A + B = (A - B) \cup (B - A)$ (= the symmetric difference $A \oplus B$);
    (b) $AB = A \cap B$.

    For subsets A, B, C of S, prove the following relations.

    (a) $(A + B) + C = A + (B + C)$;
    (b) $A(B + C) = AB + AC$;
    (c) $AA = A$;
    (d) $A + A = \varphi$;
    (e) If $A + B = A + C$, then $B = C$.

    (Also refer to Question 11 above.)

15. Given sets $A = \{1, 2, 3\}$, $B = \{3, 5, 7\}$, find $A - B$ and $B - A$. Also find $A \times B$ and $B \times A$.

16. For any sets A, B, C, prove that

    (a) $A \times (B \cap C) = (A \times B) \cap (A \times C)$;
    (b) $A \times (B \cup C) = (A \times B) \cup (A \times C)$;
    (c) $(A \times B) - (A \times C) = A \times (B - C)$.

17. If A, B are two sets, prove that $\wp(A \cap B) = \wp(A) \cap \wp(B)$.

18. Give an example to show that

    (a) $\wp(A \cup B) \neq \wp(A) \cup \wp(B)$;
    (b) $\wp(A - B) \neq \wp(A) - \wp(B)$.

19. Show that the set A does not exist when

    $$A = \{S \mid S \text{ is a set such that } S \notin S\} \text{(Russells paradox)}.$$

20. If S is a finite set, prove that any map: $S \to S$ which is one–one is onto and any map: $S \to S$ which is onto is one–one.

21. Give an example to show that the two results in Question 20 are not true if S is an infinite set.

22. Give an example to show that if $\sigma : A \to B$, $\tau : B \to C$ are maps such that

    (a) $\tau \circ \sigma \colon A \to C$ is onto, it is not necessary that σ and τ are onto;
    (b) $\tau \circ \sigma \colon A \to C$ is one–one, it is not necessary that σ and τ are one–one.

23. For the set $\mathbb{Z}$ of integers, prove that the map $f : \mathbb{Z} \times \mathbb{Z} \to \mathbb{Z}$ defined by $f(m, n) = m + n$, m, $n \in \mathbb{Z}$, is onto but is not one–one.

24. Prove that the map $f : \mathbb{R} \to \mathbb{R}$ defined by $f(x) = e^x$, $x \in \mathbb{R}$, is one–one. Is it onto? Justify.

25. Let $f : X \to Y$ be a map. Prove that $f(A \cap B) = f(A) \cap f(B)$ for all subsets A, B of X if and only if f is one–one.

26. Prove or disprove that $\lceil x + y \rceil = \lceil x \rceil + \lceil y \rceil$ for all real numbers x and y.

27. Prove or disprove that $\lfloor x + y \rfloor = \lfloor x \rfloor + \lfloor y \rfloor$ for all real numbers x and y.

28. For the functions f and g from $\mathbb{R}$ to $\mathbb{R}$ defined by $f(x) = x^2 + 1$ and $g(x) = x + 2$, $x \in \mathbb{R}$, find $f \circ g$ and $g \circ f$. Also find $f + g$ and fg.

29. If f, g are functions such that f and $f \circ g$ are both one–one, is g necessarily one–one? Justify.

30. If f, g are functions such that f and $f \circ g$ are both onto, is g necessarily onto? Justify.

31. Consider the functions f, g where $f(x) = ax + b$, $g(x) = cx + d$ with a, b, c, d real constants. Find the values of the constants a, b, c, d so that $f \circ g = g \circ f$.

32. Determine the number of bytes required to encode n bits of data when n equals

    (i) 5; (ii) 10; (iii) 501; (iv) 2999; (v) 3000; (vi) 6; (vii) 16; (viii) 166; (ix) 29900; (x) 999; (xi) 1002.

33. Data are transmitted over a particular network in blocks of 1500 octets (blocks of 8 bits). Determine the number of blocks required to transmit the amounts of data over this Network. (Note that here a byte is a synonym for an octet, a kilobyte is 1,000 bytes and a megabyte is 1,000,000 bytes.)

    (i) 149 kilobytes of data, (ii) 150 kilobytes of data,
    (iii) 274 kilobytes of data, (iv) 384 kilobytes of data,
    (v) 1455 megabytes of data, (vi) 43.6 megabytes of data.

Chapter 2

PROPOSITIONAL CALCULUS AND LOGIC

Logic is the science of reasoning. Using logic it is possible to decide the validity or otherwise of a given statement. A statement may normally have two values associated with it and the process of argument or reasoning which leads to one of the two possible outcomes is called logic, or mathematical or symbolic logic. Symbols are associated with propositions and when there is more than one proposition, they are combined using set procedures. Tables, called truth tables, are constructed for a combination of propositions and a look at the table decides if the combined proposition is valid or otherwise. This chapter is devoted to (i) introducing this propositional calculus and (ii) the construction of truth tables. Several examples on the use of logic or propositional calculus are given.

2.1. Propositions

A **proposition** is a statement or a directive statement that is either true or false. Following are some examples of propositions.

(a) It is not a holiday today.
(b) It is raining today.
(c) We shall have a dessert at dinner tonight.
(d) The attendance of students in the class is thin today.
(e) India's prime minister during the years 1971–1980 was a lady.
(f) Every integer $n >$ the integer 2 is not a prime.
(g) There exists an integer n such that $2n + 4 = 5$.
(h) Every integer $n > 1$ is divisible by a prime.

41

(i) A square matrix of any order with real entries having a non-zero determinant is always invertible.

The following statements are not propositions.

(a) What date is it today?
(b) Is it raining?
(c) Is it sunny today?
(d) Please submit your answer sheets by 11 a.m.
(e) Do you have a discrete mathematics lecture today?

Propositions may be divided into two categories. A statement which is definitely true or definitely false is also a proposition. For example, 3 divides 9; 5 always divides a number, the digit in the unit's place of which is either 0 or 5; a number in which the digit in the unit's place is 1 is divisible by 5.

A proposition which is true under all circumstances is called a **tautology** and a proposition that is false under all circumstances is called a **contradiction**.

Propositions that are of special interest to us are the propositions that describe the properties of elements of a given universal set X (say). For example, if the universal set under consideration is the set Z of integers then '2 is not divisible by 3 and is not a power of 2' and '$t^2 - 5t + 4 = 0$' are propositions over Z.

The universal set over which a particular proposition is being considered shall be clear from the context.

We, for the time being, shall return to propositions in general and not necessarily refer to a universal set.

A proposition is generally referred to by certain symbol. For example, let P denote the proposition 'Every student in the class passed the final examination with distinction'. That the proposition P is true or false can be easily decided by looking at the result of the final examination. The two possible outcomes of a proposition are called the two possible values that P can take. The two possible values of P are generally taken as T or F depending on whether P is true or false, respectively.

Let p be the proposition 'It will rain tomorrow' and let q be the proposition 'Ram will not go to school'. Now consider the proposition 'If it rains tomorrow Ram will not go to school', which we may call r. Observe that if p is true then certainly q is true. However, Ram may or may not go to school tomorrow even if it does not rain tomorrow. Thus if p is true then

q is true but q may be true even when p is false. Thus r is true if q is true and p may or may not be true. However, r is vacuously true if both p and q are false. Thus r is true in three cases, namely when q is true and when both p and q are false.

Let p, q be two propositions. In view of the above example we say that 'p implies q' or 'if p then q' (denoted by $p \to q$). This is a proposition which is true when q is true or when both p, q are false. Propositions p and q are said to be equivalent (written as $p \leftrightarrow q$) if, when p is true, then q is also true and conversely, when q is true then p is also true. Examples of pairs of equivalent propositions are:

(a) p: water froze this morning; q: the temperature this morning was below $0^0 c$;

(b) p: Ram is Shivani's brother; q: Shivani is Ram's sister;

(c) p: a number m is divisible by 3; q: the sum of the digits forming a number m is divisible by 3; and,

(d) p: a number m is divisible by 11; q: the sum and difference of the digits forming the number m taken alternatively is divisible by 11.

The following pairs of propositions are not pairs of equivalent propositions.

(a) p: Dr. O.P. Bajpai is Kurukshetra University's Vice-Chancellor; q: a male person is Kurukshetra University's Vice-Chancellor.

(b) p: m is a prime number; q: m is not divisible by 3.

2.2. Compositions of Propositions

Definition. Let p and q be two propositions.

(a) We define the **disjunction** of p and q, which is denoted by $p \vee q$ to be a proposition which is true when one or both of p and q are true and which is false when both p and q are false.

(b) We define the **conjunction** of p and q denoted by $p \wedge q$ to be a proposition which is true when both p and q are true and is false when either one or both of p and q are false.

(c) We define the **negation** of p denoted by p^- or $-p$ to be the proposition which is true when p is false and is false when p is true.

Definition. A proposition which is obtained from a combination of other propositions is called a **combined proposition**, and a proposition which is not a combination of other propositions is called an **atomic proposition**.

In symbols we can describe combined propositions through tables, which give values of the combined proposition for all possible values of propositions making the combined proposition. For example, $p \vee q$ (disjunction), $p \wedge q$ (conjunction) and $-p$ (negation) are represented by

Table 2.1.

(i) $p \vee q$			(ii) $p \wedge q$			(iii) p^-	
p	q	$p \vee q$	p	q	$p \wedge q$	p	p^-
F	F	F	F	F	F	F	T
T	F	T	F	T	F	T	F
F	T	T	T	F	F		
T	T	T	T	T	T		

respectively. The tables as above are called truth tables of the combined propositions.

Following are some examples of combined propositions.

1. A car is fitted with a central electronic lock and a siren/alarm is sounded as soon as someone tries to open it without the proper key. Consider the following propositions.

   p: A burglar touches the car door lock without the proper key,
   q: The alarm will be sounded.

   We can consider the proposition 'If the burglar touches the car door lock, then the alarm will be sounded'.

2. Most companies manufacture refrigerators that shut off as soon as the transmission voltage increases beyond a certain acceptable limit, which is normally 260 volts. We can consider the following pair of propositions.

   p: The transmission voltage increases 260 volts,
   q: The refrigerator shuts off.

   We can combine these two propositions as 'If the transmission voltage increases 260 volts, then the refrigerator shuts off'.

   To take care of such situations, we have already considered combining propositions in general. Let p and q be two propositions. We have defined two combined propositions, namely 'If p then q' or 'p implies q' and 'p if and only if q'. Using the definitions of these combined propositions, we write the truth tables of these combined propositions.

Table 2.2.

(i) $p \to q$			(ii) $p \leftrightarrow q$		
p	q	$p \to q$	p	q	$p \leftrightarrow q$
F	F	T	F	F	T
F	T	T	F	T	F
T	F	F	T	F	F
T	T	T	T	T	T

2.3. Truth Tables and Applications

As for atomic propositions, we can also combine compound propositions. Moreover, if P, Q are combined propositions then, as for atomic propositions, we can talk of P if and only if Q. If P and Q are propositions such that P is true if and only if Q is true then we say that P and Q are equivalent propositions. Thus the propositions P and Q are identical or equivalent if and only if the truth tables of P and Q are identical. Therefore, if we need to check whether P and Q are equivalent or not we only need to compare the truth tables of the two propositions and if the two truth tables are identical, then the two propositions are equivalent. If p, q, r, s are propositions, we may consider combined propositions such as:

(a) $((p \wedge q) \vee (p^- \wedge q^-))$,   (b) $((p \wedge q) \vee (p^- \wedge q^-)) \to p$,
(c) $((p \wedge q) \vee (p \wedge r)) \to s$,   (d) $(p^- \vee (q^- \wedge r^-)) \vee s$.

In this section we consider obtaining the truth tables of some such combined propositions and applications of truth tables and propositional calculus. Consider a proposition p, its negation p^- and the negation $(p^-)^-$ of p^-. Observe that if p is true, then p^- is false and, therefore, $(p^-)^-$ is true. On the other hand, if p is false, then p^- is true and therefore $(p^-)^-$ is false. Hence, it follows that p is true if and only if $(p^-)^-$ is true. Thus the propositions p and $(p^-)^-$ are equivalent.

Example 2.1. Let p, q be two propositions. We construct the truth tables of the propositions $(p \vee q)^-$, $(p \wedge q)^-$, $p^- \vee q^-$ and $p^- \wedge q^-$ and find a relationship between these.

Table 2.3. Equivalence of $(p \vee q)^-$ & $p^- \wedge q^-$ and $(p \wedge q)^-$ & $p^- \vee q^-$

p	q	p^-	q^-	$p \vee q$	$p \wedge q$	$(p \vee q)^-$	$(p \wedge q)^-$	$p^- \wedge q^-$	$p^- \vee q^-$
F	F	T	T	F	F	T	T	T	T
F	T	T	F	T	F	F	T	F	T
T	F	F	T	T	F	F	T	F	T
T	T	F	F	T	T	F	F	F	F

Looking at this table we find that the seventh and ninth columns are identical and so are the eighth and tenth columns. Therefore, the propositions $(p \vee q)^-$ and $p^- \wedge q^-$ are equivalent and so are the propositions $(p \wedge q)^-$ and $p^- \vee q^-$.

Example 2.2. If p, q, r, s are propositions, construct the truth tables of the following propositions.

(a) $((p \wedge q) \vee (p^- \wedge q^-))$,   (b) $((p \wedge q) \vee (p^- \wedge q^-)) \to p$,

(c) $((p \wedge q) \vee (p \wedge r)) \to s$,   (d) $(p^- \vee (q^- \wedge r^-)) \vee s$.

Solution

(a)

Table 2.4.1.

p	q	$p \wedge q$	p^-	q^-	$p^- \wedge q^-$	$(p \wedge q) \vee (p^- \wedge q^-)$
F	F	F	T	T	T	T
F	T	F	T	F	F	F
T	F	F	F	T	F	F
T	T	T	F	F	F	T

(b)

Table 2.4.2.

p	q	$p \wedge q$	p^-	q^-	$p^- \wedge q^-$	$(p \wedge q) \vee (p^- \wedge q^-)$	$((p \wedge q) \vee (p^- \wedge q^-)) \to p$
F	F	F	T	T	T	T	F
F	T	F	T	F	F	F	T
T	F	F	F	T	F	F	T
T	T	T	F	F	F	T	T

(c)

Table 2.4.3.

p	q	r	s	$p \wedge q$	$p \wedge r$	$(p \wedge q) \vee (p \wedge r)$	$((p \wedge q) \vee (p \wedge r)) \to s$
F	F	F	F	F	F	F	T
F	F	F	T	F	F	F	T
F	F	T	F	F	F	F	T
F	T	F	F	F	F	F	T
F	F	T	T	F	F	F	T

(*Continued*)

Table 2.4.3. (*Continued*)

p	q	r	s	$p \wedge q$	$p \wedge r$	$(p \wedge q) \vee (p \wedge r)$	$((p \wedge q) \vee (p \wedge r)) \to s$
F	T	F	T	F	F	F	T
F	T	T	F	F	F	F	T
F	T	T	T	F	F	F	T
T	F	F	F	F	F	F	T
T	F	F	T	F	F	F	T
T	F	T	F	F	T	T	F
T	T	F	F	T	F	T	F
T	F	T	T	F	T	T	T
T	T	F	T	T	F	T	T
T	T	T	F	T	T	T	F
T	T	T	T	T	T	T	T

(d)

Table 2.4.4.

p	q	r	s	p^-	q^-	r^-	$q^- \wedge r^-$	$p^- \vee (q^- \wedge r^-)$	$(p^- \vee (q^- \wedge r^-)) \vee s$
F	F	F	F	T	T	T	T	T	T
F	F	F	T	T	T	T	T	T	T
F	F	T	F	T	T	F	F	T	T
F	T	F	F	T	F	T	F	T	T
F	F	T	T	T	T	F	F	T	T
F	T	F	T	T	F	T	F	T	T
F	T	T	F	T	F	F	F	T	T
F	T	T	T	T	F	F	F	T	T
T	F	F	F	F	T	T	T	T	T
T	F	F	T	F	T	T	T	T	T
T	F	T	F	F	T	F	F	F	F
T	T	F	F	F	F	T	F	F	F
T	F	T	T	F	T	F	F	F	T
T	T	F	T	F	F	T	F	F	T
T	T	T	F	F	F	F	F	F	F
T	T	T	T	F	F	F	F	F	T

When we compare the truth tables of the propositions as in (c) and (d), we find that the two are identical. Hence, the propositions $((p \wedge q) \vee (p \wedge r)) \to s$ and $(p^- \vee (q^- \wedge r^-)) \vee s$ are equivalent.

Example 2.3. There are two fruit shops close to each other. One has a sign that says 'Good fruit is not cheap', while the other has a sign that says 'Cheap fruit is not good'. Going only by the sign boards, which shop should one prefer?

Solution. Let g be the proposition 'Fruit is good' and c be the proposition 'Fruit is cheap'. In terms of the propositions g and c, the first sign says if g is true then c is not true. In terms of symbols it says '$g \to c^-$'. Similarly, in terms of g and c, the second sign says '$c \to g^-$'. Thus in order to see if the two signs say the same thing or not, we need to compare their truth tables. The truth tables of the two propositions are as below.

Table 2.5.

g	c	g^-	c^-	$g \to c^-$	$c \to g^-$
F	F	T	T	T	T
F	T	T	F	T	T
T	F	F	T	T	T
T	T	F	F	F	F

Observe that the fifth and sixth columns in this truth table are identical. Therefore, the two propositions '$g \to c^-$' and '$c \to g^-$' are equivalent. Hence going purely by the sign boards, one may choose either of the two shops.

Example 2.4. Let p denote the proposition 'The weather is nice' and q denote the proposition 'We shall have a picnic'. Write the following in symbolic form and also write their truth tables.

(a) The weather is nice but we do not have a picnic.
(b) We shall have a picnic if and only if the weather is nice.
(c) If we do not have a picnic, the weather is not nice.
(d) Translate into English the proposition $(p^- \vee q)^- \vee (p \wedge q^-)$.

Solution. The three propositions as in (a), (b), (c) in symbolic language are: (a) $p \wedge q^-$; (b) $p \leftrightarrow q$; (c) $q^- \to p^-$. The truth tables of these propositions are given by:

Table 2.6.

p	q	p^-	q^-	$p \wedge q^-$	$p \leftrightarrow q$	$q^- \to p^-$
F	F	T	T	F	T	T
F	T	T	F	F	F	T
T	F	F	T	T	F	F
T	T	F	F	F	T	T

In view of the observations we have made about the negation of a conjunction and negation of negations, we find that $(p^- \vee q)^-$ is equivalent to $(p^-)^- \wedge q^-$, which in turn is equivalent to $p \wedge q^-$.

Therefore, the proposition $(p^- \vee q)^- \vee (p \wedge q^-)$ is equivalent to the proposition $(p \wedge q^-) \vee (p \wedge q^-)$, which in turn is equivalent to $p \wedge q^-$. However, this translates into: 'The weather is nice and we do not have a picnic'. The 'and' in this statement is in the sense of 'but'. Therefore, the proposition $(p^- \vee q)^- \vee (p \wedge q^-)$ translates into 'The weather is nice but we do not have a picnic'.

Example 2.5. Let p and q be the propositions:

p: it is below freezing point in Shimla today;
q: it is snowing in Shimla today.

Write the proposition using p and q and logical connectives:
    'That it is below freezing point in Shimla today is necessary and sufficient for it to be snowing in Shimla today'.

Solution. Observe that the given proposition is a combination of two propositions: if it is below freezing point, then it is snowing and if it is snowing, then it is below freezing point. In symbols it is the combination of the propositions 'if p then q' and 'if q then p', i.e. '$p \to q$' and '$q \to p$'. Thus the given proposition is '$p \leftrightarrow q$'. The truth table of this combined proposition has already been constructed.

Example 2.6. The truth value of the negation of a proposition in fuzzy logic is one minus the truth value of the proposition. Given that the truth value of the proposition 'Ram is happy' is 0.8 and the truth value of the proposition 'Krishan is happy' is 0.4, then the truth value of the proposition 'Ram is not happy' is equal to $1 - 0.8 = 0.2$ and the truth value of the statement Krishan is not happy is $= 1 - 0.4 = 0.6$.

Example 2.7. Show that the following are inconsistent.

$$P \to Q, \quad P \to R, \quad Q \to R^- \quad \text{and} \quad P.$$

Proof. Suppose that all four propositions are simultaneously true. Since P is true, it follows from the first two statements that both Q and R are true. Again, the proposition Q being true, it follows from the third given statement that R^- is true. Thus both R and R^- are simultaneously true. This is a contradiction as a proposition and its negation cannot be simultaneously true. Hence the given four statements are not consistent.

Example 2.8. Conclude D from the premises:

$$(A \to B) \wedge (A \to C), \quad (B \wedge C)^-, \quad D \vee A.$$

Proof. If D is true, we have nothing to prove. So, suppose that D is false. Since $D \vee A$ is true, either D is true or A is true. But D having been assumed to be false, A is true. Now $(A \to B) \wedge (A \to C)$, being true, both $A \to B$ and $A \to C$ are true. Also A being true, these two imply that both B and C are true. Therefore, $B \wedge C$ is true. However, $(B \wedge C)^-$ being given to be true, $B \wedge C$ is false. This is a contradiction, as $B \wedge C$ cannot be simultaneously true as well as false. Hence our assumption that D is false is not correct and so D is true.

Example 2.9. Translate into symbolic language and test the validity of the argument: 'If 6 is even, then 2 does not divide 7. Either 5 is not a prime or 2 divides 7. But 5 is prime. Therefore, 6 is odd'.

Solution. Let p, q, r be the propositions:

p: 6 is even,
q: 2 divides 7,
r: 5 is prime.

Then the given statements in symbolic form are: $p \to q^-$, $q \vee r^-$, and r. We have to prove that if these three propositions are true, then p is false. Suppose, on the contrary, that p is true. Then $p \to q^-$ implies that q^- holds; i.e. q is false. Then it follows from $q \vee r^-$ being true that r^- holds; i.e. r is false. But r is given to be true. Thus we have a contradiction. This proves that p is false. Observe that we had to prove the proposition $(p \to q^-) \wedge (q \vee r^-) \wedge r \to p^-$. We now prove it by constructing its truth table.

Table 2.7.

p	q	r	p^-	q^-	r^-	$p \to q^-$	$q \vee r^-$	$(p \to q^-) \wedge (q \vee r^-)$	$(p \to q^-) \wedge (q \vee r^-) \wedge r$	$(p \to q^-) \wedge (q \vee r^-) \wedge r \to p^-$
F	F	F	T	T	T	T	T	T	F	T
F	F	T	T	T	F	T	F	F	F	T
F	T	F	T	F	T	T	T	T	F	T
T	F	F	F	T	T	T	T	T	F	T
F	T	T	T	F	F	T	T	T	T	T
T	F	T	F	T	F	T	F	F	F	T
T	T	F	F	F	T	F	T	F	F	T
T	T	T	F	F	F	F	T	F	F	T

Thus the proposition $(p \to q^-) \wedge (q \vee r^-) \wedge r \to p^-$ holds. In particular, p^- holds when $(p \to q^-) \wedge (q \vee r^-) \wedge r$ is true and the argument is logically valid.

Example 2.10. Ram made the following statements.
(a) I adore Sita (b) If I adore Sita, I also adore Geeta.
Given that Ram either told the truth or lied in both cases, determine whether Ram really adores Sita.

Solution. (a) Let p, q denote the propositions:
  p: 'Ram adores Sita' q: 'Ram adores Geeta'.
Ram's statement as at (b) says 'if p then q', i.e. $p \to q$. The given hypothesis then is that either both p and $p \to q$ are true or both p and $p \to q$ are false. Consider the truth table.

Table 2.8.

p	q	$p \to q$
F	F	T
F	T	T
T	F	F
T	T	T

This table shows that both p and $p \to q$ cannot be simultaneously false. Hence, both the statements p and $p \to q$ must be true. In particular, p is true, i.e. Ram adores Sita.

Example 2.11. An island has two tribes. Any native from the first tribe always tells the truth, while any native from the second tribe always lies. A traveller arrives and asks a native if there is gold on the island. The native answers, 'There is gold on the island if and only if I always tell the truth'. Which tribe is he from? Is there gold on the island?

Solution Let p, q be the propositions:
p: He (the native) always tells the truth,
q: There is gold on the island.

Then the answer provided by the tribesman is $p \leftrightarrow q$. Construct the truth table.
Suppose that p is true, i.e. he always tells the truth. Then the answer provided by him is true, i.e. $p \leftrightarrow q$ is true. The table below shows that p and

Table 2.9.

p	q	$p \rightarrow q$	$q \rightarrow p$	$p \leftrightarrow q$
F	F	T	T	T
F	T	T	F	F
T	F	F	T	F
T	T	T	T	T

$p \leftrightarrow q$ are simultaneously true only in the case when q is true. Now suppose that p is false, i.e. he always tells a lie. Then the answer provided by him is not correct, i.e. $p \leftrightarrow q$ is false. Thus p and $p \leftrightarrow q$ are simultaneously false. Again, the above table shows that this can happen only in the case when q is true. Thus in either case, i.e. whether he always tells the truth or always tells a lie, q is true. Hence there is gold on the island. Also q can be true in both cases when p is true or p is false. Thus we are unable to decide whether he always tells the truth or he always lies.

Example 2.12. Write a compound statement that is true when none, one or two of the three statements p, q and r are true. Justify your answer.

Solution. Consider the statement '$p \wedge q \wedge r$'. This proposition is true when all the three propositions, p, q and r, are true and is false in all other cases, i.e. when none of the three statements is true or when any one of the three statements is true or when two of the three statements are true. Therefore, its negation is the proposition which is true when none, one or two of the three statements are true and is false when all the three statements are true. Hence the required statement is $(p \wedge q \wedge r)^-$. Alternatively, we can construct the truth table.

Table 2.10.

p	q	r	$p \wedge q \wedge r$	$(p \wedge q \wedge r)^-$
F	F	F	F	T
F	F	T	F	T
F	T	F	F	T
T	F	F	F	T
F	T	T	F	T
T	F	T	F	T
T	T	F	F	T
T	T	T	T	F

The statement made above is clear from the truth table.

Example 2.13. Write a compound statement that is true when exactly two of the three statements p, q, r are true.

Solution. Consider the proposition $p \wedge q \wedge r^-$, which is obviously true only when both p, q are true and r is false. Also, the proposition $p \wedge q^- \wedge r$ is true only when p, r are true and q is false and $p^- \wedge q \wedge r$ is the proposition which is true only when q, r are true but p is false. The proposition $(p \wedge q \wedge r^-) \vee (p \wedge q^- \wedge r) \vee (p^- \wedge q \wedge r)$ is true when either $p \wedge q \wedge r^-$ or $p \wedge q^- \wedge r$ or $p^- \wedge q \wedge r$ is true. Let the propositions $p^- \wedge q \wedge r$, $p \wedge q^- \wedge r$ and $p \wedge q \wedge r^-$ be denoted by P, Q and R respectively. If propositions P and Q are true together, then the propositions p and p^- are true together. But this is not possible. Hence P and Q cannot hold together. Similarly, P and R cannot hold together and nor can Q and R. Thus no two of P, Q, R can be simultaneously true. Hence when $P \vee Q \vee R$ holds, then exactly one of P, Q, R holds. Therefore, $P \vee Q \vee R$ is a proposition that is true when exactly two of the three propositions p, q, r are true.

This result can be further confirmed by the following truth table, which shows that $P \vee Q \vee R$ is a proposition that is true exactly when two of the propositions p, q, r are true and the third is false.

Table 2.11.

p	q	r	p^-	q^-	r^-	$p \wedge q$	$q \wedge r$	$p \wedge r$	P	Q	R	$P \vee Q \vee R$
F	F	F	T	T	T	F	F	F	F	F	F	F
F	F	T	T	T	F	F	F	F	F	F	F	F
F	T	F	T	F	T	F	F	F	F	F	F	F
T	F	F	F	T	T	F	F	F	F	F	F	F
F	T	T	T	F	F	F	T	F	T	F	F	T
T	F	T	F	T	F	F	F	T	F	T	F	T
T	T	F	F	F	T	T	F	F	F	F	T	T
T	T	T	F	F	F	T	T	T	F	F	F	F

Example 2.14. A certain country is inhabited only by people who always tell the truth or always tell lies and who will respond to questions only with a 'yes' or a 'no'. A tourist comes to a fork in the road, where one branch leads to the capital and the other does not. There is no sign indicating which branch to take, but there is an inhabitant, Mr. Z, standing at the fork. What single question should the tourist ask him to determine which branch to take?

Solution. Let p, q be the propositions:

p: Mr. Z always tells the truth,
q: The left branch leads to the capital.

As per the requirement, we need to formulate a question or a combined proposition which is equivalent to q. Consider the following truth table.

Table 2.12.

p	q	p^-	$p \to q$	$p^- \to q$	$(p \to q) \wedge (p^- \to q)$
T	T	F	T	T	T
T	F	F	F	T	F
F	T	T	T	T	T
F	F	T	T	F	F

Thus $(p \to q) \wedge (p^- \to q)$ is equivalent to q. This proposition translates to the statement or question 'Whether you (Mr. Z) always tell the truth or otherwise, the left branch leads to the capital'. If the answer to the question is yes, then $p \to q$ and $p^- \to q$ hold together which means that whether Mr. Z tells the truth or otherwise, q is true. Hence the left road leads to the capital.

If, on the other hand, the answer to the question is 'no', then either $p \to q$ is false, which happens when p is true but q is false, or $p^- \to q$ is false, which happens when p and q are both false. Hence q is false in any case and the road to the left does not lead to the capital.

Thus, if Mr. Z answers the question with a 'yes', the tourist will decide to go along the left road and if Mr. Z answers the question with a 'no', the tourist will decide to go along the road to the right.

Example 2.15. Ram, Rahim and Krishan belong to the Himalaya club. Every club member is either a skier or a mountain climber, or both. No mountain climber likes rain and all skiers like snow. Rahim dislikes whatever Ram likes and likes whatever Ram dislikes. Ram likes rain and snow. Is there a member of the club who is a mountain climber but not a skier?

Solution. Since Ram likes rain and snow, Rahim dislikes both rain and snow. Moreover, all skiers like snow. Therefore, Rahim is not a skier. As every club member is either a mountain climber or a skier, Rahim, not being a skier, must be a mountain climber. Thus, Rahim is a mountain climber but is not a skier.

2.4. Some Further Applications of Logic

In this section we give some other typical applications of logic where truth tables may not be of much help.

Example 2.16. Four employees have been identified as suspects for unauthorized access into the computer system of a company. They made the following statements to the investigating authorities:

Anil said, 'Chetan did it';
Balkar said,'I did not do it';
Chetan said, 'Davender did it';
Davender said, 'Chetan lied when he said that I did it'.

(a) If the authorities know that one of the four suspects is telling the truth, who did it? Explain your reasoning.
(b) If the authorities know that one of them is lying, who did it? Explain your reasoning.

Solution. For the sake of convenience, we write A, B, C and D respectively for Anil, Balkar, Chetan and Davender. Suppose that both C and D are telling a lie. C telling a lie means that D did not do it. D telling a lie means that 'C did not tell a lie when he said that I did it'. This amounts to D accepting that he did it. But 'D did it' and 'D did not do it' cannot be simultaneously true. Hence, at most, one of the two is telling a lie. If both D and C are telling the truth, again it amounts to saying that D did it and D did not do it. Hence, one of C or D is telling the truth and the other is telling a lie.

(a) When one of the four is telling the truth, then in view of the above argument, it is either C or D who is telling the truth. If C is telling the truth, then B, D telling a lie implies that both B and D did it. If, on the other hand, D is telling the truth but the other three are lying, it amounts to saying that B did it.
(b) If A, B, C are telling the truth but D is telling a lie, then both C and D did it. If, on the other hand, A, B, D are telling the truth but C is telling a lie, then C did it.

Example 2.17. The zebra puzzle, Albert Einstein: five men with different nationalities and different jobs live in consecutive houses on a street. These houses are painted in different colours. The men have different pets and different favourite drinks. Determine who owns a zebra and whose favourite drink is mineral water (which is one of the favourite drinks), given these clues: the Englishman lives in the red house; the Spaniard owns a dog; the Japanese man is a painter; the Italian drinks tea; the Norwegian lives in the first house on the left; the green house is on the right of the white house; the photographer breeds snails; the diplomat lives in the yellow house; milk is drunk in the middle house; the owner of the green house drinks coffee; the Norwegian's house is next to the blue one; the violinist drinks orange juice; the fox is in a house next to that of the physician; the horse is in a house next to that of the diplomat.

Solution. We solve this puzzle by completing the table where the rows represent nationalities of men and the columns represent the colour of their houses, their favourite drinks, their pets and their jobs. We write B, S, J, I and N for the Englishman, the Spaniard, the Japanese, the Italian and the Norwegian, respectively. List the houses in a row — the first one being occupied by N. The green house is to the right of the white one and N lives in a house next to the blue one. Since N's house is next to the blue one, N's house cannot be the white one. Also, it cannot be green because there is no house to the left of the one occupied by N. The red house is already occupied by B. Therefore, N lives in the yellow house. Since the diplomat lives in the yellow house, N is a diplomat. The yellow and blue houses are adjacent to each other and so are the white and green. The red house must be in the middle, or it must be the last one to the right. If the red house is on the extreme right, then the white house will be in the middle.

Case (i) Suppose that red house is in the middle. Then the favourite drink of B is milk. The given data also shows that, 'photographer' and 'snails' must come together and 'violinist' and 'juice' must also come together. There are two choices for the photographer and snails to come together. These go either with B or with I.

(a) Suppose that 'photographer and snails' go with B. To complete the table containing the data so far, observe that the order of the houses is as follows.

N   B

Yellow   blue   red   white   green

Since the horse is in the house next to that of the diplomat and the house next to that of the diplomat is blue, 'blue' and 'horse' must come together. Let us construct the table with the information we have collected so far.

Table 2.13.

Men	Colour of house	F. drink	Pets	Job
B	Red	Milk	Snails	Photographer
S		Juice	Dog	Violinist
J				Painter
I		Tea		
N	Yellow			Diplomat

There is now only one choice for 'violinist' and 'juice' to come together, i.e. these two together go with S. 'Green' and 'coffee' come together and, looking at the incomplete table constructed above, we find that there is only one way for these to come together. It is with J. There is only one vacant position each in the second and the fifth columns. Now we fill these positions with the only choices possible. The only choice possible for the vacant position in the fifth column is to fill it with 'physician'. The horse is in the house next to that of the diplomat's and this is blue. Therefore 'blue' and 'horse' must come together. But there is only one choice left for this to happen, namely with I. There is only one choice left for the white house, and it goes with S. Since the fox is in the house next to that of the physician and the physician is in blue house, the fox must be in the yellow house (it cannot be in the green house — the only other choice left). There is then only one choice for the zebra: it must be in the green house.

Table 2.14.

Men	Colour of house	F. drink	Pet	Job
B	Red	Milk	Snails	Photographer
S	White	Juice	Dog	Violinist
J	Green	Coffee	Zebra	Painter
I	Blue	Tea	Horse	Physician
N	Yellow	Water	Fox	Diplomat

Since the table has been completed without any ambiguity, the zebra is owned by J and mineral water is the favourite drink of N.

(b) Now suppose that the photographer and the snails go with I. Completing the table thus far, we get

Table 2.15.

Men	Colour of house	F. drink	Pet	Job
B	Red	Milk		Physician
S		Juice	Dog	Violinist
J	Blue		Horse	Painter
I		Tea	Snails	Photographer
N	Yellow			Diplomat

Since the horse is in the house next to that of the diplomat, and the house next to that of the diplomat is blue, 'blue' and 'horse' must go together. There is only one choice for these to go together and it is with J. 'Violinist' and 'juice' come together. There is only one choice for these to go together and they go together with S. Then there is only one choice for B's job: he must be a physician. The physician, i.e. B, is in the middle house and the next house is the white one, while on the other side it is blue. Thus, either the fox is in the blue house or it is in the white house. Since in the blue house there is the horse, 'white' and 'fox' must come together. But the only two vacancies for the fox are already fixed as a colour other than white. Hence the right choice for the snails and the photographer is to go with B.

Case (ii) We will next discuss the possibility of the red house being on the extreme right. Suppose that the red house is to the extreme right, so that the white house is in the middle. We tabulate the data as being already understood that N is a diplomat and lives in the yellow house.

Table 2.16.

Men	Colour of house	F. drink	Pet	Job
B	Red		Snails	Photographer
S		Juice	Dog	Violinist
J				Painter
I		Tea		Physician
N	Yellow			Diplomat

The following combinations are there: 'white and milk', 'green and coffee', 'violinist and juice' and 'photographer and snails'. 'Photographer and snails' can go either with B or with I.

(c) Suppose that 'photographer and snails' go with B. Then there is only one choice for 'violinist and juice'. In that case, I must be a physician. There is then only one possibility that colour and favourite drink may match. There being two possibilities left, namely 'white and milk' and 'green and coffee', it is not possible to complete the table.

(d) Thus, we have the other alternative that 'snails and photographer' go with I. Constructing the table with the data available so far, we have:

Table 2.17.

Men	Colour of house	F. drink	Pet	Job
B	Red			
S			Dog	
J				Painter
I		Tea	Snails	Photographer
N	Yellow			Diplomat

Since the 'white and milk' and 'green and coffee' combinations are there, there are two choices for each. In view of that, it follows that I will be in the blue house. Thus, I is in the house next to that of the diplomat and, so, his pet should be a horse — a choice which is not available. Hence, 'snails and photographer' cannot go with I.

The above then completes the proof that the red house cannot be on the extreme right and the solution to the puzzle is provided by (a) above.

Example 2.18. A detective has interviewed four witnesses to a crime. From the stories of the witnesses, the detective has concluded that: if the butler is telling the truth then so is the cook; the cook and the gardener cannot both be telling the truth; the gardener and the porter are not both lying; and, if the porter is telling the truth then the cook is lying. For each of the four witnesses, can the detective determine whether that person is telling the truth or lying? Explain your reasoning.

Solution. Let b, c, g and p stand respectively for the butler, the cook, the gardener and the porter telling the truth. Then the given hypotheses are: $b \to c$, $(c \wedge g)^- = (c^-) \vee (g^-)$, $((g^-) \wedge (p^-))^- = g \vee p$ and $p \to c^-$. If c is true, then $(c^-) \vee (g^-)$ is given, which implies that g^- is true, i.e. g is not true. Then $g \vee p$ is given, which implies that p is true and, finally, $p \to c^-$ is given, which implies that c^- holds, i.e. c is false. This is a contradiction and it follows that c is false. However, $b \to c$ implies that b is also false. Hence both the butler and the cook are telling a lie.

Now c^- and $(c^-) \vee (g^-)$ both hold, so g may or may not be true. Then $g \vee p$ holds, which implies that p may or may not be true. Also, in either case

$p \rightarrow c^-$ holds, which does not lead to a contradiction. Hence the detective is unable to decide whether the gardener is telling the truth or not. Similarly, he is unable to decide whether the porter is telling the truth or otherwise.

Example 2.19. Sanjay would like to determine the relative salaries of three colleagues using three facts. First, he knows that if Farid is not the highest paid of the three, then Jaya is. Second, he knows that if Jaya is not the lowest paid, then Manjit is paid the most. Third, no two of the three receive the same salary. Is it possible to determine the relative salaries of Farid, Manjit and Jaya from what Sanjay knows? If so, who is paid the most and who the least? Explain the reasoning.

Solution. Let F, J and M be the salaries drawn by Farid, Jaya and Manjit, respectively. Then the given information is $F \neq M$, $F \neq J$, $M \neq J$ and
If $F \not> \max\{M, J\}$, then $J > \max\{M, F\}$;
If $J \not< \min\{F, M\}$, then $M > \max\{J, F\}$.

The first among the above pieces of information says that either F or J is largest. The second piece of information says that either J is the least or M is the maximum. If we now let F^*, M^*, J^* stand for 'F is the most', 'M is the most' and 'J is the most', respectively, and let A_* stand for 'A is the least', then the above pieces of information translate to the two statements, '$F^* \vee J^*$' and '$J_* \vee M^*$' being simultaneously true. If we assume that J^* holds, then J_* does not hold and M^* holds. Thus both J^* and M^* hold, which is not the case. Therefore, J^* does not hold. Hence F^* holds. If J_* does not hold then M^* holds. Thus both F^* and M^* hold together which is again not the case. Hence J_* holds. This proves that J is the least and F is the most. Therefore, we must have $J < M < F$, and the salary structures of Jaya, Manjit and Farid increase in that order, i.e. Jaya has the lowest salary, Manjit has a higher salary than Jaya but a lower one than Farid, who has the highest salary.

Example 2.20. An island has two kinds of inhabitants, knights who always tell the truth, and their opposites, knaves who always lie. A traveller comes across two people; A and B. What are A and B if:

(a) A says, 'At least one of us is a knave', and B says nothing;
(b) A says, 'The two of us are both knights', and B says, 'A is a knave';
(c) A says, 'I am a knave or B is a knight', and B says, 'A is a knight';
(d) Both A and B say, 'I am a knight';
(e) A says, 'We are both knaves', and B says nothing.

Solution. If X is a knight we express it as X and if X is a knave we write X^-.

(a) A says, '$A^- \vee B^-$ is true'. If A is a knight, then A tells the truth and so $A^- \vee B^-$ is true. But A being a knight, A^- is not true. Therefore, B^- is true, i.e. B is a knave. Thus A is a knight and B is a knave.

If, on the other hand, A is a knave, then A lies. Therefore, $A^- \vee B^-$ is false, i.e. $(A^- \vee B^-)^-$ is true or that $A \wedge B$ is true. Thus, both A and B are knights. But A is a knave and, hence, a contradiction. Therefore, we have that A is a knight and B is a knave.

(b) A says, '$A \wedge B$ is true', and B says, 'A^- is true'. If A is a knight, then A tells the truth. Then $A \wedge B$ is indeed true and A, B are both knights. Then B also tells the truth and, therefore, A is a knave. This is a contradiction. Therefore, A is not a knight, i.e. A^- is true. Since B says, 'A^- is true', which is the case, B tells the truth. Hence B is a knight. Thus A is a knave and B is a knight.

(c) A says, '$A^- \vee B$ is true'. If A is a knight, he tells the truth. So $A^- \vee B$ is indeed true. But A^- is not true. Therefore, B is true. Hence, both A and B are knights.

If A were a knave, then $A^- \vee B$ is false, i.e. neither A^- is true nor B is true. Thus, neither A is a knave nor B is a knight. This is a contradiction, as A was assumed to be a knave. Thus the previous conclusion holds and both A and B are knights.

(d) A says, 'A is true' and B says 'B is true'. If A is indeed a knight, he tells the truth and there is no contradiction in his statement. If A were a knave, he tells a lie and, therefore, his statement, 'he is a knight' is not true. Again, there is no contradiction. Hence, A may be a knight or may be a knave. Similarly, B could be a knight or a knave.

(e) A says, '$A^- \wedge B^-$ is true'. If A were a knight, then the statement 'A is a knave and B is a knave' is true. Thus, on the one hand, A is a knight and on the other, he is a knave — a contradiction. Therefore, A is a knave. But then the statement $A^- \wedge B^-$ is false, i.e. either A^- is false or B^- is false. But A^- is true. Therefore, B^- is false, so B is true. Hence A is a knave and B is a knight.

Example 2.21. The police have three suspects for the murder of Mr. Chetan, Mr. Sumit, Mr. Janardhan and Mr. Walia. Sumit, Janardhan and Walia each declare that they did not kill Chetan. Sumit also states that Chetan was a friend of Janardhan and that Walia disliked him. Janardhan also states that he did not know Chetan and that he was out of town the day Chetan was killed. Walia also states that he saw both Sumit and Janardhan with Chetan on the day of the killing and that either Sumit or Janardhan must have killed him. Can you determine who the murderer was if:

(a) one of the three men is guilty, the two innocent men are telling the truth, but the statements of the guilty man may or may not be true;
(b) innocent men do not lie;
(c) only one person is telling the truth and the other two are lying?

Solution. Let C, S, J, W stand for Chetan, Sumit, Janardhan and Walia, respectively. Then the statements made by the three suspects are:

S: 'I did not kill C',   'C was a friend of J', and,
                                'W disliked C';
J: 'I did not kill C',   'I did not know C', and,
                                 'I was out of town on the fateful day';
W: 'I did not kill C',   'I saw both S and J with C on the fateful
                                 day', and,
                                 'Either S or J must have killed C'.

The statements on the left made by the three are simple denials and, therefore, we will concentrate on the statements on the right made by the three suspects. The statements by S and J cannot both be correct because S says C was a friend of J and J says he did not know C. The statements by J and W also cannot both be correct as J says he did not know C, while W says that he saw S and J with C on the fateful day. There is no contradiction in the first part of the statements by S and W. Therefore,

(a) if one of the men is guilty and the two others are innocent, then J, i.e. Janardhan is the guilty person;
(b) as already pointed out, J is singled out, as J and S cannot both be correct and J and W cannot both be correct. If J were telling the truth, then both S and W are lying, so neither of them is innocent. Therefore, either J killed C or S and W together killed C.
(c) As pointed out in (b) above, either only J is telling the truth and both S and W are lying or only J is lying and both S and W are telling the truth. Since only one person is telling the truth, it must be J. It is then confirmed that J did not know C and that W did not dislike C. W telling a lie means that neither S nor J killed C. So, the only possible culprit could be W. Since, 'did not dislike C' is not sufficient enough to

take away a motive for murder, W might have killed C. There is also nothing to suggest that S may not have killed C. Therefore, both S and W might have killed C.

2.5. Functionally Complete Set of Connectives

In the compound propositions we have used five connectives, namely $\rightarrow$, $\leftrightarrow, \vee, \wedge$ and the negation of a proposition. We have denoted the negation of a proposition p by $-p$ or by p^-. For notational convenience we shall now denote the negation of a proposition p by $\neg p$. Thus the five connectives used are $\rightarrow$, $\vee$, $\wedge$, $\leftrightarrow$ and $\neg$.

Definition. A subset A of the set $S = \{\rightarrow, \vee, \wedge, \leftrightarrow, \neg\}$ of connectives is said to be a **functionally complete** set of connectives if any proposition involving the five connectives is equivalent to a proposition involving connectives in the subset A. A connective in the subset A is said to be **redundant** if any proposition involving this connective is equivalent to a proposition involving connectives from A, except this connective. A subset A of the set S of connectives is called a **minimal functionally complete** set of connectives if A is a functionally complete set of connectives and A contains no redundant connective.

In this section we intend to find some minimal functionally complete sets of connectives.

Observe that if p, q are two propositions, then $p \leftrightarrow q$ is equivalent to the statements $p \rightarrow q$ and $q \rightarrow p$ simultaneously or that $p \leftrightarrow q$ is equivalent to $(p \rightarrow q) \wedge (q \rightarrow p)$. Thus any proposition involving the connective $\leftrightarrow$ can be expressed as an equivalent proposition involving the remaining four connectives, namely $\rightarrow$, $\vee$, $\wedge$ and $\neg$. In fact, in any set of connectives containing $\rightarrow$, $\leftrightarrow$ and $\wedge$, the connective $\leftrightarrow$ is redundant.

Next, we prove that, in the presence of the connectives $\vee$ and $\neg$, the connective $\rightarrow$ is redundant. For this, let p, q be two propositions. We construct the truth table.

Table 2.18.

p	q	$\neg p$	$p \rightarrow q$	$(\neg p) \vee q$
F	F	T	T	T
F	T	T	T	T
T	F	F	F	F
T	T	F	T	T

Observe that the fourth and fifth columns in the above truth table are identical. Therefore, the propositions $p \rightarrow q$ and $(\neg p) \vee q$ are equivalent. Thus in the set of connectives containing $\rightarrow$, $\vee$ and $\neg$, the connective $\rightarrow$ is redundant.

Example 2.22. Let us consider, for example, three propositions p, q and r and a combined proposition $p \wedge (q \leftrightarrow r)$. As proved above, $q \leftrightarrow r$ is equivalent to the proposition $(q \rightarrow r) \wedge (r \rightarrow q)$. Also the proposition $q \rightarrow r$ is equivalent to $(\neg q) \vee r$ and $r \rightarrow q$ is equivalent to $(\neg r) \vee q$. Therefore, the proposition $p \wedge (q \leftrightarrow r)$ is equivalent to $p \wedge ((\neg q) \vee r) \wedge ((\neg r) \vee q)$.

In the above paragraphs, we have proved the following.

Proposition 2.1. The set $\{\vee, \wedge, \neg\}$ of connectives is functionally complete.

We will next prove the following.

Proposition 2.2. The set of connectives $\{\wedge, \neg\}$ is a functionally complete set of connectives and so is the set $\{\vee, \neg\}$ of connectives.

Proof. Let p and q be two propositions. We have proved that the proposition $\neg(p \wedge q)$ is equivalent to the proposition $(\neg p) \vee (\neg q)$ and the proposition $\neg(\neg p)$ is equivalent to the proposition p. Therefore, the proposition $p \vee q$ is equivalent to the proposition $(\neg(\neg p)) \vee (\neg(\neg q))$, which in turn is equivalent to the proposition $\neg((\neg p) \wedge (\neg q))$. In view of this, we find that any combined proposition involving the connectives $\vee$, $\wedge$ and $\neg$ can be shown to be equivalent to a combined proposition involving the connectives $\wedge$ and $\neg$. Therefore, in the set of connectives $\{\vee, \wedge, \neg\}$ the connective $\vee$ is redundant. Already we have proved that the set of connectives $\{\vee, \wedge, \neg\}$ is functionally complete. Hence the set of connectives $\{\wedge, \neg\}$ is functionally complete. Again, using the fact that the proposition $p \wedge q$ is equivalent to the proposition $\neg((\neg p) \vee (\neg q))$, we find that the set of connectives $\{\vee, \neg\}$ is functionally complete. Indeed, the functionally complete sets of connectives $\{\wedge, \neg\}$ and $\{\vee, \neg\}$ are minimal as well.

2.6. The Connectives NAND and NOR

We will now define two new connectives and prove that each of them is functionally complete. The new connectives are NAND and NOR, where **NAND** is the combination of negation (NOT) and conjunction (AND) while **NOR** is the combination of negation (NOT) and disjunction (OR). The connective NAND is denoted by $\uparrow$ and NOR is denoted by $\downarrow$. Thus, if p, q are two propositions, we say that

$$p \uparrow q \quad \text{if and only if } \neg(p \wedge q) \text{ and}$$
$$p \downarrow q \quad \text{if and only if } \neg(p \vee q).$$

We have already proved that each one of the sets $\{\wedge, \neg\}$ and $\{\vee, \neg\}$ of connectives is functionally complete. To show that each one of the connectives $\uparrow$ and $\downarrow$ is functionally complete, we need to prove that each one of the connectives $\wedge$, $\neg$ and $\vee$ can be expressed in terms of $\uparrow$ and also in terms of $\downarrow$.

If P and Q are two propositions, let us write $P \Leftrightarrow Q$ to mean that P and Q are equivalent propositions. Let p, q be two propositions. Observe that:

$$\neg p \Leftrightarrow (\neg p) \vee (\neg p) \Leftrightarrow \neg(p \wedge p) \Leftrightarrow p \uparrow p.$$
$$p \wedge q \Leftrightarrow \neg(\neg(p \wedge q)) \Leftrightarrow \neg(p \uparrow q) \Leftrightarrow (p \uparrow q) \uparrow (p \uparrow q)$$
$$p \vee q \Leftrightarrow (\neg(\neg p)) \vee (\neg(\neg q)) \Leftrightarrow \neg((\neg p) \wedge (\neg q)) \Leftrightarrow (\neg p) \uparrow (\neg q)$$
$$\Leftrightarrow (p \uparrow p) \uparrow (q \uparrow q).$$

Again, we have

$$\neg p \Leftrightarrow (\neg p) \wedge (\neg p) \Leftrightarrow \neg(p \vee p) \Leftrightarrow p \downarrow p$$
$$p \wedge q \Leftrightarrow (\neg(\neg p)) \wedge (\neg(\neg q)) \Leftrightarrow \neg((\neg p) \vee (\neg q)) \Leftrightarrow (\neg p) \downarrow (\neg q)$$
$$\Leftrightarrow (p \downarrow p) \downarrow (q \downarrow q)$$
$$p \vee q \Leftrightarrow \neg(\neg(p \vee q)) \Leftrightarrow \neg(p \downarrow q) \Leftrightarrow (p \downarrow q) \downarrow (p \downarrow q).$$

Thus the negation $\neg$, disjunction $\vee$ and conjunction $\wedge$ can all be expressed in terms of the connective $\uparrow$ and also in terms of the connective $\downarrow$. Therefore, any combined proposition can be expressed by using only the connective $\uparrow$ and also by using only the connective $\downarrow$.

Exercise 2.6.1.

1. Prove that the propositions

   (a) $p \vee (q \wedge r)$ and $(p \vee q) \wedge (p \vee r)$ are equivalent;
   (b) $p \wedge (q \vee r)$ and $(p \wedge q) \vee (p \wedge r)$ are equivalent.

2. Determine whether each of the following propositions is a tautology or a contradiction.

(a) $(p \vee q) \wedge (p \to r) \wedge (q \to s) \to r \vee s$;

(b) $(p \wedge q) \to p$;

(c) $(p \wedge (p \to q)) \to q$;

(d) $q \vee (p \wedge q^-) \vee (p^- \wedge q^-)$;

(e) $p \wedge q^- \leftrightarrow p^- \wedge q$;

(f) $((p \to (q \to p)) \leftrightarrow (p^- \to (p \to q))$;

(g) $(p \to (q \vee r)) \leftrightarrow ((p \to q) \vee (p \to r))$;

(h) $q \vee (p \wedge q^-) \vee (p^- \wedge q^-)$;

(i) $(p \wedge q) \to (p \vee q)$;

(j) $(p \wedge (p \to q)) \to q$;

(k) $((p \vee q) \wedge (p \to r) \wedge (q \to r)) \to r$;

(l) $(p^- \wedge (p \to q)) \to q^-$;

(m) $(q^- \wedge (p \to q)) \to p^-$.

3. Deduce $r \vee s$ from the premises

$$C \vee D, (C \vee D) \to H^-, \quad H \to (A \wedge B^-), \quad (A \wedge B^-) \to r \vee s.$$

4. Prove that q follows logically from $p \wedge (p \to q)$.

5. Prove that $(p \leftrightarrow q)$ and $(p \wedge q) \vee (p^- \wedge q^-)$ are logically equivalent.

6. Let $P(n)$ be the proposition '$8^n - 3^n$ is a multiple of 5'. Prove that $P(n)$ is a tautology over N.

7. Construct the truth tables of the following propositions.

(a) $p \to (q \to p)$;

(b) $(q \wedge p^-) \leftrightarrow r$;

(c) $q \vee (q^- \wedge p)$;

(d) $q \to (q \to p)$;

(e) $(q \wedge p) \vee (q \wedge p^-)$.

8. Show that $p \leftrightarrow q^-$ does not logically imply $p \to q$.

9. Does p follow logically from the premises $p \to q$, q? Justify.

10. If $p \to q$ is false, determine the truth value of

(a) $(p \wedge q)^- \to q$; (b) $p^- \vee (p \leftrightarrow q)$;

giving a complete argument.

11. If $p \to q$ is true, determine the truth value of

(a) $(p \wedge q) \to q^-$; (b) $(p \to q)^- \wedge p^-$.

12. If p, q are true propositions and propositions r and s are false, find the truth value of each of the following.

(a) $p^- \to r$,

(b) $(p \to r) \wedge (r \to s)$,

(c) $s \to p^-$,

(d) $p^- \to (r \to (r \to (q \vee s)))$,

(e) $(p \to (r \to s)) \wedge ((q \to s) \to r^-)$,

(f) $(r \wedge s) \to (p \vee q)$.

13. Without using truth tables, show that $(p \vee (p^- \wedge q))^-$ and $p^- \wedge q^-$ are logically equivalent.

14. If p, q are two propositions, we define the **exclusive or** of p and q to be the proposition which is true if one of the p, q is true and the other is false and it is false otherwise. We write $p \oplus q$ for the exclusive or of p and q. It is clear from the definition that $p \oplus q = q \oplus p$. Construct the truth table of $p \oplus q$.

15. Show that each of the following pairs of propositions are logically equivalent.

    (a) $p^- \leftrightarrow q$ and $p \leftrightarrow q^-$;
    (b) $p \rightarrow q$ and $q^- \rightarrow p^-$;
    (c) $(p \oplus q)^-$ and $p \leftrightarrow q$;
    (d) $(p \rightarrow r) \wedge (q \rightarrow r)$ and $(p \vee q) \rightarrow r$;
    (e) $(p \rightarrow r) \vee (q \rightarrow r)$ and $(p \wedge q) \rightarrow r$;
    (f) $p \leftrightarrow q$ and $p^- \leftrightarrow q^-$.

16. Show that each of the following is a tautology.

    (a) $(p \vee q) \wedge (p^- \vee r) \rightarrow (q \vee r)$;
    (b) $(p \rightarrow q) \wedge (q \rightarrow r) \rightarrow (p \rightarrow r)$.

17. Show that $(p \rightarrow q) \rightarrow r$ and $p \rightarrow (q \rightarrow r)$ are not equivalent.

The **dual** of a compound proposition which contains only the logical connectives $\vee$, $\wedge$ and $\neg$ is the proposition obtained from the given proposition by replacing each $\vee$ by $\wedge$, each $\wedge$ by $\vee$, each T by F and each F by T. The last part of the above statement amounts to saying replace each p by p^- and each p^- by p. The dual of a proposition p is denoted by p^*. For example, the dual of the proposition

(a) $p \wedge q^- \wedge r^-$ is $p^- \vee (q^-)^- \vee (r^-)^-$ or $p^- \vee q \vee r$;
(b) $(p \wedge q \wedge r) \vee s$ is $(p^- \vee q^- \vee r^-) \wedge s^-$;
(c) $(p \vee F) \wedge (q \vee T)$ is $(p^- \wedge T) \vee (q^- \wedge F)$.

Observe that if s is a compound statement, then each $\vee$ in s is replaced by $\wedge$ in s^* and then this $\wedge$ in s^* is replaced by $\vee$ in $(s^*)^*$ and each $\wedge$ in s is replaced by $\vee$ in s^* and then this $\vee$ in s^* is replaced by $\wedge$ in $(s^*)^*$. Thus each $\vee$ in s remains $\vee$ in $(s^*)^*$ and each $\wedge$ in s remains $\wedge$ in $(s^*)^*$. Also a p in s gets replaced by p^- in s^* and, then this p^- in s^* gets replaced by p in $(s^*)^*$. It thus follows that $(s^*)^*$ is the same as s. We have proved that $(s^*)^* = s$.

18. Show that the propositions $p \uparrow q$ and $q \uparrow p$ are equivalent. Are the propositions $p \downarrow q$ and $q \downarrow p$ equivalent? Justify.

19. Prove, if possible, the results of Examples 2.7, 2.8 and 2.9 using truth tables.

20. Use truth tables, if possible, to solve Examples 2.16 and 2.18.

Chapter 3

MORE ON SETS

We have discussed sets and some of their properties in the first chapter. In this chapter, we discuss the principles of inclusion and exclusion, and the pigeonhole principle. Both these principles have many interesting and useful applications, some of which are discussed here. Binary relations, particularly equivalence relations, in sets are discussed. Reflexive, symmetric and transitive closures of binary relations are discussed.

3.1. The Principle of Inclusion and Exclusion

Let A, B be finite sets. When we consider the union $A \cup B$ of A and B, we observe that the elements of $A \cap B$ occur as elements of A and also as elements of B. But elements in a set are all distinct. Therefore, when we count the elements in $A \cup B$, we first count the elements of A, then count the elements of B, and the number of elements of $A \cap B$ which have been counted twice will have to be subtracted from $o(A) + o(B)$ to get the number of elements in $A \cup B$. Thus we get

$$o(A \cup B) = o(A) + o(B) - o(A \cap B). \tag{3.1}$$

This relation is called **the principle of inclusion and exclusion.**

If A, B, C are three finite sets, using the above principle of inclusion and exclusion and the distributive laws we have

$$o(A \cup B \cup C) = o(A \cup B) + o(C) - o((A \cup B) \cap C)$$
$$= o(A \cup B) + o(C) - o((A \cap C) \cup (B \cap C))$$

$$= o(A) + o(B) - o(A \cap B) + o(C) - \{o(A \cap C) + o(B \cap C)$$
$$- o(A \cap C \cap B \cap C)\}$$
$$= o(A) + o(B) + o(C) - o(A \cap B) - o(B \cap C) - o(A \cap C)$$
$$+ o(A \cap B \cap C).$$

We can generalize this result further to a finite number of finite sets and prove

Theorem 3.1. If $A_1, A_2, \ldots, A_n$, $n \geq 2$ are finite sets then

$$o(A_1 \cup A_2 \cup \cdots \cup A_n) = \sum_{1 \leq i \leq n} o(A_i) - \sum_{1 \leq i < j \leq n} o(A_i \cap A_j)$$
$$+ \sum_{1 \leq i < j < k \leq n} o(A_i \cap A_j \cap A_k)$$
$$- \cdots + (-1)^{n-1} o(A_1 \cap A_2 \cap \cdots \cap A_n).$$

We can prove this result by induction on the number n of finite sets. The proof will be presented in a later section on the principle of mathematical induction.

We will now consider some applications of the inclusion and exclusion principle.

Example 3.1. A survey was conducted among 1,000 people. Of these, 595 are Republicans, 595 wear glasses, 550 like coffee, 395 are Republicans who wear glasses, 350 are Republicans who like coffee, 400 wear glasses and like coffee, and 250 are Republicans who wear glasses and like coffee. How many of them

(a) are not Republicans, do not wear glasses and do not like coffee?
(b) are Republicans who do not wear glasses and do not like coffee?

Solution. Let S denote the set of people who took part in the survey, A be the set of those people who, among them, are Republicans, B the set of those who wear glasses, and C be the set of those who like coffee. Then A' is the set of those who are not Republicans, B' is the set of those who do not wear glasses and C' is the set of those who do not like coffee.
Then

The set of Republicans who wear glasses $= A \cap B$;
The set of Republicans who like coffee $= A \cap C$;

The set of those who wear glasses and like coffee $= B \cap C$;
The set of Republicans who wear glasses and like coffee $= A \cap B \cap C$.

We have to find the order of
(a) $A' \cap B' \cap C'$ and (b) $A \cap B' \cap C' = (A' \cup B \cup C)'$.
Now

$$o(A \cup B \cup C) = o(A) + o(B) + o(C) - o(A \cap B) - o(B \cap C) - o(A \cap C)$$
$$+ o(A \cap B \cap C)$$
$$= 595 + 595 + 550 - 395 - 350 - 400 + 250$$
$$= 845.$$

Also $A' \cap B' \cap C' = (A \cup B \cup C)'$ and, therefore, the number of people who are not Republicans, do not wear glasses and do not like coffee $= 1,000 - 845 = 155$.

To find the number of Republicans who do not wear glasses and do not like coffee, we first find the number of people who are either not Republicans or wear glasses or like coffee. Thus, we find $o(A' \cup B \cup C)$. Now (using Theorem 1.3(a)) we find that

$$o(A' \cup B \cup C) = o(A') + o(B) + o(C) - o(A' \cap B) - o(A' \cap C) - o(B \cap C)$$
$$+ o(A' \cap B \cap C)$$
$$= o(A') + o(B) + o(C) - o(B - A) - o(C - A) - o(B \cap C)$$
$$+ o((B \cap C) - A)$$
$$= o(A') + o(B) + o(C) - \{o(B) - o(A \cap B)\}$$
$$- \{o(C) - o(A \cap C)\}$$
$$- o(B \cap C) + \{o(B \cap C) - o(A \cap B \cap C)\}$$
$$= o(A') + o(A \cap B) + o(A \cap C) - o(A \cap B \cap C)$$
$$= (1,000 - 595) + 395 + 350 - 250 = 900.$$

Hence $o(A \cap B' \cap C') = o((A' \cup B \cup C)') = 1,000 - 900 = 100$.

Example 3.2. Thirty cars are assembled in a factory. The options available are a radio, an air conditioner and radial tyres. It is known that 14 of the cars have radios, 11 of them have air conditioners, 8 of them have radial tyres and 5 of them have all three options. At least how many of these cars do not have any of these option at all?

Solution. Let R denote the set of cars that have a radio, AC the set of cars that have an air conditioner and RT the set of cars that have radial tyres. Then

$$o(R) = 14, \quad o(AC) = 11, \quad o(RT) = 8 \quad \text{and} \quad o(R \cap AC \cap RT) = 5.$$

The set of cars that have at least one of the options is $R \cup AC \cup RT$ and

$o(R \cup AC \cup RT)$

$\quad = o(R) + o(AC) + o(RT) - o(R \cap AC) - o(R \cap RT) - o(AC \cap RT)$

$\qquad + o(R \cap AC \cap RT)$

$\quad = 14 + 11 + 8 + 5 - o(R \cap AC) - o(R \cap RT) - o(AC \cap RT).$

Since

$$o(R \cap AC) \geq o(R \cap AC \cap RT) = 5$$

$$o(R \cap RT) \geq o(R \cap AC \cap RT) = 5$$

$$o(AC \cap RT) \geq o(R \cap AC \cap RT) = 5,$$

the number of cars that have at least one of the three options is $\leq 38 - 5 \times 3 = 23$.

Hence the number of cars that do not have any of the three options is at least $30 - 23 = 7$.

Example 3.3. Determine the number of integers between one and 300 that are divisible by any of the numbers two, three, five and seven.

Solution. Let A, B, C, D denote the sets of integers between one and 300 that are divisible by two, three, five and seven, respectively. We are then interested in finding the order of the set $A \cup B \cup C \cup D$. Now

$o(A) = \lfloor 300/2 \rfloor = 150, \quad o(B) = \lfloor 300/3 \rfloor = 100, \quad o(C) = \lfloor 300/5 \rfloor = 60,$

$o(D) = \lfloor 300/7 \rfloor = 42,$

where $\lfloor x \rfloor$ denotes the largest integer $\leq x$. Also

$$o(A \cap B) = \lfloor 300/2 \times 3 \rfloor = 50, \quad o(A \cap C) = \lfloor 300/2 \times 5 \rfloor = 30,$$

$$o(A \cap D) = \lfloor 300/2 \times 7 \rfloor = 21, \quad o(B \cap C) = \lfloor 300/3 \times 5 \rfloor = 20,$$

$$o(B \cap D) = \lfloor 300/3 \times 7 \rfloor = 14, \quad o(C \cap D) = \lfloor 300/5 \times 7 \rfloor = 8,$$

$$o(A \cap B \cap C) = \lfloor 300/2 \times 3 \times 5 \rfloor = 10,$$

$$o(A \cap B \cap D) = \lfloor 300/2 \times 3 \times 7 \rfloor = 7,$$

$$o(A \cap C \cap D) = \lfloor 300/2 \times 5 \times 7 \rfloor = 4,$$

$$o(B \cap C \cap D) = \lfloor 300/3 \times 5 \times 7 \rfloor = 2,$$

$$o(A \cap B \cap C \cap D) = \lfloor 300/2 \times 3 \times 5 \times 7 \rfloor = 1.$$

Therefore, the number of integers between one and 300 that are divisible by at least one of the numbers two, three, five and seven is equal to

$$o(A \cup B \cup C \cup D) = o(A) + o(B) + o(C) + o(D) - o(A \cap B)$$

$$- o(A \cap C) - o(A \cap D) - o(B \cap C) - o(B \cap D)$$

$$- o(C \cap D) + o(A \cap B \cap C) + o(A \cap B \cap D)$$

$$+ o(A \cap C \cap D) + o(B \cap C \cap D) - o(A \cap B \cap C \cap D)$$

$$= 150 + 100 + 60 + 42 - (50 + 30 + 21 + 20 + 14 + 8)$$

$$+ (10 + 7 + 4 + 2) - 1$$

$$= 352 - 143 + 23 - 1 = 231.$$

Example 3.4. Determine the number of integers between one and 300 that are not divisible by three, nor by five, nor by seven. Also find how many of them are divisible by three but not by five, nor by seven.

Solution. Let A, B, C be the sets of numbers between one and 300 that are divisible by three, five and seven, respectively. Then

$$o(A) = \lfloor 300/3 \rfloor = 100, \quad o(B) = \lfloor 300/5 \rfloor = 60, \quad o(C) = \lfloor 300/7 \rfloor = 42,$$

$$o(A \cap B) = \lfloor 300/3 \times 5 \rfloor = 20, \quad o(A \cap C) = \lfloor 300/3 \times 7 \rfloor = 14,$$

$$o(B \cap C) = \lfloor 300/5 \times 7 \rfloor = 8, \quad o(A \cap B \cap C) = \lfloor 300/3 \times 5 \times 7 \rfloor = 2.$$

Therefore,

$$o(A \cup B \cup C) = o(A) + o(B) + o(C) - o(A \cap B) - o(A \cap C)$$

$$- o(B \cap C) + o(A \cap B \cap C)$$

$$= 100 + 60 + 42 - (20 + 14 + 8) + 2 = 162.$$

(a) The number of numbers that are not divisible by three, nor by five, nor by seven equals the order of the set $A' \cap B' \cap C'$, where A', etc. denotes the complement of the set A in the set of all integers between one and 300. Therefore,

$$o(A' \cap B' \cap C') = o((A \cup B \cup C)') = 300 - o(A \cup B \cup C) = 300 - 162 = 138.$$

(b) The number of numbers which are divisible by three but not by five nor by seven is

$$= \text{order of the set } A \cap B' \cap C'$$
$$= \text{order of the set } A \cap (B \cup C)'$$
$$= \text{order of the set } (A \backslash (B \cup C))$$
$$= o(A) - o(A \cap (B \cup C))$$
$$= o(A) - o((A \cap B) \cup (A \cap C))$$
$$= o(A) - o(A \cap B) - o(A \cap C) + o(A \cap B \cap C)$$
$$= 100 - 20 - 14 + 2 = 68.$$

Example 3.5. It is known that at the university, 60 per cent of the professors play tennis, 50 per cent of them play bridge, 70 per cent jog, 20 per cent play tennis and bridge, 30 per cent play tennis and jog and 40 per cent play bridge and jog. If someone claimed that 20 per cent of the professors jog and play bridge and tennis, do you accept this claim as correct? Justify.

Solution. Let T, B and J denote the set of professors who play tennis, play bridge and jog, respectively. Let us assume that the total number of professors in the university is 100. Then, as per the given data

$$o(T) = 60, \quad o(B) = 50, \quad o(J) = 70, \quad o(T \cap B) = 20, \quad o(T \cap J) = 30,$$
$$\text{and} \quad o(B \cap J) = 40.$$

Now $T \cup B \cup J$ is the set of professors who either play tennis, or play bridge, or jog. (It is assumed that every one of the professors takes part in at least one of the three activities). Then

$$100 = o(T \cup B \cup J)$$
$$= o(T) + o(B) + o(J) - o(T \cap B) - o(T \cap J) - o(B \cap J)$$
$$+ o(T \cap B \cap J)$$

$$= 60 + 50 + 70 - (20 + 30 + 40) + o(T \cap B \cap J)$$
$$= 180 - 90 + o(T \cap B \cap J),$$

which gives $o(T \cap B \cap J) = 10$.

Thus, only 10 per cent of the professors play tennis, play bridge and jog. Therefore, the claim that 20 per cent of the professors play tennis, play bridge and jog is not valid or correct.

Example 3.6. Among 100 students, 32 study mathematics, 20 study physics, 45 study biology, 15 study mathematics and biology, 7 study mathematics and physics, 10 study physics and biology and 30 do not study any of the three subjects.

(a) Find the number of students studying all the three subjects.
(b) Find the number of students studying one of the three subjects.

Solution. Let M, P and B be the sets of students studying mathematics, physics and biology respectively. Then, as per the given data

$$o(M) = 32, \quad o(P) = 20, \quad o(B) = 45, \quad o(M \cap B) = 15, \quad o(M \cap P) = 7,$$

$$o(P \cap B) = 10 \quad \text{and} \quad o(M' \cap P' \cap B') = o((M \cup P \cup B)') = 30.$$

(a) The set of students who study all the three subjects is $M \cap P \cap B$. Now

$$o(M \cup P \cup B) + o((M \cup P \cup B)') = 100.$$

Therefore,

$$o(M \cup P \cup B) + 30 = 100 \quad \text{and} \quad \text{so } o(M \cup P \cup B) = 70.$$

By the principle of inclusion and exclusion

$$70 = o(M \cup P \cup B)$$
$$= o(M) + o(P) + o(B) - o(M \cap B) - o(M \cap P) - o(P \cap B)$$
$$+ o(M \cap P \cap B)$$
$$= 32 + 20 + 45 - 7 - 15 - 10 + o(M \cap P \cap B)$$
$$= 65 + o(M \cap P \cap B)$$

which gives $o(M \cap P \cap B) = 5$.

Thus, there are five students who study all three subjects.

(b) The number of students who study mathematics alone is

$$o(M) - o(M \cap B) - o(M \cap P) + o(M \cap P \cap B).$$

Since the subset $M \cap P \cap B$ is included in both $M \cap P$ and $M \cap B$, when the elements of $M \cap P$ and $M \cap B$ are removed from M, the elements of $M \cap P \cap B$ have been removed twice and hence, have had to be added back once. Similarly, the number of students who study physics alone is

$$o(P) - o(M \cap P) - o(P \cap B) + o(M \cap P \cap B)$$

and the number of students who study biology alone is

$$o(B) - o(M \cap B) - o(P \cap B) + o(M \cap P \cap B).$$

Thus, the number of students who study one of the three subjects is

$$= o(M) + o(P) + o(B) - 2\{o(M \cap B) + o(M \cap P) + o(P \cap B)\}$$
$$+ 3o(M \cap P \cap B)$$
$$= 32 + 20 + 45 - 2\{7 + 15 + 10\} + 3 \times 5 = 97 - 64 + 15 = 48.$$

Example 3.7. (a) Among 100 students in a class, 52 got an A in the first examination and 42 got an A in the second examination. If 34 students did not get an A in either examination, find the number of students who got an A in both examinations.

(b) The number of students who got an A in the first examination is equal to the number of students who got an A in the second examination, the total number of students who got an A in one examination is 80 and 8 students did not get an A in either examination. Determine the number of students who got

(i) A in the first examination only;
(ii) A in the second examination only;
(iii) A in both examinations.

Solution. Let P, Q be the sets of students who got an A in the first and the second examination respectively.

(a) Then $o(P) = 52$, $o(Q) = 42$ and $o(P' \cap Q') = 34$.

The set of students who got an A in both the examinations is $P \cap Q$. Now

$$100 = o(P \cup Q) + o((P \cup Q)')$$
$$= o(P) + o(Q) - o(P \cap Q) + o(P' \cap Q')$$
$$= 52 + 42 - o(P \cap Q) + 34.$$

Therefore, $o(P \cap Q) = 94 + 34 - 100 = 28$.

Thus 28 students got an A in both the examinations.

(b) In this case

$$o(P) = o(Q), \quad o(P' \cap Q') = 8.$$

The number of students who got an A in one of the two examinations is

$$= o(P) - o(P \cap Q) + o(Q) - o(P \cap Q)$$
$$= o(P) + o(Q) - 2o(P \cap Q)$$
$$= 2(o(P) - o(P \cap Q)) = 80,$$

which gives

$$o(P) = 40 + o(P \cap Q).$$

Now

$$100 = o(P \cup Q) + o((P \cup Q)') = o(P \cup Q) + 8.$$

Thus

$$92 = o(P \cup Q) = o(P) + o(Q) - o(P \cap Q) = 2o(P) - o(P \cap Q)$$
$$= 40 + o(P),$$

which gives $o(P) = o(Q) = 52$. Thus the number of students who got an A in the first examination, which equals the number of students who got an A in the second examination, is equal to 52.

The number of students who got an A in both examinations is

$$= o(P \cap Q) = o(P) - 40 = 12.$$

Example 3.8. At a CPM (congress party meeting) of 30 persons, 17 are loyal to Nehru, 16 are close to Patel and 5 are close to neither Nehru nor Patel. How many of the 30 persons are close to both Nehru and Patel?

Solution. Let N denote the set of persons who are close to Nehru and let P denote the set of persons who are close to Patel. Also let S denote the set of all the 30 persons. Then

$$o(S) = 30, \quad o(N) = 17, \quad o(P) = 16 \quad \text{and} \quad o(N' \cap P') = 5.$$

The set of persons who are close to both Nehru and Patel is $N \cap P$. Now

$$30 = o(N \cup P) = o(N) + o(P) - o(N \cap P) = 17 + 16 - o(N \cap P),$$

which gives $o(N \cap P) = 33 - 30 = 3$.

Example 3.9. The 6,000 fans who attended a reception to the homecoming cricket game bought up all the paraphernalia for their cars. Altogether, 2,000 bumper stickers, 3,600 window decals, and 1,200 key rings were sold. We know that 5,200 fans bought at least one item and no one bought more than one of a given item. Also, 600 fans bought decals and key rings, 900 bought both decals and bumper stickers, and 500 bought both key rings and bumper stickers.

(a) How many fans bought all three items?

(b) How many fans bought one item?

(c) Someone questions the accuracy of total number of purchasers as being 5,200 (however, all the other numbers have been confirmed to be correct). This person claimed the total number of purchasers to be either 6,000 or 4,500. How do you dispel the claim?

Solution. Let D, S, K denote the set of fans who purchased decals, stickers and key rings respectively. By hypothesis

$$o(D \cup S \cup K) = 5{,}200 \quad \text{and} \quad o(D \cap K) = 600, \quad o(D \cap S) = 900,$$

$$o(K \cap S) = 500.$$

Then

$$5{,}200 = o(D \cup S \cup K)$$

$$= o(D) + o(S) + o(K) - o(D \cap K) - o(D \cap S) - o(K \cap S)$$

$$+ o(D \cap K \cap S)$$

$$= 3{,}600 + 2{,}000 + 1{,}200 - 600 - 900 - 500 + o(D \cap K \cap S)$$

$$= 6{,}800 - 2{,}000 + o(D \cap K \cap S)$$

$$= 4{,}800 + o(D \cap K \cap S).$$

Therefore, the number of fans who purchased all the items

$$= o(D \cap K \cap S) = 400.$$

(a) Number of fans who purchased only stickers

$$= o(S) - o(D \cap S) - o(K \cap S) + o(D \cap K \cap S)$$

$$= 2{,}000 - 500 - 900 + 400 = 1{,}000.$$

Number of fans who purchased only decals

$$= o(D) - o(D \cap K) - o(D \cap S) + o(D \cap K \cap S)$$

$$= 3{,}600 - 900 - 600 + 400 = 2{,}500.$$

Number of fans who purchased only key rings

$$= o(K) - o(D \cap K) - o(K \cap S) + o(D \cap K \cap S)$$

$$= 1{,}200 - 600 - 500 + 400 = 500.$$

(b) The number of fans who purchased exactly one item

$$= 1{,}000 + 2{,}500 + 500 = 4{,}000.$$

(c) If the number of 5,200 were not correct, let x be the number of fans who purchased at least one item. Then, as above

$$x = 4{,}800 + o(D \cap K \cap S).$$

Now $0 \le o(D \cap K \cap S) \le o(K \cap S) = 500$, so that $x \le 5{,}300$ and $x \ge 4{,}800$. Therefore, neither of the two numbers $x = 6{,}000$ and $x = 4{,}500$ is the correct position.

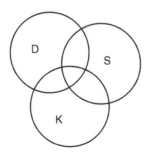

Fig. 3.1.

Exercise 3.1.

1. In a company, 80 persons applied for jobs. The establishment clerk categorized the applications and provided the following data: 30 were minors, 38 were males, 38 were Indian citizens, 11 were male minors, 15 were male Indian citizens, 13 were minor Indian citizens, 3 were male minor Indian citizens and 12 were neither minor, nor male, nor Indian citizens. Is the categorization provided by the clerk correct? Justify.

2. Prove that for all subsets A, B, C of a set S,

$$o(A \cap B' \cap C') = o(A) - o(A \cap B) - o(A \cap C) + o(A \cap B \cap C).$$

3. From among a set of 16 defective integrated circuit chips, 10 possess defect a, 6 possess defect b, 8 possess defect c, 3 possess defects a and b, 3 possess defects b and c, and 4 possess defects a and c. How many of them possess all the three defects?

4. How many bit strings of length 8 either end with a bit 0 or start with the two bits 11?

5. How many bit strings of length 6 either start with a bit 1 or end with the two bits 00?

6. Find the number of integers between one and 500 that are

   (a) divisible by 7 or 11;
   (b) neither divisible by 7 nor by 11.

7. There are 1,807 students in the 10 + 1 class in a certain school. Of these, 453 study a course in computer science, 567 study mathematics and 299 study both computer science and mathematics. Find the number of students who do not study computer science or mathematics.

8. A total of 1,232 students have taken a course in English, 879 have taken a course in French and 114 have taken a course in German. Furthermore, 103 have taken courses in both English and French, 23 have taken courses in both English and German and 14 have taken courses in both French and German. If 1,092 students have taken at least one of English, French and German, find the number of students who have taken a course in all three subjects.

9. Find the number of positive integers not exceeding 200 that are

   (a) not divisible by 3 or 5 or 7;
   (b) divisible by 3, 5 and 7;
   (c) not divisible by 3, 5 or 7.

10. Find the number of terms in the formula for the number of elements in the union of (a) 10 sets, (b) n sets, given by the principle of inclusion and exclusion.

11. Find the number of onto functions from a set with 6 elements to a set with 3 elements.

12. Find all primes not exceeding (a) 50; (b) 60; (c) 100; (d) 200; (e) 300.

13. (a) Among 60 students, 32 got an A in the first test and 26 got an A in the second test. If 19 students did not get an A in either test, find the number of students who got an A in both tests.

    (b) If the number of students who got an A in the first test is equal to the number of students who got an A in the second test, and if the total number of students who got an A in exactly one test is 44, and if 8 students did not get an A in either test, find the number of students who got an A in the first test only, who got an A in the in the second test only, and who got an A in both tests.

14. Out of a total of 150 students, 70 are wearing caps, 59 are wearing scarves and 36 are wearing both caps and scarves. Of the 60 students who are wearing sweaters, 28 are wearing caps, 27 are wearing scarves, and 14 are wearing both caps and scarves. Everyone wearing neither a cap nor a scarf is wearing gloves.

    (a) Determine the number of students wearing gloves.

    (b) Determine the number of students not wearing a sweater who are wearing caps but not a scarf.

    (c) Determine the number of students not wearing a sweater who are neither wearing a cap nor a scarf.

3.2. The Pigeonhole Principle

The pigeonhole principle is a well-known proof technique that is also known as the **shoebox argument** or the **Dirichlet drawer principle**. It roughly or in an informal way amounts to saying that if there are many pigeons and a few pigeonholes, then there is at least one pigeonhole occupied by two or more pigeons. In a formal way, it says that if A, B are finite sets with $o(A) > o(B)$, then for any map $f : A \to B$, there exist elements $a_1, a_2 \in A, a_1 \neq a_2$ such that $f(a_1) = f(a_2)$. A more general form of the pigeonhole principle states that for any map $f : A \to B$, there exist i elements $a_1, a_2, \ldots, a_i$ in A, $i = \lceil o(A)/o(B) \rceil$ (i.e. the smallest integer not less than $o(A)/o(B)$), such that $f(a_1) = f(a_2) = \cdots = f(a_i)$. Alternatively put, if $o(A) = kn + r$,

k a positive integer and $1 \leq r < n$, where $o(B) = n$, then there are $k + 1$ elements $a_1, a_2, \ldots, a_{k+1}$ in A such that $f(a_1) = f(a_2) = \cdots = f(a_{k+1})$.

Some obvious examples on the applications of the pigeonhole principle follow.

1. Among a group of 14 men, at least two of them were born in the same month of the year.
2. Among ten children, at least two will have birthdays that fall on the same day of the week.
3. If seven single shoes are selected from six pairs of shoes, there must be a pair of matched shoes among the selection.
4. Among 31 children born in the month of September, there must be at least two whose birthday falls on the same date.

3.2.1. Some typical applications of the pigeonhole principle

Example 3.10. A chess player wants to prepare for a championship match by playing some practice games in 65 days. She wants to play at least one game a day, but no more than 112 games in total. No matter how she schedules the games, there is a period of consecutive days within which she plays exactly 17 games.

Solution. Let a_i denote the total number of games she plays up to the ith day. Since she plays at least one game a day, the sequence $a_1, a_2, \ldots, a_{65}$ is a strictly increasing sequence. Also, the sequence $a_1 + 17, a_2 + 17, \ldots, a_{65} + 17$ is a strictly increasing sequence. Consider the sequence

$$a_1, a_2, \ldots, a_{65}, a_1 + 17, a_2 + 17, \ldots, a_{65} + 17, \qquad (3.2)$$

which is a sequence of 130 numbers. The total number of games she plays is, at the most, 112. Therefore, $a_{65} \leq 112$. Then, $a_{65} + 17 \leq 129$. Thus (3.2) is a sequence of integers lying between one and 129. The total number of terms in the sequence (3.2) being 130, there are at least two terms which are equal. But no two of the numbers $a_1, a_2, \ldots, a_{65}$ are equal and so no two of the numbers $a_1 + 17, a_2 + 17, \ldots, a_{65} + 17$ are equal. Therefore, there exist $i, j \leq 65$, such that $a_i = a_j + 17$. Then clearly $j < i$ and $a_i - a_j = 17$, which shows that from the $(j + 1)$th day to the ith day, exactly 17 games are played.

Given a sequence $a_1, a_2, \ldots, a_n$ of numbers, by a **subsequence** of it we mean a sequence $a_{i_1}, a_{i_2}, \ldots, a_{i_k}$ where $1 \leq i_1 < i_2 < \cdots < i_k \leq n$. The given sequence is called an **increasing sequence** if $a_1 \leq a_2 \leq \cdots \leq a_n$.

It is also called a non-decreasing sequence. Furthermore, it is called **strictly increasing** if $a_1 < a_2 < \cdots < a_n$. We can define a decreasing sequence and a strictly decreasing sequence in a similar fashion. The number of terms in a sequence is called its length.

Example 3.11. Show that in a sequence of n^2+1 distinct integers, there is either an increasing subsequence of length $n+1$ or a decreasing subsequence of length $n+1$.

Solution. For the sake of convenience, we will write $n^2 + 1 = m$. Let $a_1, a_2, \ldots, a_m$ denote the given sequence of distinct integers. We label the integer a_k with an ordered pair (x_k, y_k), where x_k is the length of the longest increasing subsequence starting with a_k, and y_k is the length of the longest decreasing subsequence starting at a_k. Suppose there is no increasing subsequence or decreasing subsequence of length $n + 1$ in the given sequence $a_1, a_2, \ldots, a_m$. This means that for every k, the values of x_k and y_k lie between 1 and n. Therefore, there are n^2 distinct ordered pairs as possible labels for the n^2+1 integers. By the pigeonhole principle, there then exist integers a_i and a_j, not equal, which are labelled with the same ordered pair, i.e. $(x_i, y_i) = (x_j, y_j)$. However, if $a_i < a_j$, then $y_i < y_j$, while if $a_i > a_j$, then $x_i < x_j$, either of which gives a contradiction to $(x_i, y_i) = (x_j, y_j)$. Hence either there is an increasing subsequence or a decreasing subsequence of length $n + 1$ in the sequence $a_1, a_2, \ldots, a_m$ of distinct integers.

Example 3.12. Show that among six persons, either there are three who are mutual friends or there are three who are complete strangers to each other.

Solution. Let A be a person in the given group of six people. By the pigeonhole principle, among the remaining five people, either there are at least three persons who are friends with A, or there are at least three persons who are strangers to A. Suppose that there are three persons who are strangers to A. Let B, C, D be the three persons who are strangers to A. If any two of B, C, D are strangers to each other, then these two persons together with A are three persons who are strangers to each other. If no two of the three persons B, C, D are strangers, then B, C, D are mutual friends. The case when there are three persons who are friendly with A can be argued similarly.

Example 3.13. A man hiked for 11 days and covered a total distance of 58 km. It is known that he hiked 8 km in the first day and only 5 km in the last day. Show that he must have hiked at least 11 km within a certain period of two consecutive days.

Solution. Let a_i be the number of kilometres hiked by the man up to (and including) the ith day. Then we have an increasing sequence

$$a_1 \le a_2 \le \cdots \le a_{11} \quad \text{with } a_{11} = 58.$$

Then a_2 is the number of kilometres the man hiked in the first two days, $a_3 - a_1$ is the number of kilometres he hiked in the second and the third days, $a_4 - a_2$ is the number of kilometres he hiked during the third and fourth days and so on, so that $a_{11} - a_9$ is the number of kilometres he hiked during the 10th and 11th days. Suppose that in every period of two consecutive days the man hiked for, at most, 10 kilometres. Then, we have

$$a_2 + (a_3 - a_1) + (a_4 - a_2) + \cdots + (a_{11} - a_9) \le 10 \times 10 = 100.$$

This gives

$$(a_2 + a_3 + \cdots + a_{11}) - (a_1 + a_2 + \cdots + a_9) \le 100 \quad \text{or } a_{10} + a_{11} - a_1 \le 100.$$

But $a_{11} = 58$, $a_{11} = a_{10} + 5$ and $a_1 = 8$. Therefore, $53 + 58 - 8 \le 100$, which is a contradiction. Therefore, there is at least one pair of consecutive days during which the man hiked for at least 11 km.

Example 3.14. If there are 101 persons of different heights standing in a line, show that it is possible to find 11 persons, in the order they are standing in the line, with heights that are either increasing or decreasing.

Solution. The heights of the persons all being distinct and in whole numbers of centimetres, we have a sequence $a_1, a_2, \ldots, a_{101}$ of positive integers, where a_is denote the heights of the persons. We then have to prove that either there is an increasing or a decreasing subsequence of this sequence of length 11. Suppose to the contrary that there is no increasing or decreasing subsequence of length 11 (i.e. every such subsequence is of length 10 at the most). With each term a_k of the sequence associate an ordered pair (i_k, d_k), where i_k is the length of the longest increasing subsequence starting at a_k, d_k is the length of the longest decreasing subsequence starting at a_k. Then, for every $k, 1 \le k \le 101$, we have

$$1 \le i_k \le 10 \quad \text{and} \quad 1 \le d_k \le 10.$$

But the number of ordered pairs (i_k, d_k) is at most $10 \times 10 = 100$. Therefore, with two terms, say a_k and $a_l, k \ne l$, the associated ordered pairs are equal.

This means that

$$(i_k, d_k) = (i_l, d_l), \quad \text{which leads to } i_k = i_l \quad \text{and} \quad d_k = d_l.$$

Since $k \neq l, a_k \neq a_l$, therefore, either $a_k < a_l$ or $a_k > a_l$. If $a_k < a_l$, an increasing subsequence of length $i_l + 1 = i_k + 1$ can be built, starting with a_k which is a contradiction, as the longest such subsequence is of length $i_k = i_l$. If, on the other hand, $a_k > a_l$, then a decreasing subsequence of length $d_l + 1 = d_k + 1$ can be built, starting with a_k, which is again a contradiction.

Example 3.15. (a) If seven integers are selected from the first 10 positive integers, show that there must be at least two pairs of these integers with the sum 11.

(b) Is the above conclusion true if six integers are chosen instead of seven?

Solution. Partition the set $S = \{1, 2, \ldots, 10\}$ of the first ten integers into two ordered subsets A, B such that $S = A \cup B, A \cap B = \varphi, o(A) = o(B)$ and if $A = \{a_1, a_2, a_3, a_4, a_5\}$ and $B = \{b_1, b_2, b_3, b_4, b_5\}$, then for every $i, 1 \leq i \leq 5, a_i + b_i = 11$. For example, we may take $A = \{1, 3, 5, 7, 9\}$, $B = \{10, 8, 6, 4, 2\}$ or $A = \{1, 2, 3, 4, 5\}$, $B = \{10, 9, 8, 7, 6\}$. There are several other ways of taking A and B.

(a) When seven elements are chosen from S, we can have a maximum of five elements of A and a minimum of two elements of B. If the elements chosen from B are b_i, b_j, then by the choice of A and B there are elements $a_i, a_j \in A$ already chosen and we have two pairs of matching integers a_i, b_i and a_j, b_j with the property that $a_i + b_i = 11 = a_j + b_j$. If only four elements are chosen from A, say the element a_5 is not chosen, then three elements have to be chosen from B. The three elements chosen from B may include b_5 or may not include b_5. If b_5 is included in the choice, then the other two bs are b_i, b_j with $i \neq 5, j \neq 5$ and we have two pairs a_i, b_i and a_j, b_j again with $a_i + b_i = 11 = a_j + b_j$. If b_5 is not included among the three elements chosen, the three elements are b_i, b_j, b_k, with five not in the set $\{i, j, k\}$ and we have three pairs, namely $a_i, b_i; a_j, b_j; a_k, b_k$ among the seven chosen elements and $a_i + b_i = a_j + b_j = a_k + b_k = 11$.

If three or fewer elements are chosen from A, then at least four elements are chosen from B and we can repeat the above argument with A and B interchanged. Hence there are at least two pairs of elements $a_i, b_i; a_j, b_j$ among the elements chosen with $a_i + b_i = 11 = a_j + b_j$.

(b) When six elements are chosen from S, we can have a maximum of five elements of A and at least one element from B. If the only element chosen from B is b_i, then we have one pair of elements a_i, b_i with $a_i + b_i = 11$.

If only four elements are chosen from A, say the element a_5 is not chosen, then two elements have to be chosen from B. These two elements chosen from B may include b_5 or may not include b_5. If b_5 is included in the choice, let b_i be the other element chosen from B. Then our six elements include only one pair a_i, b_i with $a_i + b_i = 11$. If, on the other hand, b_5 is not included in the choice of two elements of B, let these chosen elements be b_i, b_j and we have two pairs $a_i, b_i; a_j, b_j$ among our choice of six elements with $a_i + b_i = 11 = a_j + b_j$.

If only three elements are chosen from A, let the chosen elements be a_1, a_2, a_3. In this case, three elements are chosen from B and these three elements must include at least one of the elements b_1, b_2, b_3. Thus in this case, we have at least one pair of elements a_i, b_i with $a_i + b_i = 11$.

When two or less elements are chosen from A, we can repeat the above argument with A and B interchanged. Hence, in the case when six elements are chosen from S, then there is chosen at least one pair of integers which add up to 11.

Example 3.16. If nine integers are chosen from the integers one to 16, then there shall be chosen a pair of integers, one of which divides the other.

Solution. Given any positive integer, we can express it uniquely in the form $2^k m$, where $k \geq 0$ and m is odd. From the numbers one to 16, there are exactly eight odd numbers. Therefore, when nine integers are chosen from one to 16, at least two of these have the same odd part. Let these two numbers be $2^k m$ and $2^l m$, where k, l are non-negative integers. Since the two integers $2^k m$ and $2^l m$ chosen are distinct, $k < l$ or $l < k$. If $k < l$, then $2^l m = 2^{l-k}(2^k m)$, while if $k > l$, then $2^k m = 2^{k-l}(2^l m)$. Thus, either $2^k m$ divides $2^l m$ or $2^l m$ divides $2^k m$.

Example 3.17. Find the number of students in a class to be sure that three of them are born in the same month. •

Solution. The generalized pigeonhole principle states: 'If n pigeonholes are occupied by $kn + 1$ or more pigeons, where k is a positive integer, then at least one pigeonhole is occupied by $k + 1$ or more pigeons.'

Since we want three students to be born in the same month, we have $k + 1 = 3$ or $k = 2$. Also, a pigeonhole here is a month of the year. There being 12 months in a year, the number n of pigeonholes is 12. Therefore, by the generalized pigeonhole principle, the minimum number of students is $12k + 1 = 12 \times 2 + 1 = 25$.

Example 3.18. Sania Mirza prepares for the French open championship by playing some practice games over a period of 70 days. She plays at least one game a day but not more than 123 games altogether. Prove that there is a period of consecutive days during which she plays exactly 16 games.

Solution. Let a_i denote the number of games played by Sania up to and including the ith day. Since she plays at least one game every day, we have a strictly increasing sequence

$$a_1 < a_2 < a_3 < \cdots < a_{70} \tag{3.3}$$

with $a_1 \geq 1$ and $a_{70} \leq 123$. Then we also have a strictly increasing sequence

$$a_1 + 16 < a_2 + 16 < a_3 + 16 < \cdots < a_{70} + 16. \tag{3.4}$$

Observe that $a_{70} + 16 \leq 123 + 16 = 139$. The total number of positive integers in the two sequences (3.3) and (3.4) is 140, the least among these is ≥ 1 and the largest is ≤ 139. Then by the pigeonhole principle, there are at least two terms in the sequences (3.3) and (3.4) which are equal. Also no two terms of the sequence (3.3) are equal and no two terms of the sequence (3.4) are equal. Therefore, there is a term in the sequence (3.3) that equals a term in the sequence (3.4). Then, there exist integers $i, j, 1 \leq i, j \leq 70$ such that $a_j = a_i + 16$. Since the a_k's are strictly increasing, j must be $> i$ and we have $a_j - a_i = 16$. However, $a_j - a_i$ denotes the total number of games played from the $(i + 1)$th day to the jth day. So, in this period of consecutive days she plays exactly 16 games.

Example 3.19. Prove that, at a party where there are at least two persons, there are two persons who know the same number of other persons present there.

Solution. If there are only two people in the party, either they know each other or they are strangers to each other. Thus the two know the same number of other people which is either 0 or 1. Suppose that there are n people in the party with $n > 1$. Since we have already proved the result for $n = 2$, we assume that $n > 2$. Let $A_1, A_2, \ldots, A_n$ denote the people present and for any $i, 1 \leq i \leq n$, let a_i denote the number of other people known to A_i. If no two persons in the party know the same number of other persons, then $a_1, a_2, \ldots, a_n$ are all distinct. If there is a person present who does not know any one else (there will have to be only one, for otherwise the result is trivially true), then the largest among the numbers $a_1, a_2, \ldots, a_n$ will have to be less than $n - 1$. If everyone present knows at least one other person, then no a_i is zero and the largest among these is less than n. By renaming the persons, if necessary, we can assume that $1 \leq a_1 < a_2 < \cdots < a_n \leq n-1$ or $0 \leq a_1 < a_2 < \cdots < a_n \leq n - 2$.

In either case we have a contradiction, as no n distinct integers can lie between 0 and $n - 2$ or between 1 and $n - 1$. Hence there are at least two persons in the party who know exactly the same number of other persons.

Exercise 3.2.

1. If 16 integers are chosen from the integers one to 30, then there is chosen at least one pair of integers such that one of the two divides the other.

2. A wrestler is the champion for a period of 75 days. The wrestler had at least one match a day, but no more than 125 total matches. Show that there is a period of consecutive days during which the wrestler had exactly 24 matches.

3. A chess player wants to prepare for a championship match by playing some practice games in 77 days. She wants to play at least one game a day but not more than 132 games altogether. No matter how she schedules the games, there is a period of consecutive days within which she plays exactly 21 games.

4. A man hiked for 10 days and covered a total distance of 45 km. It is known that he hiked 6 km in the first day and only 3 km in the last day. Show that he must have hiked at least 9 km within a certain period of two consecutive days.

5. Let A, B be finite sets. Prove that there is no map from A to B which is (a) one–one if $o(A) > o(B)$; (b) onto if $o(A) < o(B)$.

6. A man hiked for 10 days and covered a distance of 56 km. It is known that he hiked 8 km in the first day and only 5 km in the last day. Show that he must have hiked at least 11 km within a certain period of two consecutive days.

7. Proof of Example 3.13 given does not make a use of the pigeonhole principle. Attempt a proof, if possible, of this example which makes explicit use of the pigeonhole principle.

3.3. Binary Relations

3.3.1. *Relations*

Definition. Let A, B be non-empty sets. A **binary relation** from A to B is a subset R of the Cartesian product $A \times B$. If $(a, b) \in R$, then we say that a is related to b by the relation R and write $a\,R\,b$. If a is not related to b, we express it by $a\,\check{R}\,b$. By a **binary relation on** a non-empty set A we mean a binary relation from A to A.

Example 3.20. We may define a binary relation R on the set Z of integers by

(a) $(a, b) \in R$ if 2 divides $a + b$;
(b) $(a, b) \in R$ if 3 divides $a + b$;
(c) $(a, b) \in R$ if 3 divides $a - b$;
(d) $(a, b) \in R$ if $a - b \geq 0$;
(e) $(a, b) \in R$ if a divides b;
(f) $(a, b) \in R$ if g.c.d. $(a, b) = 1$;
(g) $(a, b) \in R$ if $a \leq b$;
(h) $(a, b) \in R$ if $a < b$.

Example 3.21. Let $A = \{0, 1\}$ and $B = N$, the set of all natural numbers. We may define a binary relation R from A to B as follows.

For $a \in A, b \in B$ say that $(a, b) \in R$ if and only if $b = 2m + a, m \in N$. Observe that

$$R = \{(0, 2m)|m \in N\} \cup \{(1, 2m + 1)|m \in N\}.$$

Example 3.22. For non-empty sets A, B, every map from A to B is a relation from A to B. (Prove it!)

3.3.2. *Equivalence relations*

Definition. Let R be a binary relation defined on a non-empty set A. The relation R is said to be

(i) **reflexive** if $(a, a) \in R$ for every $a \in A$,
(ii) **symmetric** if $(a, b) \in R$ implies $(b, a) \in R$,
(iii) **anti-symmetric** if $(a, b) \in R$ and $(b, a) \in R$ imply $a = b$,
(iv) **transitive** if $(a, b) \in R, (b, c) \in R$ imply $(a, c) \in R$.

Definition. A relation R on A is called an **equivalence relation** if the relation R is reflexive, symmetric and transitive, while R is called a **partial order relation** if R is reflexive, anti-symmetric and transitive.

Let A be a non-empty set. Recall that by a partition of A we mean a collection of non-empty subsets $\{A_i\}_{i \in I}$ of A such that $A = \cup A_i$ and $A_i \cap A_j = \varphi$, for every $i, j \in I, i \neq j$. For example, if $A = \{1, 2, 3, 4\}$, then $A_1 = \{1\}, A_2 = \{2, 3, 4\}$ provide a partition of A whereas B_1, B_2 with $B_1 = \{1, 2\}, B_2 = \{2, 3, 4\}$ is not a partition of A. Also $C_1 = \{1\}, C_2 = \{2\}$ and $C_3 = \{4\}$ do not provide a partition of A.

Definition. Let A be a non-empty set and R be an equivalence relation defined on A. An **equivalence class** of the relation R or determined by the relation R means a non-empty subset of A such that any two elements in the subset are related to each other and no element not in the subset is related to any element in the subset.

Alternatively, we may define an **equivalence class determined by an element** $a \in A$ denoted by $C(a)$ or $cl(a)$ to be the subset of A defined by $C(a) = \{x \in A \,|\, xRa\}$.

Since the relation R is an equivalence relation, aRa, so, $a \in C(a)$. In particular, it proves that $C(a) \neq \varphi$.

Theorem 3.2. If $C(a)$ and $C(b)$ are two equivalence classes determined by an equivalence relation R, then either $C(a) = C(b)$ or $C(a) \cap C(b) = \varphi$.

Proof. If $C(a) \cap C(b) = \varphi$, we have nothing to prove. Suppose that $C(a) \cap C(b) \neq \varphi$. Then there exists an element $c \in C(a) \cap C(b)$. Then $c \in C(a)$ and $c \in C(b)$. Therefore, cRa and cRb. Let $x \in C(a)$. Then xRa. Also cRa implies that aRc. Now xRa and aRc imply xRc. Thus xRc and cRb which together imply that xRb and $x \in C(b)$. Hence $C(a) \subseteq C(b)$. Similarly, $C(b) \subseteq C(a)$ and, therefore, $C(a) = C(b)$.

Exercise 3.3. If R is an equivalence relation defined on a non-empty set A and $a \in A$, then $b \in C(a)$ if and only if $C(a) = C(b)$.

Theorem 3.3. Let R be an equivalence relation defined on a non-empty set A. The set of all equivalence classes of R provides a partition of the set A.

Proof. Take the collection of all distinct equivalence classes of R. By definition, none of the equivalence classes is empty. Also, if $C(a), C(b)$ are two distinct equivalence classes, then $C(a) \neq C(b)$ and, therefore, $C(a) \cap C(b) = \varphi$. Finally, let $a \in A$. Then $a \in C(a)$, i.e. every element of A belongs to some equivalence class. Thus the collection of all distinct equivalence classes determined by R gives a partition of the set A. The converse of the above result is also true.

Theorem 3.4. Let $\{A_i\}_{i \in I}$ be a partition of a non-empty set A. Then there exists an equivalence relation R defined on A and the collection $\{A_i\}$ is the collection of all distinct equivalence classes of R.

Proof. For $a, b \in A$, say that $a \, R \, b$ if there exists an $i \in I$ such that $a, b \in A_i$. Let $a \in A$. Since $A = \cup A_i$, there exists an $i \in I$ such that $a \in A_i$. Then $a, a \in A_i$ and, by definition, $a \, R \, a$ showing that R is reflexive. Let $a, b \in A$ and suppose that $a \, R \, b$. Then there exists an $i \in I$ such that $a, b \in A_i$ or $b, a \in A_i$, which implies that $b \, R \, a$. Thus R is symmetric. Next, let $a, b, c \in A$ and suppose that $a \, R \, b$ and $b \, R \, c$. Then there exist $i, j \in I$ such that $a, b \in A_i$ and $b, c \in A_j$. Observe that $b \in A_i \cap A_j$. Since $A_i \cap A_j = \varphi$ if $i \neq j$, we get $i = j$. In particular, $a, c \in A_i$ and $a \, R \, c$. This proves that the relation R is transitive and, hence, R is an equivalence relation. Let $a \in A$. Then $a \in A_i$ for some $i \in I$. For any $x \in A_i, a, x \in A_i$ and so $x \, R \, a$. This means that $x \in C(a)$ and $A_i \subseteq C(a)$. On the other hand, if $y \in C(a)$ then $y \, R \, a$, which implies the existence of a $j \in I$ such that $y, a \in A_j$. Already $a \in A_i$. Thus $a \in A_i \cap A_j$, which implies that $i = j$. Thus $y \in A_i$, i.e. $C(a) \subseteq A_i$. Hence $C(a) = A_i$ showing that every A_i equals an equivalence class. We have also almost proved that every equivalence class of R equals an A_i for some i.

3.3.3. *Union, intersection and inverse of relations*

Let A, B be two non-empty sets. As defined earlier, a relation from A to B is a subset of $A \times B$. Therefore, the power set $Þ(A \times B)$ of $A \times B$ gives all possible relations from A to B. If both A and B are finite sets, then $A \times B$ is also a finite set and $o(A \times B) = o(A) \, o(B)$. Therefore, there are only a finite number of subsets of $A \times B$ and, hence, there are only a finite number of binary relations from A to B.

Exercise 3.4. If $o(A) = m$ and $o(B) = n$, prove that the number of relations from A to B is 2^{mn}.

The empty set φ is a subset of $A \times B$ and gives a relation from A to B in which no element of A is related to any element of B. On the other extreme is the relation R consisting of all pairs (a, b), where $a \in A, b \in B$.

It is the relation in which every element of A is related to every element of B.

Definition. For any relation R from A to B, its **complementary relation** R' is defined by $R' = \{(a, b) \in A \times B | (a, b) \notin R\}$, i.e. it consists of all those ordered pairs $(a, b), a \in A, b \in B$, which are not in R. The relation R' is also written as $\bar{R}$ or $-R$.

Definition. Let R and S be two relations from A to B. By the **union** $R \cup S$ and **intersection** $R \cap S$ of R and S we mean the relations from A to B defined by

$$R \cup S = \{(a, b) \, \varepsilon \, A \times B | (a, b) \in R \quad \text{or} \quad (a, b) \in S\} \quad \text{and}$$
$$R \cap S = \{(a, b) \, \varepsilon \, A \times B | (a, b) \in R \quad \text{and} \quad (a, b) \in S\}.$$

Also, we define

$$R - S = \{(a, b) \in A \times B | (a, b) \in R \text{ but } (a, b) \notin S\}.$$

The **converse** (or also called **inverse**) **of a relation** R from A to B is a relation $\tilde{R}$ or R^{-1} from B to A defined by

$$\tilde{R} = \{(b, a) \in B \times A | (a, b) \in R\}.$$

To put it alternatively, we say that for $a \in A, b \in B$,

$$a(R \cup S)b \quad \text{if } a \, R \, b \quad \text{or} \quad a \, S \, b,$$
$$a(R \cap S)b \quad \text{if } a \, R \, b \quad \text{and} \quad a \, S \, b,$$
$$b\tilde{R}a \quad \text{if } a \, R \, b.$$

Example 3.23. Let $A = B = Z$ and $R = \{(a, |a|) | a \in Z\}$,

$$S = \{(a, 2a + 1) | a \in Z\}.$$

(a) Let $(a, b) \in R \cap S$. Then, by definition, $b = |a|$ and $b = 2a + 1$. If $a \geq 0$, then $b = a$ and $b = 2a + 1$, which together imply $a = -1$ which in turn contradicts $a \geq 0$. If $a < 0$, then $b = -a$ and $b = 2a + 1$ and the two imply that $a = -\frac{1}{3}$. Therefore, $R \cap S = \{(-\frac{1}{3}, \frac{1}{3})\}$.

(b) For every $a \in Z$, $a \geq 0$ implies $(a, a) \in R$ and $a < 0$ implies $(a, -a) \in R$.

Therefore,

$$\tilde{R} = \{(a,a)|a \in Z, a \geq 0\} \cup \{(-a,a)|a \in Z, a < 0\}$$
$$= \{(a,a)|a \in Z, a \geq 0\} \cup \{(a,-a)|a \in Z, a > 0\}.$$

Also $\tilde{S} = \{(2a+1,a)|a \in Z\}$.

Moreover, for $a,b \in Z, a(R \cup S)b$ if either $b = |a|$ or $b = 2a + 1$.

Example 3.24. Let $A = \{1,2,3,4\}$. Let

$$R = \{(a,b) \in A \times A | a - b \text{ is a non-zero integral multiple of } 2\}$$

$$S = \{(a,b) \in A \times A | a - b \text{ is a non-zero integral multiple of } 3\}.$$

Then

$$R = \{(1,3),(3,1),(2,4),(4,2)\}, \quad S = \{(1,4),(4,1)\} \quad \text{and}$$
$$R \cup S = \{(1,3),(3,1),(2,4),(4,2),(1,4),(4,1)\}, \quad R \cap S = \varphi,$$
$$R - S = \{(1,3),(3,1),(2,4),(4,2)\} = R,$$
$$S - R = \{(1,4),(4,1)\} = S.$$

3.3.4. *Composition of relations*

Definition. Let A, B, C be non-empty sets and R be a relation from A to B and S be a relation from B to C. Then the **composition** or **composite** of S and R is the relation SoR from A to C defined by

$$SoR = \{(a,c)| \text{ there exists an element } b \in B \text{ with } (a,b) \in R \text{ and } (b,c) \in S\}$$

i.e. $(a,c) \in SoR$ if and only if there exists an element $b \in B$ such that $(a,b) \in R$ and $(b,c) \in S$.

Example 3.25. It may happen that composite SoR of two relations R from A to B and S from B to C may be empty even when neither R nor S is empty. For example,

1. Let $A = \{1,2,3\}$, $B = \{4,5,6,7\}$, $C = \{a,b\}$, $R = \{(1,4),(2,6)\}$, $S = \{(5,a),(7,b)\}$. Then neither R nor S is empty but SoR is empty.
2. Let $A = B = C = Z$, the set of integers and $R = \{(a,2a+1)|a \in Z\}$, $S = \{(2a,a)|a \in Z\}$. Then neither R nor S is empty but SoR is empty. However, observe that $RoS = (2a,2a+1)|a \in Z\} \neq \varphi$.

3. Let $A = Z^+ = \{1, 2, 3, \ldots\}$ and R, S be defined by

$$R = \{(a, b) \in A \times A | a - b \text{ is a non-zero integral multiple of } 2\}$$

$$S = \{(a, b) \in A \times A | a - b \text{ is a non-zero integral multiple of } 3\}.$$

Then

$$R \cap S = \{(a, b) \in A \times A | b \neq a \text{ and both } 2, 3 \text{ divide } a - b\}$$

$$= \{(a, b) \in A \times A | b \neq a \text{ and } 6 \text{ divides } a - b\}$$

$$R - S = \{(a, b) \in A \times A | b \neq a, 2 \text{ divides } a - b \text{ but } 3 \text{ does not divide } a - b\}$$

$$S - R = \{(a, b) \in A \times A | b \neq a, 3 \text{ divides } a - b \text{ but } 2 \text{ does not divide } a - b\}$$

$$R \cup S = \{(a, b) \in A \times A | b \neq a, \text{ either } 2 \text{ divides } a - b \text{ or } 3 \text{ divides } a - b\}.$$

4. Let $A = N$, be the set $\{0, 1, 2, 3, \ldots\}$ of natural numbers and consider the relations on A defined by

   (a) $a\,R\,b$,   if $b = a^2$, $a \in A$;
   (b) $a\,S\,b$ if $b = 2a$, $a \in A$;
   (c) $a\,T\,b$ if $a \leq b$, $a, b \in A$;
   (d) $a\lambda b$ if a divides b, $a, b \in A$.

Since $a^2 = 2a$ implies $a = 0$ or 2, $R \cap S = \{(0, 0), (2, 4)\}$. Clearly, $\lambda \subseteq T$. As $2 \leq 3$ but 2 does not divide 3, $(2, 3) \in T - \lambda$ and $T \neq \lambda$. Also $T \cap \lambda = \lambda$ and $T \cup \lambda = T$.

Theorem 3.5. The composition of relations is associative.

Proof. Let A, B, C, D be non-empty sets and R, S, T be relations from A to B, B to C and from C to D, respectively. Then, we have to prove that $(ToS)oR = To(SoR)$.

Let $(a, d) \in (ToS)oR$. Then there exists an element $b \in B$, such that $(a, b) \in R$ and $(b, d) \in ToS$. By definition of ToS, it follows that there exists an element $c \in C$ such that $(b, c) \in S$ and $(c, d) \in T$. Thus, we have $(a, b) \in R, (b, c) \in S, (c, d) \in T$. Then $(a, c) \in SoR$ and $(c, d) \in T$ which together imply that $(a, d) \in To(SoR)$. This proves that $(ToS)oR \subseteq To(SoR)$. A similar argument shows that $To(SoR) \subseteq (ToS)oR$ and hence $To(SoR) = (ToS)oR$.

Definition. Let R be a relation from a set A to a set B. Then the subset $D(R) = \{a \in A| \text{ there exists an element } b \in B \text{ with } (a, b) \in R\}$ of A is called the **domain** of the relation R and the subset $R(R) = \{b \in B| \text{ there exists an element } a \in A \text{ with } (a, b) \in R\}$ of B is called the **range** of the relation R.

For the relation, $R = \{(a, |a|)|a \in Z\}$, the domain is the set Z of all integers and the range is the set of all non-negative integers.

For the relation $R = \{(a, 2a + 1)|a \in Z\}$ the domain is again the set Z of all ·integers where as the range is the set $2Z + 1$ of all odd integers.

On the set N of all natural numbers, let R, S be the relations defined by: for $a, b \in N$, aRb if $b = a^2$ and aSb if $b = 2a$. Then the range

$$R(R) = \{0, 1, 4, 9, \ldots\} = \{m^2|m \in N\} \quad \text{and}$$

$$R(S) = \{0, 2, 4, 8, \ldots\} = \{2m|m \in N\}.$$

If R is a binary relation defined on a non-empty set A, we define for $n \geq 2$, the powers R^n of R as follows. $R^2 = RoR$ and for $m \geq 2$ with R^m already defined, define $R^{m+1} = R^m oR$. Thus $R^n = RoRo\ldots oR$ is the composition of R with itself taken n times.

3.3.5. *The matrix of a relation*

Definition. Let $A = \{a_1, a_2, \ldots, a_m\}$ and $B = \{b_1, b_2, \ldots, b_n\}$ be finite ordered sets. As already mentioned, with every relation R from A to B there can be associated an $m \times n$ matrix $M_R = (m_{ij})$ where $m_{ij} = 1$ if $(a_i, b_j) \in R$ and $m_{ij} = 0$ if $(a_i, b_j) \notin R$. The matrix M_R is called the **relation matrix of the relation** R. Some authors call it simply **matrix of the relation**.

On the other hand, if we are given an $m \times n$ matrix the entries of which are in the field $B = \{0, 1\}$ of two elements, we can always define a relation R from A to B such that the given matrix is the relation matrix of the relation defined. Let $M = (m_{ij})$ be an $m \times n$ matrix where every $m_{ij} \in B = \{0, 1\}$, i.e. every $m_{ij} = 1$ or 0. We define a relation R from A to B as follows.

For $(a_i, b_j) \in A \times B$, we say that $(a_i, b_j) \in R$ if and only if $m_{ij} = 1$, i.e. $(a_i, b_j) \in R$ if $m_{ij} = 1$ and $(a_i, b_j) \notin R$ if $m_{ij} = 0$. It is then clear that the relation matrix M_R of R is the given matrix M itself.

Example 3.26. Let $A = \{1, 2, 3\}$ and $B = \{a, b\}$. Consider the relation $R = \{(1, b), (2, a), (3, b)\}$ from A to B. Then the relation matrix M_R of R is given by

$$M_R = \begin{pmatrix} 0 & 1 \\ 1 & 0 \\ 0 & 1 \end{pmatrix}.$$

Example 3.27. Let $A = \{1, 2, 3, 4, 5\}$. Define a relation R on A by: aRb if $a < b$. Then

$$R = \{(1, 2), (1, 3), (1, 4), (1, 5), (2, 3), (2, 4), (2, 5), (3, 4), (3, 5), (4, 5)\}$$

and the matrix of the relation is

$$M_R = \begin{pmatrix} 01111 \\ 00111 \\ 00011 \\ 00001 \\ 00000 \end{pmatrix},$$

which is an upper triangular matrix with every diagonal entry 0. If we modify this relation to say that aRb if $a \leq b$, then the relation matrix of this relation will be the above matrix with every diagonal entry equal to 1 instead of 0, i.e.

$$M_R = \begin{pmatrix} 11111 \\ 01111 \\ 00111 \\ 00011 \\ 00001 \end{pmatrix}.$$

The matrix M_R associated with a binary relation R is also called the **adjacency matrix** of R. The entries of adjacency matrix of a relation R are in the field of two elements.

Definition. In the set $B = \{0, 1\}$ of two elements we define **complementation**, and **Boolean compositions** by

$$0' = 1, 1' = 0, 0 \cup 0 = 0, 0 \cup 1 = 1 \cup 0 = 1 \cup 1 = 1,$$

$$0 \cap 0 = 0 \cap 1 = 1 \cap 0 = 0, 1 \cap 1 = 1.$$

Alternatively, we may also rewrite these (the Boolean compositions) as

$$0 + 0 = 0, 1 + 0 = 0 + 1 = 1 + 1 = 1, 0.0 = 0.1 = 1.0 = 0, 1.1 = 1.$$

Observe the difference of the Boolean operations from the binary operations, which is there only in one case, i.e. in binary operation, whereas $1 + 1 = 0$, in Boolean operation, $1 + 1 = 1$.

Definition. If $M = (a_{ij}), N = (b_{ij})$ are two $m \times n$ matrices over the set $B = \{0, 1\}$ of two elements, we define the matrices

$M^c = (c_{ij})$, where $c_{ij} = a_{ij}^c$ (the complement of a_{ij}), called the **complementation** of M,

$$M \cup N = (c_{ij}), \quad \text{where } c_{ij} = a_{ij} \cup b_{ij},$$

and

$$M \cap N = (c_{ij}), \quad \text{where } c_{ij} = a_{ij} \cap b_{ij}.$$

With these definitions, it is fairly easy to prove the following theorem.

Theorem 3.6. If R, S are relations from a set A to a set B with $o(A) = m, o(B) = n$, then the relation matrices of $R', R \cup S, R \cap S$ are given by

$$M_{R'} = (M_R)^c, M_{R \cup S} = M_R \cup M_S \quad \text{and} \quad M_{R \cap S} = M_R \cap M_S.$$

Proof. Let $M_R = (r_{ij})$ and $M_S = (s_{ij})$.

(a) Since $(a, b) \in A \times B$ is in R' (or R^c) if and only if $(a, b) \notin R$, (i, j)th entry of M_R^c is 1 if and only if $r_{ij} = 0$.

(b) Let $M_{R \cup S} = (x_{ij})$ and $M_R \cup M_S = (t_{ij})$. If $x_{ij} = 1$, then $(a_i, b_j) \in R \cup S$, which implies that $(a_i, b_j) \in R$ or $(a_i, b_j) \in S$, which in turn implies that $r_{ij} = 1$ or $s_{ij} = 1$. Thus $r_{ij} \cup s_{ij} = 1$, i.e. $t_{ij} = 1$.

On the other hand, $x_{ij} = 0$ implies $(a_i, b_j) \notin R \cup S$ which means that $(a_i, b_j) \notin R$ and $(a_i, b_j) \notin S$. Thus $r_{ij} = 0$ and $s_{ij} = 0$, which implies that $r_{ij} \cup s_{ij} = 0$, i.e. $t_{ij} = 0$. Hence $x_{ij} = t_{ij}$ for all i, j and $M_{R \cup S} = M_R \cup M_S$.

(c) Let $M_{R \cap S} = (y_{ij})$. Then $y_{ij} = 1$ implies that $(a_i, b_j) \in R \cap S$, which in turn shows that $(a_i, b_j) \in R$ and $(a_i, b_j) \in S$. Hence $r_{ij} = 1$ and $s_{is} = 1$ so that $r_{ij} \cap s_{ij} = 1$. Again, $y_{ij} = 0$ implies that $(a_i, b_j) \notin R \cap S$, which in turn implies that either $(a_i, b_j) \notin R$ or $(a_i, b_j) \notin S$. Thus either $r_{ij} = 0$ or $s_{ij} = 0$. Therefore, $r_{ij} \cap s_{ij} = 0$. Hence $y_{ij} = r_{ij} \cap s_{ij}$ for all i, j and $M_{R \cap S} = M_R \cap M_S$.

Theorem 3.7. If R is a relation from A to B and S is a relation from B to C, where A, B, C are finite sets with $o(A) = m, o(B) = n$ and $o(C) = t$, then the relation matrices satisfy $M_{SoR} = M_R M_S$.

Proof. As SoR is a relation from A to C, M_{SoR} is an $m \times t$ matrix. Let $M_R = (r_{ij}), M_S = (s_{ij})$ where

$$r_{ij} = 1 \text{ if } (a_i, b_j) \in R \text{ while it is } 0 \text{ if } (a_i, b_j) \notin R \quad \text{and}$$

$$s_{ij} = 1 \text{ if } (b_i, c_j) \in S \text{ while it is } 0 \text{ if } (b_i, c_j) \notin S.$$

Let $M_{SoR} = (x_{ij})$. Then $x_{ij} = 1$ if $(a_i, c_j) \in SoR$, while it is 0 if $(a_i, c_j) \notin SoR$.

For any $i, 1 \leq i \leq m$, and any $j, 1 \leq j \leq t$, let $(a_i, c_j) \in SoR$. Then there exists a $b_k, 1 \leq k \leq n$, such that $(a_i, b_k) \in R$ and $(b_k, c_j) \in S$. Therefore, $r_{ik} = 1$, $s_{kj} = 1$ so that $r_{ik}s_{kj} = 1$. Then by the definition of Boolean operations, $\sum_l r_{il}s_{lj} = 1$. However, $\sum_l r_{il}s_{lj}$ is the (i,j)th entry of M_{RoMS}. On the other hand, if the (i,j)th entry of $M_R M_S$ is 1, then $\sum_l r_{il}s_{lj} = 1$. It follows from the definition of Boolean operations that there is at least one 1 in the summation so that $r_{il}s_{lj} = 1$, which implies that $r_{il} = 1$ and $s_{lj} = 1$. Therefore, $(a_i, b_l) \in R$ and $(b_l, c_j) \in S$, which together imply that $(a_i, c_j) \in SoR$. Hence $x_{ij} = 1$. This completes the proof that $x_{ij} = \sum_l r_{il}s_{lj}$ for all i and j. Hence, we have $M_{SoR} = M_R M_S$. In particular, $M_{R^2} = M_R^2$ and in general $M_{R^n} = M_R^n$ for all $n > 1$.

Definition. Consider two relations R and S from a set A to a set B with $o(A) = m, o(B) = n$. Let the relation matrices of R and S be $M_R = (r_{ij})$ and $M_S = (s_{ij})$. Recall that $R \subseteq S$ if and only if $(a, b) \in R$ implies $(a, b) \in S$. If we want to introduce an ordered relation on the set of relation matrices, then we need to say that $M_R \leq M_S$ if and only if $R \subseteq S$. This is the same as saying that $M_R \leq M_S$ if and only if $r_{ij} \leq s_{ij}$ for all i, j. This suggests the following order relation on the Boolean set $\{0, 1\}$ of two elements. $0 \leq 0$, $0 \leq 1$, $1 \leq 1$ whereas 1 is not ≤ 0.

With this order relation on $\{0, 1\}$, for two matrices of the same order with the entries in $\{0, 1\}$, we define $M = (m_{ij}) \leq N = (n_{ij})$ if $m_{ij} \leq n_{ij}$ for all i, j.

3.3.6. *Closure operations on relations*

Definition. Let A be a finite set of order n and R be a relation on A.

(a) A relation R^s is called **symmetric closure** of R if R^s is the smallest relation on A with $R \subseteq R^s$ and R^s is symmetric.

(b) A relation R^+ is called **transitive closure** of R if R^+ is the smallest relation on A satisfying $R \subseteq R^+$ and R^+ is transitive.

(c) A relation R^0 is called **reflexive closure** of R if R^0 is the smallest relation on A that is reflexive and $R \subseteq R^0$.

Definition. Diagonal relation on a set A is the relation $\Delta = \{(a, a) | a \in A\}$. Observe that the symmetric closure of a relation R on A is $R \cup R^{-1}$, whereas the reflexive closure of R is the relation $R \cup \Delta$.

Let R be a relation on a set A. For $(a, b) \in A \times A$, we say that there is a path between a and b in R if there exist elements $(a, a_1), (a_1, a_2), \ldots, (a_k, b)$ in R.

The connectivity relation R^* consists of the pairs (a, b) such that there is a path between a and b in R. The relation R^n is the set of all pairs (a, b) such that there is a path of length n between a and b. Then $R^* = \bigcup_{n \geq 1} R^n$.

Example 3.28. Let $A = \{1, 2, 3, 4\}$. Consider the relation $R = \{(1, 2), (2, 3), (3, 4)\}$. Since in a reflexive relation on A, all the elements $(1, 1)$, $(2, 2)$, $(3, 3)$ and $(4, 4)$ should be in the relation, the reflexive closure of R is $R^0 = \{(1, 2), (2, 3), (3, 4), (1, 1), (2, 2), (3, 3), (4, 4)\}$.

To get symmetric closure of R, since $(1, 2)$, $(2, 3)$, $(3, 4)$ are in R, we must adjoin the elements $(2, 1)$, $(3, 2)$ and $(4, 3)$. Hence

$$R^s = \{(1, 2), (2, 3), (3, 4), (2, 1), (3, 2), (4, 3)\}.$$

Finally, we would like to get the transitive closure of R. For this, observe that R^2 must be $\subseteq R^+$. However, $R^2 = \{(1, 3), (2, 4)\}$. This is so because $(1, 2)$ and $(2, 3)$ are in R and $(2, 3)$, $(3, 4)$ are in R. Therefore, $R \cup R^2 = \{(1, 2), (2, 3), (3, 4), (1, 3), (2, 4)\} \subseteq R^+$. Since $(1, 3)$ and $(3, 4)$ are in $R \cup R^2$ but $(1, 4)$ is not in $R \cup R^2$, the relation $R \cup R^2$ is not transitive. For transitivity, we know that $R^3 = R^2 R \subseteq R^+$. As $(1, 3)$ is in R^2 and $(3, 4)$ is in R, $(1, 4) \in R^3$. Also, $(1, 3)$ is not the predecessor of any other element of R. Moreover, $(2, 4)$ is not the predecessor of any element of R. Therefore, there are no more elements in R^3 and we have $R^3 = \{(1, 4)\}$. Then $R \cup R^2 \cup R^3 = \{(1, 2), (2, 3), (3, 4), (1, 3), (2, 4), (1, 4)\} \subseteq R^+$. It is clear that the relation $R \cup R^2 \cup R^3$ is transitive. Hence

$$R^+ = R \cup R^2 \cup R^3.$$

Example 3.29. Let $A = \{1, 2, 3, 4, 5\}$ and consider a relation

$$R = \{(1, 4), (2, 1), (2, 2), (2, 3), (3, 2), (4, 3), (4, 5), (5, 1)\}.$$

As $(1, 4), (4, 5) \in R$ but $(1, 5) \notin R$, the relation R is not transitive. Let us agree to define a sort of product in $A \times A$ by saying that $(a, b)(c, d) = (a, d)$ if $b = c$ and $(a, b)(c, d)$ is not defined if $b \neq c$.

With this convention, we find that

$$R^2 = \{(1, 4)(4, 3), (1, 4)(4, 5), (2, 2)(2, 1), (2, 2)(2, 3), (2, 3)(3, 2),$$

$$(2, 1)(1, 4), (3, 2)(2, 1), (3, 2)(2, 2), (3, 2)(2, 3), (4, 3)(3, 2),$$

$$(4,5)(5,1),(5,1)(1,4)\}$$

$$= \{(1,3),(1,5),(2,1),(2,2),(2,3),(2,4),(3,1),(3,2),(3,3),$$

$$(4,2),(4,1),(5,4)\}.$$

As the formal product of every two elements of R is in R^+, $R^2 \subseteq R^+$. Now $(1,3),(3,1) \in R^2 \subseteq R^+$ but $(1,3)(3,1) = (1,1) \notin R \cup R^2$. Therefore, $R \cup R^2$ is not transitive. Again, the product of any element of R^2 with any element of R is in R^+. Therefore, we find R^3. Now

$$R^3 = \{(1,3)(3,2),(1,5)(5,1),(2,1)(1,4),(2,3)(3,2),(3,1)(1,4),$$

$$(3,2)(2,2),(3,2)(2,1),(3,2)(2,3),(3,3)(3,2),(4,2)(2,1),$$

$$(4,2)(2,2),(4,2)(2,3),(4,1)(1,4),(5,4)(4,3),(5,4)(4,5),$$

$$(2,2)(2,1),(2,2)(2,2),(2,2)(2,3),(2,4)(4,3),(2,4)(4,5)\}$$

$$= \{(1,1),(1,2),(2,1),(2,2),(2,3),(2,4),(2,5),(3,1),(3,2),$$

$$(3,3),(3,4),(4,1),(4,2),(4,3),(4,4),(5,3),(5,5)\}.$$

By the reasoning already given, $R^3 R = R^4 \subseteq R^+$ and

$$R^4 = \{(1,2)(2,1),(1,2)(2,2),(1,2)(2,3),(1,1)(1,4),(2,4)(4,3),$$

$$(2,4)(4,5),(2,2)(2,1),(2,2)(2,2),(2,2)(2,3),(3,4)(4,3),$$

$$(3,4)(4,5),(3,2)(2,1),(3,2)(2,2),(3,2)(2,3),(3,3)(3,2),$$

$$(4,1)(1,4),(4,2)(2,1),(4,2)(2,2),(4,2)(2,3),(4,3)(3,2),$$

$$(4,4)(4,3),(4,4)(4,5),(5,5)(5,1),(2,1)(1,4),(2,3)(3,2),$$

$$(2,5)(5,1),(3,1)(1,4),(5,3)(3,2)\}$$

$$= \{(1,1),(1,2),(1,3),(1,4),(2,1),(2,2),(2,3),(2,4),(2,5),(3,1),$$

$$(3,2),(3,3),(3,4),(3,5),(4,1),(4,2),(4,3),(4,4),(4,5),(5,1),$$

$$(5,2)\}.$$

Now

$$R \cup R^2 = \{(1,3),(1,4),(1,5),(2,1),(2,2),(2,3),(2,4),$$

$$(3,1),(3,2),(3,3),(4,1),(4,2),(4,3),(4,5),$$

$$(5,1),(5,4)\},$$

$$R \cup R^2 \cup R^3 = \{(1,1),(1,2),(1,3),(1,4),(1,5),(2,1),(2,2),(2,3),$$
$$(2,4),(2,5),(3,1),(3,2),(3,3),(3,4),(4,1),(4,2),$$
$$(4,3),(4,4),(4,5),(5,1),(5,2),(5,3),(5,4),(5,5)\}.$$
$$R \cup R^2 \cup R^3 \cup R^4 = \{(1,1),(1,2),(1,3),(1,4),(1,5),(2,1),(2,2),$$
$$(2,3),(2,4),(2,5),(3,1),(3,2),(3,3),(3,4),$$
$$(3,5),(4,1),(4,2),(4,3),(4,4),(4,5),(5,1),$$
$$(5,2),(5,3),(5,4),(5,5)\}.$$

Thus we find that $(a,b) \in R \cup R^2 \cup R^3 \cup R^4$ for all $a, b, 1 \leq a, b \leq 5$. Therefore, if (a,b), $(b,c) \in R \cup R^2 \cup R^3 \cup R^4$, then $(a,b)(b,c) = (a,c)$ also belongs to $R \cup R^2 \cup R^3 \cup R^4$. Hence $R^+ = R \cup R^2 \cup R^3 \cup R^4 = \{(a,b) \in A \times A \,|\, a, b \in A\}$, which is the transitive closure of R. Observe that R^+ is reflexive as well as symmetric.

Theorem 3.8. If R is a relation on a finite non-empty set A and order of A is n, then the transitive closure $R^+ = R \cup R^2 \cup R^3 \cup \cdots \cup R^n$.

Proof. We will first prove by induction on $m \geq 2$ that $R^m \subseteq R^+$. Let $(a,b) \in R^2$. Then, by definition of R^2, there exists a c in A such that (a,c), (c,b) are in R. Since the relation R^+ is transitive and $R \subseteq R^+$, the product

$$(a,c)(c,b) = (a,b) \in R^+. \text{ Thus } R^2 \subseteq R^+.$$

Now suppose that $m \geq 2$, $m < n$ and that $R^m \subseteq R^+$. Any element of R^{m+1} is of the form $(a,b) = (a,c)(c,b)$ where $(a,c) \in R^m$ and $(c,b) \in R$. Again, R^+ being transitive and (a,c), $(c,b) \in R^+$ imply that $(a,c)(c,b) = (a,b) \in R^+$. This proves that $R^{m+1} \subseteq R^+$ and the induction is complete. Hence $R \cup R^2 \cup R^3 \cup \cdots \cup R^n \subseteq R^+$.

Let $(a,b) \in R^{n+1}$. Then there exist elements $a_1, a_2, \ldots, a_n$ in A such that $(a,a_1), (a_1,a_2), \ldots, (a_n,b)$ are in R. Since $a, a_1, a_2, \ldots, a_n \in A$ and A has n elements, these elements cannot all be distinct. Then, either (a) there exists an $i \geq 1$ such that $a_i = a$ or (b) there exist $i, j, 1 \leq i < j \leq n$ such that $a_i = a_j$. In the case of (a), we have the string of elements $(a,a_{i+1}) = (a_i,a_{i+1}), (a_{i+1},a_{i+2}), \ldots, (a_n,b)$ in R having $n - i + 1$ elements from which it follows that $(a,b) \in R^{n-i+1}$ with $n - i + 1 \leq n$. In the case of (b), we have a string of elements $(a,a_1), \ldots, (a_{i-1},a_i)$, $(a_j = a_i, a_{j+1}), \ldots, (a_n,b)$ and since $i < j$, this string again has $\leq n$ elements. Therefore, in this case also (a,b) belong to $R \cup R^2 \cup \cdots \cup R^n$. Hence

$R^* = R \cup R^2 \cup \cdots \cup R^n$. Since R^* is clearly transitive with $R \subseteq R^*$ and R^+, by definition, is the smallest transitive relation with $R \subseteq R^+$, we have $R^+ \subseteq R^*$. Hence $R \cup R^2 \cup \cdots \cup R^n \subseteq R^+ \subseteq R \cup R^2 \cup \cdots \cup R^n$ which proves that $R^+ = R \cup R^2 \cup \cdots \cup R^n$.

Theorem 3.9. The transitive closure of a symmetric relation on a set A is again symmetric.

Proof. Let R be a symmetric relation on a set A. Since the transitive closure of R is $R^+ = R^* = \bigcup_{m \geq 1} R^m$, it is enough to prove that R^m is symmetric for all $m \geq 1$. As for $m = 1$, $R^1 = R$ is trivially symmetric, we need to prove that R^m is symmetric for all $m > 1$.

Let $m > 1$ and $(a, b) \in R^m$. Then there exist elements $a_1, a_2, \ldots, a_{m-1} \in A$ such that $(a, a_1), (a_1, a_2), \ldots, (a_{m-2}, a_{m-1}), (a_{m-1}, b) \in R$. Since the relation R is symmetric, $(a_1, a), (a_2, a_1), \ldots, (a_{m-1}, a_{m-2}), (b, a_{m-1}) \in R$, which then imply that $(b, a) \in R^m$. This completes the proof that R^m is symmetric and, hence, $R^+ = R^*$ is symmetric.

Remark. When a relation R on a set A is symmetric, the transitive closure R^+ is also symmetric, but is not always reflexive. For example, consider a relation

$$R = \{(1,2), (2,1), (1,3), (3,1)\} \text{ on the set } A = \{1,2,3,4\}. \text{ Then}$$

$$R^2 = \{(1,2)(2,1), (2,1)(1,2), (2,1)(1,3), (1,3)(3,1), (3,1)(1,2),$$
$$(3,1)(1,3)\}$$
$$= \{(1,1), (2,2), (2,3), (3,2), (3,3)\}.$$
$$R^3 = \{(1,1)(1,2), (1,1)(1,3), (2,2)(2,1), (2,3)(3,1), (3,2)(2,1),$$
$$(3,3)(3,1)\}$$
$$= \{(1,2), (1,3), (2,1), (3,1)\} = R.$$

Therefore, $R^n = R$ when n is odd and $R^n = R^2$ when n is even. Then

$$R \cup R^2 \cup R^3 = R \cup R^2 = \{(1,1), (1,2), (1,3), (2,1), (2,2), (2,3),$$
$$(3,1), (3,2), (3,3)\},$$

which is clearly the transitive closure of R. However, R^+ is not reflexive, as $(4, 4)$ is not in R^+.

Example 3.30. Let Z be the set of all integers and R be a relation defined on Z by

(a) aRb if $a + 1 = b$. Then $R = \{(a, a+1)|a \in Z\}$.
(b) aRb if $|a - b| = 2$. Then

$$R = \{(a, a+2)|a \in Z\} \cup \{(a, a-2)|a \in Z\}.$$

In the case of (a), observe that $R^n = \{(a, a+n)|a \in Z\}$. Therefore, the transitive closure $R^+ = R^* = \{(a, a+n)|a \in Z, n \in Z, n > 0\}$.

In the case of (b), $R^n = \{(a, a+2n)|a \in Z\} \cup \{(a, a-2n)|a \in Z\}$ and, therefore,

$$R^+ = R^* = \{(a, a+2n)|a \in Z, n \in Z\}.$$

Example 3.31. Let R be a symmetric relation defined on a finite set A such that for every a in A, there exists an element b in A such that either (a, b) is in R or (b, a) is in R. As already proved, the transitive closure R^+ of R is again symmetric. We explain that R^+ is reflexive as well.

For any a in A, there exists an element b in A such that either (a, b) is in R or (b, a) is in R. The relation R being symmetric, both (a, b) and (b, a) are in R. The transitive closure R^+ being a transitive relation with $R \subseteq R^+, (a, b), (b, a) \in R^+$ imply that $(a, a) \in R^+$. Hence R^+ is reflexive.

Exercise 3.5.

1. Consider the following relations on the set Z of integers.

$$R = \{(a, b)|a \leq b\}$$

$$R = \{(a, b)|a = b\}$$

$$R = \{(a, b)|a = -b\}$$

$$R = \{(a, b)|a = b \text{ or } a = -b\}$$

$$R = \{(a, b)|a = b + 2\}$$

$$R = \{(a, b)|a \leq b + 4\}$$

$$R = \{(a, b)|a < b\}$$

$$R = \{(a, b)|a \text{ divides } b, a \neq 0\}$$

$$R = \{(a, b)|g.c.d.(\text{a, b}) = 1\}.$$

Check the above relations for the properties: (a) reflexive, (b) symmetric, (c) anti-symmetric, (d) transitive.

2. Find the number of relations on a set with n elements.

3. Find the (a) reflexive closure, (b) symmetric closure, (c) transitive closure, of each of the relations in Question 1 above.

4. Let A be a finite ordered set of order m. Let R be a binary relation defined on A and $M_R = (x_{a,b})$ be the adjacency matrix of the relation R. This matrix is called **the matrix associated with or corresponding to the relation** R. Observe that (a) the matrix M_R is symmetric if and only if the relation R is symmetric, (b) every diagonal entry of M_R is 1 if and only if the relation R is reflexive. What can you say about the matrix M_R if the relation R is transitive?

5. Let A be a non-empty set and X be the set of all maps from A into R the set of all real numbers. For any $f, g \in X$, define $f \le g$ if and only if $f(a) \le g(a)$ for all $a \in A$. Show that '$\le$' is a partial order relation on X.

6. For a set A, any map from A to A is a relation on A. Give an example to show that every relation on A need not be a map from A to A.

7. Let $M_2(R)$ be the set of all square matrices of order 2 with real entries. For any $A \in M_2(R)$, let $\det(A)$ denote the determinant of A. For $A, B \in M_2(R)$ define $A \sim B$ if $\det(A) = \det(B)$. Prove that $\sim$ is an equivalence relation on $M_2(R)$.

8. If A, B are finite sets of orders m, n, respectively, determine the number of distinct relations from A to B. (Answer: 2^{mn}).

9. Characterize the relation matrices and graphs of the identity relation and the universal relation on the set $A = \{a_1, a_2, \ldots, a_n\}$.

10. A relation R defined on a set A is called **irreflexive** if no element of A is related to itself under R. If a relation R defined on a set A is transitive and irreflexive, prove that it must also be asymmetric. (Recall that a relation R defined on A is **asymmetric** if $(a, b) \in R$ implies $(b, a) \notin R$).

11. If R, S are symmetric relations on a set A, prove that $RoS \subseteq SoR$ implies that $RoS = SoR$.

12. Let R be a relation on a non-empty set A. Prove that $(R^+)^+ = R^+$.

13. On the set Z of all integers, define a relation R as follows: For $a, b \in Z$, say that aRb if five divides $a - b$. Prove that R is an equivalence relation. Also find all the equivalence classes determined by this relation.

14. If R is a relation defined on a non-empty set A and R is symmetric as well as anti-symmetric, what precisely is R?

15. In the set Z of all integers define a relation R by saying that for $a, b \in Z$, aRb if three divides $a + b$. Is the relation

  (i) reflexive?

 (ii) symmetric?

(iii) anti-symmetric?

 (iv) transitive?

  (v) irreflexive?

 (vi) asymmetric?

Chapter 4

SOME COUNTING TECHNIQUES

The principle of mathematical induction, which is based on the Peano axioms, is one of the best possible computational techniques, or proof techniques, known so far. We discuss this in this chapter. Among several applications of this, arithmetic, geometric and arithmetic-geometric series are discussed. Permutations and combinations, which form an important part of combinational techniques, are also looked at. Additionally, algorithms for generating permutations and combinations of finite sets are discussed.

4.1. The Principle of Mathematical Induction

Let S be a subset of the set N of natural numbers. For an $a \in N$, the integer $a+1$ is called the successor of a and is denoted by a^+. Suppose that $1 \in S$ and that for every $a \in S, a^+ \in S$. Then $S = N$. This principle is the basis for the proof of many varied results in mathematics.

Let P be a property which depends on the natural number n. We indicate this by writing $P(n)$. We say that $P(n)$ holds or is true if the property P holds for n. To prove that $P(n)$ holds for all integers $n \geq 1$, we need to take the following steps:

1. Prove that the property P holds for $n = 1$ or that $P(1)$ holds.
2. Prove that if $P(k)$ holds for some $k \geq 1$, then $P(k+1)$ holds.

The above stated principle shows that then $P(n)$ holds for all $n \geq 1$. We will consider some examples.

Example 4.1. Prove that the sum of the first n positive even integers is $n(n+1)$.

Proof. We have to prove that if $n \geq 1$, then
$$2 + 4 + \cdots + 2n = n(n + 1).$$
For $n = 1$, the left-hand side of the above relation is 2 and the right-hand side is $1(1 + 1) = 2$. So the result holds for $n = 1$.

Suppose that $k \geq 1$ is an integer and that the result holds for $n = k$, i.e. $2 + 4 + \cdots + 2k = k(k + 1)$.

For $n = k + 1$,
$$2 + 4 + \cdots + 2(k + 1)$$
$$= (2 + 4 + \cdots + 2k) + 2(k + 1)$$
$$= k(k + 1) + 2(k + 1)$$
$$= (k + 1)(k + 2).$$
Thus the result holds for $n = k + 1$. This completes induction and $2 + 4 + \cdots + 2n = n(n + 1)$ for all $n \geq 1$.

Example 4.2. Prove that $n \leq 2^{n-1}$ for all $n \geq 1$.

Proof. Let $P(n)$ be the property '$n \leq 2^{n-1}$'.

$P(1)$ is true because $1 \leq 2^{1-1} = 1$.

Suppose that $k \geq 1$ is an integer and that $P(k)$ holds. Thus $k \leq 2^{k-1}$. Then, for $n = k + 1$,
$$k + 1 \leq 2^{k-1} + 1 \leq 2^{k-1} + 2^{k-1}, \text{ as } k \geq 1 \text{ so that } 2^{k-1} \geq 1$$
$$= 2 \cdot 2^{k-1} = 2^k = 2^{(k+1)-1}.$$
This proves that $P(k + 1)$ holds. This completes induction and, hence, $n \leq 2^{n-1}$ for all $n \geq 1$.

Example 4.3. Use mathematical induction to prove that $n^5 - n$ is divisible by 5 for all $n \geq 1$.

Proof. Let $P(n)$ be the proposition '5 divides $n^5 - n$'.

For $P(1)$, $1^5 - 1 = 0$ and 5 divides 0. Thus $P(1)$ holds.

Suppose that $k \geq 1$ is an integer and that $P(k)$ holds. Thus 5 divides $k^5 - k$. For $n = k + 1$,
$$(k + 1)^5 - (k + 1) = (k^5 - k) + (5k^4 + 10k^3 + 10k^2 + 5k)$$
$$= (k^5 - k) + 5(k^4 + 2k^3 + 2k^2 + k).$$
Since 5 divides $k^5 - k$ and 5 divides $5(k^4 + 2k^3 + 2k^2 + k)$, 5 divides their sum, i.e. 5 divides $(k + 1)^5 - (k + 1)$. Thus $P(k + 1)$ holds. This completes induction and, hence, for every $n \geq 1$, $n^5 - n$ is divisible by 5.

Example 4.4. Prove by induction on n that for all $n \geq 1$,

$$\frac{1}{1.2} + \frac{1}{2.3} + \frac{1}{3.4} + \cdots + \frac{1}{n(n+1)} = \frac{n}{(n+1)}.$$

Proof. Let $P(n)$ be the proposition

$$`\frac{1}{1.2} + \frac{1}{2.3} + \frac{1}{3.4} + \cdots + \frac{1}{n(n+1)} = \frac{n}{(n+1)}',$$

$P(1)$ holds, since $\frac{1}{1.2} = \frac{1}{2} = \frac{1}{1+1}$.

Suppose that $k \geq 1$ is an integer and that $P(k)$ holds. Then

$$\frac{1}{1.2} + \frac{1}{2.3} + \frac{1}{3.4} + \cdots + \frac{1}{k(k+1)} = \frac{k}{(k+1)}.$$

Now

$$\frac{1}{1.2} + \frac{1}{2.3} + \frac{1}{3.4} + \cdots + \frac{1}{(k+1)(k+2)}$$

$$= \frac{1}{1.2} + \frac{1}{2.3} + \frac{1}{3.4} + \cdots + \frac{1}{k(k+1)} + \frac{1}{(k+1)(k+2)}$$

$$= \frac{k}{k+1} + \frac{1}{(k+1)(k+2)} = \frac{k(k+2)+1}{(k+1)(k+2)} = \frac{k^2+2k+1}{(k+1)(k+2)}$$

$$= \frac{(k+1)^2}{(k+1)(k+2)} = \frac{k+1}{k+2}.$$

Thus $P(k+1)$ holds. This completes induction and, hence, for all $n \geq 1$,

$$\frac{1}{1.2} + \frac{1}{2.3} + \frac{1}{3.4} + \cdots + \frac{1}{n(n+1)} = \frac{n}{n+1}.$$

Example 4.5. Prove by induction on n that if S is a finite set of order $n \geq 1$, then S has 2^n subsets.

Proof. Let S be a set having only one element a (say). Then the subsets of S are φ and $S = \{a\}$. Thus S has two subsets and the proposition $P(n)$: 'If S is a set having $n \geq 1$ elements, then S has 2^n subsets' holds for $n = 1$. Let $k \geq 1$ be an integer and suppose that $P(k)$ holds. Let T be a set having $k + 1$ elements. Then $T = S \cup \{a\}$, where $a \in T$ but $a \notin S$ and S is a subset of T having k elements. Any subset of T is either a subset S' of S or is of the form $S'' \cup \{a\}$, where S'' is a subset of S. The order of S being k, it follows by induction hypothesis that the number of subsets S' of S is 2^k and the number of subsets S'' of S is also 2^k. Therefore, the number of subsets of T is $2^k + 2^k = 2^k \times 2 = 2^{k+1}$. This completes induction and, hence, the number of subsets of a set having n elements is 2^n.

Example 4.6. Prove by induction on n that the sum of first n integers, which leave remainder 1 on division by 3 is $\frac{n(3n-1)}{2}$.

Proof. We have to prove that

$$1 + 4 + 7 + \cdots + (3n - 2) = \frac{n(3n - 1)}{2} \quad \text{for all } n \geq 1.$$

Let $P(n)$ denote the proposition: for all $n \geq 1$,

$$1 + 4 + 7 + \cdots + (3n - 2) = \frac{n(3n - 1)}{2}.$$

For $n = 1$, left-hand side is 1 and the right-hand side is $\frac{1(3-1)}{2} = 1$.

Thus $P(1)$ holds. Suppose that $k \geq 1$ and that $P(k)$ holds. For proving that $P(k + 1)$ holds, we need to prove that

$$1 + 4 + 7 + \cdots + (3k + 1) = \frac{(k + 1)(3k + 2)}{2}.$$

Now

$$1 + 4 + 7 + \cdots + (3k - 2) + 3k + 1$$

$$= \{1 + 4 + 7 + \cdots + (3k - 2)\} + 3k + 1 = \frac{k(3k - 1)}{2} + 3k + 1$$

$$= \frac{3k^2 - k + 6k + 2}{2} = \frac{3k^2 + 5k + 2}{2} = \frac{3k^2 + 3k + 2k + 2}{2}$$

$$= \frac{(k + 1)(3k + 2)}{2} = \frac{(k + 1)(3(k + 1) - 1)}{2}.$$

This completes induction and, hence, $P(n)$ holds for all $n \geq 1$.

Example 4.7. Prove that the sum of the first n odd integers is n^2.

Proof. We have to prove that for all $n \geq 1$, $1 + 3 + 5 + \cdots + (2n - 1) = n^2$.
Let $P(n)$ be the proposition: for all $n \geq 1$, '$1 + 3 + 5 + \cdots + (2n - 1) = n^2$'.
For $n = 1$, on the left-hand side is 1 and on the right-hand side is $1^2 = 1$.
Therefore, $P(1)$ holds. Suppose that $k \geq 1$ and that the proposition $P(k)$ holds. Then

$$1 + 3 + \cdots + (2k - 1) + (2k + 1) = \{1 + 3 + \cdots + (2k - 1)\} + (2k + 1)$$

$$= k^2 + 2k + 1 = (k + 1)^2.$$

Thus $P(k + 1)$ holds. This completes induction and, hence, for all $n \geq 1$, the sum of the first n odd integers is n^2.

Example 4.8. Prove that a number composed of 9^n identical digits for $n \geq 1$ is divisible by 9^n.

Proof. Suppose that a number is composed of $9^1 = 9$ identical digits. Then in the decimal representation it looks like $a\,a\,a\,a\,a\,a\,a\,a\,a$, where a is any integer with $0 \leq a \leq 9$. The sum of digits forming the number is $9a$, which is divisible by 9. This implies that the number $a\,a\,a\,a\,a\,a\,a\,a\,a$ itself is divisible by 9.

Suppose that $k \geq 1$ is an integer and that any number formed by 9^k identical digits is divisible by 9^k. Let x be a number which is composed of 9^{k+1} identical digits. It is easily seen (from the process of multiplication) that $x = yz$, where y is composed of 9^k identical digits and $z = 10 \ldots 010 \ldots 01 \ldots 10 \ldots 01$, the number of ones being 9 and the number of zeros between two consecutive ones in the expression for z being $9^k - 1$. The sum of digits forming the number z is 9. Therefore, z is divisible by 9. Also, by induction hypothesis, y is divisible by 9^k. Therefore, $yz = x$ is divisible by $9^k \times 9 = 9^{k+1}$. This completes induction. Hence every number composed of 9^n identical digits, for $n \geq 1$, is divisible by 9^n.

Example 4.9. When n couples arrived at a party, they were greeted by the host and the hostess at the door. After rounds of hand shaking, the host asked the guests, as well as his wife, to indicate the number of hands each had shaken. He got $2n + 1$ different answers. Given that no one shook hands with his or her own spouse, how many hands had the hostess shaken?

Proof. Let x be the host and y be the hostess. If only one couple arrived for the party, let it be a, b; a being the man and b his spouse. Since husband and wife do not shake hands with each other, no one of the four persons x, y, a, b can shake more than two hands. Therefore, the three different answers given by the persons y, a, b are $0, 1, 2$, i.e. one of them does not shake hands at all, one has only one handshake and the third has two handshakes. If y shook hands with both a and b, then there is no one among y, a, b who did not shake any hands. Suppose that it is b who has shaken two hands. Then b has shaken hands with x as well as y. Therefore, a must be the person who did not shake hands at all and y has shaken only one hand. If instead of b it is a who has shaken two hands, then b has not shaken any hands and y again has shaken only one hand.

Suppose that $k \geq 1$ is an integer and that y has k handshakes when k couples arrived at the party. Suppose that $k + 1$ couples $(a_1, b_1), (a_2, b_2), \ldots, (a_{k+1}, b_{k+1})$ arrived at the party. Then everyone of $a_1, a_2, \ldots, a_{k+1}, b_1, b_2, \ldots, b_{k+1}$ and y can shake at most $2(k+1)$ hands. The answers provided by $a_1, a_2, \ldots, a_{k+1}, b_1, b_2, \ldots, b_{k+1}$ and y being all distinct and being $2(k + 1) + 1$ in number, they have to be $0, 1, 2, \ldots, 2(k + 1)$. If the number given by y is $2(k + 1)$, then y has shaken hands with every one of $a_1, a_2, \ldots, a_{k+1}, b_1, b_2, \ldots, b_{k+1}$ and then there is no one who can provide the number 0. Therefore, y cannot shake hands with each and every person who arrived. Hence the number $2(k + 1)$ is provided by one $a_1, a_2, \ldots, a_{k+1}, b_1, b_2, \ldots, b_{k+1}$.

Suppose that it is a female who provides the number $2(k+1)$ (we only need change the names, if it is a male person). By renumbering the as and the bs, if necessary, we can assume that b_{k+1} gave the number $2(k + 1)$, i.e. she shook hands with every one of $x, y, a_1, a_2, \ldots, a_k, b_1, b_2, \ldots, b_k$. But then none of $y, a_1, a_2, \ldots, a_k, b_1, b_2, \ldots, b_k$ can provide the answer 0. Thus, it is a_{k+1} who did not shake any hands. Ignoring this couple, i.e. (a_{k+1}, b_{k+1}), we have the couples $(x, y), (a_1, b_1), \ldots, (a_k, b_k)$ and everyone of $y, a_1, \ldots, a_k, b_1, \ldots, b_k$ has shaken hands with b_{k+1}. Also, the numbers provided by these persons, but not in order, are $1, 2, \ldots, 2(k + 1) - 1 = 2k + 1$. Now, ignoring the one handshake that everyone had with b_{k+1}, we get the number of handshakes that $(x, y), (a_1, b_1), \ldots, (a_k, b_k)$ had among themselves. The answers then provided (but not in order) by $y, a_1, \ldots, a_k, b_1, \ldots, b_k$ are $0, 1, 2, \ldots, 2k$.

However, for k couples arriving at the party, y has k handshakes. Hence the hands that y had shaken for all the $k+1$ couples is $k+1$ (adding the one handshake that y had with b_{k+1}). This completes induction and, therefore, y, the hostess, shakes n hands when n couples arrived at the party.

So far we have considered examples of propositions $P(n)$ which are valid for all $n \geq 1$. Not all propositions are valid for all $n \geq 1$. For example, if we consider the numbers 2^n and n^3, then we find that $3^1 > 1^4, 3^2 < 2^4, 3^3 < 3^4, 3^4 < 4^4, 3^5 < 5^4, 3^6 < 6^4, 3^7 < 7^4$, whereas $3^8 = 6561 > 8^4 = 4096$. However, we can have the following:

Example 4.10. For all $n \geq 8$, $3^n > n^4$.

Proof. Let $P(n)$ denote the proposition '$3^n > n^4$'. As seen above, $3^8 = 6561 > 4096 = 8^4$. Thus $P(n)$ is true for $n = 8$.

Suppose k is an integer ≥ 8 and that $P(k)$ is true. Now

$$3^{k+1} = 3^k.3 > 3^k \left(1 + \frac{1}{8}\right)^4$$

$$> k^4 \left(1 + \frac{1}{8}\right)^4 \quad \text{(by induction hypothesis)}$$

As

$$k \geq 8, \frac{1}{8} \geq \frac{1}{k} \quad \text{and, so,} \quad \left(1 + \frac{1}{8}\right)^4 \geq \left(1 + \frac{1}{k}\right)^4.$$

Thus

$$3^{k+1} > k^4 \left(1 + \frac{1}{8}\right)^4 \geq k^4 \left(1 + \frac{1}{k}\right)^4 = (k+1)^4.$$

Thus $P(n)$ is true for $n = k+1$. This completes induction and, hence, $P(n)$ holds for all $n \geq 8$ or that $3^n \geq n^4$ for all $n \geq 8$.

Example 4.11. The dispatch clerk of the college has only two kinds of postage stamps, namely Rs. 3 each and Rs. 5 each. Prove that for all letters needing stamps worth Rs. 8 or more, the clerk can make combinations and is able to put stamps on every such letter for the exact amount required.

Proof. If a letter requires stamps worth Rs. 8, then the clerk can put on the letter one stamp of each of the two kinds. Suppose that k is an integer ≥ 8 and the clerk is able to supply stamps for Rs. k. Let the stamps supplied for Rs. k be a stamps of Rs. 3 each and b stamps of Rs. 5 each. Then, $k = 3a + 5b$ and $k + 1 = 3a + 5b + 1$. Observe that $a \geq 0$ and $b \geq 0$.

(i) $b > 0$. Then $k + 1 = 3(a + 2) + 5(b - 1)$, so that for a letter requiring stamps worth $k+1$ rupees, we may put $a + 2$ stamps of Rs. 3 each and $b - 1$ stamps of Rs. 5 each.

(ii) $b = 0$. Then $k = 3a$ and $k + 1 = 3a + 1$. As $k \geq 8$, $3a \geq 8$ and, so, $a \geq 3$ which means that $a - 3 \geq 0$. Then $k + 1 = 3a + 1 = 3(a - 3) + 9 + 1 = 3(a - 3) + 5.2$ and the clerk needs to put $a - 3$ stamps of Rs. 3 each and two stamps of Rs. 5 each. This proves that stamps can be put on the letter requiring stamps worth $k + 1$ Rs. This completes induction. Hence the exact amount of stamps can be put on any letter requiring stamps for any amount not less than Rs. 8.

Example 4.12. Prove that a bus conductor can give tickets to passengers for every amount of Rs. 12 or more using just two kind of tickets, namely Rs. 4 and Rs. 5.

Proof. For a passenger needing a ticket for Rs. 12, the conductor can give three tickets of Rs. 4 each. Suppose that k is an integer ≥ 12 and that for k rupees, the conductor can make a ticket by giving a tickets of Rs. 4 each and b tickets of Rs. 5 each. Then $k = 4a + 5b$, with $a, b \geq 0$. We have $k + 1 = 4a + 5b + 1$.

(i) $a \geq 1$. Then $k + 1 = 4a + 5b + 1 = 4(a - 1) + 5(b + 1)$.
(ii) $a = 0$. Then $k = 5b \geq 12$ implies that $b \geq 3$. Therefore, $k + 1 = 5b + 1 = 5(b - 3) + 16 = 4 \times 4 + 5(b - 3)$.

Hence in either case a ticket can be issued for Rs. $(k + 1)$. This completes induction. Therefore, for any amount of Rs. 12 or more, a ticket can be issued using Rs. 4 and Rs. 5 tickets.

4.2. Strong Induction

So far we have considered applications of mathematical induction where a given proposition $P(n)$ was proved for some starting value; $n = n_0 \geq 1$, and by assuming the validity of the proposition for some; $k \geq n_0$. There are certain problems which cannot or may not be solved by just assuming the proposition to be true for some $n = k$ and then proving it for $n = k + 1$. For example, suppose that we want to prove that every integer $n \geq 2$ is either a prime or is a product of primes. This result is fine if $n = 2$, as 2 is a prime. Suppose that $k \geq 2$ and that k is either a prime or a product of primes. With this assumption, we need to check that either $k + 1$ is a prime or is a product of primes. If $k + 1$ is a prime, we are, again, done. If $k + 1$ is not a prime, we cannot use the fact that k is a product of primes to prove that $k + 1$ is a product of primes. For example, 11 is a prime. We cannot see how to use this information to prove that 12 is a product of primes. We can circumvent this difficulty using what is called the principle of strong induction. To use this principle, we will take the following steps:

1. Prove the proposition $P(n)$ for the starting point $n = n_0$.
2. Let k be an integer $\geq n_0$ and assume that the proposition $P(n)$ is true for $n = n_0, n_0 + 1, \ldots, k$ where $k \geq n_0$.
3. Using the hypothesis as at (2) above, prove that $P(n)$ holds for $n = k + 1$.

Then $P(n)$ holds for all $n \geq n_0$. The basis for this is: If S is a non-empty subset of the set N of natural numbers such that $n_0 \in S$, and whenever $n_0, n_0 + 1, n_0 + 2, \ldots, n_0 + k - 1$ are in S, then $n_0 + k \in S$, then $n \in S$ for all $n \geq n_0$.

We now consider a few examples on the use of the above principle called the **principle of strong mathematical induction.**

Example 4.13. Prove that every natural number greater than 1 is either a prime or is a product of primes.

Proof. We know that 2 is a prime. Therefore, that the proposition $P(n) : n$ is either a prime or a product of primes is true for $n = 2$. Suppose that k is an integer at least 2 and that the proposition holds for all $n, 2 \leq n \leq k$ i.e. $P(2), P(3), \ldots, P(k)$ are all true. Now consider the number $k + 1$. Then $k+1$ is either a prime or is composite. If $k+1$ is not a prime, then $k+1 = rs$, where $2 \leq r \leq k, 2 \leq s \leq k$. By induction hypothesis, either r, s are primes or are products of primes. Writing the decompositions of r, s as products of primes in $k + 1 = rs$, we get a decomposition of $k + 1$ as a product of primes. We have thus proved that either $k + 1$ is a prime or is a product of primes. This completes induction. Hence every integer $n \geq 2$ is either a prime or is a product of primes.

A jigsaw puzzle consists of a number of pieces in which pieces with matching boundaries are to be put together to get a final complete picture. Two or more pieces with matched boundaries, when put together, form a big piece. A single piece, or a number of pieces with matched boundaries when put together to form a big piece, is called a **block.**

Thus blocks with matched boundaries can be put together to form the final picture. We say that the jigsaw puzzle is solved if all pieces have been put together and one single block forming the final picture is obtained. The process of putting together two blocks with matched boundaries is called a **move.**

Example 4.14. Prove that a jigsaw puzzle with n pieces can always be solved by making $n - 1$ moves.

Proof. When $n = 1$, there is already one piece or block and no move is needed to solve the puzzle. Suppose that $k \geq 1$ and that any puzzle with m pieces, $1 \leq m \leq k$ can be solved by making $m - 1$ moves. Now consider a jigsaw puzzle with $k + 1$ pieces. Just before the last move, we must have already obtained exactly two blocks. Let the two blocks obtained have n_1, n_2 pieces with $n_1 \geq 1, n_2 \geq 1$ and $n_1 + n_2 = k + 1$. Then both n_1 and n_2 are $\leq k$. Therefore, by induction hypothesis, the number of moves made to obtain the two blocks is $n_1 - 1, n_2 - 1$ and, therefore, the total number of moves is $n_1 + n_2 - 2$. Hence, to solve the puzzle, the total number of moves is $n_1 + n_2 - 2 + 1 = n_1 + n_2 - 1 = k + 1 - 1 = k$. This completes induction. Therefore, the number of moves required to solve a jigsaw puzzle with n pieces, $n \geq 1$, is $n - 1$.

Example 4.15. It is human nature to trace one's family tree to identify one's ancestors. This is done mainly by talking to elders, or sometimes by talking to pandits, who try to maintain records of families in a particular area or region. It happens quite often that incomplete family records prohibit us from going back beyond a few generations. To simplify things, we assume that for any of our ancestors, we can either trace both of his or her parents, or we can trace neither of them. A person in a family tree is referred to as a 'leaf' if we cannot go on to trace his or her parents while otherwise we refer to him or her as an 'internal node'. Thus, if the parents of a person can be traced then that person is called an internal node. (We shall come to this terminology when we study graphs and trees). If in a family tree there are p leaves and q internal nodes, then $p = q + 1$.

Proof. We prove this result by induction on n, the number of people in the family tree. When $n = 1$, there is only one person in the family tree. Therefore, this person must be a leaf and there is no internal node.

Suppose that $k \geq 1$ is an integer and that the result is true for all family trees with m people, where $1 \leq m \leq k$. Consider a family tree having $k + 1$ people. Let A be a person whose family tree is being traced. Since there are at least two persons in the tree, the parents of A are traced. Let the parents of A be B and C. Now consider the family trees of B and C. Both these trees have $< k$ people. Let the number of leaves in the family trees of B and C be p_1, p_2 respectively, and the number of internal nodes in the two trees be q_1, q_2. Then, by induction hypothesis, $p_1 = q_1 + 1, p_2 = q_2 + 1$. Therefore, $p_1 + p_2 = q_1 + q_2 + 2$. Now every leaf in the tree of B or C is also a leaf in the tree of A and every internal node in the family tree of B or C is an internal node in the family tree of A. Let p, q be the number of leaves and internal nodes respectively in the family tree of A. Then $p = p_1 + p_2$, $q = q_1 + q_2 + 1$ (because besides the nodes q_1, q_2 of B and C, A itself is also an internal node in the tree of A). Therefore, $p = p_1 + p_2 = q_1 + q_2 + 2 = (q_1 + q_2 + 1) + 1 = q + 1$.

This completes induction. Hence, the number of leaves in the family tree of any person is one more than the number of internal nodes in the tree of the person.

We next consider some more examples on applications of the induction and the strong induction principles.

Example 4.16. Prove that for all $n \geq 1, n! \geq 2^{n-1}$.

Proof. For $n = 1, n! = 1$ and $2^{n-1} = 2^0 = 1$. Therefore, the result holds for $n = 1$. Let $k \geq 1$ be an integer and suppose that the result holds for $n = k$ i.e. $k! \geq 2^{k-1}$. Then $(k+1)! = k!(k+1) \geq 2^{k-1}(k+1)$. As $k \geq 1$, $k+1 \geq 2$. Therefore, $(k+1)! \geq 2^{k-1}(k+1) \geq 2^{k-1} \cdot 2 = 2^k$. This completes induction and, therefore, $n! \geq 2^{n-1}$ for all $n \geq 1$.

Example 4.17. Show that for all $n \geq 1$, and a given integer $k \geq 1$,

$$\frac{1}{1 \cdot (k+1)} + \frac{1}{(k+1)(2k+1)} + \cdots + \frac{1}{(kn-k+1)(kn+1)} = \frac{n}{(kn+1)}.$$

Solution. The result is clearly true for $n = 1$. Suppose that $m \geq 1$ is an integer, and that the result holds for $n = m$. Thus

$$\frac{1}{1 \cdot (k+1)} + \frac{1}{(k+1)(2k+1)} + \cdots + \frac{1}{(mk+1)(mk-k+1)} = \frac{m}{(km+1)}.$$

Then

$$\frac{1}{1 \cdot (k+1)} + \frac{1}{(k+1)(2k+1)} + \cdots + \frac{1}{(mk+1)(mk-k+1)}$$

$$+ \frac{1}{(mk+1)(mk+k+1)}$$

$$= \left\{ \frac{1}{1 \cdot (k+1)} + \frac{1}{(k+1)(2k+1)} + \cdots + \frac{1}{(mk+1)(mk-k+1)} \right\}$$

$$+ \frac{1}{(mk+1)(mk+k+1)}$$

$$= \frac{m}{mk+1} + \frac{1}{(mk+1)(mk+k+1)}$$

$$= \frac{m(mk+k+1)+1}{(mk+1)(mk+k+1)}$$

$$= \frac{m(mk+1)+mk+1}{(mk+1)(mk+k+1)}$$

$$= \frac{(mk+1)(m+1)}{(mk+1)(mk+k+1)}$$

$$= \frac{m+1}{mk+k+1}.$$

This completes induction and, hence, the result holds for all $n \geq 1$.

Example 4.18. Prove by induction that for all $n \geq 1$,

(a) $1.1! + 2.2! + \cdots + n\,n! = (n+1)! - 1$.

(b) $1.2 + 2.3 + \cdots + n(n+1) = \frac{n(n+1)(n+2)}{3}$.

Proof. (a) The result is obviously true for $n = 1$. Suppose that $k \geq 1$ is an integer and that the result holds for $n = k$.
   Then

$$1.1! + 2.2! + \cdots + k \cdot k! = (k+1)! - 1.$$

For $n = k+1$,

$$1.1! + 2.2! + \cdots + k\,k! + (k+1)(k+1)!$$
$$= \{1.1! + 2.2! + \cdots + k\,k!\} + (k+1) \cdot (k+1)!$$
$$= \{(k+1)! - 1\} + (k+1) \cdot (k+1)!$$
$$= (k+1)!(1+k+1) - 1$$
$$= (k+1)!(k+2) - 1 = (k+2)! - 1.$$

This completes induction and, hence, the result holds for all $n \geq 1$.
   (b) For $n = 1$, $\frac{n(n+1)(n+2)}{3} = \frac{1.2.3}{3} = 2 = 1.2 = $ the left-hand side. Therefore, the result holds for $n = 1$. Suppose that $k \geq 1$ is an integer and that the result holds for $n = k$. For $n = k+1$,

$$1.2 + 2.3 \ldots + (k+1)(k+2)$$
$$= \{1.2 + 2.3 + \cdots + k \cdot (k+1)\} + (k+1)(k+2)$$
$$= \frac{k(k+1)(k+2)}{3} + (k+1)(k+2)$$
$$= (k+1)(k+2)\left(\frac{k}{3}+1\right) = \frac{(k+1)(k+2)(k+3)}{3}.$$

This completes induction and, hence, the result holds for all $n \geq 1$.

Example 4.19. In a certain kingdom, the king wanted to test the intelligence of the best mathematicians in his kingdom. For this purpose, he summoned all the best and most well-known mathematicians to his palace. In a dark room he put caps on the heads of each of them.

Some of the caps were black and some were white. The lights in the hall were then switched on and the king informed them that he had put some white caps and some black ones on their heads. The king asked them to look at each other but not to talk. The king left the hall by saying that he would return at the end of every hour and would like to know if those wearing the white caps had all guessed correctly. At the nth hour, every one of the n mathematicians who were given white caps informed the king that he knew that he or she was wearing a white cap. How did it happen? (You may assume that the king never lies.)

Solution. We can prove the result by induction on n — the number of people wearing white caps. Since the king said that he had put white caps on the heads of some and the king never lies, there must be at least one person wearing a white cap. For $n = 1$, there is only one person wearing a white cap. The one person wearing the white cap looks around at all the other mathematicians (as does every one else) and finds that everyone around is wearing a black cap. Since there is at least one person wearing a white cap, he or she argues that he or she must be the one and can tell the king on his very first visit.

Assume that $k \geq 1$ is an integer and that if k are wearing white caps then they are all able to tell the king about it exactly on the kth hour.

Now, suppose that there are $k + 1$ mathematicians wearing white caps. Every mathematician wearing a white cap finds that k of his/her colleagues are wearing white caps. He/she argues: if he/she were wearing a black cap then there would be exactly k people wearing white caps and they must have realized this and informed the king about it on the kth hour (by induction hypothesis). The very fact that they did not do so shows that there are more than k mathematicians who are wearing white caps. Except for him/her, there are exactly k persons wearing white caps. Hence, he/she argues that there are exactly $k + 1$ persons wearing white caps and that the $(k + 1)$th person must be him/her. Then, on the $(k + 1)$th hour, he/she along with all the others wearing white caps would inform the king of their conclusion. This completes induction.

Example 4.20. Show that for any integer $n \geq 0, (12)^{n+2} + (13)^{2n+1}$ is divisible by 157.

Solution. We prove the result by induction on n. For $n = 0, (12)^{n+2} + (13)^{2n+1} = (12)^2 + 13 = 144 + 13 = 157$ which is divisible by 157. Suppose that $k \geq 0$ and that $(12)^{k+2} + (13)^{2k+1}$ is divisible by 157. Let $(12)^{k+2} + (13)^{2k+1} = 157l$ where l is a positive integer.

For $n = k + 1$,

$$(12)^{k+3} + (13)^{2k+3} = (12)^{k+3} + (13)^{2k+1}(13)^2$$
$$= (12)^{k+3} + (157 \times l - (12)^{k+2})(13)^2$$
$$= 157 \times l(13)^2 + (12)^{k+2}(12 - (13)^2)$$
$$= 157l(13)^2 - 157 \times (12)^{k+2}$$
$$= 157\{l(13)^2 - (12)^{k+2}\}$$

which is divisible by 157. This completes induction and hence $(12)^{n+2} + (13)^{2n+1}$ is divisible by 157 for all $n \geq 0$.

Remark. In general we can prove:

For any positive integer a and non- negative integer $n, a^{n+2} + (a+1)^{2n+1}$ is divisible by $a^2 + a + 1$.

Example 4.21. Prove that $3n^5 + 5n^3 + 7n$ is divisible by 15 for every positive integer n.

Proof. For $n = 1, 3n^5 + 5n^3 + 7n = 3 + 5 + 7 = 15$, which is divisible by 15. Suppose that k is an integer ≥ 1 and that $3k^5 + 5k^3 + 7k$ is divisible by 15. Let $3k^5 + 5k^3 + 7k = 15l$, where l is a positive integer.

Now

$$3(k + 1)^5 + 5(k + 1)^3 + 7(k + 1)$$
$$= 3\{k^5 + 5k^4 + 10k^3 + 10k^2 + 5k + 1\}$$
$$+5\{k^3 + 3k^2 + 3k + 1\} + 7(k + 1)$$
$$= (3k^5 + 5k^3 + 7k) + 15(k^4 + 2k^3 + 2k^2 + k)$$
$$+15(k^2 + k) + (3 + 5 + 7)$$
$$= 15l + 15(k^4 + 2k^3 + 2k^2 + k) + 15(k^2 + k) + 15,$$

which is clearly divisible by 15, k being an integer. This completes induction and, hence, $3n^5 + 5n^3 + 7n$ is divisible by 15 for every positive integer n.

Example 4.22. Prove that for $n \geq 1$,

$$1^2 + 2^2 + 3^2 + \cdots + n^2 = \frac{n(n+1)(2n+1)}{6}.$$

Proof. For $n = 1$, the left-hand side is $1^2 = 1$ and the right-hand side is $\frac{1(1+1)(2+1)}{6} = 1$. Thus the result is true for $n = 1$. Suppose that $k \geq 1$ is an integer and that

$$1^2 + 2^2 + \cdots + k^2 = \frac{k(k+1)(2k+1)}{6}.$$

Then for $n = k + 1$,

$$1^2 + 2^2 + \cdots + (k+1)^2 = (1^2 + 2^2 + \cdots + k^2) + (k+1)^2$$

$$= \frac{k(k+1)(2k+1)}{6} + (k+1)^2$$

$$= (k+1)\left(\frac{k(2k+1)}{6} + k + 1\right)$$

$$= (k+1)\frac{2k^2 + k + 6k + 6}{6}$$

$$= (k+1)\frac{2k^2 + 7k + 6}{6}$$

$$= \frac{(k+1)(k+2)(2k+3)}{6}.$$

Thus the result holds for $n = k + 1$. This completes induction and, hence,

$$1^2 + 2^2 + \cdots + n^2 = \frac{n(n+1)(2n+1)}{6} \quad \text{for all } n \geq 1.$$

Example 4.23. Prove that for all $n \geq 1$,

$$1^3 + 2^3 + \cdots + n^3 = \frac{n^2(n+1)^2}{4}.$$

Proof. For $n = 1$ the left-hand side is $1^3 = 1$ and the right-hand side is $\frac{1^2(1+1)^2}{4} = 1$. This proves that the result holds for $n = 1$. Suppose k is an integer ≥ 1 and that the result holds for $n = k$, i.e.

$$1^3 + 2^3 + \cdots + k^3 = \frac{k^2(k+1)^2}{4}.$$

Now, for $n = k + 1$,

$$1^3 + 2^3 + \cdots + (k+1)^3 = (1^3 + 2^3 + \cdots + k^3) + (k+1)^3$$
$$= \frac{k^2(k+1)^2}{4} + (k+1)^3$$
$$= (k+1)^2 \left(\frac{k^2}{4} + k + 1 \right)$$
$$= (k+1)^2 \left(\frac{k^2 + 4k + 4}{4} \right)$$
$$= (k+1)^2 \frac{(k+2)^2}{4}.$$

Thus the result holds for $n = k+1$. This completes induction and, therefore, $1^3 + 2^3 + \cdots + n^3 = \frac{n^2(n+1)^2}{4}$ for all $n \geq 1$.

Example 4.24. Show that $2^n < n!$ for $n \geq 4$.

Solution. Let $P(n)$ be the proposition '$2^n < n!$'. For $n = 1$, $2^1 = 2 > 1! = 1$; for $n = 2, 2^2 = 4 > 2! = 2$; for $n = 3, 2^3 = 8 > 3! = 6$. Thus $P(n)$ does not hold for $n = 1, 2, 3$. For $n = 4, 2^4 = 16$ and $4! = 24$. Therefore $2^4 < 4!$ i.e. $P(4)$ holds. Suppose that $k \geq 4$ is an integer and that $P(k)$ holds. For proving that $P(k+1)$ holds, we have

$$2^{k+1} = 2.2^k < 2.k!, \quad \text{(by induction hypothesis)}$$
$$< (k+1)k! \quad \text{(as } k \geq 4, k+1 \geq 5 \text{ and, so, } 2 < k+1)$$
$$= (k+1)!$$

Thus $P(k+1)$ holds. This completes induction and, therefore, $P(n)$ holds for all $n \geq 4$.

Example 4.25. Formulate and prove by induction a general formula from the observations that $1^3 = 1$, $2^3 = 3 + 5$, $3^3 = 7 + 9 + 11$, $4^3 = 13 + 15 + 17 + 19$.

Solution. Observe that the first observation has only one term on the right, the second has two terms, the third has three and the fourth has four terms. Thus we observe that the expression for n^3 shall have n terms on the right, each one of which will be an odd integer and every term is two more than the previous one. Also observe that

$$1 = 1(1 - 1) + 1$$

$$3 + 5 = \{2(2 - 1) + 1\} + \{2(2 - 1) + 3\}$$

$$7 + 9 + 11 = \{3(3 - 1) + 1\} + \{3(3 - 1) + 3\} + \{3(3 - 1) + 5\}$$

$$13 + 15 + 17 + 19 = \{4(4 - 1) + 1\} + \{4(4 - 1) + 3\} + \{4(4 - 1) + 5\}$$

$$+ \{4(4 - 1) + 7\}.$$

We guess that the general formula for the above is

$$n^3 = \{n(n - 1) + 1\} + \{n(n - 1) + 3\} + \cdots + \{n(n - 1) + (2n - 1)\}.$$

Observe that for any $n \geq 1$,

$$\{n(n - 1) + 1\} + \{n(n - 1) + 3\} + \cdots + \{n(n - 1) + (2n - 1)\}$$

$$= n \times n(n - 1) + \{1 + 3 + \cdots + (2n - 1)\}$$

$$= n^2(n - 1) + \{1 + 2 + 3 + \cdots + (2n - 1) + 2n\} - \{2 + 4 + \cdots + 2n\}$$

$$= n^2(n - 1) + \frac{2n(2n + 1)}{2} - 2\{1 + 2 + \cdots + n\}$$

$$= n^2(n - 1) + n(2n + 1) - n(n + 1)$$

$$= n^3 - n^2 + 2n^2 + n - n^2 - n = n^3.$$

Thus our guess is correct. We will now prove this result by induction on n. For $n = 1, 2, 3, 4$, it is already given to be true. So we suppose that $n \geq 1$ and that the result is true for n, i.e. we assume that

$$\{n(n - 1) + 1\} + \{n(n - 1) + 3\} + \cdots + \{n(n - 1) + (2n - 1)\} = n^3.$$

For $n + 1$, we have

$$\{(n + 1)n + 1\} + \{(n + 1)n + 3\} + \cdots + \{(n + 1)n + 2(n + 1) - 1\}$$

$$= \{n(n - 1) + 1 + 2n\} + \{n(n - 1) + 3 + 2n\} + \cdots + \{n(n - 1)$$

$$+ 2n - 1 + 2n\} + \{(n + 1)n + 2n + 1\}$$

$$= [\{n(n - 1) + 1\} + \{n(n - 1) + 3\} + \cdots + \{n(n - 1) + 2n - 1\}]$$

$$+ \{2n + \cdots + 2n\} + \{n(n + 1) + 2n + 1\}$$

$$= n^3 + 2n \times n + \{n^2 + 3n + 1\} \quad \text{(by induction hypothesis)}$$
$$= n^3 + 3n^2 + 3n + 1 = (n+1)^3.$$

This completes induction and, thus, the result is proved for every $n \geq 1$.

Example 4.26. For any sets $A, B_1, B_2, \ldots, B_n, n \geq 2$, we have

(a) $A \cap (B_1 \cup B_2 \cup \cdots \cup B_n) = (A \cap B_1) \cup (A \cap B_2) \cup \cdots \cup A \cap B_n)$;
(b) $A \cup (B_1 \cap B_2 \cap \cdots \cap B_n) = (A \cup B_1) \cap (A \cup B_2) \cap \cdots \cap (A \cup B_n)$.

Proof. For $n = 2$, these are the distributive laws, which we have proved already. Suppose that $k \geq 2$ and that the two results hold for $n = k$, i.e.

$$A \cap (B_1 \cup B_2 \cup \cdots \cup B_k) = (A \cap B_1) \cup (A \cap B_2) \cup \cdots \cup (A \cap B_k);$$

and

$$A \cup (B_1 \cap B_2 \cap \cdots \cap B_k) = (A \cup B_1) \cap (A \cup B_2) \cap \cdots \cap (A \cup B_k).$$

Now, for $n = k + 1$, regarding $B_1 \cap B_2 \cap \cdots \cap B_k$ as one set,

$$A \cup (B_1 \cap B_2 \cap \cdots \cap B_{k+1})$$
$$= (A \cup (B_1 \cap B_2 \cap \cdots \cap B_k)) \cap (A \cup B_{k+1})$$
$$= ((A \cup B_1) \cap (A \cup B_2) \cap \cdots \cap (A \cup B_k)) \cap (A \cup B_{k+1})$$
$$= (A \cup B_1) \cap (A \cup B_2) \cap \cdots \cap (A \cup B_k) \cap (A \cup B_{k+1}).$$

This completes induction so far as the second relation is concerned. Hence,

$$A \cup (B_1 \cap B_2 \cap \cdots \cap B_n)$$
$$= (A \cup B_1) \cap (A \cup B_2) \cap \cdots \cap (A \cup B_n) \quad \text{for all } n \geq 2.$$

Again,

$$A \cap (B_1 \cup B_2 \cup \cdots \cup B_{k+1})$$
$$= A \cap ((B_1 \cup B_2 \cup \cdots \cup B_k) \cup B_{k+1})$$
$$= (A \cap (B_1 \cup B_2 \cup \cdots \cup B_k)) \cup (A \cap B_{k+1})$$

$$= ((A \cap B_1) \cup (A \cap B_2) \cup \cdots \cup (A \cap B_k)) \cup (A \cap B_{k+1})$$

$$= (A \cap B_1) \cup (A \cap B_2) \cup \cdots \cup (A \cap B_{k+1}).$$

This completes induction and, hence,

$$A \cap (B_1 \cup B_2 \cup \cdots \cup B_n)$$

$$= (A \cap B_1) \cup (A \cap B_2) \cup \cdots \cup (A \cap B_n), \quad \text{for all } n \geq 2.$$

Example 4.27. (Generalized Principle of Inclusion and Exclusion). For any finite subsets $A_1, A_2, \ldots, A_n$ of a universal set X,

$$o(A_1 \cup A_2 \cup \cdots \cup A_n)$$

$$= \sum_{1 \leq i \leq n} o(A_i) - \sum_{1 \leq i < j \leq n} o(A_i \cap A_j) + \sum_{1 \leq i < j < k \leq n} o(A_i \cap A_j \cap A_k) - \cdots$$

$$+ (-1)^{n-2} \sum_{1 \leq i \leq n} o(A_1 \cap \cdots \cap A_i^{\wedge} \cap \cdots \cap A_n)$$

$$+ (-1)^{n-1} o(A_1 \cap A_2 \cap \cdots \cap A_n),$$

where $A_i^{\wedge}$ means that A_i is omitted.

Proof. For $n = 2$, it is the principle of inclusion and exclusion which we have proved already. Suppose that $n \geq 2$ and that the result holds for n. Then, for $n + 1$, (for $1 \leq i \leq n$, we write B_i for $A_i \cap A_{n+1}$)

$$o(A_1 \cup A_2 \cup \cdots \cup A_{n+1})$$

$$= o((A_1 \cup A_2 \cup \cdots \cup A_n) \cup A_{n+1})$$

$$= o(A_1 \cup A_2 \cup \cdots \cup A_n) + o(A_{n+1}) - o((A_1 \cup A_2 \cup \cdots \cup A_n) \cap A_{n+1}).$$

$$= \sum_{1 \leq i \leq n+1} o(A_i) - \sum_{1 \leq i < j \leq n} o(A_i \cap A_j)$$

$$+ \sum_{1 \leq i < j < k \leq n} o(A_i \cap A_j \cap A_k) + \cdots$$

$$+ (-1)^{n-2} \sum_{1 \leq i \leq n} o(A_1 \cap \cdots \cap A_i^{\wedge} \cap \cdots \cap A_n)$$

$$+ (-1)^{n-1} o(A_1 \cap \cdots \cap A_n)$$

$$- \{o((A_1 \cap A_{n+1}) \cup (A_2 \cap A_{n+1}) \cup \cdots \cup (A_n \cap A_{n+1}))\}$$

$$= \sum_{1 \leq i \leq n+1} o(A_i) - \sum_{1 \leq i < j \leq n} o(A_i \cap A_j)$$

$$+ \sum_{1 \leq i < j < k \leq n} o(A_i \cap A_j \cap A_k) + \cdots$$

$$+ (-1)^{r-1} \sum_{1 \leq i_1 < i_2 < \cdots < i_r \leq n} o(A_{i_1} \cap A_{i_2} \cap \cdots \cap A_{i_r}) + \cdots$$

$$+ (-1)^{n-2} \sum_{1 \leq i \leq n} o(A_1 \cap \cdots \cap \hat{A}_i \cap \cdots \cap A_n)$$

$$+ (-1)^{n-1} \sum_{1 \leq i \leq n} o(A_1 \cap A_2 \cdots \cap A_n) - \left\{ \sum_{1 \leq i \leq n} o(A_i \cap A_{n+1}) + \cdots \right.$$

$$+ (-1)^{r-1} \left[\sum_{1 \leq i_1 < i_2 < \cdots < i_r \leq n} o(B_{i_1} \cap B_{i_2} \cap \cdots \cap B_{i_r}) \right] + \cdots$$

$$\left. + (-1)^{n-1} o(B_1 \cap B_2 \cap \cdots \cap B_n) \right\}$$

(the result being true by induction for any n subsets).

Using the fact that $B_i \cap B_j = A_i \cap A_j \cap A_{n+1}$ for every i, j and collecting the various terms in the above expression we get that

$$o(A_1 \cup A_2 \cup \cdots \cup A_{n+1})$$

$$= \sum_{1 \leq i \leq n+1} o(A_i) - \sum_{1 \leq i < j \leq n+1} o(A_i \cap A_j)$$

$$+ \sum_{1 \leq i < , j , < k \leq n+1} o(A_i \cap A_j \cap A_k) - \cdots$$

$$+ (-1)^{n-1} \sum_{1 \leq i \leq n+1} o(A_1 \cap \cdots \cap A_j^{\wedge} \cap \cdots \cap A_{n+1})$$

$$+ (-1)^n o(A_1 \cap A_2 \cap \cdots \cap A_{n+1}),$$

This completes induction. Hence

$$o(A_1 \cup A_2 \cup \cdots \cup A_n)$$

$$= \sum_{1 \leq i \leq n} o(A_i)$$

$$+ \sum_{1 \leq r \leq n} (-1)^{r-1} \sum_{1 \leq i_1 < i_2 < \cdots < i_r \leq n} o(A_{i_1} \cap \cdots \cap A_{i_r})$$

for all $n \geq 2$.

Example 4.28. For any positive integer m, prove that

$$^mC_0 + {}^mC_1 + \cdots + {}^mC_m = 2^m,$$

where ^mC_r is the rth binomial coefficient equal to $\frac{m!}{r!(m-r)!}$.

Proof. For $m = 1$, we have $^1C_0 + {}^1C_1 = 1 + 1 = 2 = 2^1$. Thus, the result holds for $m = 1$. Suppose that $m \geq 1$ and that the result holds for m, i.e. $^mC_0 + {}^mC_1 + \cdots + {}^mC_m = 2^m$. Now

$$^{(m+1)}C_0 + {}^{(m+1)}C_1 + {}^{(m+1)}C_2 + \cdots + {}^{(m+1)}C_i + \cdots + {}^{(m+1)}C_m + {}^{(m+1)}C_{m+1}$$

$$= {}^{(m+1)}C_0 + \{{}^mC_0 + {}^mC_1\} + \{{}^mC_1 + {}^mC_2\} + \cdots + \{{}^mC_{i-1} + {}^mC_i\}$$

$$+ \cdots + \{{}^mC_{m-1} + {}^mC_m\} + {}^{(m+1)}C_{(m+1)}$$

$$= 1 + \{{}^mC_0 + {}^mC_1 + \cdots + {}^mC_{i-1} + \cdots + {}^mC_{m-1}\} + \{{}^mC_1 + {}^mC_2$$

$$+ \cdots + {}^mC_i + \cdots + {}^mC_m\} + 1$$

$$= \{{}^mC_0 + {}^mC_1 + \cdots + {}^mC_{m-1} + {}^mC_m\} + \{{}^mC_0 + {}^mC_1 + \cdots$$

$$+ {}^mC_i + \cdots + {}^mC_m\}$$

$$= 2^m + 2^m = 2^{(m+1)}.$$

This completes induction and, hence, $^mC_0 + {}^mC_1 + \cdots + {}^mC_m = 2^m$ for all $m \geq 1$.

(In the above proof we have used the result $^{m+1}C_i = {}^mC_{i-1} + {}^mC_i$ from permutations and combinations which we shall study in detail later in this chapter and also prove the result mentioned.)

Example 4.29. Let A, B be two finite sets each of order n. Prove that the number of bijective maps from A to B is $n!$ ($= n$ factorial which is the product of first n natural numbers).

Proof. We will prove the result by induction on n- the order of A (= the order of B). Suppose that $n = 1$. Let $A = \{a_1\}, B = \{b_1\}$. Then $f : A \to B$ with $f(a_1) = b_1$ is the only map, indeed a bijective map from A to B. Thus the result holds for $n = 1$ (as $1! = 1$). Suppose that $n > 1$ and that the result holds for all finite sets C, D with $o(C) = o(D) = n - 1$. Let $A = \{a_1, a_2, \ldots, a_n\}$ and $B = \{b_1, b_2, \ldots, b_n\}$. Let $f : A \to B$ be a bijective map. By reordering the elements of B, if necessary, we may suppose that $f(a_1) = b_1$. Define a map $g : A\backslash\{a_1\} \to B\backslash\{b_1\}$ by $g(a_i) = f(a_i)$, for $2 \leq i \leq n$.

It is clear from the definition of g that g is one–one. As $o(A\backslash\{a_1\}) = o(B\backslash\{b_1\}) = n - 1$, any map : $A\backslash\{a_1\} \to B\backslash\{b_1\}$ is one–one if and only if it is onto. In particular, g is onto and, hence, bijective. By induction hypothesis, there are $(n - 1)!$ bijective maps from $A\backslash\{a_1\}$ to $B\backslash\{b_1\}$. Thus there are $(n - 1)!$ bijective maps from A to B in which a_1 gets mapped onto b_1. But b_1 could be any one of the n elements $b_1, b_2, \ldots, b_n$. Thus the number of choices of the image b_1 of a_1 being n, there are $n(n - 1)! = n!$ bijective maps from A to B. This completes induction. Hence, the number of bijective maps from A to B is $n!$ for any finite sets A, B with $o(A) = o(B) = n \geq 1$.

Example 4.30. Prove that the number of primes less than n is less than $n - 2$ for all $n \geq 5$.

Proof. For $n = 5$, the primes less than 5 are 2 and 3, so that the number of primes less than 5 is 2, which is less than $5 - 2 = 3$. Suppose that $n \geq 5$ and that the number of primes less than n is k with $k < n - 2$. We consider the number of primes which are less than $n + 1$. If n is not a prime, then the number of primes less than $n + 1$ is the same as the number of primes less than n and so is equal to k, which is less than $n - 2 < n + 1 - 2$. If n is a prime, then the number of primes less than $n + 1$ is $k + 1$. Since $k < n - 2, k + 1 < n - 2 + 1 = n + 1 - 2$. Thus in either case the number of primes less than $n + 1$ is $< n + 1 - 2$. This completes induction. Hence, for every integer $n \geq 5$, the number of primes less than n is $< n - 2$.

Observe that there are no primes less than 2 and so for $n = 2$, the result of the above example is not true, as $0 < 0$ is not valid. Similarly, it is not true for $n = 3$ or $n = 4$ as, in the first case, the number of primes is 1, which is not less than $3 - 2$ and, in the second case, the number of primes is 2 which is again not less than $4 - 2$.

Exercise 4.1.

1. Let r be a positive integer, and let $a_0, a_1, \ldots, a_j, \ldots$ be non-negative integers less than r. Then, for all $j \geq 0$, $a_o + a_1 r + \cdots + a_j r^j < r^{j+1}$.

2. A function $f : A \rightarrow A$, A any set, is called an idempotent function if $f^2 = f \circ f = f$. If f is an idempotent, prove that $f^n = f$ for all $n \geq 1$.

3. In a group of people of any order, prove that the total number of people who shake hands with an odd number of people is even.

4. Let $f : Z \rightarrow Z$ be a map defined by $f(n) = 2n + 1, n \in Z$. Obtain an expression for f^r in terms of n and r. Also, prove the validity of this expression by induction on r.

5. Prove that $n! > n$ for all $n \geq 3$.

6. Consider the proposition
   $P(n) : 1 + 5 + 9 + \cdots + (4n - 3) = (2n + 1)(n - 1)$. Prove that $P(k+1)$ holds if $P(k)$ holds. Can you say that $P(n)$ holds for all $n \geq 1$ or for all $n \geq 2$?

7. Find the least value k of n for which $(1 + n^2) < 2^n$. Then prove that $(1 + n^2) < 2^n$ for all $n \geq k$.

8. Prove by induction on n that for all $n \geq 1$,

   (a) $n^3 - n$ is divisible by 3;
   (b) $n^7 - n$ is divisible by 7;
   (c) $n^{11} - n$ is divisible by 11.

   Can you say that $n^4 - n$ is divisible by 4? Justify your answer.

9. If p is a prime and p divides $a^n, n > 1$, use induction to prove that p divides a.

10. Show by induction on n that

    (a) $n^3 + 2n$ is divisible by 3 for all $n \geq 1$;
    (b) $n^4 - 4n^2$ is divisible by 3 for all $n \geq 2$.

11. Show that for any positive integer $n > 1$,

$$\frac{1}{\sqrt{1}} + \frac{1}{\sqrt{2}} + \cdots + \frac{1}{\sqrt{n}} > \sqrt{n}.$$

12. In the post office there are available only stamps worth Rs. 3 and Rs. 4 each. Prove that for any amount $\geq$ Rs. 7 you can buy stamps with all the money you have.

13. Show that for all $n \geq 1$,

$$\frac{1}{1.3} + \frac{1}{3.5} + \frac{1}{5.7} + \cdots + \frac{1}{(2n-1)(2n+1)} = \frac{n}{2n+1}.$$

14. If a, b are real numbers, prove by induction on n, that for all $n \geq 1$,

$$(a + b)^n = \sum_{0 \leq r \leq n} {}^nC_r a^{n-r} b^r$$

$$= a^n + {}^nC_1 a^{n-1} b + \cdots + {}^nC_r a^{n-r} b^r + \cdots + b^n.$$

15. If A, B are square matrices of order 2 with real entries and $AB = BA$, prove that for all $n \geq 1$,

$$(A + B)^n = A^n + {}^nC_1 A^{n-1} B + \cdots + {}^nC_r A^{n-r} B^r + \cdots$$

$$+ {}^nC_{n-1} A B^{n-1} + B^n.$$

16. If A, B are square matrices of order 2 with real entries and A is invertible, prove by induction on n that for all $n \geq 1$,

$$(A^{-1} B A)^n = A^{-1} B^n A.$$

17. For any angle θ, prove by induction on n that

$$(\cos \theta + i \sin \theta)^n = \cos n\theta + i \sin n\theta, \quad \text{for all } n \geq 1.$$

18. For real numbers a, b and positive integer n, prove that,
    (a) $(a + b)^n = a^n + \sum_{i=1}^{n-1} c(n, i) a^{n-i} b^i + c(n, n) b^n$
    (b) $(ab)^n = a^n b^n$
    for all $n \geq 1$.

19. Prove that for every integer $n \geq 1$,
    (a) $n^2 - n$ is divisible by 2;
    (b) $n^3 - n$ is divisible by 6;
    (c) $n^3 + 2n$ is divisible by 3;
    (d) $n^4 + 2n^3 + 2n^2 + n$ is divisible by 6;
    (e) $n^5 - n$ is divisible by 30.

20. Prove that for every integer n,

$$\frac{1}{5}n^5 + \frac{1}{3}n^3 + \frac{7}{15}n \quad \text{is again an integer.}$$

21. For every positive integer n, prove that $n^{13} - n$ is divisible by each one of 2, 3, 5, 7 and 13.

22. Prove that for every positive integer n
    (a) $n^7 - n$ is divisible by 42;
    (b) $2n^3 + n$ is divisible by 3;
    (c) $n^6 + 3n^5 + 5n^4 + 5n^3 + 3n^2 + n$ is divisible by 6;
    (d) $n^5 + n^2$ is even.

23. If $a + b \neq 0$, prove that $a + b$ divides $a^{2n} - b^{2n}$ for all $n \geq 1$.
24. If $a \neq b$, prove that $a - b$ divides $a^n - b^n$ for all $n \geq 1$.
25. Prove that $n^3 + n > n^2 + 1$ for all $n > 1$.
26. Prove that for all $n \geq 1$, 6 divides

   (a) $n(n+1)(n+2)$;
   (b) $n(n+1)(2n+1)$.

27. Prove that 24 divides $n(n+1)(n+2)(n+3)$ for all $n \geq 1$.
28. Prove that for every $n > 1$, $n^4 + 4n$ is composite.
29. Prove that every $n \geq 12$ is the sum of two composite numbers.
30. If a, b are real numbers, prove by induction on n that for all $n \geq 1$,

   (a) $\begin{pmatrix} a & 0 \\ o & b \end{pmatrix}^n = \begin{pmatrix} a^n & 0 \\ o & b^n \end{pmatrix}$ (b) $\begin{pmatrix} 1 & a \\ o & 1 \end{pmatrix}^n = \begin{pmatrix} 1 & na \\ o & 1 \end{pmatrix}$

   (c) $\begin{pmatrix} a & 0 \\ b & a \end{pmatrix}^n = \begin{pmatrix} a^n & 0 \\ na^{n-1}b & a^n \end{pmatrix}$.

31. If θ is any angle, prove that

$$\begin{pmatrix} \cos\theta & \sin\theta \\ -\sin\theta & \cos\theta \end{pmatrix}^n = \begin{pmatrix} \cos n\theta & \sin n\theta \\ -\sin n\theta & \cos n\theta \end{pmatrix}$$

for all $n \geq 1$.

4.3. Arithmetic, Geometric and Arithmetic-Geometric Series

Applications of the principle of mathematical induction are more prominent in the study of arithmetic, geometric and arithmetic-geometric series than elsewhere. We study such series here. A series of the form

$$a_0 + a_1 + a_2 + a_3 + \cdots + a_n + \cdots$$

where $a_{i+1} - a_i = a_{j+1} - a_j$ for all $i, j \geq 0$ is called an **arithmetic series**.

Also, $a_{i+1} - a_i$, which is the same for every i is called the **common difference** of the series. If we write $a_0 = a, a_{i+1} - a_i = d$, then the series can be written as

$$a + (a + d) + (a + 2d) + \cdots + (a + (n-1)d) + \cdots \tag{4.1}$$

Then $a + (n-1)d$ is called the **nth term** of the series and is denoted by T_n. The sum of the first n terms of the series is denoted by S_n. Thus

$$S_n = a + (a+d) + (a+2d) + \cdots + (a+(n-1)d).$$

Theorem 4.1. For an arithmetic series (4.1), the sum S_n of the first n terms is $\frac{n}{2}\{2a + (n-1)d\}$, i.e.

$$a + (a+d) + \cdots + (a+(n-1)d) = \frac{n}{2}\{2a+(n-1)d\} \quad \text{for all } n \geq 1.$$

Proof. We will prove this result by induction on n. For $n = 1$,

$$\text{L.H.S.} = a; \quad \text{the R.H.S} = \frac{\{2a + (1-1)d\}}{2} = a.$$

Thus the result holds for $n = 1$.

Suppose that $k \geq 1$ and that the result holds for $n = k$, i.e.

$$a + (a+d) + \cdots + (a+(k-1)d) = \frac{k}{2}\{2a + (k-1)d\}.$$

For $n = k+1$,

$$a + (a+d) + \cdots + (a+kd) = \{a + (a+d) + \cdots + (a+(k-1)d)\}$$
$$+ (a+kd)$$
$$= \frac{k}{2}\{2a + (k-1)d\} + (a+kd)$$
$$= ka + a + \frac{k(k-1)}{2}d + kd$$
$$= (k+1)a + \frac{k(k-1+2)}{2}d$$
$$= (k+1)a + \frac{k(k+1)}{2}d$$
$$= \frac{(k+1)}{2}\{2a + kd\}$$
$$= \frac{(k+1)}{2}\{2a + (k+1-1)d\}$$

Thus the result holds for $n = k+1$. This completes induction and, therefore, for all $n \geq 1$,

$$S_n = a + (a+d) + \cdots + (a+(n-1)d) = \frac{n}{2}\{2a + (n-1)d\}.$$

Corollary. The sum of first n natural numbers is $\frac{n(n+1)}{2}$.

Solution. We have to prove that

$$1 + 2 + 3 + \cdots + n = \frac{n(n+1)}{2}.$$

Observe that $1 + 2 + \cdots + n + \cdots$ is an arithmetic series with $a = 1$ and $d = 1$. Therefore,

$$1 + 2 + \cdots + n = \frac{n}{2}\{2a + (n-1)d\} = \frac{n}{2}\{2.1 + (n-1)\cdot 1\} = \frac{n(n+1)}{2}.$$

A series of the form $a_0 + a_1 + a_2 + \cdots + a_n + \cdots$ where $\frac{a_{i+1}}{a_i} = \frac{a_{j+1}}{a_j}$ for all $i, j \geq 0$ is called a **geometric series**. Also the number $\frac{a_{i+1}}{a_i}$ which is the same for every i is called the **common ratio**. Writing $a_0 = a$ and $\frac{a_{i+1}}{a_i} = r$, the series takes the form

$$a + ar + ar^2 + \cdots + ar^n + \cdots. \tag{4.2}$$

As for arithmetic series, we write T_n for the nth term and S_n for the sum of first n terms of the geometric series. A simple induction argument shows that $T_n = ar^{n-1}$ for $n \geq 1$.

Theorem 4.2. For all $n \geq 1$, the sum of first n terms of the geometric series (4.2) is $\frac{a(1-r^n)}{1-r}$, provided $r \neq 1$.

Proof. For $n = 1, a + ar + ar^2 + \cdots + ar^{n-1} = a$ and

$$\frac{a(1-r^n)}{1-r} = \frac{a(1-r)}{1-r} = a.$$

Thus the result is true for $n = 1$. Suppose that $k \geq 1$ and that the result holds for $n = k$, i.e.

$$a + ar + ar^2 + \cdots + ar^{k-1} = \frac{a(1-r^k)}{1-r}. \tag{4.3}$$

For $n = k + 1$,

$$a + ar + ar^2 + \cdots + ar^k = (a + ar + ar^2 + \cdots + ar^{k-1}) + ar^k$$

$$= \frac{a(1-r^k)}{1-r} + ar^k \quad \text{by (4.3)}$$

$$= a \left\{ \frac{1 - r^k + r^k(1 - r)}{1 - r} \right\}$$

$$= \frac{a(1 - r^{k+1})}{1 - r}.$$

Thus the result holds for $n = k + 1$. This completes induction and, hence,

$$a + ar + ar^2 + \cdots + ar^{n-1} = \frac{a(1 - r^n)}{1 - r} \quad \text{for all } n \geq 1.$$

Suppose that the number r is such that $|r| < 1$. Then the absolute values of the terms of sequence $r, r^2, \ldots, r^n, \ldots$ of powers of r is a strictly decreasing sequence and r^n becomes small enough in absolute value when n is large. Thus $r^n \to 0$ when $n \to \infty$. Therefore, the sum of the infinite geometric series

$$a + ar + ar^2 + \cdots + ar^n + \cdots \quad \text{becomes} \quad \frac{a}{1 - r} \text{ when } |r| < 1.$$

For $|r| > 1$, the infinite series is a divergent series.

Next, we consider a sequence, the nth term of which is the product of the $(n - 1)$th term of an arithmetic progression and the nth power of the common ratio of a geometric progression. Such a sequence may be called an **arithmetic-geometric sequence**. The series obtained by adding the terms of an arithmetic-geometric sequence is called an arithmetic-geometric series. Thus, a general **arithmetic-geometric series** looks like

$$a + (a + d)r + (a + 2d)r^2 + \cdots + (a + (n - 1)d)r^{n-1} + \cdots \quad (4.4)$$

We now find the sum S_n of the first n terms of this arithmetic-geometric series.

Theorem 4.3. The sum S_n of the first n terms of the arithmetic-geometric series (4.4) is $\frac{a}{1-r} + \frac{dr(1-r^{n-1})}{(1-r)^2} - \frac{(a+(n-1)d)r^n}{1-r}$, provided that $r \neq 1$.

Proof. For a change, we present a proof of this result without induction. Now $S_n = a + (a + d)r + \cdots + (a + (n - 1)d)r^{n-1}$
Therefore,

$$rS_n = ar + (a + d)r^2 + \cdots + (a + (n - 2)d)r^{n-1} + (a + (n - 1)d)r^n.$$

Subtracting one from the other, we get

$$(1 - r)S_n = a + dr + dr^2 + \cdots + dr^{n-1} - (a + (n - 1)d)r^n$$

$$= a + dr(1 + r + \cdots + r^{n-2}) - (a + (n-1)d)r^n$$

$$= a + dr\frac{(1 - r^{n-1})}{(1-r)} - (a + (n-1)d)r^n.$$

Dividing both sides by $1 - r$, we get

$$S_n = \frac{a}{1-r} + \frac{dr(1 - r^{n-1})}{(1-r)^2} - \frac{(a + (n-1)d)r^n}{1-r}, \quad \text{provided that } r \neq 1.$$

If $r = 1$, the series is an arithmetic series and its sum is

$$\frac{n}{2}\{2a + (n-1)d\}.$$

When $|r| < 1$, $r^n \to 0$ when $n \to \infty$ and the sum of the infinite arithmetic-geometric series (4.4) is

$$\frac{a}{1-r} + \frac{dr}{(1-r)^2}.$$

Students will already be familiar with arithmetic, geometric and arithmetic-geometric series, having studied these during the last two years of school before joining a university/college. Therefore, we do not study these in great detail and end up giving very few examples.

Example 4.31. Sum the following series to n terms.

(a) $4 + 44 + 444 + \cdots$

(b) $.4 + .44 + .444 + \cdots$

Solution.

(a) $S_n = 4 + 44 + 444 + \cdots$ to n terms

$$= 4\{1 + 11 + 111 + \cdots \text{ to } n \text{ terms}\}$$

$$= \frac{4}{9}\{9 + 99 + 999 + \cdots \text{ to } n \text{ terms}\}$$

$$= \frac{4}{9}\{(10 - 1) + (10^2 - 1) + (10^3 - 1) + \cdots \text{ to } n \text{ terms}\}$$

$$= \frac{4}{9}\{(10 + 10^2 + 10^3 + \cdots \text{ to } n \text{ terms}) - n\}$$

$$= \frac{4}{9}\left\{\frac{10(10^n - 1)}{10 - 1} - n\right\}$$

$$= \frac{4}{81}\{10^{n+1} - 9n - 10\}.$$

(b) $S_n = .4 + .44 + .444 + \cdots$ to n terms

$= 4\{.1 + .11 + .111 + \cdots$ to n terms$\}$

$= \dfrac{4}{9}\{.9 + .99 + .999 + \cdots$ to n terms$\}$

$= \dfrac{4}{9}\{(1 - .1) + (1 - .01) + (1 - .001) + \cdots$ to n terms$\}$

$= \dfrac{4}{9}\left\{ n - \left(\dfrac{1}{10} + \dfrac{1}{10^2} + \dfrac{1}{10^3} + \cdots \text{ to } n \text{ terms} \right) \right\}$

$= \dfrac{4}{9}\left\{ n - \dfrac{1}{10} \dfrac{\left(1 - \frac{1}{10^n}\right)}{\left(1 - \frac{1}{10}\right)} \right\}$

$= \dfrac{4}{9}\left\{ n - \dfrac{1}{9}\left(1 - \dfrac{1}{10^n}\right) \right\}$

$= \dfrac{4}{81}\left\{ \dfrac{1}{10^n} + 9n - 1 \right\}.$

Example 4.32. If S is the sum, P the product and R the sum of the reciprocals of n terms in a $G.P.$, prove that

$$P^2 = \left(\frac{S}{R} \right)^2.$$

Solution. Let $a, ar, ar^2, ar^3, \ldots, ar^{n-1}$ be the given $G.P.$ Then

$$S = a + ar + ar^2 + \cdots \text{ to } n \text{ terms}$$
$$= \frac{a(1 - r^n)}{1 - r}.$$
$$P = a.ar.ar^2 \ldots .ar^{n-1}$$
$$= a^n.1.r.r^2.r^3 \ldots .r^{n-1}$$
$$= a^n r^{1+2+3+\cdots+n-1}$$
$$= a^n\, r^{\frac{n(n-1)}{2}}.$$

Also,

$$R = \frac{1}{a} + \frac{1}{ar} + \frac{1}{ar^2} + \cdots \text{ to } n \text{ terms}$$

$$= \frac{1}{a} \left\{ 1 + \frac{1}{r} + \frac{1}{r^2} + \cdots \right\}$$

$$= \frac{1}{a} \left\{ \frac{1 - (1/r)^n}{1 - 1/r} \right\}.$$

Therefore,

$$\left(\frac{S}{R} \right)^n = \left\{ \frac{a(1 - r^n)}{1 - r} \cdot \frac{ar^{n-1}(r - 1)}{r^n - 1} \right\}^n = (a^2 r^{n-1})^n = P^2.$$

Example 4.33. If a, b, c are in $A.P.$ and x, y, z are in $G.P$, prove that

$$x^b y^c z^a = x^c y^a z^b.$$

Solution. Since a, b, c are in $A.P.$, we have

$$b - a = c - b \quad \text{or} \quad c - a = 2(b - a). \tag{4.5}$$

Again x, y, z are in $G.P.$ Therefore

$$y^2 = xz. \tag{4.6}$$

By (4.5) and (4.6), we get

$$(y^2)^{c-a} = (xz)^{2(b-a)} \quad \text{or} \quad y^{2(c-a)} = x^{2(b-a)} z^{2(b-a)}.$$

Taking square roots, we get

$$y^{(c-a)} = x^{(b-a)} z^{(b-a)} = x^{c-b} z^{b-a} \quad \text{or} \quad y^c y^{-a} = x^c x^{-b} z^b z^{-a}$$

which gives

$$\frac{y^c}{y^a} = \frac{x^c}{x^b} \frac{z^b}{z^a}.$$

Therefore,

$$x^b y^c z^a = x^c y^a z^b.$$

Example 4.34. Two cars start together in the same direction from the same place. The first car goes with a uniform speed of 10 km/hr. The second car goes at a speed of 8 km/hr in the first hour and increases its speed by 0.5 km/hr each succeeding hour. After how many hours will the second car overtake the first if both the cars go non-stop?

Solution. Suppose the second car overtakes the first at the end of n hours. The first car moving at a uniform speed of $10\,\text{km/hr}$ covers a distance of $10 \times n = 10n\,\text{km}$ in n hours. The distance covered by the second car in n hours equals

$$8 + \left(8 + \frac{1}{2}\right) + \left(8 + \frac{1}{2} + \frac{1}{2}\right) + \cdots \text{ to } n \text{ terms.}$$

This is an arithmetic series with $a = 8$ and $d = \frac{1}{2}$.

Therefore, the distance covered by the second car in n hours

$$= \frac{n}{2}\{2a + (n-1)d\}$$

$$= \frac{n}{2}\left\{2*8 + (n-1)\frac{1}{2}\right\}$$

$$= \frac{n}{4}\{32 + n - 1\}$$

$$= \frac{n(n+31)}{4}.$$

Since the two cars are at the same place after n hours, they have covered the same distance. Therefore,

$$10n = \frac{n(n+31)}{4} \quad \text{or} \quad n(n + 31 - 40) = 0.$$

which gives $n = 0$ or $n = 9$. Thus the two cars are together at $n = 0$ hours, i.e. at the starting time or after 9 hours. Hence the second car overtakes the first at the end of 9 hours.

Example 4.35. An insect starts from a point and travels in a straight path covering $1\,\text{mm}$ in the first second and half of the distance covered in the previous second in the succeeding second. How long would it take to reach a point $3\,\text{mm}$ away from the starting point?

Solution. If x is the distance covered by the insect in the ith second, then the distance covered in $(i+1)$th second is $\frac{x}{2}$. Suppose the insect covers the distance of $3\,\text{mm}$ in n seconds. Then

$$3 = 1 + \frac{1}{2} + \frac{1}{2^2} + \cdots \text{ to } n \text{ terms}$$

$$= \frac{(1 - 1/2^n)}{(1 - 1/2)}$$

or

$$\frac{3}{2} = 1 - \frac{1}{2^n} \quad \text{which gives} \quad \left(\frac{1}{2}\right)^n = -\frac{1}{2}.$$

But for no positive integer n it cannot happen that $\left(\frac{1}{2}\right)^n = -\frac{1}{2}$. Hence the insect never reaches the target.

Example 4.36. Show that the sum of cubes of any number of consecutive, positive integers is divisible by the sum of those integers.

Solution. Let n be a positive integer and consider n consecutive integers $m+1, m+2, m+3, \ldots, m+n$. We have to prove that

$$S = (m+1)^3 + (m+2)^3 + \cdots + (m+n)^3,$$

the sum of the cubes is divisible by

$$s = (m+1) + (m+2) + \cdots + (m+n),$$

the sum of the integers. Now

$$S = (m+1)^3 + (m+2)^3 + \cdots + (m+n)^3$$
$$= 1^3 + 2^3 + \cdots + m^3 + (m+1)^3 + (m+2)^3 + \cdots + (m+n)^3$$
$$\quad - \{1^3 + 2^3 + \cdots + m^3\}$$
$$= \frac{(m+n)^2(m+n+1)^2}{4} - \frac{m^2(m+1)^2}{4}$$
$$= \left\{ \frac{(m+n)(m+n+1)}{2} + \frac{m(m+1)}{2} \right\}$$
$$\quad \times \left\{ \frac{(m+n)(m+n+1)}{2} - \frac{m(m+1)}{2} \right\}.$$

Also,

$$s = (m+1) + (m+2) + \cdots + (m+n)$$
$$= 1 + 2 + \cdots + m + (m+1) + (m+2) + \cdots + (m+n)$$
$$\quad - \{1 + 2 + \cdots + m\}$$
$$= \frac{(m+n)(m+n+1)}{2} - \frac{m(m+1)}{2}.$$

Since s occurs as a factor in S, s divides S and

$$\frac{S}{s} = \frac{(m+n)(m+n+1)}{2} + \frac{m(m+1)}{2}$$

and the right-hand side is indeed an integer.

Example 4.37. The sum of the first m terms of an *A.P.* is the same as the sum of its first n terms. Show that the sum of its first $m + n$ terms is zero.

Solution. Let a be the initial term and d be the common difference of the *A.P.* Then,

$$S_m = \frac{m}{2}\{2a + (m-1)d\} \quad \text{and} \quad S_n = \frac{n}{2}\{2a + (n-1)d\}.$$

Since $S_m = S_n$, we have

$$\frac{m}{2}\{2a + (m-1)d\} = \frac{n}{2}\{2a + (n-1)d\}.$$

which implies

$$m\{2a + (m-1)d\} = n\{2a + (n-1)d\}$$

or

$$2a(m-n) + (m^2 - m - n^2 + n)d = 0,$$

which gives

$$(m-n)\{2a + (m+n-1)d\} = 0.$$

Since $m \neq n.$, we get

$$2a + (m+n-1)d = 0.$$

Therefore,

$$S_{m+n} = \frac{m+n}{2}\{2a + (m+n-1)d\} = 0.$$

Example 4.38. If the pth term of an *A.P.* is $\frac{1}{q}$ and the qth term is $\frac{1}{p}$, prove that the sum of the first pq terms is $\frac{(pq+1)}{2}$.

Solution. Let a be the initial term and d be the common difference of the *A.P.* If T_n denotes the nth term of this series, then

$$T_n = a + (n-1)d.$$

Therefore,

$$T_p = a + (p-1)d = \frac{1}{q} \quad \text{and} \quad T_q = a + (q-1)d = \frac{1}{p}.$$

Subtracting one from the other, we get

$$(p-q)d = \frac{1}{q} - \frac{1}{p} = \frac{p-q}{pq} \quad \text{which gives } d = \frac{1}{pq}$$

(since $p \neq q$, as the question makes sense only in that case).
Also then

$$a + \frac{(p-1)}{pq} = \frac{1}{q}$$

which gives

$$a = \frac{1}{pq}.$$

Therefore,

$$S_{pq} = \frac{pq}{2}\{2a + (pq-1)d\} = \frac{pq}{2}\left\{\frac{2}{pq} + \frac{(pq-1)}{pq}\right\}$$

$$= \frac{pq}{2}\left\{\frac{2}{pq} + 1 - \frac{1}{pq}\right\} = \frac{pq}{2}\frac{(pq+1)}{pq} = \frac{1}{2}(pq+1).$$

Example 4.39. If the mth, nth, pth terms of any $G.P.$ are themselves in $G.P.$, then m, n and p are in $A.P.$

Solution. If any two of m, n, p are equal then all three of them are equal and the result is trivially true. So suppose that no two of m, n, p are equal. Let a be the initial term and r the common ratio of the $G.P.$ Then $T_m = ar^{m-1}$, $T_n = ar^{n-1}$ and $T_p = ar^{p-1}$. Since T_m, T_n and T_p are in $G.P.$, $T_n^2 = T_m T_p$. Therefore, $a^2 r^{2(n-1)} = a^2 r^{m+p-2}$ which gives

$$2(n-1) = m + p - 2 \quad \text{or} \quad 2n = m + p.$$

This shows that m, n and p are in $A.P.$

Example 4.40. Prove that arithmetic mean (A.M) of two positive numbers is greater than their geometric mean (G.M).

Solution. Let a, b be the two positive numbers. Then

$$A.M. = \frac{a+b}{2} \quad \text{and} \quad G.M. = \sqrt{ab}.$$

Then $A.M. \gtreqless G.M$ if and only if

$$\frac{a+b}{2} \gtreqless \sqrt{ab} \quad \text{or} \quad a + b - 2\sqrt{ab} \gtreqless 0 \quad \text{or} \quad (\sqrt{a} - \sqrt{b})^2 \gtreqless 0.$$

But for $a \neq b$, $\sqrt{a} - \sqrt{b} \neq 0$ and the square of any non-zero real number is positive. Therefore, $(\sqrt{a} - \sqrt{b})^2 > 0$ and $A.M > G.M.$ Equality will hold for these two means if and only if the two numbers are equal.

Example 4.41. For any positive integers, a, b, c prove that

$$a^2b^2 + b^2c^2 + c^2a^2 > abc(a + b + c).$$

Solution. Let a, b and c be positive integers. Then ab, bc, ca are positive integers. Since the arithmetic mean between two numbers is greater than their geometric mean,

$$\frac{ab + bc}{2} > \sqrt{ab \cdot bc}.$$

Squaring both sides we get

$$(ab + bc)^2 > 4ab^2c$$

or

$$a^2b^2 + b^2c^2 + 2ab^2c > 4ab^2c.$$

Therefore,

$$a^2b^2 + b^2c^2 > 2ab^2c.$$

Similarly,

$$b^2c^2 + c^2a^2 > 2bc^2a \quad \text{and} \quad c^2a^2 + a^2b^2 > 2ca^2b.$$

Adding the three relations, we get

$$2(a^2b^2 + b^2c^2 + c^2a^2) > 2abc(a + b + c)$$

or

$$a^2b^2 + b^2c^2 + c^2a^2 > abc(a + b + c).$$

Example 4.42. Sum up the series

(a) $1 + 2x + 3x^2 + 4x^3 + \cdots$ to n terms.

(b) $2.5 + 4.5^2 + 6.5^3 + 8.5^4 + \cdots$ to n terms.

Solution. Let S_n denote the sum of the series to n terms. Then

(a)

$$S_n = 1 + 2x + 3x^2 + 4x^3 + \cdots + nx^{n-1}$$

and

$$xS_n = x + 2x^2 + 3x^3 + \cdots + (n-1)x^{n-1} + nx^n.$$

Subtracting, we get

$$(1-x)S_n = 1 + x + x^2 + \cdots + x^{n-1} - nx^n$$

$$= \frac{1-x^n}{1-x} - nx^n, \quad \text{provided } x \neq 1.$$

$$= \frac{1 - (n+1)x^n + nx^{n+1}}{1-x}.$$

Dividing by $(1-x)$,

$$S_n = \frac{1 - (n+1)x^n + nx^{n+1}}{(1-x)^2}.$$

For $x = 1$, the series is arithmetic — in fact it is the sum of the first n natural numbers and $S_n = \frac{n(n+1)}{2}$.

(b) Although we can solve this directly exactly as in (a) above, we deduce it as special case of (a).

$$2.5 + 4.5^2 + 6.5^3 + 8.5^4 + \cdots = 2.5(1 + 2.5 + 3.5^2 + \cdots)$$

and the bracketed sum can be obtained by taking $x = 5$ in part (a) above. Thus,

$$2.5 + 4.5^2 + 6.5^3 + 8.5^4 + \cdots = \frac{10(1 - (n+1)5^n + n5^{n+1})}{(1-5)^2}$$

$$= \frac{5}{8}\left(1 - (n+1)5^n + n5^{n+1}\right).$$

Exercise 4.2.

1. If a is the first term and l is the nth term of an arithmetic series, prove that the sum of the first n terms of this series is $\frac{n}{2}(a+l)$.
2. Find the 40th term and the sum of the first 40 terms of the arithmetic series $10 + 7 + 4 + \cdots$
3. Compute the sum of the first 100 positive integers exactly divisible by (a) 7 and (b) 6.

4. How long will it take to pay off a debt of Rs. 8,800 if Rs. 250 is paid the first month, Rs. 270 the second month, Rs. 290 in the third month, and so on?

5. Determine the arithmetic series the sum of the first n terms of which $n^2 + 2n$.

6. Compute the sum of all integers between 100 and 800 that are divisible by 7.

7. Determine a formula for the arithmetic mean between two numbers p and q. (See Example 4.40 above.)

8. Insert four arithmetic means between 8 and 23.

9. Insert between 1 and 31 a number of arithmetic means so that the sum of the terms of the arithmetic sequence is 112.

10. Determine a formula for the geometric mean between two numbers a and b. Deduce that for no value x the three numbers

    (a) $x, x + 3$ and $x + 6$
    (b) $x, x + 4$ and $x + 8$

    are in geometric progression.

11. Find three numbers in a geometric sequence whose sum is 12 and whose product is 64.

12. From a tank filled with 2,400 litres of milk, 400 litres are drawn off and the tank is filled up with water. Then 400 litres of the mixture are removed and replaced with water and the process is repeated. How many litres of milk remain in the tank after (a) 4 drawings; and (b) 5 drawings of 60 litres each, are made?

13. Find three numbers in a geometric progression whose sum is 18 and whose product is 216.

14. Among four numbers, the first three are in geometric progression and the last three are in arithmetic progression with a common difference of 6. If the first number is the same as the fourth, find the four numbers.

4.4. Permutations and Combinations

4.4.1. *Rules of product and sum*

By an **experiment,** we mean a physical process that has a number of observable outcomes, for example, (i) tossing a coin, (ii) drawing a card from a pack of playing cards, (iii) throwing a dice, (iv) selecting a representative from a class. While considering the outcomes of several experiments, the following rules apply.

1. **Rule of product.** If one experiment has m outcomes and another experiment has n outcomes, then there are $m \times n$ outcomes when both the experiments take place.

2. **Rule of sum.** If an experiment has m outcomes and another experiment has n possible outcomes, then there are $m + n$ outcomes when exactly one of these experiments takes place. We will consider some examples.

Example 4.43. If there are 60 students in a second-year computer engineering class and 50 students in a second-year mechanical engineering class in a college, then a representative from the computer engineering class can be chosen in 60 ways, while a representative from the mechanical engineering class can be chosen in 50 ways. Then there are 60×50 ways of selecting a representative each from both the classes and there are $60 + 50$ ways of selecting a representative from either the computer engineering class or from the mechanical engineering class.

Example 4.44. While forming binary sequences of length r, i.e. r digit sequences, each place may be filled in two ways by taking either a 0 or a 1 in this place. Therefore, the number of r digit binary sequences is

$$2 \times 2 \times \cdots \times 2 = 2^r.$$

To count the r digit binary sequences with odd number of 1s, we have to consider r digit binary sequences, which have only one 1 or three 1s or five 1s, and so on. For the sequences with one 1, the 1 may appear in any one of the r positions. Therefore, there are r r digit binary words with one 1. In fact, for any $i, 1 \leq i \leq r$, the number of sequences with i digits equal to 1 is $\binom{r}{i}$. Therefore, the number of r digit binary sequences with odd 1s

$$= \binom{r}{1} + \binom{r}{3} + \binom{r}{5} + \cdots,$$

which can be proved to be equal to 2^{r-1}. This result, in fact, follows by induction on r. For $r = 1$, there is only one sequence of length 1 with entry 1, i.e. it is 2^{r-1}. Now suppose that $r \geq 1$ and that

$$\binom{r}{1} + \binom{r}{3} + \binom{r}{5} + \cdots = 2^{r-1}.$$

When r is odd,

$$\binom{r+1}{1} + \binom{r+1}{3} + \cdots + \binom{r+1}{r}$$

$$= \left\{ \binom{r}{0} + \binom{r}{1} \right\} + \left\{ \binom{r}{2} + \binom{r}{3} \right\} + \cdots + \left\{ \binom{r}{r-1} + \binom{r}{r} \right\}$$

$$= \binom{r}{0} + \binom{r}{1} + \binom{r}{2} + \cdots + \binom{r}{r-2} + \binom{r}{r-1} + \binom{r}{r}$$

$$= 2^r = 2^{(r+1)-1}$$

and, for r even,

$$\binom{r+1}{1} + \binom{r+1}{3} + \cdots + \binom{r+1}{r+1}$$

$$= \left\{ \binom{r}{0} + \binom{r}{1} \right\} + \left\{ \binom{r}{2} + \binom{r}{3} \right\} + \cdots + \left\{ \binom{r}{r-2} + \binom{r}{r-1} \right\} + 1$$

$$= \binom{r}{0} + \binom{r}{1} + \binom{r}{2} + \cdots + \binom{r}{r-2} + \binom{r}{r-1} + \binom{r}{r}$$

$$= 2^r = 2^{(r+1)-1}.$$

This completes induction. Therefore, the number of r digit binary sequences with odd 1s is 2^{r-1}.

Example 4.45. The number of three-digit decimal numbers is $9 \times 10 \times 10 = 9 \times 10^2$ because the left-most position is occupied by any one of the 9 non-zero numbers from 1 to 9 and either of the other two positions have 10 choices each. The number of three-digit decimal numbers without repetition is $9 \times 9 \times 8 = 9^2 \times 8$.

Example 4.46. The number of three-letter words that can be formed with the English alphabet, whether meaningful or not, is $26 \times 26 \times 26 = 26^3$ and the number of three-letter words without repeating an alphabet is $26 \times 25 \times 24$.

Example 4.47. The number of ways of getting 6 in a throw of two dice is 3, while the number of ways of getting 6 in two throws of one dice is 5. As an application of both the rule of products and the rule of sums we have the following example.

Example 4.48. Three integers are selected from the integers $1, 2, \ldots$ 1,000. In how many ways can these integers be selected such that their sum is divisible by 4?

Solution. We divide the integers from 1 to 1,000 into 4 disjoint subsets. Let A, B, C, D be the sets of integers as follows:

$$A = \{m | 1 \leq m \leq 1000 \text{ and } 4 \text{ divides } m\}$$

$$B = \{m | 1 \leq m \leq 1000 \text{ and } 4 \text{ divides } m - 1\}$$

$$C = \{m | 1 \leq m \leq 1000 \text{ and } 4 \text{ divides } m - 2\}$$

$$D = \{m | 1 \leq m \leq 1000 \text{ and } 4 \text{ divides } m - 3\}.$$

Observe that $o(A) = o(B) = o(C) = o(D) = 250$, i.e. there are 250 positive integers up to 1,000, which are divisible by 4,250 integers which leave the remainder 1 on division by 4,250 integers which leave the remainder 2 on division by 4 and 250 such integers which leave the remainder 3 on division by 4. The sum of no three integers from either B or C or D is divisible by 4. If the sum of three integers is to be divisible by 4, then either

(a) all the three integers are from A; or
(b) one integer is from each of A, B and D; or
(c) two integers are from C and one from A; or
(d) two integers are from D and one from C; or
(e) two integers are from B and one from C.

These are all the possible choices of three integers, so that their sum is divisible by 4. Therefore, the total number of choices of the three integers, the sum of which is divisible by 4, is

$$= \binom{250}{3} + \binom{250}{1}\binom{250}{2} + \binom{250}{1}\binom{250}{1}\binom{250}{1} + \binom{250}{2}\binom{250}{1}$$

$$+ \binom{250}{1}\binom{250}{2}$$

$$= \frac{250 \times 249 \times 248}{6} + 250 \times 250 \times 250 + 3 \times \frac{250 \times 250 \times 249}{2}$$

$$= 250 \times (83 \times 124 + 250 \times 250 + 3 \times 249 \times 125) = 41541750.$$

4.4.2. *Permutations*

There are three coloured balls: white, red and green, and there are two holes each capable of holding or accommodating exactly one ball. The holes are numbered 1 and 2. We want to put two balls in the two holes. The first hole can be assigned to any one of the three balls. Thus the first hole can be occupied in three ways. For each way of assigning a ball to the first hole, there are two ways of placing a ball in the second hole. Thus, altogether, there are $3 \times 2 = 6$ ways of placing two balls in the two holes. The six assignments are WR, WG, GR, GW, RG, RW, where we are writing W for white, R for red and G for green. Also, WR means that the white one is placed in the first hole and the red one in the second. As such, WR and RW are different assignments. Each one of these assignments is called a permutation of the letters W, R, G, taken two at a time. Also, each one of these assignments may be regarded as a map from the set $\{1, 2\}$ to the set $\{W, R, G\}$.

In general, by a **permutation** of n objects, say $x_1, x_2, \ldots, x_n$, taken r at a time, we mean an assignment of one of the n objects to the first place (from the left), one of the remaining $n - 1$ objects to the second position, one of the remaining $n - 2$ objects to the third position and finally, one of the remaining $n - r + 1$ objects to the rth position.

This is the same thing as saying that it is a one–one map from the set $\{1, 2, \ldots, r\}$ to the set $\{x_1, x_2, \ldots, x_n\}$. In particular, a permutation of n objects taken all at a time is a one–one map from the set $\{1, 2, \ldots, n\}$ to the set $\{x_1, x_2, \ldots, x_n\}$ It is equivalent to saying that it is a one–one map from $\{x_1, x_2, \ldots, x_n\}$ to itself: the usual way we define a permutation of a finite set. (Usually a permutation of a finite set A is a map from A to A which is both one–one and onto. However, any map from A to A is one–one if and only if it is onto). The number of ways of assigning an element out of $A = \{x_1, x_2, \ldots, x_n\}$ to the first position is n, the number of ways of assigning an element of A to the second position is $n - 1$, the number of assignments of an element of A to the third position is $n - 2$, and so on (of course without repeating an element). Finally, the number of assignments of an element to the rth position is $n - r + 1$. Therefore, by the product rule, the number of ways of assigning r elements of A (without repetition) to the r positions 1st, 2nd, $\ldots$, rth is

$$n(n - 1)(n - 2) \cdots (n - r + 1).$$

Thus the total number of permutations of n objects taken r at a time is $n(n - 1) \cdots (n - r + 1)$, which we write as $P(n, r)$ or ^nP_r. Observe that for $r = n, P(n, n) = {}^n P_n = n!$. Also note that $P(n, n - 1) = {}^n P_{n-1} = n!$. Let us consider a couple of particular cases of the above.

Example 4.49. Consider $n = 3$ and let $A = \{a, b, c\}$. The permutations of the three objects taken two at a time are ab, ac, bc, ba, ca, cb, which are

$6 = 3 \times 2 = P(3, 2)$ in number. The permutations of the three objects taken all at a time are $abc, bca, cab, acb, bac, cba$, which are again $6 = P(3, 3)$ in number. We have not considered permutations of the three objects taken one at a time. These permutations are a, b, c. This is a trivial case as it is just the collection of all the objects under consideration.

Example 4.50. Consider the case $n = 4$ and let the four objects be a, b, c, d. The permutations of these objects taken two at a time are ab, ac, ad, ba, bc, bd, ca, cb, cd, da, db, dc. The permutations of these objects taken three at a time are abc, abd, acd, acb, adb, adc, bca, cab, cba, bac, bda, dab, dba, bad, cda, dac, dca, cad, bcd, cdb, dbc, bdc, dcb, cbd. These permutations are all distinct and are $24 = P(4, 3)$ in number. Therefore, the above list exhausts all the permutations of the four objects taken three at a time. Next, we consider the permutations of the four objects taken all at a time. These are $abcd$, $bcda$, $cdab$, $dabc$, $abdc$, $bdca$, $dcab$, $cabd$ $acbd$, $dacb$, $bdac$, $dcba$, $bacd$, $dbac$, $cdba$, $acdb$, $adbc$, $adcb$, $bcad$, $badc$, $cbda$, $cbad$, $cadb$, $dbca$, which are again $24 = P(4, 4)$ in number. Therefore, these are all the possible permutations of the four objects a, b, c, d.

The amount and difficulty of work in computing permutations of n objects goes on increasing as n increases. Also, it is difficult to keep track of whether we have obtained all the permutations correctly. Later on we shall give an algorithm which will generate all the permutations of a given set of n elements in a systematic manner.

4.4.3. The arrangements of objects that are not all distinct

So far, we have considered permutations of elements which are all distinct. Permutations of n elements which are not all distinct cannot be considered as bijective maps from the set $\{1, 2, \ldots, n\}$ to the collection A of the n objects. Here A is no longer a set but is a multiset.

Example 4.51. Consider a collection A of n objects, in which q elements are equal, each equal to a (say). Let $A = \{a_1, a_2, \ldots, a_q, a_{q+1}, \ldots, a_n\}$ in which $a_1 = a_2 = \cdots = a_q = a$ and none of the remaining elements equals a. Choose q positions $i_1, i_2, \ldots, i_q$, which are occupied by the elements $a_1, a_2, \ldots, a_q$. For every placement of the elements $a_{q+1}, \ldots, a_n$, there are $q!$ ways of placing the q elements $a_1, a_2, \ldots, a_q$ in the positions $i_1, i_2, \ldots, i_q$. However, since $a_1 = a_2 = \cdots = a_q = a$, all the $q!$ permutations or ways of placement are identical. Thus the $n!$ permutations of $a_1, a_2, \ldots, a_n$, considering them all distinct, can be partitioned into blocks of size $q!$

in which all the permutations in each block are equal. Therefore, there are $\frac{n!}{q!}$ distinct ways of placement of the elements $a_1, a_2, \ldots, a_n$ in which $a_1 = a_2 = \cdots = a_q = a$ and all the other elements are distinct.

Next, suppose that among the above n elements in which q of them are equal, each equal to a, there are r elements $a_{q+1}, a_{q+2}, \ldots, a_{q+r}$ which are all equal and equal to b (say) with $b \neq a$ and none of the remaining $n - r$ elements equals b. Suppose that in a permutation the elements $a_{q+1}, \ldots, a_{q+r}$ occupy the positions $j_1, j_2, \ldots, j_r$. For a fixed choice of places for the remaining elements $a_1, \ldots, a_q, a_{q+r+1}, \ldots, a_n$, there are $r!$ ways of placing the elements $a_{q+1}, : \ldots, a_{q+r}$ in the $j_1, \ldots, j_r$ positions and all these $r!$ ways lead to only one arrangement. Thus all the $\frac{n!}{q!}$ permutations of $a_1, a_2, \cdots, a_n$ in which $a_1 = a_2 = \cdots = a_q = a$ are partitioned into blocks of size $r!$ in which all the permutations in any particular block are equal. Therefore,

(a) the number of permutations of the n objects in which q objects are equal each equal to a and r objects are equal each equal to $b \neq a$, is $\frac{n!}{q!r!}$.

We can continue the above process and obtain

(b) The number of ways of placing n balls of which q_1 are of one colour, q_2 are of second colour, q_3 are of a third colour, etc. in n numbered boxes is

$$\frac{n!}{q_1! \, q_2! \, q_3! \, \ldots}.$$

If we consider the placing of r balls instead of all the n balls, the number of ways to place r coloured balls out of a total of n balls in n numbered boxes, where q_1 of these balls are of one colour, q_2 of these balls are of a second colour, $\ldots, q_t$ of these balls are of tth colour, may appear to be $\frac{P(n,r)}{q_1!q_2!\ldots q_t!}$. However, this is not correct. Consider, for example, 3 dashes and 2 dots and we want the number of messages of length 3 by using the dots and dashes given. If the above formula were correct, then the number of messages shall be

$$= \frac{P(5,3)}{3!2!} = \frac{5 \times 4 \times 3}{6 \times 2} = 5.$$

By actually writing the messages, there is one message having 3 dashes, there are 3 messages having 2 dashes and 1 dot, namely (for the sake of clarity of writing the messages, we represent a dash by 1 and a dot by 0), 1 1 0, 1 0 1, 0 1 1, there are 3 messages having one dash and 2 dots, namely, 1 0 0, 0 1 0, 0 0 1. There are no messages with 3 dots. Thus, there are 7 messages, which is different from the result obtained by using the proposed (assumed) formula. However, the following is still true.

(c) The number of ways of placing r balls of which q_1 are of one colour, q_2 are of a second colour and so on, with q_t of the tth colour in $n \geq r$ boxes (each capable of holding one ball) is

$$\frac{P(n,r)}{(q_1!q_2!\ldots q_t!)}.$$

We will now explain the above for some small values of n.

Example 4.52. Consider a multiset $A = \{a, a, a, b\}$. Let us take $A = \{a_1, a_2, a_3, b\}$ with $a_1 = a_2 = a_3 = a$. In any permutation of a_1, a_2, a_3, b, the element b may occupy the first position, the second position, the third position or the fourth position. We will consider the four placements separately.

(a) b occupies the first place. Then a_1, a_2, a_3 can be arranged in six ways in the second, third and fourth places as:
$ba_1a_2a_3$, $ba_2a_3a_1$, $ba_3a_1a_2$, $ba_2a_1a_3$, $ba_3a_2a_1$, $ba_1a_3a_2$,
which all lead to only one method of placement, i.e. *baaa.*

(b) For b occupying the second position, the placements are
$a_1ba_2a_3$, $a_2ba_3a_1$, $a_3ba_1a_2$, $a_2ba_1a_3$, $a_3ba_2a_1$, $a_1ba_3a_2$,
which all lead to only one method of placement, i.e. *abaa.*

(c) For b occupying the third place, the placements are
$a_1a_2ba_3, a_2a_3ba_1, a_3a_1ba_2$, $a_2a_1ba_3$, $a_3a_2ba_1$, $a_1a_3ba_2$,
which all give only one placement, namely *aaba.*

(d) Let b occupy the fourth place. Then the corresponding placements of the four objects are
$a_1a_2a_3b$, $a_2a_3a_1b$, $a_3a_1a_2b$, $a_2a_1a_3b$, $a_1a_3a_2b$, $a_3a_2a_1b$,
all of which give only one placement, i.e. *aaab.*
Therefore, there are four possible arrangements of these objects and $4 = \frac{4!}{3!}$.

Example 4.53. Consider the number of different messages that can be formed by using four 1 s and three 0 s. For this we need sequences of length 7 using four 1 s (i.e. four elements all equal) and three 0 s (i.e. three equal elements). Therefore, the number of messages is $\frac{7!}{4!3!} = 35$.

Example 4.54. Consider the word 'ELEPHANT', which is a word of length 8, i.e. a word using 8 letters, namely E, L, E, P, H, A, N and T. The letters are distinct except for the letter E, which occurs twice. The total number of words (meaningful or otherwise) of length 8 that can be formed by using these alphabet equals $\frac{8!}{2!} = 20160$.

Example 4.55. Consider the word 'COMMITTEE', which is a word of length 9. Among these letters, each of M, T and E occurs twice, while each of the remaining three occurs only once. Therefore, the number of words formed by using all the letters occurring in the word 'COMMITTEE' equals $\frac{9!}{2!2!2!} = 45360$.

4.4.4. *Combinations*

Suppose we want to choose two elements out of a set $A = \{1, 2, 3\}$. We may choose the first to be any one of a, b, c. Then we are left with two elements and the second may be chosen in two ways. Thus, altogether, there are six ways of choosing two elements. However, we may have chosen a in the first step and b in the second step or vice versa. The net result is that we have chosen the two elements as a, b. Thus the two choices lead to one choice of a and b. Similarly, there is one choice for the elements b, c and only one choice for the elements a, c. Therefore, the choices of two elements out of the three are $a, b; b, c; a, c$.

Consider, in general, a set $A = \{a_1, a_2, \ldots, a_n\}$ of n elements. We want to choose r elements out of A. We can choose the first element in n ways, the second in $n-1$ ways and so on, and the rth element in $n-r+1$ ways. Suppose that the elements chosen are $a_{i_1}, a_{i_2}, \ldots, a_{i_r}$. These r elements could be chosen in $r!$ ways but all these ways finally lead to the choice of the same elements. Therefore, the number of ways we can choose r elements out of the n elements is

$$\frac{P(n, r)}{r!} = \frac{n!}{(n-r)!r!}.$$

Each choice of r objects out of the n (necessarily distinct) objects is called a **combination** of n objects taken r at a time. Since the given objects are all distinct and the order in which the r elements are chosen is immaterial, each combination of n objects taken r at a time is just a subset of order r of the set A. The number $\frac{n!}{(n-r)!r!}$ of combinations of n objects taken r at a time is denoted by ^nC_r *or* $C(n, r)$. Thus the number of combinations of n objects taken r at a time is

$$^nC_r = C(n, r) = \frac{n!}{(n-r)!r!}.$$

Theorem 4.4. Prove that for any positive integer n and integer $r, 0 \le r \le n, C(n, r) = C(n, n-r)$.

Proof. Consider a set A of n objects. Let A_1 be a subset of A of order r and define $A_1' = \{a \in A | a \notin A_1\} = A - A_1$.

Since $o(A) = n$ and $o(A_1) = r$, it follows that $o(A_1') = n - r$. Let A_1, A_2 be two subsets of A with $o(A_1) = o(A_2) = r$ and $A_1 \neq A_2$. Then there exists an element $a_1 \in A_1$ but $a_1 \notin A_2$ and there exists an $a_2 \in A_2, a_2 \notin A_1$. Now, $a_1 \notin A_2$ implies that $a_1 \in A_2'$. Similarly $a_2 \in A_1'$. Also, $a_1 \in A_1$ implies that $a_1 \notin A_1'$. Thus $a_1 \in A_2'$ but $a_1 \notin A_1'$. Hence $A_1' \neq A_2'$. This proves that to every subset A_1 of order r in A there corresponds one and only one subset A_1' of order $n - r$ in A. Therefore, the number of subsets of A of order r equals the number of subsets of A of order $n - r$ i.e. ${}^n C_r = {}^n C_{n-r}$.

Remark. We define $0! = 1$.
Therefore

$$C(n, 0) = \frac{n!}{(n-0)!0!} = \frac{n!}{n!0!} = \frac{n!}{n!} = 1$$

and

$$C(n, n) = C(n, n - n) = C(n, 0) = 1.$$

Example 4.56. Among 11 MLAs (Members of Legislative Assembly of a state) there are $C(11, 4) = {}^{11}C_4$ ways to form a committee of four members. There are $C(10, 3) = {}^{10}C_3$ ways of forming a committee of four members so that a particular MLA is always in the committee. There are $C(10, 4) = {}^{10}C_4$ ways of forming a committee of four members in which a particular MLA is never taken.

Example 4.57. Consider the number of ways of forming a committee of four MLAs out of the total 11 MLAs so that at least one of the two MLAs, say A and B, is always included.

Solution. The number of ways is equal to the number of ways of forming a committee so that (i) both A and B are chosen, (ii) A is chosen but B is not chosen and (iii) B is chosen but A is not. The number of ways when both A and B are chosen is equal to

$$C(9, 2) = {}^9C_2 = \frac{9 \times 8}{2} = 36;$$

A is chosen but B is not is equal to

$$C(9, 3) = {}^9C_3 = \frac{9 \times 8 \times 7}{6} = 84;$$

B is chosen but A is not, again, equals $C(9,3) = 84$. Therefore, the total number of ways of forming a committee in which at least one of A and B is always chosen equals $36 + 84 + 84 = 204$.

Alternatively, we could tackle this problem by finding the total number of committees of four MLAs without any restriction, minus the number of committees in which both A and B are absent. This number is equal to

$$C(11,4) - C(9,4) = \frac{11 \times 10 \times 9 \times 8}{1 \times 2 \times 3 \times 4} - \frac{9 \times 8 \times 7 \times 6}{1 \times 2 \times 3 \times 4}$$

$$= 330 - 126 = 204.$$

We next consider a couple of examples for finding the number of combinations of r things out of n when repetitions are allowed.

Example 4.58. A boy stands first in the class in the annual school examination. His father offers to buy him three of the given five items, such as some storybooks, pens of different inks, science fiction books, some wall posters of international sports players and books of historical importance. He could choose more than one of the same kind (but the total number in any case being three). Let the five items be a, b, c, d and e. The man asks the boy if he would like to have item a. The boy may either say yes (Y) or no (N). If the boy's answer is Y, the man again asks the boy if he would like to have item a. The answer may again be Y or N. If the answer is Y, the man repeats the same question. If the answer is Y again, the questioning stops as the boy has already chosen three articles of item a. If, at any stage, the answer is N, the man asks the boy if he would like to have item b. If the boy has not chosen three articles even when the questioning about b has to stop (it stops the first time the answer is N), he asks the same question with c and then d and finally e, in order to allow him to choose three articles from the five items. The process of questioning stops as soon as three Y answers are obtained. The maximum number of questions will be $4 + 3 = 7$, which happens when the boy would not like to have any one of the four items a, b, c, d and would like to have three of item e. Otherwise, the number of questions will be less than 7. Thus there are 7 places available to be filled by Y or N, with exactly three Y. In case the sequence of Y and N having three Y is of length less than 7, the remaining places are filled with B to indicate a blank (i.e. no question). Thus, the three Y answers could be in any of the three places out of the seven. The sequences could be $N\,N\,N\,Y\,N\,Y\,Y$, $N\,Y\,N$ $Y\,Y\,B\,B$, $Y\,N\,N\,Y\,N\,Y\,B$, etc. which indicate the choices $\{d, e, e\}$, $\{b, c, c\}$, $\{a, c, d\}$ made by the boy. The total number of choices made is $C(7,3)$.

Example 4.59. When two dice are rolled, the possible pairs of numbers that appear on the faces of the dice facing upwards are 1, 1 or 1, 2 or 1, 3 or 2, 3 or 2, 4 or 2, 6 etc. All possible choices for the outcome could be obtained by the following process. Is the number 1 there? If the answer provided is *Yes*, the same question is repeated, while if the answer is *No*, the question is repeated with 1 replaced by 2. If the answer provided is *Yes*, the question with number 2 is repeated but if the answer is *No*, the question is repeated with the number 3 in place of 2. The process stops only when we have got the two *Yes* answers (two because of there being two dice only two numbers will show up). The questioning will continue right up to the number 6 if two *Yes* answers have not already been obtained. If we have already got 5 *No* answers, the next two answers will have to be *Yes* as the numbers appearing on a dice are from 1 to 6. Thus the maximum number of questions is 7. Again, if the sequence of Y and N is of length less than 7, the remaining place(s) will be filled with B to indicate blank. Thus, we need to choose two places out of the 7 possible to put two Y. The total number of choices is thus $C(7, 2)$.

The argument used in the above two examples can be generalized to get the following.

Theorem 4.5. The number of choices of r objects out of n objects (there being unlimited supply of each object) with repetitions allowed is

$$C(n + r - 1, r).$$

Example 4.60. Suppose three six faced unbiased dice are rolled. The possible sequences of numbers that show up are

```
6 6 6   6 5 5   6 4 4   6 3 3   6 2 2   6 1 1   5 5 5   5 4 4   5 3 3   5 2 2
6 6 5   6 5 4   6 4 3   6 3 2   6 2 1           5 5 4   5 4 3   5 3 2   5 2 1
6 6 4   6 5 3   6 4 2   6 3 1                   5 5 3   5 4 2   5 3 1
6 6 3   6 5 2   6 4 1                           5 5 2   5 4 1
6 6 2   6 5 1                                   5 5 1
6 6 1
5 1 1   4 4 4   4 3 3   4 2 2   4 1 1   3 3 3   3 2 2   3 1 1   2 2 2   2 1 1   1 1 1
        4 4 3   4 3 2   4 2 1           3 3 2   3 2 1           2 2 1
        4 4 2   4 3 1                   3 3 1
        4 4 1
```

which are

$$6 + 5 + 4 + 3 + 2 + 1 + 5 + 4 + 3 + 2 + 1 + 4 + 3 + 2 + 1 + 3 + 2$$
$$+ 1 + 2 + 1 + 1 = 21 + 15 + 10 + 6 + 3 + 1 = 56$$

in number and $56 = C(8,3) = C(6 + 3 - 1, 3)$.

4.4.5. *Generation of permutations and combinations*

Recall that if $a_1 a_2 \ldots a_n, b_1 b_2 \ldots b_n$ are two permutations of the numbers $1, 2, \ldots, n$, we say that $a_1 a_2 \ldots a_n$ comes before $b_1 b_2 \ldots b_n$ in lexicographic order if for some m, $1 \leq m < n, a_1 = b_1, \ldots, a_{m-1} = b_{m-1}$ and $a_m < b_m$. For example, among the permutations 1 2 4 6 5 3 and 1 2 4 5 6 3, the permutation 1 2 4 5 6 3 comes before 1 2 4 6 5 3. We want to describe a procedure using lexicographic ordering to generate all the permutations of $\{1, 2, \ldots, n\}$. For this we consider the following:

Algorithm 4.1

Suppose we are given a permutation $a_1 a_2 \ldots a_n$. We want to write down the next permutation in the lexicographic order. Take the next permutation, say, $b_1 b_2 \ldots b_n$ as follows:

Step 1. Choose the largest possible m such that a_m is less than at least one of $a_{m+1}, \ldots, a_n$.

Step 2. Choose $b_i = a_i$ for $i = 1, 2, \ldots, m - 1$ and choose b_m to be the smallest number among $a_{m+1}, \ldots, a_n$ which is larger than a_m.

Step 3. Arrange the remaining a_i's as $b_{m+1}, \ldots, b_n$ subject to $b_{m+1} < b_{m+2} < \cdots < b_n$.

Theorem 4.6. Given a permutation $a_1 a_2 \ldots a_n$ of the numbers $1, 2, \ldots, n$ the largest m for which $a_m <$ at least one of $a_{m+1}, \ldots, a_n$ is the largest m for which $a_m < a_{m+1}$ and conversely.

Proof. Let m be the largest index such that

$$a_m < \text{at least one of } a_{m+1}, \ldots, a_n.$$

Then there exists a $j, 1 \leq j \leq n - m$ such that $a_m < a_{m+j}$. If $j \neq 1$ i.e. if a_m is not less than a_{m+1}, then $a_{m+1} \leq a_m < a_{m+j}$ which contradicts the maximality of m as $a_{m+1} < a_{m+j}$ and, so,

$$a_{m+1} < \text{ at least one of } a_{m+2}, \ldots, a_n.$$

Therefore $a_m < a_{m+1}$.

If m is not the largest with $a_m < a_{m+1}$, then there is an $i \geq 1$ such that $a_{m+i} < a_{m+i+1}$. But then $a_{m+i} < $ at least one of $a_{m+i+1}, \ldots, a_n$. This contradicts the assumption that m is the largest with $a_m < $ at least one of $a_{m+1}, \ldots, a_n$. Thus, if m is the largest with $a_m < $ at least one of $a_{m+1}, \ldots, a_n$, then m is the largest subject to $a_m < a_{m+1}$.

Conversely, suppose that m is the largest subject to $a_m < a_{m+1}$. If m is not the largest with

$$a_m < \text{ at least one of } a_{m+1}, \ldots, a_n$$

then there exists an $i \geq 1$ such that $m + i$ is the largest with

$$a_{m+i} < \text{ at least one of } a_{m+i+1}, \ldots, a_n.$$

By the first part already proved above $a_{m+i} < a_{m+i+1}$ which contradicts the maximality of m with $a_m < a_{m+1}$. Therefore m is the largest such that $a_m < $ at least one of $a_{m+1}, \ldots, a_n$.

Hence the largest m for which $a_m < $ at least one of $a_{m+1}, \ldots, a_n$ is the same as the largest m such that $a_m < a_{m+1}$.

In view of the above, Step 1 in the above procedure may be replaced by

Step $1'$. Choose the largest possible m such that $a_m < a_{m+1}$.

Example 4.61. Let a permutation 1 2 4 6 3 5 of the numbers $1, 2, 3, 4, 5, 6$ be given. To write the next permutation in lexicographic order, let us think of the given permutation as $a_1 \, a_2 \, a_3 \, a_4 \, a_5 \, a_6$ so that $a_1 = 1, a_2 = 2, a_3 = 4, a_4 = 6, a_5 = 3$ and $a_6 = 5$. Then the largest m such that $a_m < a_{m+1}$ is 5 as $a_5 = 3 < 5 = a_6$. There is only one element in $a_{m+1}, \ldots, a_n$ and it is a_6. Therefore $b_5 = a_6 = 5$. Also $b_i = a_i$, for $i = 1, 2, \ldots, m - 1$ so that the next permutation looks like $a_1 a_2 a_3 a_4 a_6$ x or 1 2 4 6 5 x. There is only one choice for the last entry namely, it has to be 3. Thus the next permutation is 1 2 4 6 5 3.

Let us next find the next permutation in lexicographic order. For this we represent the permutation 1 2 4 6 5 3 as $a_1 \, a_2 \, a_3 \, a_4 \, a_5 \, a_6$. Observe that the largest m for which $a_m < a_{m+1}$ is 3 as $a_3 = 4 < 6 = a_4$ while a_4 is not less than a_5 and a_5 is not less than a_6. The smallest among a_4, a_5, a_6 which is larger than a_3 is a_5. Therefore $b_3 = 5$ and $b_1 = a_1, b_2 = a_2$. The remaining numbers are a_3, a_4, a_6 i.e. 3, 4, 6 and, therefore $b_4 = 3, b_5 = 4$ and $b_6 = 6$. Thus the next permutation is 1 2 5 3 4 6.

To find the next permutation, we again represent the permutation 1 2 5 3 4 6 as $a_1 a_2 a_3 a_4 a_5 a_6$. As $a_5 = 4 < 6 = a_6$, the largest m with $a_m < a_{m+1}$ is 5. There is only one element beyond a_5 and it is a_6. Therefore $b_5 = a_6 = 6$. As $b_i = a_i$ for $i = 1, 2, 3, 4$, there is only one choice for b_6 and it is $b_6 = a_5 = 4$. Hence the next permutation is 1 2 5 3 6 4.

We now find the permutation next to 1 2 5 3 6 4. Call this permutation $a_1 a_2 a_3 a_4 a_5 a_6$. The largest m for which $a_m < a_{m+1}$ is 4 as 3 < 6 but 6 is not less than the next entry which is 4. Then out of the entries next to a_4 the entry 4 is >3 and is smaller of the two. Therefore $b_4 = 4$. Then the next two entries are 3, 6 in that order. Hence the next permutation is 1 2 5 4 3 6. We can check that the next three permutations are 1 2 5 4 6 3, 1 2 5 6 3 4, 1 2 5 6 4 3 in that order.

Exercise 4.3. Using the algorithm given, find 10 permutations next to 1 2 6 3 4 5 in lexicographic order.

In order to generate all permutations of $\{1, 2, \ldots, n\}$ we have to have a starting point. Given any permutation $a_1 a_2 \ldots a_n$ of $1, 2, \ldots, n$, observe that the permutation $12 \ldots n$ is less than $a_1 a_2 \ldots a_n$ in lexicographic order. Thus to generate all permutations we start with the smallest among them namely $12 \ldots n$ and apply the above procedure. Since the total number of permutations is finite i.e. $n!$, this process must come to an end. As soon as we arrive at a permutation when we cannot find a permutation next to it, in lexicographic order, we stop, as all the possible permutations have been obtained.

We now use the procedure to get all the permutations in the case of $n = 3$ and $n = 4$.

Example 4.62. Find all possible permutations of the numbers 1, 2, 3.

Solution. As mentioned above we start with the permutation 1 2 3. (Any permutation for which we need to write the next in lexicographic order shall be assumed to be $a_1 a_2 a_3$. Also the next permutation being written shall be assumed to be $b_1 b_2 b_3$.). As 2 < 3, the largest m for which $a_m < a_{m+1}$ is $m = 2$. Then there is only one choice for b_2 and it equals a_3. Hence the next permutation in lexicographic order is 1 3 2.

Now the largest m for which $a_m < a_{m+1}$ is $m = 1$. The choices for b_1 are 3, 2 and the smaller of these being 2 and it being $> a_1, b_1 = 2$. Then the next two entries are 1, 3 in that order. Thus the next permutation is

2 1 3. Now the largest m for which $a_m < a_{m+1}$ is $m = 2$. Then there is only one choice for b_2 namely $b_2 = 3$. Therefore the next permutation is 2 3 1. Next, the largest m for which $a_m < a_{m+1}$ is $m = 1$. Then b_1 is the smaller of 1, 3 which is greater than 2. Therefore $b_1 = 3$. Hence the next permutation is 3 1 2. To find the next permutation, we find that the largest m for which $a_m < a_{m+1}$ is $m = 2$. Then there is only one choice for b_2 and, so, $b_2 = 2$. Therefore the next permutation in lexicographic order is 3 2 1. Now there being no m for which $a_m < a_{m+1}$, the process terminates and we have obtained all the permutations of 1, 2, 3. These are

$$123 \to 132 \to 213 \to 231 \to 312 \to 321.$$

Example 4.63. Apply the procedure given to generate all the permutations of 1, 2, 3, 4.

Solution. As in the last example, the permutation for which we have to find the next in lexicographic order we shall write as $a_1 \, a_2 \, a_3 \, a_4$. Also the starting permutation is 1 2 3 4. We obtain the permutations in the following table. An arrow between two permutations means that the permutation on the right is next to the permutation on the left in lexicographic order. Also m below an arrow is the value of the largest possible m with $a_m < a_{m+1}$ used to get the next permutation.

$1\,2\,3\,4 \xrightarrow{3} 1\,2\,4\,3 \xrightarrow{2} 1\,3\,2\,4 \xrightarrow{3} 1\,3\,4\,2 \xrightarrow{2} 1\,4\,2\,3 \xrightarrow{3} 1\,4\,3\,2 \xrightarrow{1}$

$2\,1\,3\,4 \xrightarrow{3} 2\,1\,4\,3 \xrightarrow{2} 2\,3\,1\,4 \xrightarrow{3} 2\,3\,4\,1 \xrightarrow{2} 2\,4\,1\,3 \xrightarrow{3} 2\,4\,3\,1 \xrightarrow{1}$

$3\,1\,2\,4 \xrightarrow{3} 3\,1\,4\,2 \xrightarrow{2} 3\,2\,1\,4 \xrightarrow{3} 3\,2\,4\,1 \xrightarrow{2} 3\,4\,1\,2 \xrightarrow{3} 3\,4\,2\,1 \xrightarrow{1}$

$4\,1\,2\,3 \xrightarrow{3} 4\,1\,3\,2 \xrightarrow{2} 4\,2\,1\,3 \xrightarrow{3} 4\,2\,3\,1 \xrightarrow{2} 4\,3\,1\,2 \xrightarrow{3} 4\,3\,2\,1.$

Our aim now is to give a procedure to generate all k-subsets of the set $\{1, 2, \ldots, n\}$ of order n. This procedure will automatically generate all k-subsets of any set $\{a_1, a_2, \ldots, a_n\}$ of order n- there being only a difference of symbols. Since the order of listing the elements of a set is immaterial, we agree to represent each subset of $\{1, 2, \ldots, n\}$ as a sequence with the elements in the subset arranged in increasing order. Then we arrange the sequences in lexicographic order.

Consider a subset $a_1 a_2 \ldots a_k$ of $A = \{1, 2, \ldots, n\}$ with $a_1 < a_2 < \ldots < a_k$. If $a_i = n - k + i$ for $i = 1, 2, \ldots, k$, then this k-subset of A is $n - k + 1, n - k + 2, \ldots, n$ and there is no k-subset of A which is next to this subset in lexicographic order.

Algorithm 4.2

Procedure for finding a k-subset next to $a_1a_2 \ldots a_k$ in lexicographic order.

If $a_i = n - k + i$ for every $i, 1 \leq i \leq k$, we can do nothing and so suppose that $a_i \neq n - k + i$ for some $i, 1 \leq i \leq k$.

Step 1. Choose the largest m with $a_m \neq n - k + m$. Then $a_m < n - k + m$.
Step 2. Let $b_i = a_i$ for every $i, 1 \leq i \leq m - 1$.
Step 3. Set $b_m = a_m + 1$.
Step 4. For $k \geq j > m$, set $b_j = b_{j-1} + 1 = a_m + j + 1 - m$.

Observe that as $a_m < n - k + m, b_k = a_m + k + 1 - m < n - k + m + k + 1 - m = n + 1$ so that we do stay within the set A.

Theorem 4.7. Let $a_1a_2 \ldots a_k$ be a k-subset of A with $a_i \neq n - k + i$ for some $i, 1 \leq i \leq k$ and $b_1b_2 \ldots b_k$ be a subset of A constructed by using the above procedure. Then $a_1a_2 \ldots a_k < b_1b_2 \ldots b_k$ in lexicographic order and there is no other subset of A which comes next to $a_1a_2 \ldots a_k$ and prior to $b_1b_2 \ldots b_k$.

Proof. That the subset $a = a_1a_2 \ldots a_k$ comes before $b = b_1b_2 \ldots b_k$ in lexicographic order is clear from the construction of $b = b_1b_2 \ldots b_k$. Suppose that $c = c_1c_2 \ldots c_k$ is a subset of A which comes before b in lexicographic order and after a. Since m is the largest with $a_m \neq n - k + m$, for every $j > m, a_j = n - k + j$. Now we have $(a_k = n - k + k = n)$

$$a: a_1 < \cdots < a_{m-1} < a_m < n - k + 1 + m < n - k + 2 + m < \cdots < n = a_k$$

$$c: c_1 < \cdots < c_{m-1} < c_m < c_{m+1} < c_{m+2} < \cdots < c_k$$

$$b: a_1 < \cdots < a_{m-1} < a_m + 1 < a_m + 2 < a_m + 3 < \cdots < a_m + k - m + 1.$$

Since $a < c < b$, it follows from the above that

$$a_i \leq c_i \leq a_i \quad \text{for all } i, 1 \leq i \leq m - 1.$$

Hence $c_1 = a_1, c_2 = a_2, c_{m-1} = a_{m-1}$.

Again, because of $a < c < b$, we have either $a_m < c_m$ or $a_m = c_m$ and either $c_m < a_m + 1$ or $c_m = a_m + 1$.

Since $a_m < c_m < a_m + 1$ cannot happen nor can $a_m = c_m = a_m + 1$ happen, we have either $a_m < c_m = a_m + 1$ or $a_m = c_m < a_m + 1$.

Case (i). Suppose that $a_m < c_m = a_m + 1$.

Now $c < b$ implies that there exists a $j \geq 0$ such that

$$c_{m+j} = a_m + j + 1 \quad \text{and} \quad c_{m+j+1} < a_m + j + 2.$$

But $a_m + j + 1 = c_{m+j} < c_{m+j+1} < a_m + j + 2$ cannot happen which implies that c_{m+j+1} cannot be less than $a_m + j + 2$. Hence

$$c_{m+j} = a_m + j + 1 \quad \text{for all } j \geq 0 \quad \text{and} \quad c = b.$$

Case (ii). Suppose that $c_m = a_m$. Again, since $a < c$, there exists an $i \geq 1$. such that $n - k + i < c_{m+i}$. Then $n - k + i + 1 < c_{m+i} + 1 \leq c_{m+i+1}$, which in turn implies that $n - k + i + 2 < c_{m+i+2}$. Continuing in this fashion (or otherwise by using induction) we find that $n = a_k < c_k$ which is a contradiction. Hence $n - k + i = c_{m+i}$ for all $i \geq 1$ and $c = a$.

We have thus proved that either $c = a$ or $c = b$. Hence there is no subset of A which comes before b and after a in lexicographic order.

We now use the above algorithm to find k-subsets of $A = \{1, 2, \ldots, n\}$ for some small values of n and k.

Example 4.64. Consider $A = \{1, 2, 3\}$ with $n = 3$. To find all 2-subsets of A, we have $k = 2$ so that $n - k = 1$. We start with the subset 1 2. Now for $m = 2, n - k + m = 1 + 2 = 3 \neq 2 = a_2$. Therefore, the next subset is 1 3.

For $m = 2$ again, $n - k + m = 1 + 2 = 3 = a_2$ while for $m = 1$, $n - k + m = 1 + 1 = 2 > 1 = a_1$. Therefore, the next subset in order is 2 3.

Now $a_2 = 3 = 1 + 2 = n - k + m$ and $a_1 = 2 = 1 + 1 = n - k + m$ so that the process terminates here. Hence the 2-subsets of A are 1 2, 1 3, 2 3 or $\{1, 2\}, \{1, 3\}, \{2, 3\}$.

Example 4.65. Consider $A = \{1, 2, 3, 4\}$

(a) To find all 2-subsets of A, we have $n - k = 4 - 2 = 2$. We start with the first subset 1 2. Now for $m = 2, n - k + m = 2 + 2 = 4 > 2 = a_2$ Therefore the next subset in order is 1 3.

Again, for $m = 2, n - k + m = 2 + 2 = 4 > 3 = a_2$ and the next subset in order is 1 4.

In this case for $m = 2, n - k + m = 2 + 2 = 4 = a_2$ and for $m = 1, n - k + m = 2 + 1 = 3 > 1 = a_1$. Thus the next subset in order is 2 3.

Here $a_2 = 3 < 2 + 2 = n - k + m$ and, so, the next subset in order is 2 4.

Now $a_2 = 4 = 2 + 2 = n - k + m$ but $a_1 = 2 < 2 + 1 = n - k + m$ and the next subset in order is 3 4.

Since $a_1 = 3 = n - k + m = 2 + 1$, and $a_2 = 4 = n - k + m = 2 + 2$, the process terminates here. Hence all the 2-subsets of A are

$$\{1, 2\}, \{1, 3\}, \{1, 4\}, \{2, 3\}, \{2, 4\}, \{3, 4\}.$$

(b) To find all 3-subsets of A, $k = 3$ and $n - k = 4 - 3 = 1$. The starting point for generating 3-subsets of A is the subset 1 2 3. Here $a_3 = 3 < 1 + 3 = n - k + m$ and, so, the next subset is 1 2 4.

In this case $a_3 = 4 = 1 + 3 = n - k + m$ but $a_2 = 2 < 1 + 2 = n - k + m$ and the next subset in order is 1 3 4.

Observe that a_3 again equals $n - k + m, a_2 = 3 = 1 + 2 = n - k + m$, but $a_1 = 1 < 1 + 1 = n - k + m$ so that the next subset in order is 2 3 4.

Since $a_1 = 2 = n - k + 1, a_2 = 3 = n - k + 2$ and $a_3 = 4 = n - k + 3$, the process terminates here. Hence all the 3-subsets of A are

$$\{1, 2, 3\}, \{1, 2, 4\}, \{1, 3, 4\}, \{2, 3, 4\}.$$

Example 4.66. Consider $A = \{1, 2, 3, 4, 5\}$. We find all 2-subsets, 3-subsets and 4-subsets of A. We obtain these in the tabular form indicating the value of largest possible m with $a_m < n - k + m$ below the arrow between two subsets which are written in lexicographic order.

(a) All 2-subsets of A are $(n - k = 5 - 2 = 3)$

$$1\,2 \ \overset{2}{\rightarrow}\ 1\,3 \ \overset{2}{\rightarrow}\ 1\,4 \ \overset{2}{\rightarrow}\ 1\,5 \ \overset{1}{\rightarrow}\ 2\,3 \ \overset{2}{\rightarrow}\ 2\,4 \ \overset{2}{\rightarrow}\ 2\,5 \ \overset{1}{\rightarrow}\ 3\,4 \ \overset{2}{\rightarrow}\ 3\,5 \ \overset{1}{\rightarrow}\ 4\,5 \quad \text{or}$$

$$\{1, 2\}, \{1, 3\}, \{1, 4\}, \{1, 5\}, \{2, 3\}\{2, 4\}, \{2, 5\}, \{3, 4\}, \{3, 5\}, \{4, 5\}.$$

(b) All 3-subsets of A are $(n - k = 5 - 3 = 2)$

$$1\,2\,3 \ \overset{3}{\rightarrow}\ 1\,2\,4 \ \overset{3}{\rightarrow}\ 1\,2\,5 \ \overset{2}{\rightarrow}\ 1\,3\,4 \ \overset{3}{\rightarrow}\ 1\,3\,5 \ \overset{2}{\rightarrow}\ 1\,4\,5 \ \overset{1}{\rightarrow}\ 2\,3\,4 \ \overset{3}{\rightarrow}$$

$$2\,3\,5 \ \overset{2}{\rightarrow}\ 2\,4\,5 \ \overset{1}{\rightarrow}\ 3\,4\,5 \quad \text{or}$$

$$\{1, 2, 3\}, \{1, 2, 4\}, \{1, 2, 5\}, \{1, 3, 4\}, \{1, 3, 5\}, \{1, 4, 5\}, \{2, 3, 4\}, \{2, 3, 5\},$$
$$\{2, 4, 5\}, \{3, 4, 5\}.$$

(c) All 4-subsets of A are $(n - k = 5 - 4 = 1)$

$$1\,2\,3\,4 \ \overset{4}{\rightarrow}\ 1\,2\,3\,5 \ \overset{3}{\rightarrow}\ 1\,2\,4\,5 \ \overset{2}{\rightarrow}\ 1\,3\,4\,5 \ \overset{1}{\rightarrow}\ 2\,3\,4\,5 \quad \text{or}$$

$$\{1, 2, 3, 4\}, \{1, 2, 3, 5\}, \{1, 2, 4, 5\}, \{1, 3, 4, 5\}, \{2, 3, 4, 5\}.$$

Example 4.67. Consider the set $A = \{1,2,3,4,5,6\}$.

(a) All 2-subsets of A are (here $n - k = 6 - 2 = 4$)

$$1\,2 \ \overset{2}{\to}\ 1\,3 \ \overset{2}{\to}\ 1\,4 \ \overset{2}{\to}\ 1\,5 \ \overset{2}{\to}\ 1\,6 \ \overset{1}{\to}\ 2\,3 \ \overset{2}{\to}\ 2\,4 \ \overset{2}{\to}\ 2\,5 \ \overset{2}{\to}\ 2\,6 \ \overset{1}{\to}$$

$$3\,4 \ \overset{2}{\to}\ 3\,5 \ \overset{2}{\to}\ 3\,6 \ \overset{1}{\to}\ 4\,5 \ \overset{2}{\to}\ 4\,6 \ \overset{1}{\to}\ 5\,6 \quad \text{or}$$

$\{1,2\}, \{1,3\}, \{1,4\}, \{1,5\}, \{1,6\}, \{2,3\}, \{2,4\}, \{2,5\}, \{2,6\}, \{3,4\},$
$\{3,5\}, \{3,6\}, \{4,5\}, \{4,6\}, \{5,6\}.$

(b) All 3-subsets of the set A are (here $n - k = 6 - 3 = 3$)

$$1\,2\,3 \ \overset{3}{\to}\ 1\,2\,4 \ \overset{3}{\to}\ 1\,2\,5 \ \overset{3}{\to}\ 1\,2\,6 \ \overset{2}{\to}\ 1\,3\,4 \ \overset{3}{\to}\ 1\,3\,5 \ \overset{3}{\to}\ 1\,3\,6 \ \overset{2}{\to}$$

$$1\,4\,5 \ \overset{3}{\to}\ 1\,4\,6 \ \overset{2}{\to}\ 1\,5\,6 \ \overset{1}{\to}\ 2\,3\,4 \ \overset{3}{\to}\ 2\,3\,5 \ \overset{3}{\to}\ 2\,3\,6 \ \overset{2}{\to}\ 2\,4\,5 \ \overset{3}{\to}$$

$$2\,4\,6 \ \overset{2}{\to}\ 2\,5\,6 \ \overset{1}{\to}\ 3\,4\,5 \ \overset{3}{\to}\ 3\,4\,6 \ \overset{2}{\to}\ 3\,5\,6 \ \overset{1}{\to}\ 4\,5\,6 \quad \text{or}$$

$\{1,2,3\}, \{1,2,4\}, \{1,2,5\}, \{1,2,6\}, \{1,3,4\}, \{1,3,5\}, \{1,3,6\},$
$\{1,4,5\}, \{1,4,6\}, \{1,5,6\}, \{2,3,4\}, \{2,3,5\}, \{2,3,6\}, \{2,4,5\},$
$\{2,4,6\}, \{2,5,6\}, \{3,4,5\}, \{3,4,6\}, \{3,5,6\}, \{4,5,6\}.$

(c) All 4-subsets of A are $(n - k = 6 - 4 = 2)$

$$1\,2\,3\,4 \ \overset{4}{\to}\ 1\,2\,3\,5 \ \overset{4}{\to}\ 1\,2\,3\,6 \ \overset{3}{\to}\ 1\,2\,4\,5 \ \overset{4}{\to}\ 1\,2\,4\,6 \ \overset{3}{\to}\ 1\,2\,5\,6 \ \overset{2}{\to}$$

$$1\,3\,4\,5 \ \overset{4}{\to}\ 1\,3\,4\,6 \ \overset{3}{\to}\ 1\,3\,5\,6 \ \overset{2}{\to}\ 1\,4\,5\,6 \ \overset{1}{\to}\ 2\,3\,4\,5 \ \overset{4}{\to}\ 2\,3\,4\,6 \ \overset{3}{\to}$$

$$2\,3\,5\,6 \ \overset{2}{\to}\ 2\,4\,5\,6 \ \overset{1}{\to}\ 3\,4\,5\,6 \quad \text{or}$$

$\{1,2,3,4\}, \{1,2,3,5\}, \{1,2,3,6\}, \{1,2,4,5\}, \{1,2,4,6\},$
$\{1,2,5,6\}, \{1,3,4,5\}, \{1,3,4,6\}, \{1,3,5,6\}, \{1,4,5,6\},$
$\{2,3,4,5\}, \{2,3,4,6\}, \{2,3,5,6\}, \{2,4,5,6\}, \{3,4,5,6\}.$

(d) All 5-subsets of A are (here $n - k = 6 - 5 = 1$)

$$1\,2\,3\,4\,5 \ \overset{5}{\to}\ 1\,2\,3\,4\,6 \ \overset{4}{\to}\ 1\,2\,3\,5\,6 \ \overset{3}{\to}\ 1\,2\,4\,5\,6 \ \overset{2}{\to}\ 1\,3\,4\,5\,6 \ \overset{1}{\to}$$

$$2\,3\,4\,5\,6 \quad \text{or}$$

$\{1,2,3,4,5\}, \{1,2,3,4,6\}, \{1,2,3,5,6\}, \{1,2,4,5,6\},$
$\{1,3,4,5,6\} \quad \text{and} \quad \{2,3,4,5,6\}.$

A Course in Discrete Mathematical Structures

Example 4.68. We next list all the 4-subsets of $A = \{1, 2, 3, 4, 5, 6, 7\}$. Observe that here $n - k = 7 - 4 = 3$. The required subsets are

$$1\,2\,3\,4 \overset{\rightarrow}{}^{4} \; 1\,2\,3\,5 \overset{\rightarrow}{}^{4} \; 1\,2\,3\,6 \overset{\rightarrow}{}^{4} \; 1\,2\,3\,7 \overset{\rightarrow}{}^{3} \; 1\,2\,4\,5 \overset{\rightarrow}{}^{4} \; 1\,2\,4\,6 \overset{\rightarrow}{}^{4}$$

$$1\,2\,4\,7 \overset{\rightarrow}{}^{3} \; 1\,2\,5\,6 \overset{\rightarrow}{}^{4} \; 1\,2\,5\,7 \overset{\rightarrow}{}^{3} \; 1\,2\,6\,7 \overset{\rightarrow}{}^{2} \; 1\,3\,4\,5 \overset{\rightarrow}{}^{4} \; 1\,3\,4\,6 \overset{\rightarrow}{}^{4}$$

$$1\,3\,4\,7 \overset{\rightarrow}{}^{3} \; 1\,3\,5\,6 \overset{\rightarrow}{}^{4} \; 1\,3\,5\,7 \overset{\rightarrow}{}^{3} \; 1\,3\,6\,7 \overset{\rightarrow}{}^{2} \; 1\,4\,5\,6 \overset{\rightarrow}{}^{4} \; 1\,4\,5\,7 \overset{\rightarrow}{}^{3}$$

$$1\,4\,6\,7 \overset{\rightarrow}{}^{2} \; 1\,5\,6\,7 \overset{\rightarrow}{}^{1} \; 2\,3\,4\,5 \overset{\rightarrow}{}^{4} \; 2\,3\,4\,6 \overset{\rightarrow}{}^{4} \; 2\,3\,4\,7 \overset{\rightarrow}{}^{3} \; 2\,3\,5\,6 \overset{\rightarrow}{}^{4}$$

$$2\,3\,5\,7 \overset{\rightarrow}{}^{3} \; 2\,3\,6\,7 \overset{\rightarrow}{}^{2} \; 2\,4\,5\,6 \overset{\rightarrow}{}^{4} \; 2\,4\,5\,7 \overset{\rightarrow}{}^{3} \; 2\,4\,6\,7 \overset{\rightarrow}{}^{2} \; 2\,5\,6\,7 \overset{\rightarrow}{}^{1}$$

$$3\,4\,5\,6 \overset{\rightarrow}{}^{4} \; 3\,4\,5\,7 \overset{\rightarrow}{}^{3} \; 3\,4\,6\,7 \overset{\rightarrow}{}^{2} \; 3\,5\,6\,7 \overset{\rightarrow}{}^{1} \; 4\,5\,6\,7.$$

The subsets we get number $35 = C(7, 4)$.

(Please note that some of the solved examples in this chapter are taken from the unsolved exercises in Kolman, Busby and Ross (2005), Liu (2000) and Rosen (2003/ 2005)).

Exercise 4.4.

1. Find the number of permutations of $\{a, b, c, d, e, f, g\}$ that
   (i) end with b, (ii) begin with b.
2. Find the values of each of the quantities
   (a) $P(7,3)$ (b) $P(5,4)$ (c) $P(7,7)$ (d) $P(9,6)$ (e) $C(5,2)$ (f) $C(6,3)$
   (g) $C(8,5)$ (h) $C(8,3)$ (i) $C(5,3)$.
3. Nine runners take part in 400 meter race. If all orders of a finish are possible, find the number of possibilities of the first three positions.
4. Find the number of bit strings of length 8 that contain
   (a) exactly three 1s, (b) at most four 1s, (c) at least three 1s.
5. Find the number of bit strings of length 9 that contain
   (a) exactly four 1s, (b) at most four 1s, (c) at least four 1s.
6. Find the number of ways to make n men and n women stand in a row

   (i) if the men and women alternate,
   (ii) if the men and women alternate and the first person in every row is a man,
   (iii) if the men and women alternate and the last person in every row is a man.

7. Find r if

   (i) $5\ P(4,r) = 6P(5,r-1)$, (ii) $P(5,r) = 2P(6,r-1)$,

   (iii) $P(5,r) = P(6,r-1)$, (iv) $P(19,r) = P(20,r-1)$,

   (v) $P(11,r) = P(12,r-1)$.

8. Find n if

   (i) $P(2n,3) = 60P(n,2)$, (ii) $P(2n,3) = 68P(n,2)$,

   (iii) $P(2n,3) = 92P(n,2)$, (iv) $2P(5,3) = P(n,4)$.

9. Prove that $P(n,n) = 2P(n,n-2)$ for all $n > 1$.

10. Prove that the product of any k consecutive positive integers is divisible by $k!$.

11. A polygon has 54 diagonals, find the number of its sides.

12. In an examination paper part A contains five questions and part B contains three questions. A student has to attempt five questions selecting at least one from each part. In how many ways can he attempt the questions (without regard to the order in which the questions are answered)?

13. If seven fair coins are tossed and the results recorded, how many

    (a) outcome recorded sequences are possible?

    (b) recorded sequences contain exactly 4 heads?

14. Five unbiased dice are rolled and the numbers showing up on top are recorded. Find the number of

    (a) all possible record sequences,

    (b) all possible record sequences containing exactly one six,

    (c) all possible record sequences containing exactly three fives.

15. In the discrete mathematics paper, there are four units each containing two questions. The candidates are required to answer five questions selecting at least one from each unit. In how many ways can the student answer five questions without regard to the order in which the questions are answered.

16. In a question paper part A contains eight questions and part B contains four questions. The candidates are required to answer six questions but not all from part A. Find the number of ways that the choice can be made by a candidate.

17. An urn contains 18 balls of which 10 are red and 8 are white. In how many ways can six balls be chosen so that
    (a) all six are red? (b) all six are white? (c) two are red and four are white?

(d) three are red and three are white? (e) at least three are red? (f) at most three are red?

18. An objective type paper contains 30 true/false questions. Of these 13 statements are true and the rest are false. If the questions can be written in any order, find the number of different answer keys that are possible.

19. Find the number of bit strings of length 8 that contain at least three 1s and at least four 0 s.

20. Find the number of license plates consisting of two letters followed by four digits that contain no letter or digit twice.

21. Find the number of license plates consisting of two letters followed by six digits that contain no letter twice.

22. Prove that the number of ways to select n objects from $3n$ objects of which n are identical and the rest are all different is

$$2^{2n-1} + \frac{1}{2}\frac{1}{2}\frac{(2n)!}{(n!)^2}.$$

23. Find the number of ways of seating three girls and nine boys in two vans, each having numbered seats, three in the front and four in the back. Also find the number of possible seating arrangements if three girls sit together in the back row on adjacent seats.

24. Find the number of numbers of five digits that can be formed with the digits 1, 2, 3 each of which can be used at most twice in a number.

25. Prove that the number of ways to distribute n different things between two persons one receiving p things and the other receiving q things, where

$$p + q = n \text{ is } \frac{n!}{p!q!}.$$

26. An AIEEE question paper contains 20 multiple choice questions, each with possible answers a, b, c, d. If the number of questions with a, b, c and d as their answers is 8, 3, 4 and 5 respectively, find the number of different answer keys that are possible, if the questions can be placed in any order.

Chapter 5

RECURRENCE RELATIONS

This chapter is devoted to a study of recurrence relations. Homogeneous, particular and total solutions of linear recurrence relations are discussed. For solving recurrence relations by the method of generating functions that are discussed here, we need to resolve given rational functions into partial fractions. Resolving rational functions into partial fractions is discussed in the beginning of the chapter.

5.1. Partial Fractions

5.1.1. *Rational functions*

Recall that numbers which can be expressed in the form $\frac{p}{q}$, where p, q are integers and $q \neq 0$ are called rational numbers. If p, q are functions in a variable x instead of being integers, then the fractions $\frac{p}{q}$ are called **rational functions**. Thus a rational function in the variable x is a quotient $\frac{f(x)}{g(x)}$, where $f(x)$ and $g(x)$ are polynomials in x and $g(x) \neq 0$. For example,

$$\frac{2x+1}{3x^2+4x+1}, \quad \frac{3x^2+2x+1}{4x^3+3x+2}, \quad \frac{x^3+1}{x^2+2x+1}$$

are rational functions in x.

A rational function $\frac{f(x)}{g(x)}$ is called a proper rational fraction if the degree of the numerator $f(x)$ is less than the degree of the denominator $g(x)$. On the other hand, if the degree of the numerator $f(x)$ is greater than the degree of the denominator $g(x)$, then the rational function is called an improper rational function. For example,

$$\frac{2x^2 - 3}{4x^3 + 3x + 1}, \quad \frac{2x}{3x^2 + x - 4}, \quad \frac{x^3 - 3x + 2}{x^4 - x^3 + 2x^2 - 5}$$

are proper functions, whereas

$$\frac{2x^2 - 5}{x + 2}, \quad \frac{2x^2 + 3x - 4}{x^2 + 2x}, \quad \frac{3x^5 - 4x^2 - 2x + 1}{x^4 + 3x^2 - 2x + 7}$$

are improper functions. By dividing the numerator by the denominator, an improper fraction can always be written as the sum of a polynomial in x and a proper fraction in x. Thus

$$\frac{2x^2 - 5}{x + 2} = 2x - 4 + \frac{3}{x + 2}; \quad \frac{2x^2 + 3x - 4}{x^2 + 2x} = 2 - \frac{x + 4}{x^2 + 2x};$$

$$\frac{3x^5 - 4x^2 - 2x + 1}{x^4 + 3x^2 - 2x + 7} = 3x - \frac{9x^3 - 2x^2 + 23x - 1}{x^4 + 3x^2 - 2x + 7}.$$

5.1.2. *Partial fractions*

When a proper fraction is expressed as the sum of other rational fractions, such that the degree of the numerator of the new rational fraction is strictly less than the degree of the corresponding denominator of the rational fraction, and the denominator is either an irreducible polynomial or a power of such a polynomial, the new rational fractions which add up to the original proper fraction are called partial fractions.

A basis for solving a given rational fraction into partial fractions is provided by the following theorem:

Theorem 5.1. If a polynomial of degree $\le n$ has $n + 1$ or more distinct roots, then the coefficients of all the powers of x in the polynomial are zero and the polynomial is identically zero.

Proof. Consider a polynomial $a_0 x^n + a_1 x^{n-1} + a_2 x^{n-2} + \cdots + a_{n-1} x + a_n$ where $a_0, a_1, a_2, \ldots, a_n$ are constants. Let $\alpha_1, \alpha_2, \ldots, \alpha_{n+1}$ be $n + 1$ distinct roots of this polynomial. Then we have

$$a_0 \alpha_1^n + a_1 \alpha_1^{n-1} + \cdots + a_{n-1} \alpha_1 + a_n = 0$$

$$a_0 \alpha_2^n + a_1 \alpha_2^{n-1} + \cdots + a_{n-1} \alpha_2 + a_n = 0$$

$$\cdots$$

$$a_0 \alpha_{n+1}^n + a_1 \alpha_{n+1}^{n-1} + \cdots + a_{n-1} \alpha_{n+1} + a_n = 0.$$

In matrix notation, this system of $n + 1$ equations in the variable $a_0, a_1, a_2, \ldots, a_n$ takes the form $AX = 0$, where

$$A = \begin{pmatrix} \alpha_1^n & \alpha_1^{n-1} & \cdots & \alpha_1 & 1 \\ \alpha_2^n & \alpha_2^{n-1} & \cdots & \alpha_2 & 1 \\ \cdots\cdots\cdots\cdots\cdots\cdots\cdots\cdots \\ \alpha_{n+1}^n & \alpha_{n+1}^{n-1} & \cdots & \alpha_{n+1} & 1 \end{pmatrix}, \quad X = \begin{pmatrix} a_0 \\ a_1 \\ a_2 \\ \cdot \\ \cdot \\ \cdot \\ a_n \end{pmatrix}.$$

This system of homogeneous linear equations has a non-zero solution if and only if $\det A = 0$. (We discuss systems of linear equations in Chapter 13.) Observe that $\det A = 0$ is a polynomial of degree

$$n + (n - 1) + \cdots + 2 + 1 = \frac{n(n + 1)}{2} \text{ in } \alpha_1, \alpha_2, \ldots, \alpha_{n+1}.$$

Also, taking α_j equal to α_i for any $i, i \neq j$, in $\det A$ makes two rows of the determinant identical, hence the resulting determinant is zero. Therefore, $\alpha_j - \alpha_i$ for all $i \neq j$ are factors of $\det A$. Since no two of these differences are equal, it follows that

$$\prod_{1 \leq i < j \leq n+1} (\alpha_i - \alpha_j)$$

is a factor of $\det A$. The number of factors in

$$\prod_{1 \leq i < j \leq n+1} (\alpha_i - \alpha_j) \text{ is } n + (n - 1) + \cdots + 2 + 1 = \frac{n(n + 1)}{2}.$$

Hence

$$\deg \prod_{1 \leq i < j \leq n+1} (\alpha_i - \alpha_j) = \frac{n(n + 1)}{2} = \deg \ \det A \text{ and } \prod_{1 \leq i < j \leq n+1} (\alpha_i - \alpha_j)$$

divides $\det A$. Also, the coefficient of the highest degree term in $\det A$ is one, while the coefficients of various degree terms in

$$\prod_{1 \leq i < j \leq n+1} (\alpha_i - \alpha_j)$$

are integers. Therefore,

$$\det A = \pm \prod_{1 \leq i < j \leq n+1} (\alpha_i - \alpha_j).$$

Since $\alpha_i \neq \alpha_j$ for any $l \neq j$,

$$\prod_{1 \leq i < j \leq n+1} (\alpha_i - \alpha_j) \neq 0.$$

Thus, $\det A \neq 0$ and the given system of equations has only the zero solution and $a_0 = a_1 = a_2 = \cdots = a_n = 0$.

Corollary. If two polynomials of degree n in the same variable x are equal for more than n values of the variable x, then coefficients of the like powers of x in the two polynomials are equal, it being understood that if a term is missing in either of the two polynomials, the same is considered present with coefficient zero.

Consider a given proper fraction $\frac{f(x)}{g(x)}$.
Suppose that

$$\frac{f(x)}{g(x)} = \frac{f_1(x)}{g_1(x)} + \frac{f_2(x)}{g_2(x)} + \cdots + \frac{f_k(x)}{g_k(x)}, \qquad (5.1)$$

where $\frac{f_i(x)}{g_i(x)}, 1 \leq i \leq k$, are partial fractions such that every $g_i(x)$ is a factor of $g(x)$. Multiplying both sides of (5.1) by $g(x)$, we get

$$f(x) = f_1(x)h_1(x) + f_2(x)h_2(x) + \cdots + f_k(x)h_k(x), \qquad (5.2)$$

where $h_i(x) = \frac{g(x)}{g_i(x)}$ are polynomials in x. Since $\frac{f(x)}{g(x)}$ is the sum of the partial fractions on the right-hand side of (5.1), the relation (5.1) is true for all values of x. Therefore, the relation (5.2) is also true for all values of x. Thus, if $f(x)$ and the polynomial on the right-hand side of (5.2) are of degree n, say (by introducing zero coefficients, if necessary, we can assume the two polynomials to be of the same degree — strictly speaking we cannot call the degree of a polynomial n if the coefficient of the highest power x^n of x is 0), then the two polynomials take the same value for $\geq n + 1$ values of x. Hence by the above corollary, coefficients of the like powers of x on the two sides of relation (5.2) are equal.

5.1.3. Procedure for resolving into partial fractions

We will now describe a procedure for resolving a proper fraction $\frac{f(x)}{g(x)}$ into partial fraction in several steps. We do not give a single general procedure/process.

1. (a) Suppose that $g(x)$ is the product of distinct linear factors, say $g(x) = (a_1 x + b_1)(a_2 x + b_2) \cdots (a_t x + b_t)$, then we write

$$\frac{f(x)}{g(x)} = \frac{A_1}{a_1 x + b_1} + \frac{A_2}{a_2 x + b_2} + \cdots + \frac{A_t}{a_t x + b_t},$$

where $A_1, A_2, \ldots, A_t$ are constants.

(b) When $g(x)$ is a product of linear factors not all distinct, say $g(x) = (a_1 x + b_1)^r (a_2 x + b_2)^s \cdots (a_t x + b_t)^u$, then we write

$$\frac{f(x)}{g(x)} = \frac{A_1}{a_1 x + b_1} + \frac{A_2}{(a_1 x + b_1)^2} + \cdots + \frac{A_r}{(a_1 x + b_1)^r}$$

$$+ \frac{B_1}{a_2 x + b_2} + \frac{B_2}{(a_2 x + b_2)^2} + \cdots + \frac{Bs}{(a_2 x + b_2)^s} + \cdots$$

$$+ \frac{C_1}{a_t x + b_t} + \frac{C_2}{(a_t x + b_t)^2} + \cdots + \frac{C_u}{(a_t x + b_t)^u}.$$

For example, we write

$$\frac{x^3 - 2x^2 + 4x - 5}{(x-1)(x+2)(2x+1)(3x+5)} = \frac{A}{x-1} + \frac{B}{x+2} + \frac{C}{2x+1} + \frac{D}{3x+5},$$

$$\frac{x^3 - 2x^2 + 4x - 5}{(x+1)^2(2x+3)^3} = \frac{A}{x+1} + \frac{B}{(x+1)^2} + \frac{C}{2x+3}$$

$$+ \frac{D}{(2x+3)^2} + \frac{E}{(2x+3)^3}$$

$$\frac{x^3 + 2x^2 - 3x + 7}{(-x+1)(2x+1)^2(3x-4)^2} = \frac{A}{-x+1} + \frac{B}{2x+1} + \frac{C}{(2x+1)^2}$$

$$+ \frac{D}{3x-4} + \frac{E}{(3x-4)^2}.$$

2. (a) When $g(x)$ is a product of distinct quadratic polynomials, say $g(x) = (a_1 x^2 + b_1 x + c_1)(a_2 x^2 + b_2 x + c_2) \cdots (a_t x^2 + b_t x + c_t)$, then we write

$$\frac{f(x)}{g(x)} = \frac{A_1 x + B_1}{a_1 x^2 + b_1 x + c_1} + \frac{A_2 x + B_2}{a_2 x^2 + b_2 x + c_2} + \cdots + \frac{A_t x + B_t}{a_t x^2 + b_t x + c_t}.$$

(b) When $g(x)$ is a product of quadratic polynomials not all distinct, say $g(x) = (a_1 x^2 + b_1 x + c_1)(a_2 x^2 + b_2 x + c_2)^2$, then we write

$$\frac{f(x)}{g(x)} = \frac{A_1 x + B_1}{a_1 x^2 + b_1 x + c_1} + \frac{A_2 x + B_2}{a_2 x^2 + b_2 x + c_2} + \frac{A_3 x + B_3}{(a_2 x^2 + b_2 x + c_2)^2}.$$

For example, we write

$$\frac{x^3 - 2x^2 + 4x - 5}{(x^2 + x + 1)(2x^2 + 3x + 5)} = \frac{A_1x + B_1}{x^2 + x + 1} + \frac{A_2x + B_2}{2x^2 + 3x + 5}$$

$$\frac{x^3 + x^2 - 3x + 7}{(x^2 - x + 1)(x^2 - 2x + 3)^2} = \frac{A_1x + B_1}{x^2 - x + 1} + \frac{A_2x + B_2}{x^2 - 2x + 3}$$

$$+ \frac{A_3x + B_3}{(x^2 - 2x + 3)^2}.$$

3. (a) When $g(x)$ is a product of some linear factors, some quadratic polynomials, say $g(x) = (ax+b)^r(a_1x^2+b_1x+c_1)^s(a_2x^2+b_2x+c_2)^t$, then we write

$$\frac{f(x)}{g(x)} = \frac{A_1}{ax + b} + \frac{A_2}{(ax + b)^2} + \cdots + \frac{A_r}{(ax + b)^r} + \frac{B_1x + C_1}{a_1x^2 + b_1x + c_1}$$

$$+ \frac{B_2x + C_2}{(a_1x^2 + b_1x + c_1)^2} + \cdots + \frac{B_sx + C_s}{(a_1x^2 + b_1x + c_1)^s}$$

$$+ \frac{D_1x + E_1}{a_2x^2 + b_2x + c_2} + \frac{D_2x + E_2}{(a_2x^2 + b_2x + c_2)^2}$$

$$+ \cdots + \frac{D_tx + E_t}{(a_2x^2 + b_2x + c_2)^t}.$$

(b) When $g(x)$ is a product of some linear polynomials, some quadratic polynomials, some cubic polynomials etc, say $g(x) = (ax+b)(a_1x^2+b_1x+c_1)(a_2x^3+b_2x^2+c_2x+d_2)$, then we write

$$\frac{f(x)}{g(x)} = \frac{A}{ax + b} + \frac{Bx + C}{a_1x^2 + b_1x + c_1} + \frac{Dx^2 + Ex + F}{(a_2x^3 + b_2x^2 + c_2x + d_2)}.$$

For example, we write

$$\frac{x^4 - x^3 + 3x^2 + 7}{(2x + 3)(x^2 + x + 1)(x^3 + 2x^2 - 7x + 3)}$$

$$= \frac{A}{2x + 3} + \frac{Bx + C}{x^2 + x + 1} + \frac{Dx^2 + Ex + F}{x^3 + 2x^2 - 7x + 3};$$

$$\frac{x^4 + x^3 - 3x^2 + 5}{(x + 3)^2(x^3 + x^2 + x - 1)} = \frac{A}{x + 3} + \frac{B}{(x + 3)^2} + \frac{Cx^2 + Dx + E}{x^3 + x^2 + x - 1}.$$

5.1.4. *Some solved examples*

Example 5.1. Resolve into partial fractions $\frac{12x+11}{x^2+x-6}$.

Solution. Observe that $x^2 + x - 6 = (x - 2)(x + 3)$. Therefore,

$$\frac{12x + 11}{x^2 + x - 6} = \frac{12x + 11}{(x - 2)(x + 3)}.$$

Let

$$\frac{12x + 11}{x^2 + x - 6} = \frac{A}{x - 2} + \frac{B}{x + 3}.$$

Multiplying both sides with $x^2 + x - 6$, we get

$$12x + 11 = A(x + 3) + B(x - 2).$$

which is an identity in x. Comparing the coefficients of x and x^0 on both sides, we get

$$A + B = 12, \quad 3A - 2B = 11.$$

Solving these relations for A and B we get
$2A + 2B + 3A - 2B = 24 + 11$ or $5A = 35$, i.e. $A = 7$. Then $B = 12 - A = 12 - 7 = 5$
Hence

$$\frac{12x + 11}{x^2 + x - 6} = \frac{7}{x - 2} + \frac{5}{x + 3}.$$

Example 5.2. Resolve into partial fractions $\frac{10x^2+9x-7}{(x+2)(x^2-1)}$.

Solution. Observe that $x^2 - 1 = (x - 1)(x + 1)$. Therefore,

$$\frac{10x^2 + 9x - 7}{(x + 2)(x^2 - 1)} = \frac{10x^2 + 9x - 7}{(x + 2)(x - 1)(x + 1)}.$$

Let

$$\frac{10x^2 + 9x - 7}{(x + 2)(x^2 - 1)} = \frac{10x^2 + 9x - 7}{(x + 2)(x - 1)(x + 1)}$$

$$= \frac{A}{x + 2} + \frac{B}{x - 1} + \frac{C}{x + 1}.$$

Multiplying both sides with $(x + 2)(x^2 - 1)$, we get

$$10x^2 + 9x - 7 = A(x - 1)(x + 1) + B(x + 2)(x + 1) + C(x + 2)(x - 1)$$
$$= A(x^2 - 1) + B(x^2 + 3x + 2) + C(x^2 + x - 2).$$

Since this is an identity in x, comparing the coefficients of x^2, x and the constant terms on both sides of this relation, we get

$$A + B + C = 10, \tag{5.3}$$

$$3B + C = 9, \tag{5.4}$$

$$-A + 2B - 2C = -7. \tag{5.5}$$

We have to solve these relations for A, B, C. Now

$$A + B + C + (-A + 2B - 2C) = 10 - 7,$$

$$\text{or } 3B - C = 3. \tag{5.6}$$

Adding (5.4) and (5.6),

$$3B - C + 3B + C = 12 \quad \text{or} \quad 6B = 12 \text{ i.e. } B = 2.$$

Substituting for B in (5.6), we get $C = 3$. Putting the value of B and C in (5.3), we get $A = 5$.

   Hence

$$\frac{10x^2 + 9x - 7}{(x + 2)(x^2 - 1)} = \frac{5}{x + 2} + \frac{2}{x - 1} + \frac{3}{x + 1}.$$

Example 5.3. Resolve into partial fractions $\frac{7x-2}{x^3-x^2-2x}$.

Solution. Observe that $x^3 - x^2 - 2x = x(x^2 - x - 2) = x(x + 1)(x - 2)$. Therefore,

$$\frac{7x - 2}{x^3 - x^2 - 2x} = \frac{7x - 2}{x(x + 1)(x - 2)}.$$

Let

$$\frac{7x - 2}{x^3 - x^2 - 2x} = \frac{A}{x} + \frac{B}{x + 1} + \frac{C}{x - 2}.$$

Multiplying both sides with $x^3 - x^2 - 2x$, we get

$$7x - 2 = A(x + 1)(x - 2) + Bx(x - 2) + Cx(x + 1)$$
$$= A(x^2 - x - 2) + B(x^2 - 2x) + C(x^2 + x).$$

This being an identity in x, we can compare coefficients of like powers of x on both sides and get

$$A + B + C = 0 \tag{5.7}$$

$$-A - 2B + C = 7 \tag{5.8}$$

$$-2A = -2. \tag{5.9}$$

The relation (5.9) gives $A = 1$. Putting this value of A in (5.7) and (5.8), we get

$$B + C = -1 \quad \text{and} \quad -2B + C = 8.$$

Therefore, $3B = -9$ i.e. $B = -3$. Then $C = -1 - B = -1 + 3 = 2$. Hence

$$\frac{7x - 2}{x^3 - x^2 - 2x} = \frac{1}{x} - \frac{3}{x+1} + \frac{2}{x-2}.$$

Example 5.4. Resolve into partial fractions $\frac{10x^2 + 9x - 2}{(x-1)^2(x+2)^2}$.

Solution. Let

$$\frac{10x^2 + 9x - 2}{(x-1)^2(x+2)^2} = \frac{A}{x-1} + \frac{B}{(x-1)^2} + \frac{C}{x+2} + \frac{D}{(x+2)^2}.$$

Multiply both sides with $(x-1)^2(x+2)^2$. We get

$$10x^2 + 9x - 2$$

$$= A(x-1)(x+2)^2 + B(x+2)^2 + C(x-1)^2(x+2) + D(x-1)^2$$

$$= A(x-1)(x^2 + 4x + 4) + B(x^2 + 4x + 4) + C(x^2 - 2x + 1)(x+2)$$

$$+ D(x^2 - 2x + 1)$$

$$= A(x^3 + 3x^2 - 4) + B(x^2 + 4x + 4) + C(x^3 - 3x + 2)$$

$$+ D(x^2 - 2x + 1).$$

This being an identity in x, we can compare the coefficients of like powers of x on the two sides and get

$$A + C = 0 \tag{5.10}$$

$$3A + B + D = 10 \tag{5.11}$$

$$4B - 3C - 2D = 9 \tag{5.12}$$

$$-4A + 4B + 2C + D = -2. \tag{5.13}$$

By (5.10), $C = -A$ and putting $C = -A$ in (5.12) and (5.13), we get

$$3A + 4B - 2D = 9 \tag{5.14}$$

$$-6A + 4B + D = -2 \tag{5.15}$$

$$(5.11) \text{ and } (5.15) \text{ give } 9A - 3B = 12 \quad \text{or} \quad 3A - B = 4. \tag{5.16}$$

Also (5.11) and (5.14) give

$$9A + 6B = 29. \tag{5.17}$$

Substituting for B from (5.16) in (5.17), $9A + 6(-4 + 3A) = 29$, which gives $A = \frac{53}{27}$ and $C = -\frac{53}{27}$

Now $B = 3A - 4 = \frac{53}{9} - 4 = \frac{17}{9}$ and $D = 10 - 3A - B = 10 - \frac{53}{9} - \frac{17}{9} = \frac{20}{9}$.
Hence

$$\frac{10x^2 + 9x - 2}{(x-1)^2(x+2)^2} = \frac{53/27}{x-1} + \frac{17/9}{(x-1)^2} - \frac{53/27}{x+2} + \frac{20/9}{(x+2)^2}.$$

Example 5.5. Resolve into partial fractions $\frac{5x^2+8x+21}{(x^2+x+6)(x+1)}$.

Solution. Let

$$\frac{5x^2 + 8x + 21}{(x^2 + x + 6)(x + 1)} = \frac{A}{x+1} + \frac{Bx + C}{x^2 + x + 6}.$$

Multiply both sides by $(x^2 + x + 6)(x + 1)$ to obtain

$$5x^2 + 8x + 21 = A(x^2 + x + 6) + (Bx + C)(x + 1)$$

$$= A(x^2 + x + 6) + (Bx^2 + Bx) + C(x + 1).$$

This being an identity in x, compare the coefficients of like powers of x on the two sides to obtain

$$A + B = 5 \tag{5.18}$$

$$A + B + C = 8 \tag{5.19}$$

$$6A + C = 21. \tag{5.20}$$

Now, (5.18) and (5.19) together give $C = 3$. Putting the value of C in (5.20) gives $A = 3$. Then (5.18) shows that $B = 2$. Hence

$$\frac{5x^2 + 8x + 21}{(x^2 + x + 6)(x + 1)} = \frac{3}{x+1} + \frac{2x + 3}{x^2 + x + 6}.$$

Example 5.6. Resolve into partial fractions $\frac{2x^3 - x + 3}{(x^2+4)(x^2+1)}$.

Solution. Both the factors of the denominator being quadratic and irreducible, let

$$\frac{2x^3 - x + 3}{(x^2+4)(x^2+1)} = \frac{Ax+B}{x^2+1} + \frac{Cx+D}{x^2+4}.$$

Multiplying both sides by $(x^2+1)(x^2+4)$, we get

$$2x^3 - x + 3 = (Ax+B)(x^2+4) + (Cx+D)(x^2+1)$$

$$= (A+C)x^3 + (B+D)x^2 + (4A+C)x + (4B+D).$$

Comparing the coefficients of like powers of x on the two sides of this relation, we get

$$A + C = 2 \qquad (5.21)$$

$$B + D = 0 \qquad (5.22)$$

$$4A + C = -1 \qquad (5.23)$$

$$4B + D = 3. \qquad (5.24)$$

(5.21) and (5.23) give $A = -1$, $C = 3$ and (5.22) and (5.24) give $B = 1, D = -1$.

Hence

$$\frac{2x^3 - x + 3}{(x^2+4)(x^2+1)} = \frac{-x+1}{x^2+1} + \frac{3x-1}{x^2+4}.$$

Example 5.7. Resolve into partial fractions $\frac{7x^3 + 16x^2 + 20x + 5}{(x^2+2x+2)^2}$.

Solution. $x^2 + 2x + 2$ is irreducible over the rational numbers. Therefore, let

$$\frac{7x^3 + 16x^2 + 20x + 5}{(x^2+2x+2)^2} = \frac{Ax+B}{x^2+2x+2} + \frac{Cx+D}{(x^2+2x+2)^2}.$$

Multiplying both sides by $(x^2+2x+2)^2$, we get

$$7x^3 + 16x^2 + 20x + 5 = (Ax+B)(x^2+2x+2) + Cx + D$$

$$= Ax^3 + (B+2A)x^2 + (2A+2B+C)x + (2B+D).$$

Comparing the coefficients of like powers of x on the two sides we get

$$A = 7, \quad B + 2A = 16, \quad 2A + 2B + C = 20, \quad 2B + D = 5.$$

Then, solving these equations, we get $A = 7, B = 2, C = 2$ and $D = 1$. Hence

$$\frac{7x^3 + 16x^2 + 20x + 5}{(x^2 + 2x + 2)^2} = \frac{7x + 2}{x^2 + 2x + 2} + \frac{2x + 1}{(x^2 + 2x + 2)^2}.$$

Example 5.8. Resolve into partial fractions $\frac{x^2-7}{x^4+x^2-6}$.

Solution. Regarding $x^4 + x^2 - 6$ as a quadratic in x^2, we find that $x^2 = 2$ is a root of this polynomial. Therefore, we can factorize $x^4 + x^2 - 6 = (x^2 - 2)(x^2 + 3)$ and each of the factors $x^2 - 2$ and $x^2 + 3$ is irreducible over the rational numbers. Hence, let

$$\frac{x^2 - 7}{x^4 + x^2 - 6} = \frac{Ax + B}{x^2 - 2} + \frac{Cx + D}{x^2 + 3}.$$

Multiplying both sides with $x^4 + x^2 - 6$, we get

$$x^2 - 7 = (Ax + B)(x^2 + 3) + (Cx + D)(x^2 - 2)$$

$$= (A + C)x^3 + (B + D)x^2 + (3A - 2C)x + 3B - 2D.$$

This is an identity in x. Compare the coefficients of like powers of x on the two sides. We get $A + C = 0, B + D = 1, 3A - 2C = 0, 3B - 2D = -7$. These relations give $A = C = 0$ and $3B - 2(1 - B) = -7$ or $B = -1$. Then $D = 2$. Hence

$$\frac{x^2 - 7}{x^4 + x^2 - 6} = \frac{-1}{x^2 - 2} + \frac{2}{x^2 + 3}.$$

Example 5.9. Resolve into partial fractions $\frac{x^4+3x^2+x+1}{(x+1)(x^2+1)^2}$.

Solution. Since $x^2 + 1$ is an irreducible quadratic polynomial over rational numbers, let

$$\frac{x^4 + 3x^2 + x + 1}{(x + 1)(x^2 + 1)^2} = \frac{A}{x + 1} + \frac{Bx + C}{x^2 + 1} + \frac{Dx + E}{(x^2 + 1)^2}.$$

Multiplying both sides of this relation by $(x + 1)(x^2 + 1)^2$, we get

$$x^4 + 3x^2 + x + 1$$

$$= A(x^2 + 1)^2 + (Bx + C)(x + 1)(x^2 + 1) + (Dx + E)(x + 1)$$

$$= (A + B)x^4 + (B + C)x^3 + (2A + B + C + D)x^2$$
$$+ (B + C + D + E)x + C + E + A.$$

This is an identity in x. Compare the coefficients of like powers of x on both sides. We get $A + B = 1, B + C = 0, 2A + B + C + D = 3, B + C + D + E = 1$, $A + C + E = 1$.

Solve these equations for A, B, C, D and E. We get $A + B = 1, B + C = 0, 2A + D = 3, D + E = 1, A - B + E = 1$. Then $2A + E = 2$ and $2A - E = 2$ which imply that $A = 1, E = 0$. But then $D = 1, B = 0 = C$. Hence

$$\frac{x^4 + 3x^2 + x + 1}{(x + 1)(x^2 + 1)^2} = \frac{1}{x + 1} + \frac{x}{(x^2 + 1)^2}.$$

Exercise 5.1. Resolve each of the following rational functions into partial fractions.

1. $\dfrac{x^2 - 3}{(x - 2)(x^2 + 4)}$.

2. $\dfrac{x^3 - 6}{x(2x^2 + 3x + 8)(x^2 + x + 1)}$.

3. $\dfrac{5x^3 - 2}{x^3(x + 1)^2}$.

4. $\dfrac{2x - 1}{(x - 4)^2}$.

5. $\dfrac{x^2 - 3x + 1}{(x^2 + 1)^2(x^2 + x + 1)}$.

6. $\dfrac{3x^2 + 3x + 8}{(x - 2)^2(x^2 + 1)}$.

7. $\dfrac{x + 2}{2x^2 - 13x + 15}$.

8. $\dfrac{x + 2}{2x^2 - 7x - 15}$.

9. $\dfrac{2x^2 + 10x - 5}{(x + 1)(x^2 - 9)}$.

10. $\dfrac{2x^2 + 7x + 19}{(x - 1)(x + 2)^2}$.

11. $\dfrac{x^2 - 3x + 1}{x^2(x - 2)^2}$.

12. $\dfrac{x^2 + 4x - 10}{(x + 2)^3}$.

13. $\dfrac{7x^2 - 25x + 6}{(x^2 - 2x - 1)(3x - 2)}$.

14. $\dfrac{x^2 - 7}{x^4 - x^2 - 6}$.

15. $\dfrac{3x^2 - 8x + 7}{(x - 2)^3}$.

16. $\dfrac{5x^2 + 8x + 21}{(x^2 + x + 6)(x + 1)}$.

17. $\dfrac{7x^3 + 16x^2 + 20x + 5}{(x^2 + 2x + 2)^2}$.

18. $\dfrac{x + 2}{x^2 + 7x + 12}$.

19. $\dfrac{10x + 13}{x^2 + x - 6}$.

20. $\dfrac{12x + 11}{x^2 - x - 6}$.

21. $\dfrac{6 - x}{2x^2 + 3x - 2}$.

22. $\dfrac{8 - x}{2x^2 + 5x + 2}$.

23. $\dfrac{8+x}{2x^2+3x-2}$.

24. $\dfrac{5x+3}{2x^2+2x}$.

25. $\dfrac{x}{x^2+3x-18}$.

26. $\dfrac{7x+2}{x^3-x^2-2x}$.

27. $\dfrac{2x^3+x-3}{(x^2+4)(x^2+1)}$.

28. $\dfrac{x^4+3x^2+x+1}{(x+1)(x^2+1)^2}$.

29. $\dfrac{2x-9}{(2x+1)(4x^2+9)}$.

30. $\dfrac{x^3}{(x^2+3)^2}$.

31. $\dfrac{x^3}{(x^2+4)^2}$.

32. $\dfrac{3x-1}{x^3-1}$.

33. $\dfrac{3x}{x^2-1}$.

34. $\dfrac{5x^2+3x+1}{(x+1)(x^2+2)}$.

35. $\dfrac{3x^3+5x^2+13x+14}{x^2(x+3)^2}$.

36. $\dfrac{x^3}{(x^2+4)^2}$.

37. $\dfrac{5x^2+8x+21}{(x^3+x+6)(x+1)}$.

38. $\dfrac{5x^3+4x^2+7x+3}{(x^3+2x+2)(x^2-x-1)}$.

•

5.2. Recurrence Relations: Preliminaries

Suppose we want to construct a sequence of positive integers in which the first term is 1 and beyond that every term exceeds its predecessor by 2. We can construct this sequence recursively. The first term is 1, the next term is $1+2=3$, the next term is $3+2=5$, etc. We can continue this process and obtain the sequence as $1, 3, 5, \ldots$, which is the sequence of all odd positive integers. Observe that if we write a_r for the rth term of this sequence, the sequence could be constructed from the condition that $a_r = a_{r-1} + 2$ for $r > 1$ with $a_1 = 1$. As another example, let us construct a sequence of positive integers in which the first term is 1 and every term from the second onwards is 1 more that twice the previous term. Then the sequence is $1, 2 \times 1 + 1 = 3, 2 \times 3 + 1 = 7, 2 \times 7 + 1 = 15, 2 \times 15 + 1 = 31$, i.e. the sequence is $1, 3, 7, 15, 31, 63, 127, \ldots$ There are innumerable situations where sequences, not necessarily sequences of positive integers, can be obtained recursively. (Students familiar with group theory may note that powers of an element in a group, the terms of lower central series, upper central series of a group, etc. all admit of recursive definitions.)

Definition. For a numeric function $(a_0, a_1, a_2, \ldots, a_r, \ldots)$ an equation relating a_r, for any r, to one or more of the $a_i's$, $i < r$, is called a **recurrence relation** or a **difference equation**.

According to the recurrence relation, we can carry out a step-by-step computation to determine a_r from $a_{r-1}, a_{r-2}, \ldots$, then to determine a_{r+1} from $a_r, a_{r-1}, \ldots$, and so on, provided that the value of the (numeric) function at one or more points is given so that the computations can be initiated. These given values of the function for some values of r are called the 'boundary conditions'. Thus a numeric function can be described by a recurrence relation together with an appropriate set of boundary conditions. The numeric function is called the **solution of the recurrence relation**.

Definition. The recurrence relation

$$c_0 a_r + c_1 a_{r-1} + c_2 a_{r-2} + \cdots + c_k a_{r-k} = f(r) \qquad (5.25)$$

where $c_i s$ are constants, is called a **linear recurrence relation** with constant coefficients. Also (5.25) is called a kth **order** recurrence relation, provided both c_0 and c_k are non-zero. For example, $2a_r + 5a_{r-1} = 2^r$ is a first order linear recurrence relation with constant coefficients, while $a_r = a_{r-1} + a_{r-2}$ is a linear recurrence relation of order 2 with constant coefficients. Also, both

$$3a_r + 4a_{r-1} - a_{r-2} = r^2 + 3 \qquad (5.26)$$

and $a_r + 5a_{r-2} = 6r^3 + 3r - 8$ are linear recurrence relations of order 2 with constant coefficients. However, the recurrence relation $a_r + ra_{r-1} + 3a_{r-2} = 7$ is of order 2, but is not with constant coefficients. On the other hand the relation $a_r^2 + ra_{r-1} + a_{r-3} = 0$ is neither linear nor with constant coefficients.

Example 5.10. Consider the recurrence relation (5.26) as above and suppose that we are given $a_3 = 3, a_4 = 5$. Then $a_5 = \frac{1}{3}\{-4 \times 5 + 3 + 25 + 3\} = \frac{11}{3}$. Then $a_6 = \frac{1}{3}\{-\frac{44}{3} + 5 + 36 + 3\} = \frac{88}{9}$. Then we can substitute for a_5 and a_6 in the relation $a_r = \frac{1}{3}\{-4a_{r-1} + a_{r-2} + r^2 + 3\}$ for $r = 7$, then for a_6 and a_7 in the above relation for $r = 8$ and so on, and we get the values of a_r for every $r, r > 2$.

We can also find the values of a_0, a_1 and a_2 by using the relation $a_{r-2} = 4a_{r-1} + 3a_r - r^2 - 3$ taking $r = 4, 3, 2$ respectively. We get

$$a_2 = 4a_3 + 3a_4 - 16 - 3 = 12 + 15 - 16 - 3 = 8$$

$$a_1 = 4a_2 + 3a_3 - 9 - 3 = 32 + 9 - 9 - 3 = 29$$

$$a_0 = 4a_1 + 3a_2 - 4 - 3 = 116 + 24 - 4 - 3 = 133.$$

In the general case of a linear recurrence relation with constant coefficients of order k, as in (5.25), if k consecutive values of the numeric function $a_{m-k}, a_{m-k+1}, \ldots, a_{m-1}$ are known for some value m, then we get $a_m = -\frac{1}{c_0}\{c_1 a_{m-1} + c_2 a_{m-2} + \cdots + c_0 a_{m-k} - f(m)\}$. Once a_m is known, a_{m+1} is given by $a_{m+1} = -\frac{1}{c_0}\{c_1 a_m + c_2 a_{m-1} + \cdots + c_k a_{m-k+1} - f(m+1)\}$ and so on, we can find the values of $a_{m+2}, a_{m+3}, \ldots$ Also we can determine the values of $a_{m-k-1}, a_{m-k-2}, \ldots$ by first computing $a_{m-k-1} = -\frac{1}{c_k}\{c_0 a_{m-1} + c_1 a_{m-2} + \cdots + c_{k-1} a_{m-k} - f(m-1)\}$; then $a_{m-k-2} = -\frac{1}{c_k}\{c_0 a_{m-2} + c_1 a_{m-3} + \cdots + c_{k-1} a_{m-k-1} - f(m-2)\}$ and so on.

It is clear from the above process that, indeed, the values of k consecutive $a_i's$ are sufficient to determine the numeric function uniquely.

On the other hand, fewer than k values of the numeric function are not sufficient to determine uniquely the numeric function for a linear recurrence relation of order k with constant coefficients. For example, let

$$a_r + a_{r-1} + a_{r-2} = 5$$

and suppose that we are given $a_0 = 2$. Then

$$2, 0, 3, 2, 0, 3, 2, 0, \ldots \quad \text{and}$$

$$2, -2, 5, 2, -2, 5, 2, -2, \ldots$$

are two different numeric functions that satisfy the recurrence relation along with the given boundary condition.

The values of non-consecutive $a_i's$ need not always constitute an appropriate set of boundary conditions for a unique solution of a linear recurrence relation of order k with constant coefficients to exist. In the example considered above, suppose the boundary conditions are now given to be $a_0 = 2, a_3 = 2$. Then again, there are two solutions as obtained above.

We have discussed a way of finding all the terms of a recurrence relation by using the boundary condition(s). As another, similar method of finding the terms of a given recurrence relation, let us consider the recurrence relation $a_r = a_{r-1} + 2$ with $a_1 = 1$. Now

$$a_r = a_{r-1} + 2$$

$$= (a_{r-2} + 2) + 2 = a_{r-2} + 2 \times 2$$

$$= (a_{r-3} + 2) + 2 \times 2 = a_{r-3} + 3 \times 2$$

$$= (a_{r-4} + 2) + 3 \times 2 = a_{r-4} + 4 \times 2.$$

Observe that on the right-hand side, the subscript of a in a_k, plus the coefficient of 2 is $r-1+1 = r, r-2+2 = r, r-3+3 = r, r-4+4 = r$.

Thus this sum is r always. Therefore, if we continue substituting for a_k in terms of a_{k-1} we eventually arrive at a_1 plus a multiple of 2. Since $1 = r - (r-1)$, the coefficient of 2 shall be $r - 1$, i.e.

$$a_r = a_1 + (r-1) \times 2 = 1 + 2 \times r - 2 = 2r - 1.$$

Thus we have found the value of a_r for general r. Finding such a value of a_r is called 'solving the given recurrence relation', and the value so obtained is called a 'solution of the recurrence relation'. The method adopted in the above solution is called the **method of backtracking.**

Example 5.11. As another application of the method of backtracking, consider the recurrence relation $a_r = 3 \times a_{r-1} + 1$ for $r > 1$ and $a_1 = 1$. Now

$$a_r = 3 \times a_{r-1} + 1$$

$$= 3(3 \times a_{r-2} + 1) + 1 = 3^2 \times a_{r-2} + 3 + 1$$

$$= 3^2(3 \times a_{r-3} + 1) + 3 + 1 = 3^3 \times a_{r-3} + 3^2 + 3 + 1.$$

Continuing the above procedure, we find that the solution of the recurrence relation is $a_r = 3^{r-1} \times a_1 + (3^{r-2} + 3^{r-3} + \cdots + 3 + 1) = 3^{r-1} + 3^{r-2} + 3^{r-3} + \cdots + 3 + 1 = (3^r - 1)/2$.

5.2.1. *Homogeneous solutions*

Consider a linear difference equation

$$c_0 a_r + c_1 a_{r-1} + \cdots + c_k a_{r-k} = f(r) \tag{5.27}$$

with constant coefficients. The numeric function $a^{(h)} = (a_o^{(h)}, a_1^{(h)}, \ldots, a_r^{(h)}, \ldots)$, which satisfies the recurrence relation

$$c_0 a_r + c_1 a_{r-1} + \cdots + c_k a_{r-k} = 0 \tag{5.28}$$

is called a **homogeneous solution** of the difference equation (5.27). If $a^{(p)} = (a_o^{(p)}, a_1^{(p)}, \ldots, a_r^{(p)}, \ldots)$ is some solution of the difference equation (5.27) obtained in any manner, then it, i.e. $a^{(p)}$, is called a **particular solution** of the difference equation (5.27). Also then $a = a^{(h)} + a^{(p)} = (a_0^{(h)} + a_0^{(p)}, a_1^{(h)} + a_1^{(p)}, \ldots, a_r^{(h)} + a_r^{(p)}, \ldots)$ is called the **total solution** of the difference equation (5.27). A homogeneous solution of the difference equation is not, in general, a solution of the difference equation (5.27). A particular solution of a difference equation may or may not satisfy the given boundary conditions. The total solution is indeed a solution of the difference equation (5.27) and the advantage of considering a total solution is that the homogeneous solution may be so adjusted that the total solution satisfies the given boundary conditions. Recall that we also talk of homogeneous solutions, particular solutions and total solutions of differential equations. However, we do not study differential equations in this text.

Suppose that $a_r = A\alpha^r$, where A is a non-zero constant, is a homogeneous solution of the difference equation (5.27). Then $Ac_0\alpha^r + Ac_1\alpha^{r-1} + \cdots + Ac_k\alpha^{r-k} = 0$, which gives

$$c_0\alpha^r + c_1\alpha^{r-1} + \cdots + c_k\alpha^{r-k} = 0 \quad \text{or} \tag{5.29}$$

$$c_0\alpha^k + c_1\alpha^{k-1} + \cdots + c_k = 0. \tag{5.30}$$

In this case, α is called a **characteristic root** and equation (5.30) is called the **characteristic equation** of the difference equation (5.27). We shall also come across characteristic roots and characteristic equations of square matrices.

Observe that if α is a characteristic root of the difference equation (5.27), then $a_r = A\alpha^r$ is a homogeneous solution of the difference equation for any non-zero constant A. Also, any linear combination of homogeneous solutions is, again, a homogeneous solution. If all the characteristic roots of the recurrence relation (5.27) are distinct, say, $\alpha_1, \alpha_2, \ldots, \alpha_k$ then $A_1\alpha_1^r + A_2\alpha_2^r + \cdots + A_k\alpha_k^r$ is a homogeneous solution of the relation (5.27).

Next, suppose that α_1 is a repeated characteristic root of the recurrence relation (5.27), say repeated twice. Then α_1 is a root of (5.30) and so is also a repeated root of (5.29). Therefore, α_1 is also a root of

$$c_0 r\alpha^r + c_1(r-1)\alpha^{r-1} + \cdots + c_k(r-k)\alpha^{r-k} = 0. \tag{5.31}$$

Adding B times the relation (5.31) to A times the relation (5.29), where A, B are non-zero constants (with α replaced by α_1), we get $c_0(Br+A)\alpha_1^r + c_1(B(r-1)+A)\alpha_1^{r-1} + \cdots + c_k(B(r-k)+A)\alpha_1^{r-k} = 0$. It follows from this that $(Br+A)\alpha_1^r$ is a homogeneous solution of the difference equation.

Suppose α_1 is a characteristic root of (5.27) repeated m times. Then differentiating (5.29) (formally, since strictly speaking we cannot talk of derivative of (5.29)) $m-1$ times, we find that $A_1 r^{m-1}\alpha_1^r, A_2 r^{m-2}\alpha_1^r, \ldots, A_{m-1}r\alpha_1^r, A_m\alpha_1^r$ are homogeneous solutions of the difference equation (5.27), where $A_1, A_2, \ldots, A_m$ are arbitrary constants. Therefore, $(A_1 r^{m-1} + A_2 r^{m-2} + \cdots + A_{m-1}r + A_m)\alpha_1^r$ is a homogeneous solution of the difference equation (5.27) corresponding to the characteristic root α_1 repeated m times. Similar homogeneous solutions can be obtained corresponding to other characteristic roots of (5.27), if any. Taking a sum of all these solutions, we get a homogeneous solution of the given difference equation (5.27).

We will consider some examples.

Example 5.12. Consider the Fibonacci sequence of numbers

$$1, 1, 2, 3, 5, 8, 13, 21, 34, \ldots$$

the corresponding recurrence relation of which is

$$a_r = a_{r-1} + a_{r-2}, \quad \text{for } r > 1 \quad \text{with } a_0 = a_1 = 1.$$

This is a recurrence relation of order 2 with $f(r) = 0$. The characteristic equation is $\alpha^2 - \alpha - 1 = 0$ the roots of which are (distinct)

$$\alpha_1 = \frac{1 + \sqrt{5}}{2}, \quad \alpha_2 = \frac{1 - \sqrt{5}}{2}.$$

Therefore any homogeneous solution is

$$a_r = A_1 \left(\frac{1 + \sqrt{5}}{2} \right)^r + A_2 \left(\frac{1 - \sqrt{5}}{2} \right)^r,$$

where A_1, A_2 are arbitrary constants. The values of the two constants A_1, A_2 are determined by the boundary conditions $a_0 = a_1 = 1$. Indeed,

$$1 = A_1 + A_2, \quad \text{and}$$

$$1 = A_1 \left(\frac{1 + \sqrt{5}}{2} \right) + A_2 \left(\frac{1 - \sqrt{5}}{2} \right)$$

$$= A_1 \left(\frac{1 + \sqrt{5}}{2} \right) + (1 - A_1) \left(\frac{1 - \sqrt{5}}{2} \right)$$

$$= A_1 \sqrt{5} + \frac{1 - \sqrt{5}}{2}$$

which gives $A_1 = \left(\frac{1+\sqrt{5}}{2\sqrt{5}} \right)$ Then $A_2 = \left(\frac{-1+\sqrt{5}}{2\sqrt{5}} \right)$

Example 5.13. Consider the recurrence relation $a_r = 2a_{r-1} + 2a_{r-2}$, for $r > 1$ with $a_0 = a_1 = 1$. This is a recurrence relation of order 2 with $f(r) = 0$. The characteristic equation is $\alpha^2 - 2\alpha - 2 = 0$ the roots of which are (distinct) $\alpha_1 = 1 + \sqrt{3}, \alpha_2 = 1 - \sqrt{3}$. Therefore, any homogeneous solution is $a_r = A_1(1 + \sqrt{3})^r + A_2(1 - \sqrt{3})^r$, where A_1, A_2 are arbitrary constants. The values of the two constants A_1, A_2 are determined by the boundary conditions $a_0 = a_1 = 1$. Indeed,

$$1 = A_1 + A_2, \quad \text{and}$$

$$1 = A_1(1 + \sqrt{3}) + A_2(1 - \sqrt{3})$$
$$= A_1(1 + \sqrt{3}) + (1 - A_1)(1 - \sqrt{3})$$
$$= 2A_1\sqrt{3} + (1 - \sqrt{3})$$

which gives $A_1 = \frac{1}{2}$ and then $A_2 = \frac{1}{2}$

Example 5.14. Consider the recurrence relation $a_r + 6a_{r-1} + 9a_{r-2} = 3$. The characteristic equation is $\alpha^2 + 6\alpha + 9 = 0$, giving -3 as a repeated characteristic root. Therefore, $(A_1r + A_2)(-3)^r$ is a homogeneous solution of the given recurrence relation.

Example 5.15. Consider the difference equation

$$2a_r - 9a_{r-1} + 12a_{r-2} - 4a_{r-3} = 0.$$

This is again a recurrence relation with $f(r) = 0$. Therefore any homogeneous solution of this recurrence relation is indeed a solution of the relation. The degree of the relation being 3, there are 3 characteristic roots of the relation. The characteristic equation of the difference equation is $2\alpha^3 - 9\alpha^2 + 12\alpha - 4 = 0$ which may be rewritten as $2\alpha^2(\alpha - 2) - 5\alpha(\alpha - 2) + 2(\alpha - 2) = 0$ or $(\alpha - 2)(2\alpha^2 - 5\alpha + 2) = 0$ giving $\alpha = \frac{1}{2}, 2, 2$ as characteristic roots.

Therefore an arbitrary solution of the given difference equation is

$$a_r = (A_1r + A_2)2^r + A_3\left(\frac{1}{2}\right)^r,$$

where A_1, A_2, A_3 are arbitrary constants.

Example 5.16. Solve the recurrence relation

$$a_r + 9a_{r-1} + 27a_{r-2} + 27a_{r-3} = 0, \quad r > 2.$$

Solution. The characteristic equation is $\alpha^3 + 9\alpha^2 + 27\alpha + 27 = 0$ which may be rewritten as

$$\alpha^3 + 3\alpha^2 + 6\alpha^2 + 18\alpha + 9\alpha + 27 = 0 \quad \text{or}$$

$$(\alpha + 3)(\alpha^2 + 6\alpha + 9) = 0 \quad \text{or} \quad (\alpha + 3)^3 = 0.$$

This shows that -3 is a characteristic root of multiplicity 3. Hence a homogeneous solution and, therefore, a solution of the recurrence relation is

$$a_r = (A_1 r^2 + A_2 r + A_3)(-3)^r,$$

where A_1, A_2, A_3 are arbitrary constants.

Example 5.17. Solve the following recurrence relation. $a_r - 2a_{r-1} + 2a_{r-2} - a_{r-3} = 0, r > 2$, given that $a_0 = 2, a_1 = 1$ and $a_2 = 1$.

Solution. Characteristic equation of the recurrence relation is $\alpha^3 - 2\alpha^2 + 2\alpha - 1 = 0$, which may be rewritten as $(\alpha - 1)(\alpha^2 + \alpha + 1) - 2\alpha(\alpha - 1) = 0$ or $(\alpha - 1)(\alpha^2 - \alpha + 1) = 0$. The roots of this equation are

$$1, \frac{1 - \sqrt{-3}}{2} \quad \text{and} \quad \frac{1 + \sqrt{-3}}{2}.$$

Therefore, a homogeneous solution and, hence, a solution of the given recurrence relation is

$$a_r = A1^r + B\left(\frac{1 - \sqrt{-3}}{2}\right)^r + C\left(\frac{1 + \sqrt{-3}}{2}\right)^r$$

$$= A + B\left(\frac{1 - \sqrt{-3}}{2}\right)^r + C\left(\frac{1 + \sqrt{-3}}{2}\right)^r.$$

Using the given boundary conditions $a_0 = 2, a_1 = 1$ and $a_2 = 1$, we get three equations in terms of A, B, C, which can then be solved to get the values of the constants A, B, C. (Suggested exercise: find the values of A, B, C.)

5.2.2. *Particular solutions*

There is no general method of finding a particular solution of a difference equation and one has to be obtained by inspection. However, following are some points which help us in finding a particular solution of a difference equation by inspection. Consider the recurrence relation as given in (5.27).

(i) When $f(r)$ is a polynomial of degree m, then the corresponding particular solution will be of the form $g(r) = P_0 + P_1 r + P_2 r^2 + \cdots + P_m r^m$.

(ii) When $f(r)$ is of the form α^r and α is not a characteristic root of (5.27), then a particular solution of (5.27) is of the form $P\alpha^r$. Furthermore, when $f(r)$ is of the form $g(r)\alpha^r$, where $g(r)$ is a polynomial of degree

m in r and α is not a characteristic root of (5.27), then a particular solution of (5.27) is of the form $(P_0 + P_1 r + \cdots + P_{m-1} r^{m-1} + P_m r^m)\alpha^r$.

(iii) When $f(r)$ is of the form $g(r)\alpha^r$, where $g(r)$ is a polynomial of degree m in r and α is a characteristic root of multiplicity n of (5.27), then the particular solution of (5.27) is of the form $r^n(P_0 + P_1 r + P_2 r^2 + \cdots + P_m r^m)\alpha^r$.

We find a particular solution of the difference equation in each of Examples 5.18 to 5.25.

Example 5.18. The difference equation $a_r + 7a_{r-1} + 12a_{r-2} = 5r^2$.

Solution. The function $f(r)$ in this case being of degree 2 in r, let a particular solution be

$$P_1 r^2 + P_2 r + P_3. \tag{5.32}$$

Substituting for a_r from (5.32) in the given difference equation, we get

$$
\begin{aligned}
5r^2 &= P_1 r^2 + P_2 r + P_3 + 7\{P_1(r-1)^2 + P_2(r-1) + P_3\} \\
&\quad + 12\{P_1(r-2)^2 + P_2(r-2) + P_3\} \\
&= r^2(P_1 + 7P_1 + 12P_1) + r(P_2 - 14P_1 + 7P_2 - 48P_1 + 12P_2) \\
&\quad + (P_3 + 7P_1 - 7P_2 + 7P_3 + 48P_1 - 24P_2 + 12P_3) \\
&= r^2(20P_1) + r(20P_2 - 62P_1) + (20P_3 - 31P_2 + 55P_1).
\end{aligned}
$$

Comparison of the like powers of r on the two sides of this equation gives

$$
P_1 = \frac{1}{4}, \quad P_2 = \frac{(31P_1)}{10} = \frac{31}{40}, \quad P_3 = \frac{\{31 \times 31/40 - 55/4\}}{20} = \frac{411}{800}.
$$

Thus a particular solution is

$$
a_r^{(p)} = \frac{1}{4}r^2 + \frac{31}{40}r + \frac{411}{800}.
$$

Example 5.19. The difference equation

$$a_r + 7a_{r-1} + 12a_{r-2} = 5r^2 - 4r + 3.$$

Solution. Let a particular solution be $a_r^{(p)} = P_1 r^2 + P_2 r + P_3$. Substituting for a_r from the particular solution expression in the given difference equation, we get

$$5r^2 - 4r + 3 = P_1 r^2 + P_2 r + P_3 + 7\{P_1(r-1)^2 + P_2(r-1) + P_3\}$$
$$+ 12\{P_1(r-2)^2 + P_2(r-2) + P_3\}.$$
$$= 20P_1 r^2 + (20P_2 - 62P_1)r + (20P_3 - 31P_2 + 55P_1).$$

Comparing the coefficients of like powers of r on the two sides we get $P_1 = \frac{1}{4}, 20P_2 = 62P_1 - 4 = \frac{31}{2} - 4 = \frac{23}{2}$ i.e. $P_2 = \frac{23}{40}, 20P_3 = \frac{31 \times 23}{40} - \frac{55}{4} + 3 = \frac{713 - 550 + 120}{40} = \frac{283}{40}$ or $P_3 = \frac{283}{800}$. Thus a particular solution is

$$a_r^{(p)} = \frac{1}{4}r^2 + \frac{23}{40}r + \frac{283}{800}.$$

Example 5.20. The difference equation $a_r + a_{r-1} = 2r3^r$.

Solution. Three is not a root of the characteristic equation $\alpha + 1 = 0$. Thus general form of a particular solution is $(P_1 r + P_2)3^r$. Substituting from this for a_r in the given difference equation, we get

$$2r3^r = (P_1 r + P_2)3^r + \{P_1(r-1) + P_2\}3^{r-1}.$$

Cancelling the common factor 3^{r-1} from the two sides, we get

$$6r = 3(P_1 r + P_2) + (P_1 r + P_2 - P_1)$$

which on comparison of like powers of r on the two sides gives $6 = 3P_1 + P_1$ and $0 = 3P_2 + P_2 - P_1$. Thus $P_1 = \frac{3}{2}, P_2 = \frac{3}{8}$. Hence a particular solution is

$$a_r^{(p)} = \left(\frac{r}{2} + \frac{1}{8}\right)3^{r+1}.$$

Example 5.21. The difference equation $a_r - 3a_{r-1} = 2.3^r$.

Solution. Here, three is a simple characteristic root of this difference equation, so the general form of a particular solution is $a_r = Pr3^r$. Substituting for a_r from the particular solution in the given difference equation, we get $rP3^r - 3P(r-1)3^{r-1} = 2.3^r$, which gives $rP - (r-1)P = 2$, or that $P = 2$. Thus a particular solution is $a_r^{(p)} = 2r3^r$.

Example 5.22. The difference equation $a_r - 6a_{r-1} + 9a_{r-2} = (r+1)3^r$.

Solution. The characteristic equation of this recurrence relation is $\alpha^2 - 6\alpha + 9 = 0$, giving $\alpha = 3$ as a root of multiplicity 2. General form of a particular solution is, therefore, $a_r^{(p)} = r^2(P_1 r + P_2)3^r$. Substituting this for a_r in the difference equation, we get

$$(r+1)3^r = r^2(P_1 r + P_2)3^r - 6(r-1)^2\{P_1(r-1) + P_2\}3^{r-1}$$

$$\cdot + 9(r-2)^2\{P_1(r-2) + P_2\}3^{r-2}.$$

On cancelling the common factor 3^r from the two sides, we get

$$r+1 = r^2(P_1 r + P_2) - 2\{P_1(r-1)^3 + P_2(r-1)^2\} + \{P_1(r-2)^3 + P_2(r-2)^2\}.$$

Comparison of coefficients of like powers of r on the two sides gives $P_1 - 2P_1 + P_1 = 0, P_2 + 6P_1 - 2P_2 - 6P_1 + P_2 = 0, -6P_1 + 4P_2 + 12P_1 - 4P_2 = 1$ and $2P_1 - 2P_2 - 8P_1 + 4P_2 = 1$, which lead to

$$P_1 = \frac{1}{6} \quad \text{and} \quad P_2 = 1.$$

A particular solution of the difference equation is

$$a_r^{(p)} = r^2\left(1 + \frac{r}{6}\right)3^r.$$

Example 5.23. Consider the difference equation $a_r = a_{r-2} + 5$.

Solution. Since 1 is a characteristic root of the difference equation of multiplicity 1 (the other root being -1 again of multiplicity 1) and 5 can be written as 5.1^r, the general form of the particular solution is Pr. Substitution of this for a_r in the difference equation gives $rP - (r-2)P = 5$, from which it follows that $P = 5/2$. Hence a particular solution of the recurrence relation is $a_r^{(p)} = (5/2)r$.

Not writing 5 as 5.1^r, the general form of a particular solution will have to be taken as just P. If that is accepted as a particular solution, we get $P = P + 5$ leading to $5 = 0$ — an absurdity.

Example 5.24. Consider the recurrence relation

$$a_r - 4a_{r-1} + 5a_{r-2} - 2a_{r-3} = 5.$$

Solution. The situation here is similar to that in the last example. The characteristic equation of this difference equation is $\alpha^3 - 4\alpha^2 + 5\alpha - 2 = 0$, giving 1 as a characteristic root of the difference equation with multiplicity 2 (the other root being 2 of multiplicity 1). Thus we need to consider the general form of the particular solution as Pr^2. Substitution of this for a_r in the difference equation gives

$$Pr^2 - 4P(r-1)^2 + 5P(r-2)^2 - 2P(r-3)^2 = 5$$

or $-4P + 20P - 18P = 5$ so that $P = -5/2$ and a particular solution is

$$a_r^{(p)} = \left(\frac{-5}{2}\right)r^2.$$

Again, if we had not considered 5 as 5.1^r and had assumed the form of a particular solution simply as P or even as Pr, then we shall get $5 = 0$ in both the cases, which is an absurdity.

Example 5.25. Consider the difference equation $a_r - 7a_{r-1} + 12a_{r-2} = 3^r + r$.

Solution. In this case, $f(r)$ is neither of the form $g(r)$, nor of the form $g(r)\alpha^r$. Rather, it is of the form $g(r)\alpha^r + h(r)$, where $g(r)$ and $h(r)$ are certain polynomials in r. In such a case, we need to take a particular solution as a sum of two solutions, one corresponding to $g(r)\alpha^r$ and the other corresponding to $h(r)$. The characteristic equation in the present case is $\alpha^2 - 7\alpha + 12 = 0$, which has 4, 3 as characteristic roots both of multiplicity 1. The general form of a particular solution shall be a combination of $P_1 r 3^r$ (since 3 is a characteristic root) and $P_2 r + P_3$. Thus we take

$$a_r^{(p)} = P_1 r 3^r + P_2 r + P_3.$$

Substitution of this for a_r in the given difference equation gives

$$3^r + r = P_1 r 3^r + P_2 r + P_3 - 7\{P_1(r-1)3^{r-1} + P_2(r-1) + P_3\}$$
$$+ 12\{P_1(r-2)3^{r-2} + P_2(r-2) + P_3\}$$
$$= P_1(3r - 7(r-1) + 4(r-2))3^{r-1} + \{P_2 r + P_3 - 7P_2(r-1)$$
$$- 7P_3 + 12P_2(r-2) + 12P_3\}$$
$$= -P_1 3^{r-1} + \{6P_2 r - 17P_2 + 6P_3\}.$$

This relation being true for all values of r, by taking $r = 0, 1, 2$ successively in the above relation, we get the three equations

$$1 = \frac{-P_1}{3} + 6P_3 - 17P_2;$$

$$4 = -P_1 + 6P_3 - 17P_2 + 6P_2;$$

$$11 = -3P_1 + 6P_3 - 17P_2 + 12P_2.$$

Solving these for P_1, P_2, P_3 we get
$P_1 = -3, P_2 = \frac{1}{6}, P_3 = \frac{17}{36}$. Thus the corresponding particular solution is

$$a_r^{(p)} = -r3^{r+1} + \frac{r}{6} + \frac{17}{36}.$$

Example 5.26. Determine a particular solution for each of the recurrence relations

(a) $a_r - 3a_{r-1} + 2a_{r-2} = 2^r$,
(b) $a_r - 4a_{r-1} + 4a_{r-2} = 2^r$.

Solution. (a) The characteristic equation of the difference equation is $\alpha^2 - 3\alpha + 2 = 0$, the roots of which are 1, 2 each of multiplicity 1. Therefore, a particular solution of the difference equation is of the form $a_r^{(p)} = Ar2^r$. Substituting from this for a_r in the given difference equation, we get $Ar2^r - 3A(r-1)2^{r-1} + 2A(r-2)2^{r-2} = 2^r$, which, on cancelling the common factor 2^{r-1} from the two sides gives $2Ar - 3A(r-1) + A(r-2) = 2$, which gives $A = 2$. Hence a particular solution of the recurrence relation is $a_r^{(p)} = r2^{r+1}$.

(b) Here the characteristic equation is $\alpha^2 - 4\alpha + 4 = 0$, the roots of which are 2 and 2. Therefore, a particular solution is of the form $a_r^{(p)} = Ar^2 2^r$. Substituting this for a_r in the given recurrence relation, we get $Ar^2 2^r - 4A(r-1)^2 2^{r-1} + 4A(r-2)^2 2^{r-2} = 2^r$, which, on cancelling the common factor 2^r from both sides gives $A\{r^2 - 2(r^2 - 2r + 1) + (r^2 - 4r + 4)\} = 1$. This gives $A = \frac{1}{2}$ and, hence, a particular solution of the recurrence relation is $a_r = r^2 2^{r-1}$.

Example 5.27. Solve the following recurrence relations

(a) $a_r - 5a_{r-1} + 6a_{r-2} = 1$;
(b) $a_r + 5a_{r-1} + 6a_{r-2} = 42. \, 4^r$;
(c) $a_r - 4a_{r-1} + 4a_{r-2} = (r+1)2^r$.

Solution. (a) For a homogeneous solution, the characteristic equation is $\alpha^2 - 5\alpha + 6 = 0$, which has 2, 3 as its roots. Thus a homogeneous solution is $a_r^{(0)} = A2^r + B3^r$.

Since $f(r) = 1$ is a constant and 1 is not a characteristic root, a particular solution is of the form P, where P is a constant. Substituting P for a_r in the given recurrence relation gives $P = \frac{1}{2}$. Hence a total solution is $a_r = A2^r + B3^r + \frac{1}{2}$ where A, B are constants, which could be determined if we were given some boundary conditions.

(b) The characteristic equation is $\alpha^2 + 5\alpha + 6 = 0$, which has -2 and -3 as roots. Therefore, a homogeneous solution is $a_r^{(h)} = A(-2)^r + B(-3)^r$, where A, B are constants.

In this case, since 4 is not a characteristic root, a particular solution is of the form $P4^r$, where P is a constant. Putting $a_r = P4^r$ in the given recurrence relation gives $P4^r + 5P4^{r-1} + 6P4^{r-2} = 42.4^r$ leading to $(16 + 20 + 6)P = 42 \times 16$, so that $P = 16$. Hence a particular solution is $a_r^{(p)} = 16.4^r$.

Therefore, a total solution is $a_r = A(-2)^r + B(-3)^r + 16.4^r$.

(c) The characteristic equation here is $\alpha^2 - 4\alpha + 4 = 0$, having 2 as a root of multiplicity 2. Then a particular solution of the difference equation is, therefore, of the form $a_r^{(p)} = (Ar + B)r^2 2^r$. On substituting this value of a_r in the given recurrence relation, we get

$$(r+1)2^r = (Ar + B)r^2 2^r - 4(A(r-1) + B)(r-1)^2 2^{r-1}$$
$$+ 4(A(r-2) + B)(r-2)^2 2^{r-2}.$$

On cancelling the common factor 2^r from the two sides, we get

$$Ar^3 + Br^2 - 2(Ar - A + B)(r-1)^2 + (Ar - 2A + B)(r-2)^2 = r + 1.$$

Comparison of the coefficients of r and the constant terms on the two sides gives

$$-2A - 4(A - B) + 4A + 4(2A - B) = 1 \quad \text{and} \quad 2(A - B) - 4(2A - B) = 1,$$

which together give $A = \frac{1}{6}$ and $B = 1$. Therefore $a_r^{(p)} = (\frac{r}{6} + 1)r^2 2^r$. Also the homogeneous solution is $a_r^{(h)} = (Ar + B)2^r$. Hence a total solution is $a_r = (Ar + B)2^r + (\frac{r}{6} + 1)r^2 2^r$, where A, B are arbitrary constants.

5.2.3. Solution by the method of generating functions

Instead of solving a difference equation for finding an expression for a_r by finding a homogeneous solution and a particular solution of the recurrence relation, we may find a solution with the help of what is called a generating function of the recurrence relation.

Definition. Given a recurrence relation, as in (5.27), a formal series $A(z) = a_0 + a_1 z + a_2 z^2 + \cdots + a_r z^r + \cdots = \sum_{r=0}^{\infty} a_r z^r$ is called a **generating function** of the recurrence relation.

Observe that if a generating function of a recurrence relation is obtained, then the coefficient of z^r gives the value of a_r and, hence, a solution of the difference equation is obtained. We will explain this procedure through some examples.

Example 5.28. Consider

$$a_r = 2a_{r-1} + 3, \quad r \geq 1, \tag{5.32}$$

with the boundary condition $a_0 = 1$. Multiplying both sides of (5.32) by z^r and taking the sum over r, we get

$$\sum_{1}^{\infty} a_r z^r = 2 \sum_{1}^{\infty} a_{r-1} z^r + 3 \sum_{1}^{\infty} z^r. \tag{5.33}$$

Let $\sum_0^{\infty} a_r z^r = A(z)$ so that $\sum_1^{\infty} a_r z^r = A(z) - a_0$, $\sum_1^{\infty} a_{r-1} z^r = z \sum_1^{\infty} a_{r-1} z^{r-1} = zA(z)$ and $\sum_1^{\infty} z^r = \frac{z}{1-z}$. Substituting these relations in (5.33), we get

$$A(z) - a_0 = 2zA(z) + \frac{3z}{1-z} \quad \text{or} \quad (1 - 2z)A(z) = 1 + \frac{3z}{1-z} = \frac{1 + 2z}{1-z}$$

which implies that

$$A(z) = \frac{1 + 2z}{(1-z)(1-2z)} = \frac{4}{1-2z} - \frac{3}{1-z}. \tag{5.34}$$

Consequently, $a_r = 4.2^r - 3$. Thus (5.34) is the generating function of the given difference equation.

Now consider the general difference equation of order k as in (5.27) which is valid for $r \geq s$ where $s \geq k$. Multiplying both sides of (5.27) by z^r and summing over $r = s$ to ∞,

$$\sum_{r=s}^{\infty} (c_0 a_r + c_1 a_{r-1} + \cdots + c_k a_{r-k}) z^r = \sum_{r=s}^{\infty} f(r) z^r \quad \text{or}$$

$$c_0 \sum_{r=s}^{\infty} a_r z^r + c_1 z \sum_{r=s}^{\infty} a_{r-1} z^{r-1} + \cdots + c_k z^k \sum_{r=s}^{\infty} a_{r-k} z^{r-k} = \sum_{r=s}^{\infty} f(r) z^r. \tag{5.35}$$

Now

$$\sum_{r=s}^{\infty} a_r z^r = A(z) - a_0 - a_1 z - \cdots - a_{s-1} z^{s-1},$$

$$\sum_{r=s}^{\infty} a_{r-1} z^{r-1} = A(z) - a_0 - a_1 z - \cdots - a_{s-2} z^{s-2},$$

$$\cdots$$

$$\sum_{r=s}^{\infty} a_{r-k} z^{r-k} = A(z) - a_0 - a_1 z - \cdots - a_{s-k-1} z^{s-k-1}.$$

Substituting these values in (5.35), we get

$$A(z) = \frac{1}{c_0 + c_1 z + \cdots + c_k z^k}$$

$$\times \left\{ \sum_{r=s}^{\infty} f(r) z^r + c_0 (a_0 + a_1 z + \cdots + a_{s-1} z^{s-1}) \right.$$

$$+ c_1 z (a_0 + a_1 z + \cdots + a_{s-2} z^{s-2})$$

$$\left. + \cdots + c_k z^k (a_0 + a_1 z + \cdots + a_{s-k-1} z^{s-k-1}) \right\}$$

which gives the required generating function.

Example 5.29. Solve the difference equation

$$a_r - 7a_{r-1} + 12a_{r-2} = 3^r + r, \quad r \geq 2 \tag{5.36}$$

by first finding its generating function, with the boundary conditions $a_0 = 1, a_1 = 2$.

Solution. Multiply both sides of (5.36) by z^r and sum over $r = 2$ to ∞. We get

$$\sum_{r=2}^{\infty} a_r z^r - 7z \sum_{r=2}^{\infty} a_{r-1} z^{r-1} + 12z^2 \sum_{r=2}^{\infty} a_{r-2} z^{r-2} = \sum_{r=2}^{\infty} 3^r z^r + \sum_{r=2}^{\infty} r z^r$$

or

$$A(z) - a_0 - a_1 z - 7z(A(z) - a_0) + 12z^2 A(z) = \frac{9z^2}{1 - 3z} + z \left[\frac{1}{(1-z)^2} - 1 \right]$$

or

$$A(z)(1 - 7z + 12z^2)$$

$$= \frac{9z^2}{1 - 3z} + z\left[\frac{1}{(1 - z)^2} - 1\right] + 1 - 5z$$

$$= \frac{9z^2}{1 - 3z} + \frac{z(2z - z^2)}{(1 - z)^2} + (1 - 5z)$$

$$= \frac{9z^2(1 - 2z + z^2) + (z - 3z^2)(2z - z^2) + (1 - 5z)(1 - 2z + z^2)(1 - 3z)}{(1 - z)^2(1 - 3z)}$$

$$= \frac{9z^2 - 18z^3 + 9z^4 + (1 - 7z + 13z^2 - 6z^3)(1 - 3z)}{(1 - z)^2(1 - 3z)}$$

$$\times \frac{1 - 10z + 43z^2 - 63z^3 + 27z^4}{(1 - z)^2(1 - 3z)}.$$

Therefore

$$A(z) = \frac{1 - 10z + 43z^2 - 63z^3 + 27z^4}{(1 - 3z)^2(1 - z)^2(1 - 4z)}$$

$$= \frac{11/36}{1 - z} + \frac{1/6}{(1 - z)^2} - \frac{21/4}{1 - 3z} - \frac{3}{(1 - 3z)^2} + \frac{79/9}{1 - 4z}.$$

Consequently $a_r = $ coefficient of z^r in the generating function

$$= \frac{11}{36} + \left\{\frac{r + 1}{6}\right\} - \frac{7}{4} \cdot 3^{r+1} - 3(r + 1)3^r + \frac{79 \cdot 4^r}{9}$$

$$= \frac{17}{36} + \frac{r}{6} - \frac{11}{4} \cdot 3^{r+1} - r \cdot 3^{r+1} + \frac{79.4^r}{9}.$$

Example 5.30. Solve the recurrence relation $a_r = a_{r-1} + a_{r-2}$ for $r \geq 3$, with $a_0 = 1, a_1 = 2, a_2 = 3$.

Solution. Multiply both sides of the given recurrence relation by z^r and add for $r = 3$ to ∞. we get

$$\sum_{r=3}^{\infty} a_r z^r = z\sum_{r=3}^{\infty} a_{r-1} z^{r-1} + z^2\sum_{r=3}^{\infty} a_{r-2} z^{r-2}.$$

Set $\sum_{r=0}^{\infty} a_r z^r = A(z)$. We get
$$(A(z) - a_0 - a_1 z - a_2 z^2) = z(A(z) - a_0 - a_1 z) + z^2(A(z) - a_0) \text{ or}$$
$$A(z)(1 - z - z^2) = a_0 + a_1 z + a_2 z^2 - z(a_0 + a_1 z) - a_0 z^2 = 1 + z.$$

Therefore,

$$A(z) = \frac{1+z}{1-z-z^2} = \frac{1+z}{\left(\frac{\sqrt{5}-1}{2}-z\right)\left(\frac{\sqrt{5}+1}{2}+z\right)}$$

$$= \frac{A}{\left(\frac{\sqrt{5}-1}{2}-z\right)} + \frac{B}{\left(\frac{\sqrt{5}+1}{2}+z\right)}$$

where $A = \frac{\sqrt{5}+1}{2\sqrt{5}}$, $B = \frac{1-\sqrt{5}}{2\sqrt{5}}$. Set $a = \frac{\sqrt{5}-1}{2}$, $b = \frac{\sqrt{5}+1}{2}$. Then $A(z) = \frac{A}{a}(1-\frac{z}{a})^{-1} + \frac{B}{b}(1+\frac{z}{b})^{-1}$. Therefore, $a_r = \frac{A}{a^{r+1}} + \frac{B(-1)^r}{b^{r+1}}$, where A, B, a, b are as above.

5.2.4. *Some typical examples*

Example 5.31. Let $4a_r + C_1 a_{r-1} + C_2 a_{r-2} = f(r), r > 1$ be a second order linear recurrence relation with constant coefficients. For some boundary conditions a_0 and a_1, the solution of the recurrence relation is $1 - 2r + 3 \cdot 2^r$. Determine a_0, a_1, C_1, C_2 and $f(r)$.

Solution. Since $1 - 2r + 3 \cdot 2^r$ is a solution of the given recurrence relation, we have

$$4(1 - 2r + 3 \cdot 2^r) + C_1(1 - 2(r-1) + 3 \cdot 2^{r-1}) + C_2(1 - 2(r-2)$$
$$+ 3 \cdot 2^{r-2}) = f(r)$$

which implies that

$$(4 + 3C_1 + 5C_2) - r(8 + 2C_1 + 2C_2) + 3(16 + 2C_1 + C_2)2^{r-2} = f(r).$$

Therefore, $f(r) = a + 2br + 3c \cdot 2^{r-2}$, where

$$4 + 3C_1 + 5C_2 = a, \quad 4 + C_1 + C_2 = -b, \quad 16 + 2C_1 + C_2 = c.$$

Let us suppose that $a = b = 0$. Then $4 + 3C_1 + 5C_2 = 0$ and $4 + C_1 + C_2 = 0$ which lead to $C_1 = -8, C_2 = 4$. With these values of C_1, C_2, we get $c = 4$. Hence $f(r) = 3 \cdot 2^r$. The given recurrence relation then becomes $a_r - 8a_{r-1} + 4a_{r-2} = 3 \cdot 2^r, r > 1$. Since $a_r = 1 - 2r + 3 \cdot 2^r$, taking $r = 0$, 1 gives $a_0 = 4, a_1 = 5$.

Observe that by giving another set of values to a, b we get another set of values of C_1, C_2 and $f(r)$ — thus giving another difference equation of second order with the same solution and the same boundary conditions.

Example 5.32. Solve the recurrence relation $a_r - 5a_{r-1} + 6a_{r-2} = f(r)$, where $f(r) = 0$ for $r = 0, 1$ and $f(r) = 6$ otherwise, given that $a_0 = a_1 = 0$.

Solution. The characteristic equation is $\alpha^2 - 5\alpha + 6 = 0$ which has 2 and 3 as its roots. Therefore, a homogeneous solution is $a_r^{(h)} = A2^r + B3^r$, where A, B are certain constants. Since $f(r) = 0$ for $r = 0, 1$ and $f(r) = 6$ for $r > 1$, the recurrence relation becomes $a_r - 5a_{r-1} + 6a_{r-2} = 6$, $r > 1$, with $a_0 = a_1 = 0$. A particular solution of the recurrence relation is $a_r^{(p)} = P$, where P is a constant. Substituting this in the given relation gives $P - 5P + 6P = 6$ or that $P = 3$. A total solution of the recurrence relation is $a_r = A2^r + B3^r + 3$. Since $a_0 = a_1 = 0$, taking $r = 0, 1$ in succession in the above relation gives $A + B + 3 = 0$, $2A + 3B + 3 = 0$ and solving for A, B, we get $B = 3$ and $A = -6$.

Hence a solution of the recurrence relation is

$$a_r = -6(2)^r + 3(3)^r + 3$$

$$= -6(2)^r + 3^{r+1} + 3, \quad \text{for all } r.$$

Exercise. Solve the above recurrence relation by the method of generating functions.

Example 5.33. Show that the Fibonacci numbers satisfy the recurrence relation $f_n = 5f_{n-4} + 3f_{n-5}$ for $n = 5, 6, 7$, together with the initial conditions $f_0 = 0, f_1 = 1, f_2 = 1, f_3 = 2$ and $f_4 = 3$. Use this recurrence relation to show that f_{5n} is divisible by 5 for $n = 1, 2, 3 \ldots$.

Solution. As already seen, the Fibonacci numbers satisfy the recurrence relation

$$a_n = a_{n-1} + a_{n-2}, \quad n > 2 \tag{5.37}$$

with the initial conditions $a_1 = 1, a_2 = 1$. Since $f_1 = f_2 = 1$ as given, we have $f_1 = a_1, f_2 = a_2$. Then $a_3 = a_1 + a_2 = 1 + 1 = 2 = f_3$ and $a_4 = a_2 + a_3 = 1 + 2 = 3 = f_4$.

Now, suppose that $n > 4$ and that f_k for $k < n+1$ satisfy the recurrence relation (5.37). Then we need to check that $f_{n+1} = a_{n+1}$. Now

$f_{n+1} = 5f_{n-3} + 3f_{n-4} = 5a_{n-3} + 3a_{n-4}$ (by induction hypothesis)

$= 5a_{n-3} + 3(a_{n-2} - a_{n-3}) = 2a_{n-3} + 3a_{n-2} = 2a_{n-3} + 3(a_n - a_{n-1})$

$= 2(a_{n-1} - a_{n-2}) + 3a_n - 3a_{n-1} = 3a_n - a_{n-1} - 2a_{n-2}$

$= 3a_n - a_{n-1} - 2(a_n - a_{n-1})$

$= a_n + a_{n-1} = f_n + f_{n-1}.$

This completes induction and, hence, $f_n, n > 4$, satisfy the recurrence relation for Fibonacci numbers.

For $n = 1, f_{5n} = f_5 = 5f_1 + 3f_0 = 5 \times 1 + 3 \times 0 = 5$ which is divisible by 5. Suppose that $n > 0$ and that f_{5k} is divisible by 5 for every $k < n + 1$. Now $f_{5(n+1)} = f_{5n+5} = 5f_{5n+1} + 3f_{5n}$ and $5f_{5n+1}$ is divisible by 5 trivially while f_{5n} is divisible by 5 (by induction hypothesis). Hence $f_{5(n+1)}$ is divisible by 5. This completes induction and, therefore, f_{5n} is divisible by 5 for all $n > 0$.

Next we will consider some typical examples for formation of recurrence relations.

Example 5.34. Gossip is spread among r people via telephone. In particular, in a telephone conversation between A and B, A tells B all the gossip he has heard, and B reciprocates. Let a_r denote the minimum number of telephone calls among r people so that all gossip will be known to every one. Show that

(a) $a_2 = 1, a_3 = 3$ and $a_4 = 4$;

(b) $a_r = a_{r-1} + 2$;

(c) $a_r < 2r - 3$, for $r > 3$.

Solution. (a) For $r = 2$, there are two persons, A and B, and when A telephones B or B telephones A, they share all the gossip known to them. Therefore, $a_2 = 1$.

For $r = 3$, there are three persons A, B and C. Firstly, A and B can share the gossip with each other by making one telephone call. The gossip shared by A and B needs to be shared with C by making one telephone call, either from A or B to C, or from C to A or B. Suppose that a conversation has taken place between A and C. Then A and C share all the gossip known to all the three. However, B is ignorant about the gossip revealed by C to A. To know this gossip, B must talk to either A or to C or vice versa. Hence at least three telephone calls are necessary for the three persons so that everyone knows all the gossip.

For $r = 4$, let the four persons be A, B, C and D. One telephone call is necessary for A, B to share the common gossip and another telephone call is necessary for C, D to share the common gossip. Now one telephone call from A to either C or D is necessary for the two, A, C or A, D to be able to share all the gossip that is known to A, B, C, D. Suppose A, C know all the gossip known to all four of them. Then there must be a telephone call between B and D for them to share all the gossip known to all four of them. Thus $a_4 = 4$.

(b) Let $A_1, A_2, \ldots, A_r$ be all the r persons. Let there be a call between A_1 and A_2, so that the two know all the gossip known to each other. Consider now the $r - 1$ people $A_2, A_3, \ldots, A_r$. For them to share all the gossip known to every one of them, the minimum number of calls made is a_{r-1}. Since the gossip known to A_1 is already known to A_2, the persons $A_2, A_3, \ldots, A_r$ share all the gossip known to all the r people. Now, for A_1 to know all the gossip known to $A_3, \ldots, A_r$, there has to be one telephone call between A_1 and any one of $A_2, A_3, \ldots, A_r$. Hence there are exactly $a_{r-1} + 2$ calls necessary for $A_1, A_2, \ldots, A_r$ to share all the gossip known to all of them. Therefore, $a_r = a_{r-1} + 2$, for all $r > 2$.

(c) Since $a_4 = 4 < 2 \cdot 4 - 3$, we have the starting point for induction. Now, by (b) above $a_{r+1} = a_r + 2 < 2 \cdot r - 3 + 2 = 2(r + 1) - 3$, which completes induction.

Example 5.35. The Tower of Hanoi problem. There are three pegs mounted on a board, together with disks of different sizes. Initially these disks are placed on the first peg in order of size, with the largest on the bottom. The rules of the puzzle allow the disks to be moved one at a time from one peg to another, as long as a disk is never placed on top of a smaller disk. The goal of the puzzle is to find the number of moves to have all the disks on the second peg in order of size, with the largest on the bottom.

Solution. Let H_n denote the number of moves needed to solve the Tower of Hanoi problem with n disks. We will set up a recurrence relation for the sequence $\{H_n\}$. We begin with n disks on peg 1. We can transfer the top $n - 1$ disks, following the rules of the puzzle, to peg 3 using H_{n-1} moves. While these moves are made, we keep the largest disk fixed. Use one move to transfer the largest disk from peg 1 to peg 2. Now, suppose the largest disk is fixed on the second peg. Transfer the $n-1$ disks on peg 3 to peg 2, placing these disks on top of the largest disk and we use H_{n-1} moves for doing so. Therefore, the total number of moves for the whole transfer is $2H_{n-1} + 1$. Hence $H_n = 2H_{n-1} + 1$, which is the required recurrence relation for the transfer of disks.

We will now use the method of backtracking to solve this problem. Using iteration, we get

$$
\begin{aligned}
H_n &= 2H_{n-1} + 1 \\
&= 2\{2H_{n-2} + 1\} + 1 = 2^2 H_{n-2} + 2 + 1 \\
&= 2^2\{2H_{n-3} + 1\} + 2 + 1 = 2^3 H_{n-3} + 2^2 + 2 + 1
\end{aligned}
$$

$$= 2^3 \{2H_{n-4} + 1\} + 2^2 + 2 + 1 = 2^4 H_{n-4} + 2^3 + 2^2 + 2 + 1$$

$$\cdots$$

$$= 2^{n-1} H_1 + 2^{n-2} + 2^{n-3} + \cdots + 2^2 + 2 + 1$$

$$= 2^{n-1} + 2^{n-2} + 2^{n-3} + \cdots + 2^2 + 2 + 1 = \frac{(2^n - 1)}{(2 - 1)} = 2^n - 1.$$

Remark. The above puzzle was invented by the French mathematician Édouard Lucas in the late nineteenth century. There is a myth associated with the puzzle. It tells that in a tower in Hanoi, monks transferred 64 gold disks from one peg to another, according to the rules of the puzzle. The myth says that the world will end when the monks finish the transfer of discs. How long after the monks started will the world end if the monks take one second to move a disk?

From the above formula for transferring n disks, we find that it will take the monks $2^{64} - 1 = 18,446,744,073,709,551,615$ moves to transfer the disks. As they make one move per second, it will take them more than 500 billion years to solve the puzzle, thus the world will survive a while longer than it already has. (Refer to Rosen, 2003, for some more myths associated with the problem.)

Example 5.36. Code word enumeration. A computer system considers a string of decimal digits a valid code word if it contains an even number of 0 digits. Let a_n denote the number of valid n-digit code words. Find a recurrence relation for a_n.

Solution. For $n = 1$, there are no valid code words as no string of length 1 can have an even number of zeros. For $n = 2$, there is only one valid code word, namely 00, since the other sequences of length 2 have either no zero or one zero. Thus $a_1 = 0, a_2 = 1$. For $n = 3$, the number of valid code words equals the number of strings of length 3 with two zeros. Therefore, $a_3 = C(3, 2) \times 9 = 3 \times 9 = 27 = 8a_2 + (10)^2 - 9^2$.

Now suppose that $n > 2$. Let b_n denote the number of strings of length n of decimal digits having an odd number of zeros. Then, there being a total number of $(10)^n$ strings of length n and 9^n being the number of strings of length n having no zeros, we have $b_n = (10)^n - a_n - 9^n$.

Let c_{n+1} denote the last entry of a code word of length $n+1$. If $c_{n+1} = 0$, the valid code word results from a string of length n in b_n and if c_{n+1} is non-zero, the valid code word results from a valid code word of length n. On the other hand, attaching a non-zero entry at the end of a code word of length n and attaching a zero entry at the end of a string in b_n leads to a code word of length $n + 1$. Hence

$$a_{n+1} = 9a_n + b_n = 9a_n + (10)^n - a_n - 9^n = 8a_n + (10)^n - 9^n.$$

Therefore, $a_{n+1} = 8a_n + (10)^n - 9^n$, for all $n > 2$.

Since it has already been obtained for $n = 2$, the result is valid for all $n > 1$ with the boundary conditions $a_1 = 0$ and $a_2 = 1$ (proved already).

Example 5.37. Compound interest. A person deposits Rs. 10,000 in a savings account at a bank yielding 5% interest, the interest being compounded yearly. How much will the person get after 20 years?

Solution. Let A_n denote the amount that the person gets at the end of the nth year. Then A_{n-1} is the amount due to the man at the end of the $(n-1)$th year. The amount due at the end of the nth year is A_{n-1} plus the interest on this amount for one year. Letting P denote the amount invested and R the annual rate of interest paid by the bank, which is compounded annually, then the amount receivable at the end of the nth year is given by

$$A_n = A_{n-1} + \left(\frac{A_{n-1} \times R}{100} \right) = A_{n-1} \times \left(\frac{100 + R}{100} \right).$$

Using the method of backtracking, which is the same as using iteration, we get

$$A_n = A_0 \times \frac{(100 + R)^n}{100^n} = P \times \frac{(100 + R)^n}{100^n}.$$

For the particular question under reference, $P = 10,000$ and $R = 5$. Also $n = 20$. Therefore,

$$A_{20} = 10000 \times \left(\frac{21}{20} \right)^{20} \quad \text{which equals Rs. } 26,500.$$

Example 5.38. Rabbits and the Fibonacci numbers. We have seen that the sequence of Fibonacci numbers satisfies the recurrence relation $a_n = a_{n-1} + a_{n-2}$ for $n > 2$ with $a_0 = a_1 = 1$ and $a_2 = 2$. The sequence of Fibonacci numbers and the associated recurrence relation could be generated as follows.

A young pair of rabbits, one of each sex, is placed on an island. A pair of rabbits does not breed until they are two months old. After they are two months old, each pair of rabbits produces another pair each month. Let a_n

denote the number of pairs of rabbits at the end of the nth month. Then a_0 is the number of pairs of rabbits initially put on the island and so it is 1. Since no new pairs of rabbits are produced during the first month, the number of pairs of rabbits at the end of the first month is the same as the number of pairs put on the island initially. Therefore, $a_1 = a_0 = 1$. At the end of the second month the pair of rabbits put initially on the island produce one pair of rabbits and so at the end of the second month the number of pairs of rabbits is $1 + a_0 = 1 + 1 = 2$. Suppose that $n > 2$ and that $a_n =$ the number of pairs of rabbits at the end of the nth month $=$ the number of pairs of rabbits at the end of the previous month $+$ the number of pairs of rabbits that are produced during (in fact at the end of) the nth month (it being assumed that no rabbits ever die). The number of rabbits at the end of the previous month is a_{n-1}. Also it is the number of pairs of rabbits at the end of the $(n-2)$th month that produce one pair each at the end of the nth month. Thus the number of pairs of rabbits produced at the end of the nth month is a_{n-2}. Hence $a_n = a_{n-1} + a_{n-2}$ which is the required recurrence relation valid for all $n > 2$. It is already seen to be valid for $n = 2$. Therefore, the recurrence relation is valid for all $n > 1$ with the initial conditions $a_0 = a_1 = 1$.

Example 5.39. Find a recurrence relation for c_n, the number of ways to parenthesize the product of $n + 1$ symbols $x_0, x_1, x_2, \ldots, x_n$ without changing the order.

Solution. Before we take up the general case, we will consider the problem for the values $n = 1, 2, 3, 4$. For $n = 1$, there are two symbols, x_0, x_1 and there is only one way to parenthesize the product as $x_0 x_1$.

For $n = 2$, the symbols are x_0, x_1, x_2 and we can consider these as two symbols in two ways, such as $x_0 x_1, x_2$ and $x_0, x_1 x_2$ and there is exactly one way to parenthesize the two symbols in each case. Therefore, $c_2 = 1 + 1 = 2$.

For $n = 3$, the symbols used are x_0, x_1, x_2, x_3 and to parenthesize these is to parenthesize the symbols in the following sets

$$x_0 x_1, x_2, x_3, \quad x_0, x_1 x_2, x_3, \quad x_0, x_1, x_2 x_3.$$

There is no common way to parenthesize the symbols in the first two sets whereas, while parenthesizing the symbols in the third set, we get some common arrangements with the parenthesizing of the first set. The common arrangements are precisely the arrangements obtained on parenthesizing the symbols $x_0 x_1, x_2 x_3$. The number of such arrangements is precisely c_1. Also, the number of arrangements obtained for the three sets is c_2 in each case. Therefore, $c_3 = 3c_2 - c_1$.

For $n = 4$, the symbols are x_0, x_1, x_2, x_3, x_4. The number of ways for parenthesizing these is precisely the number of ways for parenthesizing the symbols in the sets

$$x_0x_1, x_2, x_3, x_4; \quad x_0, x_1x_2, x_3, x_4; \quad x_0, x_1, x_2x_3, x_4; \quad x_0, x_1, x_2, x_3x_4.$$
$$(5.38)$$

The number of ways of parenthesizing the symbols in each of these sets is c_3. However, the number of ways of parenthesizing the symbols x_0x_1, x_2x_3, x_4 is common to the number of ways of parenthesizing the elements in the first and the third lists in (5.38) and the number of ways of parenthesizing the symbols x_0x_1, x_2, x_3x_4 is common to the number of ways of parenthesizing the symbols in the first and the fourth list in (5.38), while the number of ways of parenthesizing the symbols x_0, x_1x_2, x_3x_4 is common to the number of ways of parenthesizing the symbols in the second and the fourth list in (5.38). The number of ways of parenthesizing the symbols in each of the three lists $x_0x_1, x_2x_3, x_4; x_0x_1, x_2, x_3x_4; x_0, x_1x_2, x_3x_4$ is c_2. Therefore, the number of ways of parenthesizing the five symbols x_0, x_1, x_2, x_3, x_4 is $c_4 = 4c_3 - 3c_2$.

With this much motivational preparation, we now consider the general case and prove that

$$c_n = nc_{n-1} - \left(\frac{(n-1)(n-2)}{2} \right) c_{n-2} \quad \text{for all } n > 1 \qquad (5.39)$$

with $c_0 = c_1 = 1$.

We have already established this result for $n = 2$. Suppose that $n > 1$ and that the result (5.39) holds for n. Consider the number of ways of parenthesizing the product of $n + 2$ symbols: $x_0, x_1, x_2, \ldots, x_{n+1}$. This number is precisely the number of ways of parenthesizing the symbols in the $n + 1$ sets of symbols

$$x_0, x_1, \ldots, x_{i-1}, y_i, x_{i+2}, \ldots, x_{n+1} \qquad (5.40)$$

for $i = 0, 1, 2, \ldots, n$ where $y_i = x_i x_{i+1}$.

By induction hypothesis, the number of ways to parenthesize the symbols in each of the sets (5.40) is c_n. Therefore, the total number of ways to parenthesize the symbols in these $n + 1$ sets of symbols is $(n + 1)c_n$. However, all these ways of parenthesizing are not distinct. Consider two of the sets from (5.40), say

$$x_0, x_1, \ldots, x_{i-1}, y_i, x_{i+2}, \ldots, x_{n+1}, \qquad (5.41)$$

$$x_0, x_1, \ldots, x_{j-1}, y_j, x_{j+2}, \ldots, x_{n+1}, \qquad (5.42)$$

with $i + 1 < j$ or $j + 1 < i$. In parenthesizing symbols in these two sets, we are led to a set of symbols (assuming that $i + 1 < j$)

$$x_0, x_1, \ldots, x_{i-1}, \quad y_i, \ldots, y_j, \ldots, x_{n+1}. \qquad (5.43)$$

Once we come to the set (5.43), the number of ways of parenthesizing the symbols in (5.43) is common to the number of ways of parenthesizing the symbols in (5.41) and (5.42). The number of ways of parenthesizing the symbols in (5.43) is c_{n-1}. Since either $i + 1 < j$ or $j + 1 < i, i, j$ are to be chosen from the n integers $0, 1, \ldots, i, i + 2, \ldots, j, j + 2, \ldots, n + 1$. The number of choices for i and j is $\frac{n(n-1)}{2}$ which is the number of choices of the sets (5.41) and (5.42) from among the sets (5.40). Therefore, the number of distinct ways of parenthesizing the symbols in the set (5.40) is $(n + 1)c_n - (\frac{n(n-1)}{2})c_{n-1}$. Hence

$$c_{n+1} = (n + 1)c_n - \left(\frac{n(n - 1)}{2} \right) c_{n-1}.$$

5.2.5. *Recurrence relations reducible to linear recurrence relations*

We will next consider some examples of recurrence relations that are reducible to linear recurrence relations.

Example 5.40. Consider the simultaneous linear recurrence relations $a_r = 3a_{r-1} + 4b_{r-1}, b_r = a_{r-1} + b_{r-1}$ with the boundary conditions $a_0 = 1, b_0 = 0$.

Solution. From the second equation, $a_{r-1} = b_r - b_{r-1}$ substituting in the first equation we get $b_{r+1} - b_r = 3(b_r - b_{r-1}) + 4b_{r-1}$, or $b_{r+1} - 4b_r - b_{r-1} = 0, r > 0$. This is now a linear recurrence relation. The characteristic equation of this recurrence relation is $t^2 - 4t - 1 = 0$ having roots a, b where $a = 2 + \sqrt{5}, b = 2 - \sqrt{5}$. Therefore a solution of this recurrence relation is $b_r = Aa^r + Bb^r$, where A, B are constants. Since $b_0 = 0$, taking $r = 0$, we get $A + B = 0$ so that $B = -A$, and, therefore, $b_r = A\{a^r - b^r\}$.

Substituting this value of b_r in the first given recurrence relation, we get

$$a_r - 3a_{r-1} = 4A\{a^{r-1} - b^{r-1})\qquad(5.44)$$

which is again a linear recurrence relation. The characteristic equation of this recurrence relation is $t-3=0$ having 3 as a root. Therefore a particular solution of (5.44) is of the form $Pa^r + Qb^r$, where P, Q are constants. Substituting this in (5.44), we get $Pa^r + Qb^r - 3(Pa^{r-1} + Qb^{r-1}) = 4A(a^{r-1} - b^{r-1})$ or $(Pa - 3P - 4A)a^{r-1} + (Qb - 3Q + 4A)b^{r-1} = 0$. Since this relation is true for all values of r, $Pa - 3P - 4A = 0$ and

$$Qb - 3Q + 4A = 0 \text{ giving } P = \frac{4A}{(a-3)}, \quad Q = -\frac{4A}{(b-3)}.$$

Therefore a particular solution of (5.44) is

$$a_r^{(p)} = 4A \left\{ \frac{a^r}{(a-3)} - \frac{b^r}{(b-3)} \right\}.$$

As 3 is a simple characteristic root of (5.44), a general solution of this recurrence relation is

$$a_r = 4A \left\{ \frac{a^r}{(a-3)} - \frac{b^r}{(b-3)} \right\} + \alpha 3^r.$$

For $r = 0, a_0 = 1$ and we get $1 = 4A\frac{(b-a)}{(a-3)(b-3)} + \alpha$ giving

$$\alpha = 1 + 4A \left(\frac{1}{b-3} - \frac{1}{a-3} \right).$$

Hence

$$a_r = 4A \left\{ \frac{a^r}{a-3} - \frac{b^r}{b-3} \right\} + \left\{ 1 + 4A \left(\frac{1}{b-3} - \frac{1}{a-3} \right) \right\} 3^r.$$

Putting the values of a_r and b_r in the second difference equation given, we get

$$A(a^r - b^r) = 4A \left\{ \frac{a^{r-1}}{a-3} - \frac{b^{r-1}}{b-3} \right\} + \left\{ 1 + 4A \left(\frac{1}{b-3} - \frac{1}{a-3} \right) \right\} 3^{r-1}$$

$$+ A(a^{r-1} - b^{r-1})$$

$$= A \left(\frac{(a+1)a^{r-1}}{a-3} - \frac{(b+1)b^{r-1}}{b-3} \right)$$

$$+ \left\{ 1 + 4A \left(\frac{1}{b-3} - \frac{1}{a-3} \right) \right\} 3^{r-1}$$

or

$$Aa^{r-1}\frac{(a^2-4a-1)}{a-3} - Ab^{r-1}\frac{(b^2-4b-1)}{b-3}$$

$$= \left\{1 + 4A\left(\frac{1}{b-3} - \frac{1}{a-3}\right)\right\}3^{r-1}.$$

Since this result is true for all r, we get $\{1 + 4A(\frac{1}{b-3} - \frac{1}{a-3})\} = 0$ which finally leads to $A = \frac{(a-3)(b-3)}{4(b-a)} = \frac{(-1+\sqrt{5})(-1-\sqrt{5})}{4(-2\sqrt{5})} = \frac{1}{2\sqrt{5}}$. Putting this value of A, we get

$$a_r = \frac{2}{\sqrt{5}}\left(\frac{a^{r-1}}{\sqrt{5}-1} + \frac{b^{r-1}}{\sqrt{5}+1}\right) \quad \text{and} \quad b_r = \frac{1}{2\sqrt{5}}(a^r - b^r).$$

Example 5.41. Solve the recurrence relation $a_r^2 - 3a_{r-1}^2 = 1$, given that $a_0 = 3$.

Solution. Let $b_r = a_r^2$. Then the given recurrence relation becomes $b_r - 3b_{r-1} = 1$, with $b_0 = 9$. Multiplying both sides by z^r and summing over r for $r = 1$ to ∞, we get (with $B(z) = \sum_{r=0}^{\infty} b_r z^r$)

$$B(z) - b_0 - 3zB(z) = \sum_{r=1}^{\infty} z^r = \frac{1}{1-z} - 1 \quad \text{or} \quad (1-3z)B(z) = 8 + \frac{1}{1-z}.$$

Thus

$$B(z) = \frac{8}{1-3z} + \frac{1}{(1-3z)(1-z)} = \frac{8}{1-3z} + \frac{1}{2}\left(\frac{3}{1-3z} - \frac{1}{1-z}\right)$$

$$= \frac{19}{2}(1-3z)^{-1} - \frac{1}{2}(1-z)^{-1}.$$

Therefore, $b_r = 19 \cdot \frac{3^r}{2} - \frac{1}{2}$ and hence, $a_r = \sqrt{\frac{19 \cdot 3^r - 1}{2}}$.

Example 5.42. Solve the recurrence relation $ra_r - ra_{r-1} + a_{r-1} = 2^r$, given that $a_0 = 273$.

Solution. The given recurrence relation may be written as $ra_r - (r-1)a_{r-1} = 2^r$. Set $ra_r = b_r$, for $r = 0, 1, 2, \ldots$
 Then $b_0 = 0$ and $b_r - b_{r-1} = 2^r$.
 The characteristic equation of this relation is $\alpha - 1 = 0$, giving $\alpha = 1$ as a characteristic root. A homogeneous solution of this relation is $b_r^{(h)} = A.1^r = A$. Let $b_r^p = B2^r$ be a particular solution of the recurrence relation. Then $B.2^r - B.2^{r-1} = 2^r$ which yields $B = 2$. Therefore, a total solution

of the relation is $b_r = A + 2^{r+1}$. Thus $r a_r = A + 2^{r+1}$. Taking $r = 0$, we get $0 = A + 2$ so that $A = -2$. Hence

$$r a_r = 2^{r+1} - 2 \quad \text{or } a_r = \frac{2^{r+1}}{r} - \frac{2}{r}.$$

Observe that solution obtained for the recurrence relation is valid for $r \geq 1$ and the given boundary condition has absolutely no effect on the problem.

Example 5.43. Solve the recurrence relation $a_r^3 - 3a_{r-1} = 0$, given that $a_0 = 9$.

Solution. The given equation may be rewritten as $a_r^3 = 3a_{r-1}$. Taking logarithm of both sides with base 3, we get $3 \log a_r = 1 + \log a_{r-1}$.

Set $\log a_r = b_r$. Then the above equation reduces to $3b_r = 1 + b_{r-1}$ with $b_0 = \log a_0 = 2 \log 3 = 2$.

This is a linear recurrence relation and solving it by the method of generating functions, we get

$$3(B(z) - b_0) - zB(z) = \frac{z}{(1-z)} \quad \text{or } (3-z)B(z) = 6 + z(1-z)^{-1} \quad \text{or}$$

$$B(z) = \frac{6 - 5z}{(1-z)(3-z)} = \frac{(1-z)^{-1}}{2} + \frac{9(3-z)^{-1}}{2}.$$

Comparison of coefficients of z^r on both sides gives $b_r = \frac{1}{2} + \frac{9}{2 \cdot 3^{r+1}}$. Hence

$$a_r = 3^{\left(\frac{1}{2} + \frac{9}{2 \cdot 3^{r+1}}\right)}.$$

Example 5.44. Solve the recurrence relation

$$a_r = \sqrt[3]{a_{r-1} + \sqrt[3]{a_{r-2} + \sqrt[3]{a_{r-3} + \cdots}}}, \quad r > 0$$

given that $a_0 = 4$.

Solution. Taking cubes of both sides we get

$$a_r^3 = a_{r-1} + \sqrt[3]{a_{r-2} + \sqrt[3]{a_{r-3} + \cdots}}$$

$$= a_{r-1} + a_{r-1} = 2a_{r-1},$$

by the given relation.

Taking logarithm of both sides to the base 2 and setting $\log a_r = b_r$, we get

$$3b_r - b_{r-1} = 1, \quad r > 0.$$

Multiply both sides with z^r and sum over $r = 1$ to ∞. Writing

$$\sum_{r=0}^{\infty} b_r z^r = B(z),$$

we get

$$3(B(z) - b_0) - zB(z) = \frac{z}{1-z} \quad \text{or} \quad (3-z)B(z) = 3b_0 + \frac{z}{1-z}.$$

Since $b_0 = \log a_0 = \log 4 = 2$, we get

$$(3-z)B(z) = 6 + \frac{z}{1-z} = \frac{6-5z}{1-z} \quad \text{or}$$

$$B(z) = \frac{6-5z}{(1-z)(3-z)} = \frac{1}{2} \cdot \frac{1}{1-z} + \frac{9}{2} \cdot \frac{1}{3-z}.$$

Hence $b_r = \frac{1}{2} + \frac{3}{2} \cdot \frac{1}{3^r} = \frac{1}{2} + \frac{1}{2 \cdot 3^{r-1}}$ and so $a_r = 2^{\frac{1}{2}(1+\frac{1}{3^{r-1}})}$.

Note. A couple of solved examples are taken from unsolved exercises in Liu (2000). Examples on Tower of Hanoi problem, code enumeration problem, counting the number of parenthesizing product and the rabbits and the Fibonacci numbers have been taken from Rosen (2003).

Exercise 5.2.

1. Let a_n denote the nth Fibonacci number with $a_0 = 0$. Prove, by induction, that for every positive integer n
   (a) $a_1^2 + a_2^2 + \cdots + a_n^2 = a_n a_{n+1}$.
   (b) $a_1 + a_3 + \cdots + a_{2n-1} = a_{2n}$.
   (c) $a_0 a_1 + a_1 a_2 + \cdots + a_{2n-1} a_{2n} = a_{2n}^2$.
   (d) $a_0 - a_1 + a_2 - a_3 + \cdots - a_{2n-1} + a_{2n} = a_{2n-1} - 1$.
   (e) $a_{n-1} a_{n+1} - a_n^2 = (-1)^2$.

2. If $a_{m,n}$ is defined recursively by $a_{0,0} = 0$ and $a_{m,n} = a_{m-1,n} + 1$ if $n = 0$ and $m > 0$ while $a_{m,n} = a_{m,n-1} + 1$ if $n > 0$, prove by induction that $a_{m,n} = m + n$ for all $(m, n) \in N \times N$, where N is the set of all natural numbers.

3. Let $a_{m,n}$ be defined recursively by $a_{1,1} = 7$ and $a_{m,n} = a_{m-1,n} + 3$ if $n = 1$ and $m > 1$, while $a_{m,n} = a_{m,n-1} + 3$ if $n > 1$. Prove by induction

on $m + n$ that $a_{m,n} = 3(m + n) + 1$ for all $(m, n) \in Z^+ \times Z^+$, where Z^+ is the set of all positive integers.

4. Prove that none of the following recurrence relations admits a unique solution.

   (a) $a_n = 1 + a_{n-2}$ for $n > 1$ and $a_1 = 0$.

   (b) $a_n = 1 + a_{n-2}$ if n is odd and $a_1 = 1$.

   (c) $a_n = 1 + a_{n-1} + a_{n-3}$ for all $n > 3$ and $a_1 = 1, a_2 = 3$.

5. Use the method of backtracking to find the nth term of the following recurrence relations.

   (a) $a_n = a_{n-1} + a_{n-2}, n > 1$ and $a_0 = 1, a_1 = 2$.

   (b) $a_n = a_{n-1} + a_{n-2} + a_{n-3}$ for $n > 2$ with $a_0 = 1, a_1 = 2, a_2 = 3$.

   (c) $a_n - 2a_{n-1} = 1$ with the initial condition $a_1 = 3$.

6. Find the difference equation in each of the following cases.

   (a) $a_n = A \cdot 3^n + B \cdot 5^n$.

   (b) $a_n = (A + nB)3^n$.

   (c) a_n is the number of 2-*set* partitions of a set with n elements.

7. Solve the following recurrence relations.

   (a) $a_{n+3} = 2a_{n+2} + 5a_{n+1} - 6a_n$.

   (b) $a_{n+1} - 2a_n \sin \alpha + a_{n-1} = 0$.

   (c) $a_n = a_{n-1} + 2a_{n-2}$.

   (d) $a_n - 4a_{n-2} = 0$, for $n > 1$, given that $a_0 = 0, a_1 = 2$.

   (e) $a_{n+3} = 3a_{n+2} - 4a_n$.

   (f) $a_n = 3a_{n-2} - 2a_{n-3}$, for $n > 3$, given that $a_1 = 0, a_2 = 8$ and $a_3 = -2$.

   (g) $b_n = 3b_{n-1} - 2b_{n-2}$ given that $b_1 = 5, b_2 = 3$.

   (h) $b_n = 2b_{n-1} - b_{n-2}$ with the initial conditions $b_1 = 1$ and $b_2 = 3$.

   (i) $a_n = 7a_{n-2} + 6a_{n-3}, a_1 = 3, a_2 = 6, a_3 = 10$.

   (j) $a_n + 2a_{n-1} - 2a_{n-2} - 4a_{n-3} = 0$ with $a_1 = 0, a_2 = 2, a_3 = 8$.

   (k) $a_n = 4a_{n-1} + 5_{n-2}$, for $n > 2$, given that $a_1 = 2, a_2 = 6$.

   (l) $b_n + 3b_{n-1} + 2b_{n-2} = 0$ for $n > 2$, given that $b_1 = -2, b_2 = 4$.

   (m) $a_n = 4a_{n-1} - 4a_{n-2}$ for $n > 2$, given that $a_1 = 1, a_2 = 7$.

8. Solve the recurrence relations in Question 7 above where boundary conditions are given by the method of generating functions.

9. Solve the following recurrence relations.

   (a) $a_{n+2} - 4a_{n+1} + 3a_n = 3^n$.

   (b) $a_{n+2} - 4a_{n+1} + 3a_n = 7^n$.

(c) $a_n - 5a_{n-1} + 6a_{n-2} = 4^n$, for $n > 1$ given that $a_0 = 0, a_1 = 1$.

(d) $a_{n+2} - 6a_{n+1} + 9a_n = 3^n$.

(e) $a_{n+2} - 5a_{n+1} + 6a_n = n + 2^n$.

(f) $a_n - 2a_{n-1} + a_{n-2} = 3n - 1$.

(g) $a_n - 3a_{n-1} + 2a_{n-2} = n^2 - 2n - 1$.

(h) $a_{n+2} + 6a_{n+1} + 9a_n = n2^n + 3^n + 7$.

(i) $a_{n+2} - 5a_{n+1} + 6a_n = 4^n(n^2 - n + 5)$.

10. Solve the following simultaneous recurrence relations.

(a) $a_{n+1} - b_n = 2(n + 1), b_{n+1} - a_n = -2(n + 1)$.

(b) $a_{n+1} + b_n + c_n = 1, a_n + b_{n+1} + c_n = n, a_n + b_n + c_{n+1} = 2n$.

11. Solve the difference equation in Example 5.42 by method of generating functions.

Chapter 6

PARTIALLY ORDERED SETS

We have already defined partially ordered sets (posets). In this chapter we study such sets in more detail. Finite partially ordered sets can be represented by diagrams called Hasse diagrams, which are also discussed here. There is a close connection between the sizes of chains and antichains in finite posets and this is provided by the theorems of Mirsky and of Dilworth. These theorems are proved.

6.1. Preliminaries

We have already defined a partial order relation on a non-empty set. Recall that a relation R defined on a non-empty set S is called **a partial order relation** if the relation R is reflexive, anti-symmetric and transitive. Also, as has already been said, a non-empty set S with a partial order relation defined on it is called **a partially ordered set** or a **poset**. We may represent a poset by (S, R).

Example 6.1. The sets

(a) Z of all integers;
(b) Q of all rational numbers;
(c) $\mathcal{R}$ of all real numbers;
(d) $\mathcal{N}$ of all natural numbers,

are partially ordered sets with the relation R — the usual '$\leq$' relation defined for real numbers.

Example 6.2. On the set $\mathcal{N}$ of all natural numbers, define a relation R by: $a\,R\,b$ for $a, b \in \mathcal{N}$ if a divides b. As every natural number divides itself, the relation R is reflexive. If a, b are natural numbers and $a|b$ and $b|a$, then there exist positive integers k and l such that $b = ka$, $a = lb$. Then $a = kla$, which gives $kl = 1$. As k, l are both positive integers, we must have $k = l = 1$ and $b = ka = a$. This proves that the relation R ($=$'$|$') is anti-symmetric. If a, b, c are natural numbers and $a|b, b|c$, then $b = ka$, $c = lb$ and, so, $c = kla$, which shows that $a|c$. Hence the relation '$|$' is transitive. This completes the proof that the relation '$|$' is a partial order relation.

Example 6.3. Let S be a non-empty set and $\mathcal{P}(S)$ be the power set of S, i.e. $\mathcal{P}(S)$ is the set of all subsets of S. Consider the usual relation '$\subseteq$' defined on sets, i.e. for $A, B \in \mathcal{P}(S)$, $A \subseteq B$ if A is a subset of B. If $A \subseteq B$ and $B \subseteq A$, then $A = B$, showing that the relation is anti-symmetric. Also, every set is a subset of itself and the relation $\subseteq$ is reflexive. It is clear from the definition of a subset of a set that if $A \subseteq B$ and $B \subseteq C$, then $A \subseteq C$. Hence the relation '$\subseteq$' is transitive as well. Therefore, the relation '$\subseteq$' defined on $\mathcal{P}(S)$ is a partial order relation.

Example 6.4. The set $S = \{1, 2, 3, 4, 6, 8, 10\}$ with the relation '$|$'($=$ *divides*) is a poset.

Example 6.5. In the set $\mathcal{N} \times \mathcal{N}$ of all ordered pairs (a,b), where $a, b \in \mathcal{N}$ define an order relation '$\leq$' as follows:

For $(a_1, b_1), (a_2, b_2) \in \mathcal{N} \times \mathcal{N}$, say that $(a_1, b_1) \leq (a_2, b_2)$ if either $a_1 < a_2$ or $a_1 = a_2$ but $b_1 \leq b_2$. This relation is clearly reflexive. Let $(a_1, b_1), (a_2, b_2) \in \mathcal{N} \times \mathcal{N}$ and suppose that $(a_1, b_1) \leq (a_2, b_2)$ and $(a_2, b_2) \leq (a_1, b_1)$. Then

(a) either $a_1 < a_2$ or $a_1 = a_2$ and $b_1 \leq b_2$;
(b) $a_2 < a_1$ or $a_1 = a_2$ and $b_2 \leq b_1$

hold together. If $a_1 < a_2$, then (b) cannot hold. Therefore, we must have $a_1 = a_2$. Then (a) and (b) imply that $b_1 \leq b_2, b_2 \leq b_1$, which together give $b_1 = b_2$. Hence $(a_1, b_1) = (a_2, b_2)$ and the relation is anti-symmetric.

Next, let $(a_1, b_1), (a_2, b_2), (a_3, b_3) \in \mathcal{N} \times \mathcal{N}$ and suppose that $(a_1, b_1) \leq (a_2, b_2)$ and $(a_2, b_2) \leq (a_3, b_3)$. Then

(c) either $a_1 < a_2$ or $a_1 = a_2$ but $b_1 \leq b_2$;
(d) either $a_2 < a_3$ or $a_2 = a_3$ and $b_2 \leq b_3$.

If $a_1 < a_2$, then for both cases of (d), $a_1 < a_3$. On the other hand, suppose that $a_1 = a_2$ so that $b_1 \leq b_2$. If $a_2 < a_3$, then $a_1 < a_3$, while if $a_2 = a_3$ then $b_2 \leq b_3$ and, so, $b_1 \leq b_3$. Thus, either $a_1 < a_3$ or $a_1 = a_3$ but $b_1 \leq b_3$. Hence $(a_1, b_1) \leq (a_3, b_3)$. This proves that the relation is transitive and, therefore, is a partial order relation. The order relation defined above is called **lexicographic ordering.**

We can extend the lexicographic ordering defined above on $\mathcal{N} \times \mathcal{N}$ to $\mathcal{N}^{(k)} = \mathcal{N} \times \cdots \times \mathcal{N}$ (k-terms) in a natural way. If $(a_1, a_2, \ldots, a_k)$, $(b_1, b_2, \ldots, b_k) \in \mathcal{N}^{(k)}$ we say that $(a_1, a_2, \ldots, a_k) \leq (b_1, b_2, \ldots, b_k)$ if there exists an integer $i, 1 \leq i \leq k$, such that $a_1 = b_1, a_2 = b_2, \ldots, a_i = b_i$ but $a_{i+1} \leq b_{i+1}$. It may be proved on the lines of the above proof that the lexicographic order defined on $\mathcal{N}^{(k)}$ is a partial order relation.

Lexicographic ordering is also called **dictionary ordering.** Strictly speaking, we can define the dictionary ordering as follows: Let $a = a_1 a_2 \cdots a_m$, $b = b_1 b_2 \cdots b_n$ be two words using the English alphabet, i.e. every a_i and b_j is an English letter. Let $k = \min(m, n)$. We say that $a \leq b$ if there exists an $i \leq k$ such that $a_1 = b_1, \ldots, a_i = b_i$ and $a_{i+1} < b_{i+1}$ if $i + 1 \leq k$ while if $i = k$, then $a \leq b$ provided $m < n$, i.e. if $k = m$, then $a \leq b$ if either $a_1 a_2 \cdots a_m \leq b_1 b_2 \cdots b_m$ or $a_1 a_2 \cdots a_m = b_1 b_2 \cdots b_m$ and $m < n$. This is precisely the order used in the dictionary to give the words listed in it.

Lexicographic ordering could, in fact, also be defined on $A_1 \times A_2 \times \cdots \times A_k$ where $(A_1, \leq), (A_2, \leq), \ldots, (A_k, \leq)$ are totally ordered posets not necessarily all equal. (Definition to follow.) Observe that $(3, 6) \leq (4, 7), (3, 6) \leq (4, 5), (3, 6) \leq (3, 7)$. If we compare the words **discrete** and **discriminent**, since $e < i$ we have discrete $\leq$ discriminent while discreet $\leq$ discrete.

Consider the partially ordered set $(\mathcal{N}, |)$ with the relation '|' dividing, i.e. if $a, b \in \mathcal{N}$, then aRb if $a|b$. Observe that in this poset the elements 2 and 3 are such that neither $2|3$ nor $3|2$ i.e. neither 2 is related to 3 nor 3 to 2. Again, if $S = \{1, 2, 3\}$, then $\mathcal{P}(S) = \{\emptyset, \{1\}, \{2\}, \{3\}, \{1, 2\}, \{1, 3\}, \{2, 3\}, S\}$. In the poset $(\mathcal{P}(S), \leq)$ with $A \leq B$ if A is a subset of B, we have pairs of elements $\{1\}$, $\{2\}; \{1\}, \{3\}; \{2\}, \{3\}; \{1, 2\}, \{1, 3\}; \{1, 2\}, \{2, 3\}; \{1, 3\}, \{2, 3\}$ which have the property that neither is related to the other.

Motivated by the above observations, we make the following definition.

Definition. Given a poset $(S, \leq)$, two elements $a, b \in S$ are said to be **comparable** if either $a \leq b$ or $b \leq a$. If a, b are not comparable, we say that a, b are **incomparable.**

Observe that in the poset $(\mathcal{N}, |)$, the elements $2, 6$ are comparable, while $3, 5$ are not comparable. In fact, there are infinite pairs of elements in this poset which are comparable and infinitely many pairs of elements which are incomparable.

In the poset $\{S, |\}$ where $S = \{1, 2, 3, 4, 6, 8, 10\}, 1, 2; 1, 3; 2, 4; , 2, 6;$ $2, 8; 3, 6; 4, 8; 2, 10$ are pairs of comparable elements while the elements $2, 3$ are incomparable, and so are the pairs of elements $3, 4; 4, 10; 6, 8$.

Definition. If $(S, \leq)$ is a poset in which every two elements of S are comparable, i.e. when the law of dichotomy holds, then S is called a **totally ordered** or **linearly ordered** set. Observe that the poset $(\mathcal{N}, \leq)$ is linearly ordered while $(\mathcal{N}, |)$ is not linearly ordered. If S is a set having at least two elements and $\mathcal{P}(S)$ is the power set of S, then $(\mathcal{P}(S), \subseteq)$ is not linearly ordered.

Observe that $(\mathcal{N}, \leq)$ is a totally ordered set. We can generalize the concept of lexicographic ordering defined on $\mathcal{N}^{(k)}$, $k \geq 1$, as follows: Let A be a non-empty, linearly ordered set and A^* be the set of all words in A. The set A^* is called the **Kleene closure** of A. We define an ordering on A^* as follows:

(a) $u \leq v$, if u is the empty word and v is a non-empty word in A^*.

(b) If $u = au'$ and $v = bv'$, where $a, b \in A$ and $u', v' \in A^*$ then say that $u \leq v$ if either $a < b$ or when $a = b$, then $u' < v'$.

This is a generalization of the ordering used in the dictionary.

There is another ordering, which may be defined on A^*. For $u \in A^*$, let $|u|$ denote the length of the word u. For $u, v \in A^*$, define $u \leq v$ if $|u| < |v|$ or if $|u| = |v|$, then $u \leq v$ in lexicographic order. This ordering is called **short-lex order** or also **free semigroup order**.

6.2. Hasse Diagrams

Let S be a finite set with a relation R defined on S. We plot points corresponding to the elements of S in some manner and if $a, b \in S$ and aRb, we connect the point A to the point B by an arrow from A to B. We are writing A, B for the points which correspond to a, b respectively. In this fashion we get a graph associated with the relation R defined on S. Henceforth we denote the point corresponding to an element $x \in S$ by x itself.

If (S, R) is a poset, then in the graph associated with this there is a loop around every point of S. As a matter of convention, while drawing the

graph associated with a poset, we omit the loops around the points of S.
If $a, b, c \in S$ and aRb, bRc, then aRc. Thus if there is an arrow from a to b
and an arrow from b to c, then there is an arrow from a to c. Therefore, in
the presence of arrows from a to b and from b to c, we can omit an arrow
from a to c, it being understood in the case of a poset. Also, while drawing
a graph associated with a poset, we position the point a at a lower level
than b if aRb. Thus, when we have a lower than b in the plane with a, b
connected, there is no need to have an arrow pointing from a to b. The
final graph that is obtained by using the conventions as mentioned is called
the **Hasse diagram** of the poset. To fix our ideas, we will draw the Hasse
diagrams of some finite posets.

Example 6.6. Draw the Hasse diagram representing the partial order
relation R defined by aRb if $a|b$ on the set
(a) $\{1, 2, 3, 4, 6, 8, 10\}$; (b) $\{1, 2, 3, 4, 6, 8, 12\}$; (c) $\{2, 3, 4, 6, 8, 10\}$;
(d) $\{2, 3, 4, 6, 8, 12\}$.

Solution. Using the procedure as detailed above, the Hasse diagrams of
the four relations are given in Fig. 6.1 (a) to (d), below.

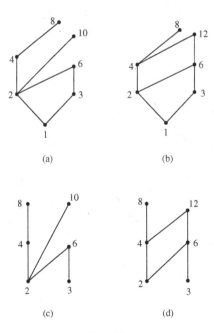

Fig. 6.1.

Example 6.7. Draw the Hasse diagram for the partial order relation R defined by ARB if $A \subseteq B$ on the power set of: (a) $\{1,2\}$; (b) $\{1,2,3\}$ (c) $\{1,2,3,4\}$.

Solution.

(a) $\mathcal{P}(S) = \{\phi, \{1\}, \{2\}, \{1,2\}\}$

(b) $\mathcal{P}(S) = \{\phi, \{1\}, \{2\}, \{3\}, \{1,2\}, \{1,3\}, \{2,3\}, \{1,2,3\}\}$

(c) $\mathcal{P}(S) = \{\phi, \{1\}, \{2\}, \{3\}, \{4\}, \{1,2\}, \{1,3\}, \{1,4\}, \{2,3\}, \{2,4\},$
$\{3,4\}, \{1,2,3\}, \{1,2,4\}, \{1,3,4\}, \{2,3,4\}, \{1,2,3,4\}\}.$

Therefore, the Hasse diagrams are given below:

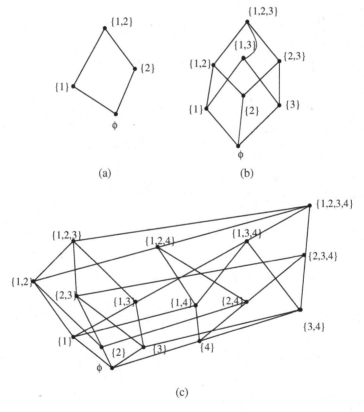

(a)　　　　　　　　(b)

(c)

Fig. 6.2.

(a) The Hasse diagram is obtained from the associated graph or diagraph by omitting all the loops; then omitting all the edges that occur because of transitivity, namely, only the edge $\{\phi, \{1, 2\}\}$.

(b) The Hasse diagram is obtained from the associated diagraph by omitting all the loops, and then omitting all the edges that occur due to transitivity, namely, $\{\phi, \{1, 2\}\}, \{\phi, \{1, 3\}\}$, $\{\{1\}, \{1, 2, 3\}\}$, $\{\{2\}\}, \{1, 2, 3\}$ and $\{\{3\}, \{1, 2, 3\}\}$.

(c) In this case again, the Hasse diagram is obtained from the associated diagraph by omitting all the loops; and then omitting the edges that occur due to transitivity, namely,

$$\{\phi, \{1, 2\}\}, \{\phi, \{1, 3\}\}, \{\phi, \{1, 4\}\}, \{\phi, \{2, 3\}\}, \{\phi, \{2, 4\}\},$$

$$\{\phi, \{3, 4\}\}; \{\{1\}, \{1, 2, 3\}\}, \{\{1\}, \{1, 2, 4\}\}, \{\{1\}, \{1, 3, 4\}\},$$

$$\{\{2\}, \{1, 2, 3\}\}, \{\{2\}, \{1, 2, 4\}\}, \{\{2\}, \{2, 3, 4\}\}; \{\{3\}, \{1, 3, 4\}\},$$

$$\{\{3\}, \{1, 2, 3\}\}, \{\{4\}, \{1, 2, 4\}\}, \{\{4\}, \{1, 3, 4\}\}, \{\{4\}, \{2, 3, 4\}\},$$

$$\{\{3\}, \{3, 2, 4\}\}, \{\{1, 2\}, \{1, 2, 3, 4\}\}, \{\{3, 4\}, \{1, 2, 3, 4\}\},$$

$$\{\{1, 3\}, \{1, 2, 3, 4\}\}, \{\{1, 4\}, \{1, 2, 3, 4\}\}, \{\{2, 3\}, \{1, 2, 3, 4\}\},$$

$$\{\{2, 4\}, \{1, 2, 3, 4\}\}.$$

Thus the Hasse diagrams are as already shown above. Because of the positioning of the points, all the arrowheads are removed.

Example 6.8. Obtain the partial order relation on the set $S = \{a, b, c, d, e\}$, the Hasse diagram associated with which is as in Fig. 6.3.

Solution. Let R be the partial order relation on S, the Hasse diagram of which is as given in Fig. 6.3. Since R is reflexive, we must have

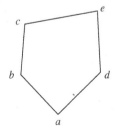

Fig. 6.3.

$(a, a), (b, b), (c, c), (d, d), (e, e)$ in R. The elements b, d being at a higher level than a and being connected with $a, (a, b), (a, d) \in R$. The element c being at a higher level than b and connected to it, $(b, c) \in R$. For the same reason $(c, e), (d, e) \in R$. Now

(a) $(a, b), (b, c) \in R$ imply $(a, c) \in R$
(b) $(a, d), (d, e) \in R$ imply $(a, e) \in R$
(c) $(b, c), (c, e) \in R$ imply $(b, e) \in R$

Thus we have

$\{(a, a), \ (b, b), \ (c, c), \ (d, d), \ (e, e), \ (a, b), \ (a, d), \ (b, c), \ (a, c), \ (b, e),$
$(a, e), (d, e), (c, e)\} \subseteq R.$
It is then clear that R is equal to

$\{(a, a), (b, b), (c, c), (d, d), (e, e), (a, b), (a, c), (a, d), (d, e), (a, e), (b, c),$
$(b, e), (c, e)\}.$

Example 6.9. Consider the set D_{30} of all positive divisors of 30 with the relation $(|)$, i.e. for $a, b \in D_{30}, a \leq b$ if a divides b. Now $D_{30} = \{1, 2, 3, 5, 6, 10, 15, 30\}$ has 8 elements. D_{30} is clearly a poset and its Hasse diagram is given in Fig. 6.4.

Observe that the Hasse diagram of $(D_{30}, |)$ is similar to the Hasse diagram of the poset as considered in Example 6.7 (b) above. Consider a mapping $f : \mathcal{P}(S) \to D_{30}$ given by $f(\phi) = 1, f(\{1\}) = 2, f(\{2\}) = 3,$ $f(\{3\}) = 5, f(\{1, 2\}) = 6, f(\{2, 3\}) = 15, f(\{1, 3\}) = 10$ and $f(\{1, 2, 3\}) = 30.$ The map is both one–one and onto and has the property that $a \leq b$ in $\mathcal{P}(S)$ if and only if $f(a) \leq f(b).$

Motivated by this, we say two posets $(A, \leq), (X, \leq)$ are said to be **isomorphic** if there exists a one–one correspondence $f : A \to X$ such that $f(a) \leq f(b)$ in X if, and only if, $a \leq b$ in A. Also, we say that $(A, \leq), (X, \leq)$ are **anti-isomorphic** if the one–one correspondence f has the property that $a \leq b$ if, and only if, $f(b) \leq f(a).$

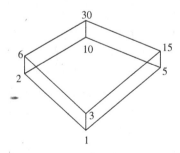

Fig. 6.4.

6.3. Chains and Antichains in Posets

Let (A, R) be a partially ordered set. If for $a, b, \in A$, aRb or that $(a, b) \in R$, we write $a \leq b$ and say that a is less than or equal to b. Also, in this case, we write the poset as $(A, \leq)$. We also say that a is less than b if $a \leq b$ but $a \neq b$. In the situation $a \leq b$, we can also say that b is greater than or equal to a. If a is less than b, we say b is greater than a.

Let $(A, \leq)$ be a partially ordered set. A subset B of A is called a **chain** if, for any two elements a, b of B, either $a \leq b$ or $b \leq a$. In terms of the earlier terminology, we say that B is a chain if every two elements of B are comparable. If no two elements of B are comparable, we say that B is an **antichain**.

Lemma. If $B = \{a_1, a_2, \ldots, a_n\}$ is a finite chain in a poset $(A, \leq)$, then there exists a unique element a_{i_1} such that $a_{i_1} \leq a$ for every $a \in B$.

Proof. We prove the existence part of this result by induction on n — the number of elements in the chain. If $n = 2$, say $B = \{a_1, a_2\}$, as any two elements of B are comparable, either $a_1 \leq a_2$ or $a_2 \leq a_1$, and we are through. So, suppose that $n \geq 2$ and that for any chain B having n elements, there exists an element $a \in B$ such that $a \leq b$ for every $b \in B$. Consider a chain $B = \{a_1, a_2, \ldots, a_{n+1}\}$ having $n + 1$ elements. Since the elements a_1, a_{n+1} are comparable, either $a_1 \leq a_{n+1}$ or $a_{n+1} \leq a_1$. For the sake of notational convenience, let us assume that $a_1 \leq a_{n+1}$. Consider $B_1 = \{a_1, a_2, \ldots, a_n\}$, which is again a chain in A. By induction hypothesis, there exists an $a \in B_1$, such that $a \leq a_i$ for every $i, 1 \leq i \leq n$. Then $a \leq a_1$, and $a_1 \leq a_{n+1}$ already. Therefore, by transitivity of the relation '$\leq$', $a \leq a_{n+1}$. Hence $a \leq a_i$ for every $i, 1 \leq i \leq n + 1$. This completes induction and, therefore, every finite chain B of A has an element $a \in B$ such that $a \leq b$ for every $b \in B$.

If $a, a' \in B$ are two elements such that $a \leq b$ for every $b \in B$ and $a' \leq b$ for every $b \in B$, then $a \leq a'$ and $a' \leq a$. Therefore $a' = a$.

In a similar fashion we can also prove

Lemma. If $B = \{a_1, a_2, \ldots, a_n\}$ is a finite chain in a poset $(A, \leq)$, then there exists a unique element $a_{ij} \in B$ such that $a \leq a_{ij}$ for every $a \in B$.

Using the two results, we find that, if $B = \{a_1, a_2, \ldots, a_n\}$ is a finite chain in a poset $(A, \leq)$ let $a_{i_1} \in B$ be such that $a_{i_1} \leq b$ for every $b \in B$. Then consider $B_1 = B \backslash \{a_{i_1}\}$. B_1 is also a chain and there exists an $a_{i_2} \in B_1$ such that $a_{i_2} \leq b$ for every $b \in B_1$. By the choice of a_{i_1} we have $a_{i_1} \leq a_{i_2}$. Next consider $B_2 = B_1 \setminus \{a_{i_2}\}$. B_2 is a chain and, so, there exists an element $a_{i_3} \in B_2$ such that $a_{i_3} \leq b$ for every $b \in B_2$. By the choice of a_{i_2} we find that $a_{i_2} \leq a_{i_3}$. Continuing this process, we have $a_{i_1} \leq a_{i_2}, a_{i_2} \leq a_{i_3}, \ldots, a_{i_{n-1}} \leq a_{i_n}$, where $\{i_1, i_2, \ldots, i_n\} = \{1, 2, \ldots, n\}$. The above relations are abbreviated simply as $a_{i_1} \leq a_{i_2} \leq \cdots \leq a_{i_n}$. Also then n is called the **length** of the chain.

Consider the partially ordered set $(A, \leq)$ where $A = \{1, 2, 3, 4, 6, 8, 12\}$ with $a \leq b$ if a divides b. Then $\{1, 2, 4, 8, \}, \{1, 2, 4\}, \{2, 4, 8\}, \{1, 4, 8\}, \{1, 2\}, \{2, 4\}, \{2, 8\}, \{4, 8\}$ are chains, while $\{2, 3\}, \{3, 4\}, \{3, 8\}, \{8, 12\}$ are antichains. Observe that there is no antichain in A having three or more elements.

In the partially ordered set $(A, \leq), A = \{a, b, c, d, e\}$ with the Hasse diagram of the relation being as given in Fig. 6.3, $\{a, b, c\}, \{a, b, c, e\}, \{a, d, e\}, \{b, c, e\}$ are chains while $\{b, d\}, \{c, d\}$ are antichains.

Definition. Let $(A, \leq)$ be a partially ordered set. An element $a \in A$ is called

(a) a **maximal** element if there is no b in $A, b \neq a$, for which $a \leq b$;
(b) a **minimal** element if there is no b in $A, b \neq a$ for which $b \leq a$.

Definition. Let $(A, \leq)$ be a partially ordered set and a, b be two elements of A. An element c of A is called

(a) an **upper bound** of a, b if $a \leq c, b \leq c$;
(b) a **lower bound** of a, b if $c \leq a, c \leq b$;
(c) a **least upper bound** of a, b if c is an upper bound of a, b and there is no other upper bound $d(\neq c)$ of a, b with $d \leq c$;
(d) a **greatest lower bound** of a, b if c is a lower bound of a, b and there is no other lower bound $d(\neq c)$ of a, b with $c \leq d$.

Some authors define least upper bound $(l.u.b.)$ and greatest lower bound $(g.l.b.)$ of elements a, b of a poset $(A, \leq)$ as follows:

(a) an element $c \in A$ is called a least upper bound of a, b if $a \leq c, b \leq c$ and if $d \in A$ with $a \leq d, b \leq d$, then $c \leq d$;
(b) an element c is a greatest lower bound of a, b if $c \leq a, c \leq b$ and if $d \in A$ with $d \leq a, d \leq b$, then $d \leq c$.

With this definition of $g.l.b.$ and $l.u.b.$ of a, b, it is quite clear that $l.u.b.$ of $a, b \in A$, if it exists, is unique and similarly, $g.l.b.$ of $a, b \in A$, if it exists, is unique. Observe that in the poset with Hasse diagram (Fig. 6.5) c, d are

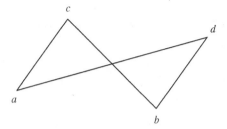

Fig. 6.5.

l.u.b. of a, b while a, b are *g.l.b.* of c, d in the sense of the earlier definition while this is not so with the above definition. However, *l.u.b.* of a, b in the sense of the above definition shall always be a *l.u.b.* in the sense of the earlier definition. The case with *g.l.b.* is similar.

We can define lower and upper bounds and greatest lower and least upper bounds of non-empty subsets of a poset in a similar fashion. Let $(P, \leq)$ be a poset and A be a non-empty subset of P. An element $x \in P$ is called

(a) a **lower bound** of A if $x \leq a$ for every $a \in A$;

(b) an **upper bound** of A if $a \leq x$ for every $a \in A$;

(c) a **greatest lower bound** of A if x is a lower bound of A and there is no lower bound y of A with $x \leq y$;

(d) a **least upper bound** of A if x is an upper bound of A and there is no upper bound y of A with $y \leq x$.

A greatest lower bound of $(P, \leq)$ if it exists, is called a **least element** or a **zero element** of the poset and a least upper bound of $(P, \leq)$, if it exists, is called a **largest** or **greatest** or a **unit element** of $(P, \leq)$.

Proposition 6.1. (a) There is exactly one greatest element of a poset, if it exists. (b) There is exactly one least element of a poset, if it exists.

Proof. (a) Consider a poset $(P, \leq)$. Suppose that a greatest element exists in P. If possible, let x, y be two greatest elements of P. By definition of greatest element $a \leq x$ for all $a \in P$ and $a \leq y$ for all $a \in P$

In particular, $x \leq y$ and $y \leq x$. Therefore, $x = y$. (b) may be proved on similar lines.

Example 6.10. Observe that in the poset with the Hasse diagram of Fig. 6.3, above, e is a maximal element and a is a minimal element. Also

e is an upper bound of c, d, indeed, a least upper bound while a is a lower bound of b, d and also of c, d. Indeed a is a greatest lower bound of b, d as well as of c, d.

Consider the partially ordered set with Hasse diagram as in Fig. 6.6. In this poset the pairs $b, c; b, a; c, d; a, e; d, e; f, g; h, i; j, k$ are incomparable. The elements, j, k are maximal while the elements b, a, e are minimal. The elements h, i, j, k are upper bounds of b, c as well as d, e, while f is the least upper bound of b, c and g is the least upper bound of d, e. Moreover, k and j are least upper bounds of h, i. Also f, g are greatest lower bounds of h, i and h, i are greatest lower bounds of j, k. The subsets $\{b, f, h, j\}, \{b, f, i, k\}, \{b, f, i, j\}, \{g, h, k\}, \{g, i, k\}, \{g, h, j\}, \{g, i, j\}$ are some of the chains.

Exercises 6.1.

1. In the poset with Hasse diagram as in Fig. 6.6: (i) Decide if $\{a, c, f, i, h, j, k\}$ is a chain. (ii) We have listed above 8 pairs of elements which are incomparable. Find all other pairs of points which are incomparable.
2. Find lower and upper bounds of the subsets $\{a, b, c\}, \{g, h\}, \{b, d, g\}$ and $\{a, c, d, f\}$ in the poset, the Hasse diagram of which is given in Fig. 6.7
3. Find the greatest lower bounds and the least upper bounds of the sets $\{3, 9, 12, 15\}$ and $\{1, 2, 4, 5, 10, 15\}$ if they exist, in the poset $(\mathcal{N}, |)$

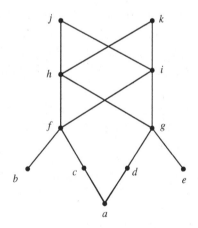

Fig. 6.6.

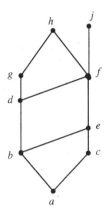

Fig. 6.7.

Example 6.11. Let $\mathcal{N}$ be the set of natural numbers and $\leq$ be the relation defined by $a \leq b$ for $a, b \in \mathcal{N}$ if a divides b. Then $(\mathcal{N}, \leq)$ is a partially ordered set. For $m, n \in \mathcal{N}$, common multiple of m, n is an upper bound of m, n, while the least common multiple of m, n is the (only) least upper bound of m, n. On the other hand, a common divisor of m, n is a lower bound of m, n and the greatest common divisor of m, n is the (only) greatest lower bound of m, n.

Example 6.12. Consider the poset $P = (\{2, 4, 5, 10, 12, 20\}, |)$. Observe that $2 \leq 4, 2 \leq 10, 2 \leq 12, 2 \leq 20, 4 \leq 12, 4 \leq 20, 5 \leq 10, 5 \leq 20, 10 \leq 20$, whereas the pairs of elements $2, 5; 4, 5; 5, 12; 4, 10; 12, 20$ are not comparable. If $a \in P$ and $a \leq 2$, then a divides 2. But the only divisor of 2 other than 2 is 1 and $1 \notin P$. Therefore, there is no $a \in P$ with $a \leq 2$, $a \neq 2$. Hence 2 is a minimal element of P. As 2 does not divide $5, 2 \nleq 5$ and, so, 2 is not a lower bound of P. Also, 5 is a minimal element of P but, again, 5 is not a lower bound of P. There are no other minimal elements in P. There is no element in P which is divisible by 12 or by 30. Therefore, 12 and 20 are maximal elements of P but these elements not being comparable, neither of them is an upper bound of P. The Hasse diagram of this poset is given in Fig. 6.8(a).

If we consider $(Q, |)$ with $Q = \{1, 2, 4, 5, 10, 12, 20\}$ then 1 is a lower bound of Q and $2, 5$ are no longer minimal elements. However, 12 and 20 are maximal elements and there is no upper bound of Q. The Hasse diagram of this poset is given in Fig. 6.8 (b) below.

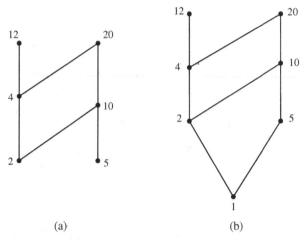

(a) (b)

Fig. 6.8.

Example 6.13. Let S be a non-empty set, $P(S)$ be the power set of S and consider the poset $(P(S), \subseteq)$. Since the empty set ϕ is a subset of every subset A of S, ϕ is a lower bound. Also, there is no subset $A \neq \phi$ of S which is a subset of ϕ. Therefore, ϕ is the least element of the poset. Every subset A of S being contained in S, S is an upper bound. In fact, S is the least upper bound or the largest element of the poset.

Example 6.14. In the poset $(\mathcal{N}, |)$, 1 divides every natural number and, so, 1 is the least element of the poset. However, there is no greatest element in this poset.

Example 6.15. Consider the subsets $A = \{3, 9, 12\}$ and $B = \{1, 2, 4, 5, 10\}$ of the poset $(\mathcal{N}, |)$. Observe that $3 \leq 9, 3 \leq 12$ and 1, 3 being the only natural numbers which divide $3, 9, 12$, the numbers $1, 3$ are the lower bounds of A. Also, 3 is the greatest lower bound of A. Indeed, 3 is the least element of A. If an integer a is an upper bound of A, then 3, 9, 12 must divide a. The integers a which have this property are the ones which are divisible by 36, the least common multiple of 3, 9, 12. Thus, there are infinitely many upper bounds of A but 36 is the least upper bound of A.

A natural number a divides $1, 2, 4, 5$ and 10 if and only if it is a divisor of the greatest common divisor of $1, 2, 4, 5$ and 10. This greatest common divisor being 1, the number 1 is a lower bound and, in fact, the greatest lower bound (and, so, also the least element) of B. The least common multiple of $1, 2, 4, 5$ and 10 is 20. Therefore, 20 is the least upper bound of B and any multiple of 20 is an upper bound.

Example 6.16. Consider the binary relation R on the set $A = \{a, b, c\}$ the associated matrix of which is

$$
\text{(a)} \begin{pmatrix} 1 & 0 & 1 \\ 1 & 1 & 0 \\ 0 & 0 & 1 \end{pmatrix}, \quad \text{(b)} \begin{pmatrix} 1 & 0 & 0 \\ 0 & 1 & 0 \\ 1 & 0 & 1 \end{pmatrix}, \quad \text{(c)} \begin{pmatrix} 1 & 1 & 0 \\ 0 & 1 & 1 \\ 1 & 0 & 1 \end{pmatrix}.
$$

(a) Let R_i be the relation defined on A the associated matrix of which is as given in (a), (b), (c) respectively. Then

$$
\begin{aligned}
R_1 &= \{(a, a), (a, c), (b, a), (b, b), (c, c)\}, \\
R_2 &= \{(a, a), (b, b), (c, a), (c, c)\}, \\
R_3 &= \{(a, a), (a, b), (b, b), (b, c), (c, a), (c, c)\}.
\end{aligned}
$$

We find that $(b, a), (a, c) \in R_1$ but $(b, c) \notin R_1$. Therefore, R_1 is not transitive and so this relation is not a partial order relation.

The relation R_2 is clearly reflexive as $(a, a), (b, b), (c, c) \in R_2$. Also, it is transitive. Since $(a, c) \notin R_2$ and $(c, a) \in R_2$, the anti-symmetry is clear. Therefore, R_2 is a partial order relation.

Observe that $(a, b), (b, c) \in R_3$ but $(a, c) \notin R_3$. Therefore, R_3 is not transitive and, hence, it is not a partial order relation.

Example 6.17. Consider the binary relations R_i defined on the set $A = \{1, 2, 3, 4\}$ the associated matrices of which are respectively

$$
\text{(a)} \begin{pmatrix} 1 & 0 & 1 & 0 \\ 0 & 1 & 1 & 0 \\ 0 & 0 & 1 & 1 \\ 1 & 1 & 0 & 1 \end{pmatrix}, \quad \text{(b)} \begin{pmatrix} 1 & 0 & 0 & 1 \\ 0 & 1 & 0 & 1 \\ 1 & 1 & 1 & 0 \\ 1 & 1 & 0 & 1 \end{pmatrix}.
$$

Then

$$
R_1 = \{(1, 1), (1, 3), (2, 2), (2, 3), (3, 3), (3, 4), (4, 1), (4, 2), (4, 4)\}
$$

and

$$
R_2 = \{(1, 1), (1, 4), (2, 2), (2, 4), (3, 1), (3, 2), (3, 3), (4, 1), (4, 2), (4, 4)\}.
$$

Both the relations are clearly reflexive. Observe that $(4, 1), (1, 3) \in R_1$ but $(4, 3) \notin R_1$. Therefore R_1 is not transitive and, therefore, it is not a partial order relation.

In the case of R_2, observe that $(3,1), (1,4) \in R_2$ but $(3,4) \notin R_2$. Hence R_2 also not being transitive cannot be a partial order relation.

Remark. Let $S = \{x_1, \ldots, x_n\}$ be a finite set of order n, R be a binary relation defined on S and $A = (a_{ij})$ be the matrix associated with the binary relation R. Recall that then

$$a_{ij} = \begin{cases} 1 \text{ if } (x_i, x_j) \in R \\ 0 \text{ if } (x_i, x_j) \notin R \end{cases}.$$

In order that the relation R be reflexive, it is necessary and sufficient that $a_{ii} = 1$, for every i. For R to be anti-symmetric, it is necessary and sufficient that for $i \neq j, a_{ij}a_{ji} = 0$. Observe that if $a_{ij}a_{ji} = 1$, then $a_{ij} = a_{ji} = 1$ which implies that both (x_i, x_j) and (x_j, x_i) are in R. Then for R being anti-symmetric, $x_i = x_j$ which is not possible for $i \neq j$. On the other hand if $a_{ij}a_{ji} = 0$, then either $a_{ij} = 0 = a_{ji}$ or one of a_{ij} is 0 and the other is one. Therefore, either neither of $(x_i, x_j), (x_j, x_i)$ is in R or one of $(x_i, x_j), (x_j, x_i)$ is in R and the other is not. Therefore, the condition of anti-symmetry is automatically met in this case.

The relation R is transitive if and only if R itself is its transitive closure. Therefore, R is transitive if and only if $R = R \cup R^2 \cup \cdots \cup R^n$ which is so if and only if $A = A + A^2 + \cdots A^n$ (under Boolean sum and product).

Definition. Let $(S, \leq)$ be a poset. We say that an element $y \in S$ **covers an element** $x \in S$ if $x < y$ and there is no element $z \in S$ such that $x < z < y$. The set $\{(x,y) \in S \times S | y \text{ covers } x\}$ is called the **covering relation** of the poset $(S, \leq)$.

Example 6.18. Consider $S = \{1, 2, 3, 4, 6, 12\}$ and the relation $\leq$ as $|$ i.e. $a \leq b$ if a divides b. Since 1 divides $2, 3, 4, 6, 1 \leq 2, 1 \leq 3, 1 \leq 4, 1 \leq 6, 2 \leq 12$ and $3 \leq 6, 3 \leq 12$ but $3 \not\leq 4$, 2 is a cover of 1, and 3 is also a cover of 1. Again, 2 divides 4, 2 divides 6, and neither 4 divides 6 nor 6 divides 4. Therefore, 4 and 6 cover 2. Similarly, we find that 6 covers 3, 12 covers 4, and 12 covers 6. Hence the covering relation of this relation is $\{(1,2), (1,3), (2,4), (2,6), (3,6), (4,12), (6,12)\}$.

Example 6.19. Let $S = \{a, b, c\}$ and $R = \{(A, B) | A \subseteq B\} \subseteq P(S) \times P(S)$.

Now the subsets of S are:

$$\phi, \{a\}, \{b\}, \{c\}, \{a, b\}, \{a, c\}, \{b, c\}, \{a, b, c\}.$$

It is then clear that each of the subsets

$\{a\}, \{b\}, \{c\}$ covers ϕ;
$\{a, c\}, \{a, b\}$ covers $\{a\}$;
$\{a, b\}, \{b, c\}$ covers $\{b\}$;
$\{b, c\}, \{a, c\}$ covers $\{c\}$;

and S is a cover of each one of $\{a, b\}, \{a, c\}, \{b, c\}$. Therefore, the covering relation of R is: $\{(\phi, \{a\}), (\phi, \{b\}), (\phi, \{c\}), (\{a\}, \{a, b\}), (\{a\}, \{a, c\}),$ $(\{b\}, \{a, b\}), (\{b\}, \{b, c\}), (\{c\}, \{a, c\}), (\{c\}, \{b, c\}), (\{a, b\}, S), (\{b, c\}, S)$ $(\{a, c\}, S)\}$.

Example 6.20. Let (x, y) belong to the covering relation of a finite poset $(S, \leq)$. Then $x \leq y$ implies that in the Hasse diagram of the poset $(S, \leq), x$ is on a lower level than y. Also, there is a chain of edges $(x, z_1), (z_1, z_2), \ldots, (z_r, y)$. However, if there is an edge (x, z_i) with x at a lower level than z_i then $x \leq z_i \leq y$ and y is no longer a cover of x. Hence there is an edge between x and y. Thus, we have: if (x, y) is in the covering relation of the finite poset $(S, \leq)$, then in the Hasse diagram of the poset, x is at a lower level than y and there is an edge between x and y.

Example 6.21. Consider a poset represented by the Hasse diagram (Fig. 6.9). Then l, m are the maximal elements, a, b, c are the minimal elements and there are no least or greatest elements. l, m, k are the upper bounds of $\{a, b, c\}$ and the least upper bound of $\{a, b, c\}$ is k. The subset $\{f, g, h\}$ has no lower bound and, therefore, no greatest lower bound of this subset exists.

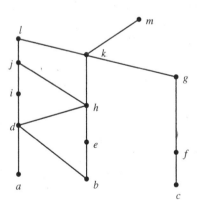

Fig. 6.9.

Example 6.22. Let $(S, \leq)$ be a poset with a (a) least element x; and a (b) greatest element y. Then (a) $x \leq a$ for every $a \in S$ and x is a minimal element of S. If z is another minimal element, then by definition of least element, $x \leq z$. But z being a minimal element, $z = x$. Hence x is the only minimal element of $(S, \leq)$. (b) By definition of greatest element,

$$a \leq y \quad \text{for all } a \in S. \tag{6.1}$$

Therefore, there is no $z \in S$ with $y < z$. Hence y is a maximal element. On the other hand, let z be another maximal element. By (6.1), $z \leq y$. Then the definition of a maximal element shows that $z = y$. Hence y is the only maximal element of $(S, \leq)$.

Example 6.23. Consider the posets, the Hasse diagrams of which are as in Fig. 6.10.

(a) Then $S = \{a, b, c, d\}$ and if R is the relation defined on S, we find that $(a, b), (b, c), (b, d) \in R$. The relation R being a partial order relation, $R = \{(a, a), (b, b), (c, c), (d, d), (a, b), (a, c), (a, d), (b, c), (b, d)\}$.

(b) Here $S = \{a, b, c, d, e\}$ and the relation R defined by the Hasse diagram contains $(a, b), (a, c), (b, e), (b, d), (c, e)$. The relation R being a partial order relation, we have

$$R = \{(a, a), (b, b), (c, c), (d, d), (e, e), (a, b), (a, c),$$
$$(b, e), (b, d), (c, e), (a, e), (a, d)\}.$$

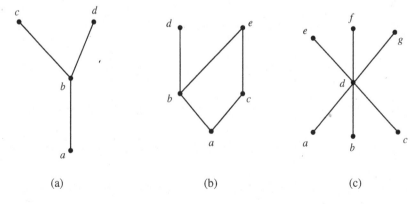

Fig. 6.10.

(c) Here $S = \{a, b, c, d, e, f, g\}$. Let R be the partial order relation defined on S by the given Hasse diagram. Then $(a, d), (b, d), (c, d),$ $(d, e), (d, f), (d, g) \in R$. Using reflexive and transitive property of R, we get

$$R = \{(a, a), (b, b), (c, c), (d, d), (e, e), (f, f), (g, g), (a, d),$$
$$(b, d), (c, d), (d, e), (d, f), (d, g), (a, e), (a, f), (a, g),$$
$$(b, e), (b, f), (b, g), (c, e), (c, f), (c, g)\}.$$

Example 6.24. Let S be a non-empty set and R be a binary relation defined on S. Suppose that R is an equivalence relation as well as a partial order relation. For $a, b \in S$, let $(a, b) \in R$. Then R being symmetric, $(b, a) \in R$. Also R is anti-symmetric. Therefore, $(a, b), (b, a) \in R$ imply that $a = b$. Hence $R = \{(a, a) | a \in S\}$.

Example 6.25. Let $S = \{a, b, c\}$ be a set of order 3. Let R be a partial order relation defined on S. Observe that if $(a, b) \in R$, then (b, a) cannot belong to R. Thus at most one of the two elements (i) $(a, b), (b, a)$; (ii) $(a, c), (c, a)$; (iii) $(b, c), (c, b)$ can belong to R. It follows that although $S \times S$ has nine elements, R has at most six elements. The relation R being reflexive, it has at least three elements. If $(a, b), (b, c) \in R$, then $(a, c) \in R$ and R has six elements. Therefore, if R is a relation having five elements, then

(a) If $(a, b) \in R$, the other element has to be one of $(a, c), (b, c),$ $(c, a), (c, b), (b, a)$. But $(b, a) \notin R$ and if (b, c) or $(c, a) \in R$, then R has to have (a, c) or (c, b) as well, and R will have six elements. Therefore, in this case there are two choices for R, namely $\{(a, a), (b, b), (c, c),$ $(a, b), (a, c)\}$ or $\{(a, a), (b, b), (c, c), (a, b), (c, b)\}$.

(b) If $(a, c) \in R$, the other element has to be one of $(a, b), (b, a),$ $(b, c), (c, b), (c, a)$. But the occurrence of any of the elements $(c, a), (b, a), (c, b)$ is ruled out and accepting (a, b) as an element gives one of the two relations obtained in (a) above. Thus, the only new relation we can get is $\{(a, a), (b, b), (c, c), (a, c), (b, c)\}$.

(c) If $(b, c) \in R$, the remaining element has to be one of $(a, b), (b, a),$ $(a, c), (c, a), (c, b)$. Occurrence of the pairs $(a, b), (c, a), (c, b)$ is ruled out and accepting the pair (a, c) gives the relation as at (b) above. Thus the only choice for R in this case is $\{(a, a), (b, b), (c, c), (b, c), (b, a)\}$.

(d) If $(b, a) \in R$, the only new poset admissible is $\{(a, a), (b, b),$ $(c, c), (b, a), (c, a)\}$. If we accept (c, a) in R, we do not get any new poset with five elements. Accepting (c, b) in R leads to a new poset with five elements, namely $\{(a, a), (b, b), (c, c), (c, b), (c, a)\}$. Thus, we have six posets with five elements each:

$$\{(a, a), (b, b), (c, c), (a, b), (a, c)\},$$
$$\{(a, a), (b, b), (c, c), (a, b), (c, b)\},$$
$$\{(a, a), (b, b), (c, c), (a, c), (b, c)\},$$
$$\{(a, a), (b, b), (c, c), (c, b), (c, a)\},$$
$$\{(a, a), (b, b), (c, c), (b, c), (b, a)\},$$
$$\{(a, a), (b, b), (c, c), (b, a), (c, a)\}.$$

There are six more posets having six element each:

$$R_1 = \{(a, a), (b, b), (c, c), (a, b), (b, c), (a, c)\},$$
$$R_2 = \{(a, a), (b, b), (c, c), (b, a), (a, c), (b, c)\},$$
$$R_3 = \{(a, a), (b, b), (c, c), (a, b), (a, c), (c, b)\},$$
$$R_4 = \{(a, a), (b, b), (c, c), (a, b), (c, a), (c, b)\},$$
$$R_5 = \{(a, a), (b, b), (c, c), (b, a), (c, a), (b, c)\},$$
$$R_6 = \{(a, a), (b, b), (c, c), (b, a), (c, a), (c, b)\}.$$

as with the elements a, b there are three posets containing (a, b) and three posets containing (b, a). This is similar to the pair of elements a, c and the pair of elements b, c. But among the six posets listed above, three already contain (a, c), three contain (c, a), three contain (b, c) and three contain (c, b). Thus the above six posets are the only ones possible having six elements.

In addition to the six posets of order five, and six posets of order six, there are six posets having four elements. These are

$$\{(a, a), (b, b), (c, c), (a, b)\}, \{(a, a), (b, b), (c, c), (b, a)\},$$
$$\{(a, a), (b, b), (c, c), (a, c)\}, \{(a, a), (b, b), (c, c), (c, a)\},$$
$$\{(a, a), (b, b), (c, c), (b, c)\}; \{(a, a), (b, b), (c, c), (c, b)\}.$$

Interchanging a and b, we find that $R_1 \leftrightarrow R_2$, $R_3 \leftrightarrow R_5$ and $R_4 \leftrightarrow R_6$, where $\leftrightarrow$ indicates interchange of the posets concerned. Therefore, but for the change of names, there are left only three posets of order six, namely R_1, R_3 and R_4. Then interchanging a and c, R_3 and R_4 are interchanged. Also, interchanging b and c interchanges R_1 and R_3. Thus effectively, so far as Hasse diagrams are concerned, these six posets lead to only one Hasse diagram. Similarly, effectively there is only one poset

Fig. 6.11.

of order four, and so only one Hasse diagram. However, for order five there are effectively two posets, namely $\{(a,a),(b,b),(c,c),(a,b),(a,c)\}$ and $\{(a,a),(b,b),(c,c),(b,a),(c,a)\}$. One of these has a least element and two maximal elements, while the other has a greatest element and two minimal elements. Effectively, we find that there are five Hasse diagrams for the posets as in Fig. 6.11 on the set S having three elements.

The first one is the Hasse diagram of a poset having three elements, the second is the Hasse diagram of a poset having four elements, the third and fourth are diagrams of posets having five elements each, while the sixth one is the Hasse diagram of a poset having six elements.

Chains and antichains in partially ordered sets are closely related. As an illustration we have Theorem 6.2.

Theorem 6.2. Let $(P, \leq)$ be a partially ordered set and n be the length of the longest chains in P. Then the elements in P can be partitioned into n disjoint antichains.

Proof. We prove the result by induction on the length n of the longest chain in P. If $n = 1$, no two elements in P are comparable and so P itself is the single antichain. Let $n \geq 2$ be the length of the longest chains in P and suppose that any partially ordered set with the length $n - 1$ of the longest chains in it can be partitioned into $n - 1$ disjoint antichains. P has at least one chain of length n. Let one such chain be

$$x_1 \leq x_2 \leq \cdots \leq x_n. \tag{6.2}$$

Then there is no x in P such that $x_n < x$, for otherwise the chain (6.2) can be extended by one step and we get a chain $x_1 \leq x_2 \leq \cdots \leq x_n < x$ of length $n + 1$, which is a contradiction. Therefore, x_n is a maximal element of P. Let M be the set of all maximal elements of P. As proved above, $M \neq \phi$. It is clear from the definition of maximal elements that no two elements of M are comparable. Therefore, M is an antichain. Also, the last element in every chain of length n in P is in M. Now consider the subset $N = P - M$. The relation '$\leq$' in P induces a relation '$\leq$' in N which is reflexive, anti-symmmetric and transitive. Thus $(N, \leq)$ is a partially ordered set. Moreover, $n - 1$ is the length of the longest chains in N. By induction hypothesis, N is partitioned into $n - 1$ disjoint antichains. The antichain M in P is disjoint from N (by the choice of N). Therefore, altogether we find that P is partitioned into n disjoint antichains.

Corollary. Let $(P, \leq)$ be a partially ordered set consisting of $mn + 1$ elements. Then either there is an antichain consisting of $m + 1$ elements or there is a chain of length $n + 1$ in P.

Proof. If there is a chain of length $\geq n + 1$ in P, then there is a chain of length $n + 1$ in P and we have nothing to prove. Suppose that every chain in P is of length $\leq n$. Suppose that the length of the longest possible chains in P is $k \leq n$. Then P is partitioned into k antichains. If every antichain in P has $\leq m$ elements, then the total number of elements in P is $\leq mk \leq mn$, which is a contradiction to the order of P being $mn + 1$. Hence there is at least one antichain in P which has $\geq m + 1$ elements. Out of the elements of this antichain, we can choose $m + 1$ elements and these elements together form an antichain consisting of $m + 1$ elements.

The dual of the result of Theorem 6.2 also holds and is known as Dilworth's theorem. We will now prove this result.

Theorem 6.3. (Dilworth's theorem). Let $(P, \leq)$ be a (finite) partially ordered set. If the size of the largest antichain in P is at most n, then the elements in P can be partitioned into n disjoint chains.

Proof. We will prove the result by induction on the order of P. If the order of P is 1, the result follows trivially. Suppose that $o(P) = m \geq 2$ and that the result holds for all posets with order $< m$. Let C be a maximal chain, say $x_1 < x_2 < \cdots < x_k$ in P. Consider the subset $P - C$ of P which is a poset under the relation induced by the relation in P. If $P - C$ has no antichain of order n, then the largest antichain in $P - C$ is of order $\leq n - 1$. By induction hypothesis, $P - C$ is a union of $n - 1$ chains and, hence, P is the union of these $n - 1$ chains and the chain C.

Let $A = \{a_1, a_2, \ldots, a_n\}$ be the largest antichain in $P - C$. Consider

$$B = \{x \in P | x \leq a_i \text{ for some } i, 1 \leq i' \leq n\} \text{ and}$$
$$D = \{x \in P | x \geq a_i \text{ for some } i, 1 \leq i' \leq n\}.$$

Clearly A is contained in both B and D. If $x \in B \cap D$, then there exist $i, j, 1 \leq i, j \leq n$, such that $a_j \leq x \leq a_i$. But a_i, a_j are not comparable if $i \neq j$. Therefore, $i = j$ and $x = a_i$. Thus $B \cap D = A$.

If $x \in P$ is not in $B \cup D$, then the antichain A can be enlarged to an antichain of order $n + 1$ in P, which is not possible. Therefore, $P = B \cup D$. Since $A \cap C = \phi$, the maximal element x_k of C does not belong to B and the minimal element x_1 of C is not in D for otherwise, in either case, the chain C can be enlarged by one step which is a contradiction to C being a maximal chain in P. Therefore, $B \neq P$ and $D \neq P$. As both B and D have a largest possible antichain A of order n, it follows by induction hypothesis that $B = \cup_{i=1}^{n} B_i$, $D = \cup_{i=1}^{n} D_i$, where B_i are chains in B and D_i are chains in D. Every element of A belongs to some B_i and no two elements of A can belong to the same B_i. By renaming the B_i, if necessary, we can assume that $a_i \in B_i$ for every $i, 1 \leq i \leq n$. Similarly, we can assume that $a_i \in D_i$ for every $i, 1 \leq i \leq n$. If a_i is not a minimal element of D_i, there exists an $x \in D_i$ such that $x < a_i$. But $x \in D_i$ implies that $a_j \leq x$ for some j. Therefore, $a_j \leq x < a_i$ and a_j, a_i become comparable. Hence a_i is the minimal element of D_i. Similarly a_i is the maximal element of B_i. The two chains B_i and D_i can, therefore, be combined to get a chain $C_i = B_i \cup D_i$ in P. Also, then $P = B \cup D = \cup_{i=1}^{n} (B_i \cup D_i) = \cup_{i=1}^{n} C_i$. This completes induction and, therefore, the proof.

Dilworth's theorem has several interesting applications. Some among these include a result of Erdos and Szekeres on monotone subsequence of sequence of real numbers, Sperner's lemma on the maximal cardinality of an antichain in the power set of a finite set, and the marriage theorem. We do not consider these here and the interested reader may refer to the book by D. Jungnickel (Jungnickel, 1999) and the relevant references cited there.

Example 6.26. A point (x, y) in the first quadrant of the $x - y$ plane determines a rectangle, the coordinates of the four vertices being $(0, 0), (x, 0), (x, y)$ and $(0, y)$. If (x_1, y_1) is another point, then it is clear that the rectangle determined by (x_1, y_1) is contained in the rectangle determined by (x, y) if and only if $x_1 \leq x, y_1 \leq y$. Consider ten rectangles determined by points with coordinates $(x_v y_i), 1 \leq i \leq 10$. Define an order on the set $S = \{(x_i, y_i) | 1 \leq i \leq 10\}$ by saying that $(x_v y_i) \leq (x_j, y_j)$ if and only if $x_i \leq x_j$ and $y_i \leq y_j$. This is clearly a partial order relation on the set S of order $10 = 3 \times 3 + 1$. It then follows from corollary to Theorem 6.2 that either there is a chain of $3 + 1 = 4$ elements or there is an antichain of four elements. Let $R_i, 1 \leq i \leq 10$, denote the rectangle determined by $(x_v y_i)$. If there is a chain of four points, say $(x_{i_1}, y_{i_1}), (x_{i_2}, y_{i_2}), (x_{i_3}, y_{i_3})$ and (x_{i_4}, y_{i_4}), then there are four rectangles $R_{i_1}, R_{i_2}, R_{i_3}$ and R_{i_4} such that $R_{i_1}, R_{i_2}, R_{i_3}$ are contained in R_{i_4}; R_{i_1}, R_{i_2} are contained in R_{i_3} and R_{i_1} is contained in R_{i_2}. If there is an antichain of length four in S, thén there are four rectangles $R_{i_1}, R_{i_2}, R_{i_3}$ and R_{i_4} such that none of these is contained in any of the other three. Observe that the rectangles determined by the points (2,3) and (3,2) are not comparable.

Example 6.27. Consider a set S of $mn + 1$ persons. For $x, y \in S$ say that x is a descendent of y in the usual sense of the word. For $x, y \in S$, say that $x \leq y$ if and only if x is a descendent of y. Assume that every person is his or her own descendent. The relation '$\leq$' defined on S is a partial order relation. It follows from corollary to Theorem 6.2 that either there is chain of length $m + 1$ in S or there is an antichain containing $n + 1$ elements of S. If there is a chain $x_1 \leq x_2 \leq \cdots \leq x_{m+1}$ in S, then there are $m + 1$ persons $x_1, x_2, \ldots, x_{m+1}$ in S such that x_1 is a descendent of x_2, x_2 is a descendent of x_3, and so on with x_m a descendent of x_{m+1}. Thus there are $m + 1$ persons such that for every pair of persons, one is descendent of the other. If there is an antichain containing $n + 1$ elements from S, then there is a subset of S having $n + 1$ persons such that none of them is a descendent of any other person in the subset.

Please note that good sources of information on this subject are books by Birkhoff and Bartee (1975), Dornhoff and Hohn (1978), Kolman, Busby and Ross (2005), Lidl and Pilz (1998), Liu (2000) and Rosen (2003/2005).

Exercises 6.2.

1. Prove that a finite poset $(P, \leq)$ has at least one maximal element and at least one minimal element.

2. Let $\mathcal{B}^n$ denote the set of all ordered sequences of 0s and 1s with n elements. Draw the Hasse diagram of (i) $\mathcal{B}^2$, (ii) $\mathcal{B}^3$, (iii) $\mathcal{B}^4$ where $\leq$ is the lexicographic order.

3. Let C denote the set of all real, continuous functions defined on the closed interval $1 \leq x \leq 2$. For $f, g \in C$, say that $f \leq g$ if $f(x) \leq g(x)$ for all x with $1 \leq x \leq 2$. Show that $(C, \leq)$ is a poset which has neither zero nor unit element.

4. Show that in the poset $(R, \leq)$ of all real numbers with the usual $\leq$ relation, no element covers any other element. Show that the same is true in the poset $(Q, \leq)$ of all rational numbers with the usual $\leq$ relation.

5. If a poset $(P, \leq)$ has the zero element z and if a covers z, then a is called an atom, while if $(P, \leq)$ has the unit u and u covers b, then b is

called a co-atom of P. Prove that if a, b are co-atoms of a poset $(P, \leq)$ and $a \leq b$, then $a = b$. (Also refer to Chapter 12).

6. Draw the Hasse diagram of the poset $(P, |)$ where P is the set of all positive integral divisors of (a) 24; (b) 36; (c) 48.

7. Determine the Hasse diagram of the relation on the set $A = \{1, 2, 3, 4, 5\}$ the associated matrix of which is

$$
\text{(a)} \begin{pmatrix} 1 & 1 & 1 & 1 & 1 \\ 0 \cdot & 1 & 1 & 1 & 1 \\ 0 & 0 & 1 & 1 & 1 \\ 0 & 0 & 0 & 1 & 1 \\ 0 & 0 & 0 & 0 & 1 \end{pmatrix}
\qquad
\text{(b)} \begin{pmatrix} 1 & 0 & 1 & 1 & 1 \\ 0 & 1 & 1 & 1 & 1 \\ 0 & 0 & 1 & 0 & 1 \\ 0 & 0 & 0 & 1 & 0 \\ 0 & 0 & 0 & 0 & 1 \end{pmatrix}.
$$

8. On the set Z of integers, define a relation R by saying that for $a, b \in Z$, aRb if $a = b^k$ for some positive integer k. Is R a partial order relation?

9. If R is a linear order on a set A, show that R^{-1} is also a linear order on A.

10. Let $A = \{2, 3, 6, 9, 12, 18, 24\}$ and $B = A \times A$ be the Cartesian product of A with itself. Define a relation $\prec$ on B by: $(a, b) \prec (a', b')$ if a divides a' and $b \leq b'$, the relation $\leq$ being the usual relation of the number system. Show that the relation $\prec$ is a partial order relation.

11. Find the greatest and the least elements, if they exist, of the following posets.

   (a) $A = \{2, 4, 6, 8, 12, 18, 24, 36, 72\}$ with the partial order relation of divisibility.

   (b) $A = \{2, 3, 4, 6, 12, 18, 24, 36\}$ with the partial order relation of divisibility.

   (c)

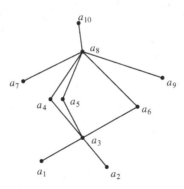

Fig. 6.12.

(d)

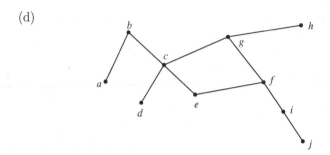

Fig. 6.13.

Also find $l.u.b.$ of $\{d, f\}, \{a, i\}$ and $g.l.b.$ of $\{b, i\}$ and $\{b, f\}$, if they exist for the poset (d).

12. If $p_1 \leq p_2 \leq \cdots \leq p_n \leq p_1$, where $p_1 \in P$, is a chain in a poset $(P, \leq)$, prove that $p_1 = p_2 = \cdots = p_n$.

13. If a and b are atoms of a poset $(P, \leq)$ and if $a \leq b$, prove that $a = b$.

14. Give an example of a poset other than $\{P(S), \leq\}$ or $\{P(S), \geq\}$ that is isomorphic to its dual.

15. Let C be the set of all complex numbers and define a relation R on C by $x + iy \leq x' + iy'$ if and only if $x \leq x'$ and $y \leq y'$. Is this relation a linear order? Does there exist a minimal or a maximal element in $(C, \leq)$?

16. Are the relations with the associated matrices as given below partial orders?

(a) $\begin{pmatrix} 1 & 0 & 1 \\ 1 & 1 & 0 \\ 0 & 0 & 1 \end{pmatrix}$, (b) $\begin{pmatrix} 1 & 0 & 0 & 0 \\ 0 & 1 & 0 & 0 \\ 0 & 0 & 1 & 1 \\ 0 & 0 & 0 & 1 \end{pmatrix}$, (c) $\begin{pmatrix} 1 & 0 & 1 & 0 \\ 0 & 1 & 1 & 0 \\ 0 & 0 & 1 & 1 \\ 1 & 1 & 0 & 0 \end{pmatrix}$.

17. Draw the Hasse diagram for the divisibility relation on the set
(a) $\{3, 5, 7, 11, 13, 16, 17\}$;   (b) $\{2, 3, 5, 10, 11, 15, 25\}$;
(c) $\{1, 2, 3, 5, 7, 11, 13\}$;   (d) $\{1, 2, 3, 6, 12, 24, 36, 48\}$;
(e) $\{1, 2, 4, 8, 16, 32, 64, 128\}$;   (f) $\{1, 3, 9, 27, 81, 243\}$.

18. Let $(S, \leq)$ be a poset and R be a relation on S defined by xRy if y is a cover of x i.e. $R = \{(x, y) | x, y \in S$ and y covers $x\}$. Find the covering relation on the poset

(a) $S = \{1, 2, 3, 4, 6, 12\}$ with the relation '$|$' divides.

(b) $S = p(A), A = (1, 2, 3\}$ with the relation $\subseteq$ (set inclusion).

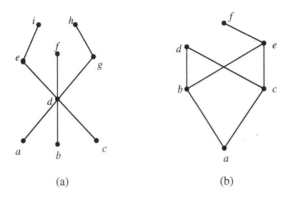

(a)   (b)

Fig. 6.14.

19. Decide if it is possible to reconstruct a finite poset from its covering relation. Justify this for an example of a set of order (a) 5; (b) 6.

20. List all ordered pairs in the poset with the Hasse diagrams given in Fig. 6.14. Also find their covering relations.

Chapter 7

GRAPHS

The subject of graph theory arose from real-life problems. This began with the solution of the Königsberg bridge problem by L. Euler in 1736. Rightly, Euler is called the father of graph theory. This chapter is devoted to presenting a brief introduction to the subject. In addition to the study of Euler and Hamiltonian paths (circuits), the problem of shortest distance from one vertex to another of a weighted graph is discussed. Also given is a brief introduction to planar graphs.

7.1. Preliminaries and Graph Terminology

A **graph** G is an ordered pair (V, E) where,

(a) V is a non-empty set,

(b) E is a set of multisets, each multiset having two elements from V.

Elements of the set V are called **vertices** (or **nodes**, **points** or **junctions**) and the elements of E are called **edges** (or **arcs** or **lines**). Also V is called the **vertex set** and E the **edge set** of the graph G. The graph may also be written as or denoted by $G(V, E)$. A graph $G(V, E)$, in which both V and E are finite sets is called a **finite graph**. Since in this study we are concerned only with finite graphs, by a graph we shall mean a finite graph.

Diagrammatically, a graph $G(V, E)$ may be described by representing elements of V as points or small circles along with lines or arcs between various points (not necessarily all) of V. An edge is said to be **incident** **with the** vertices it joins. For example, if $a, b \in V$ and $e = \{a, b\} \in E$, we say that the edge e is incident with the vertices a, b. Also, we say that the

241

vertices a, b are **adjacent**, i.e. vertices a, b are said to be adjacent if $\{a, b\}$ is an edge in E. Two distinct edges which have a common vertex are called **adjacent edges**. A vertex $a \in V$ is called an **isolated vertex** if there is no vertex $b \in V$, for which $\{a, b\} \in E$. If $a \in V$ and $\{a, a\} \in E$, then the edge $\{a, a\}$ is called a **loop**. A graph with n vertices and m edges is referred to as an (n, m) graph. An $(n, 0)$ graph is called a **trivial graph**.

A **directed graph** G may be defined as an ordered pair (V, E), where V is a non-empty set and E is a binary relation on V or that E is a subset of the Cartesian product $V \times V$. As in the case of graph defined earlier, the elements of V are called **vertices**, or **nodes** or **points**, V is called the **vertex set**, the elements of E are called **edges** or **arcs**, etc. and E is called the **edge set** of the directed graph $G(V, E)$. Diagrammatically, a directed graph is represented by a set V of points or small circles along with arrows between various points (not necessarily all) of V. These arrows constitute the set E of edges. If $e = (a, b) \in E$ is an edge in the directed graph, we say that e is **incident from** a and **incident into** b. Also a is called the **initial vertex** and b **the terminal vertex** *of* $e = (a, b)$. Moreover, the vertices a, b are said to be **adjacent** and e and a (or e and b) are said to be **incident with each other**. We may also use the terminology that a is **adjacent** to b and b is **adjacent** to a. A vertex with no edge incident with it is called an **isolated vertex**. A graph defined earlier, as opposed to a directed graph may be called an **undirected graph**. However, we use the word graph to mean an undirected graph and a directed graph shall always be referred to as such, or as a **diagraph**. Following are some examples of graphs.

Example 7.1.

1. $G = (\{a, b, c, d\} = V, E = \{\{a, b\}, \{a, c\}, \{b, d\}, \{d, d\}\})$. Or diagrammatically as in Fig. 7.1.
2. $G = (V, E)$, where V is the set of all villages in Kurukshetra district and E is the set of all roads connecting villages two-by-two, by a direct road.

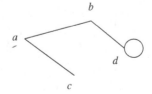

Fig. 7.1.

3. $G = (V, E)$, where V is the set of all towns in Kurukshetra district and E is the set of all roads leading from Kurukshetra to the towns (only direct roads are roads to be considered).

4. $G = (V, E)$, where V is the set of four tennis players called a, b, c, d while E contains ordered pairs (x, y) if x beats y in a direct match. This is a directed graph.

5. $G = (V, E)$, where $V = \{1, 2, 3, 4\}$ is the set of first four chapters in a book and an ordered pair $(i, j) \in E$ if the ith chapter is referred to in the jth chapter. This also gives a directed graph. If we take $E = \{(1, 2), (1, 3), (2, 4), (3, 4)\}$ the graph may be represented as in Fig. 7.2.

6. The graph $G = (V, E)$, where $V = \{v_1, v_2, v_3, v_4, v_5\}$ and

$$E = \{\{v_1, v_2\}, \{v_1, v_3\}, \{v_2, v_3\}, \{v_2, v_4\}, \{v_3, v_5\}, \{v_4, v_5\}\}$$

is represented by Fig. 7.3.

7. Some directed graphs with three vertices and three edges are given in Figs. 7.4–7.6.

8. The graph $G = (V, E)$, where $V =$ the set of all railway stations in Haryana; and $E =$ the collection of all pairs of stations connected by a rail track.

9. The graph $G = (V, E)$, where $V =$ the set of all railway stations in India, and $E =$ the set of all pairs of stations connected by a rail track.

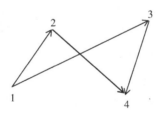

Fig. 7.2.

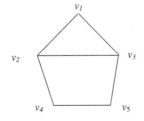

Fig. 7.3.

Fig. 7.4.

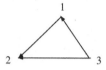

Fig. 7.5.

Fig. 7.6.

10. The graph $G = (V, E)$, where $V =$ the set of all airports in India; and $E =$ the set of all pairs of airports between which there is an air route (i.e. between which there is a direct flight).
Observe that none of the graphs 8–10 has a loop.

Exercise 7.1.

1. How many graphs are there with three vertices and three edges?
2. How many directed graphs are there with three vertices and three edges?
3. How many graphs are there with three vertices and two edges?
4. How many directed graphs are there with three vertices and two edges?

A graph with no loops, in which there is exactly one edge between every pair of distinct vertices is called a **complete graph**. A complete graph with n vertices is denoted by K_n. The notion of 'complete graph' may also be defined for directed graphs and, as such, in a complete directed graph (also called **directed complete diagraph**) there is exactly one arrow between two distinct vertices. With the same vertex set V, we can have more than one complete diagraph. For example, if $V = \{a, b, c\}$, we may take $E_1 = \{(a, b), (a, c), (b, c)\}$, $E_2 = \{(a, b), (b, c), (c, a)\}$, $E_3 = \{(a, b), (c, b), (a, c)\}$, etc. and we get three distinct diagraphs $G_1 = G(V, E_1)$,

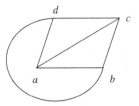

Fig. 7.7. K_4

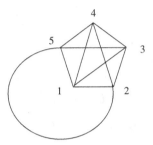

Fig. 7.8. K_5

$G_2 = G(V, E_2)$, $G_3 = G(V, E_3)$. There are still some more directed complete graphs with the vertex set V (decide these additional diagraphs). Figures 7.7 and 7.8 give complete graphs with four and five vertices.

Let G be a complete (n, m) graph. Then a given vertex a is adjacent to the remaining $n - 1$ vertices and so there are $n - 1$ edges incident with this vertex a. Next we consider a vertex $b \neq a$. The vertex b is also incident with $n - 1$ edges but one of these edges is the one which is incident with a as well. Thus there are $n - 2$ new edges incident with b. We continue in this way and find that the total number of edges is $(n - 1) + (n - 2) + \cdots + 3 + 2 + 1 = n(n - 1)/2 = {}^nC_2$. We thus have:

Theorem 7.1. If G is an (n, m) complete graph, then $m = C(n, 2) = n(n - 1)/2$.

Observe that none of the graphs in Example 7.1 (1 to 6 and 8 to 10) are complete. However, the three graphs as in Example 7.1 (7) are complete diagrams.

In an (n, m) graph $G(V, E)$, the number of edges incident with a vertex v is called the **degree** *of* v and is denoted by $\deg(v)$. Every edge $e = \{v_i, v_j\}$ contributes 1 to the degree of v_i and 1 to the degree of v_j. A loop $\{v_i, v_i\}$ contributes 2 to the degree of v_i. Therefore,

Theorem 7.2. In an (n, m) graph $\sum_{1 \leq i \leq n} \deg(v_i) = 2m$.

In a diagraph, v_i is the initial vertex and v_j is the terminal vertex of $e = (v_i, v_j)$. Therefore, instead of talking of degree of a vertex v, we define the outgoing degree of v and the incoming degree of v. The number of edges incident from v is called the **outgoing degree** of v and is denoted by outdeg(v). The number of edges incident into v (i.e. for which v is the terminal vertex) is called the **indegree** of v and is denoted by indeg(v). A loop $e = (v_i, v_i)$ contributes 1 to indeg(v_i) and 1 to outdeg(v_i).

Corollary. An undirected graph has an even number of vertices of odd degree.

Proof. Consider an undirected graph $G = (V, E)$. Partition the set V of vertices as $V = V_1 \cup V_2 \cup V_3$, where V_1 is the subset of V consisting of all vertices of odd degree in G, V_2, the subset of V consisting of all vertices of even degree in G and V_3, the set of all isolated vertices in G. The degree of every isolated vertex is 0. By the theorem

$$2o(E) = \sum_{v \in V} \deg v$$

$$= \sum_{v \in V_1} \deg v + \sum_{v \in V_2} \deg v + \sum_{v \in V_3} \deg v = \sum_{v \in V_1} \deg v + \sum_{v \in V_2} \deg v.$$

For every $v \in V_2$, $\deg v$ is even. Therefore, $\sum \deg v$ for all $v \in V_2$ is an even number (it could be zero if V_2 is an empty set, i.e. if there are no vertices of non-zero even degree). Also $2o(E)$ is an even number. Therefore, $\sum \deg v$, where $v \in V_1$ is even. Since the sum of even number of odd integers is even, whereas the sum of an odd number of odd integers is odd, it follows that V_1 has an even number of vertices. Hence G has an even number of vertices, every one of which is of odd degree.

Theorem 7.3. In an (n, m) diagraph,

$$\sum_{1 \leq i \leq n} \text{in} \deg(v_i) = \sum_{1 \leq i \leq n} \text{out} \deg(v_i) = m.$$

A graph G in which every vertex has the same degree d is called a **regular graph of degree** d or a d-**regular graph**. Figures 7.9 and 7.10 show regular graphs of degree two and three respectively. Yet another regular graph of degree three is given by Fig. 7.11. However, the graph given by Fig. 7.12 is not regular. We may define a **regular diagraph** in a similar fashion.

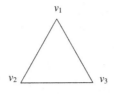

Fig. 7.9.

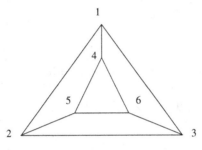

Fig. 7.10.

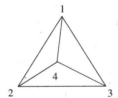

Fig. 7.11.

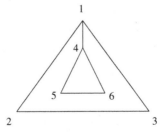

Fig. 7.12.

Let $G = (V, E)$ be a graph. A graph $G' = (V', E')$ is called a **subgraph** *of G* if V' is a subset of V and E' is a subset of E. A subgraph G' of G is called a **spanning subgraph** of G if $V' = V$, i.e. G' contains all vertices of G. Let $G' = (V', E')$ be a subgraph of $G = (V, E)$. The **complement** of the subgraph G' in G is defined to be the subgraph $G'' = (V'', E'')$ of G such that $E'' = E - E'$ and V'' contains all those vertices of G, which are incident with the edges in E'' or are isolated vertices of G not contained in V'. The complement of an (n, m) graph $G = (V, E)$ is the complement of the subgraph G in the complete graph K_n defined on the vertex set V.

In the graph $G = (V, E)$, where $V = \{a, b, c, d\}$ and $E = \{\{a, b\}, \{a, d\}, \{b, d\}\}$, $G' = (V', E')$ with $V' = \{a, b, d\}$, $E' = \{\{a, b\}, \{b, d\}\}$ is a subgraph and $G'' = (V'', E'')$, where $E'' = \{\{a, d\}\}$, $V'' = \{a, c, d\}$ is the complement of G' in G. Also, $\widetilde{G} = (\widetilde{V}, \widetilde{E})$ with $\widetilde{E} = \{\{a, c\}, \{b, c\}, \{c, d\}\}$ and $\widetilde{V} = \{a, b, c, d\} = V$ is the complement of the graph G. Moreover, the complement of the subgraph $G_1 = (V_1, E_1)$ with $V_1 = \{a, b, d\}$, $E_1 = E$ is the $(1, 0)$ graph $(\{c\}, \phi)$.

We, however, deviate from the above definition of complement of a graph. Given a simple (n, m) graph $G(V, E)$ without loops, the graph G' with vertex set V and the edge set E' consisting of edges required to make G complete is called the complement of G. This means that the graph $(V, E \cup E')$ is a complete graph on the vertex set V. Observe that if $G' = (V, E')$ is the complement of $G = (V, E)$, then the graph $G = (V, E)$ is the complement of the graph $G' = (V, E')$.

Observe that the graphs given by Figs. 7.13 and 7.14 are complements of each other and so are the graphs given by Figs. 7.15 and 7.16.

The two definitions of the complement of an (n, m) graph given above are not the same. The difference between the two definitions is made clear by the following examples. Observe that the complement of the graph of Fig. 7.17 is the graph of Fig. 7.18, according to the first definition, whereas it is graph of Fig. 7.19 in the sense of the second definition.

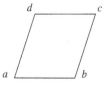

Fig. 7.13.

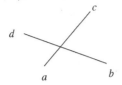

Fig. 7.14.

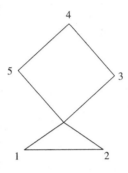

Fig. 7.15.

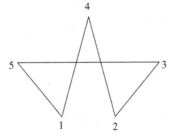

Fig. 7.16.

Fig. 7.17.

Fig. 7.18.

Fig. 7.19.

We find that the graph of Fig. 7.18 is a complete graph on the vertex set $V = \{a, c\}$ and as such we cannot talk of its complement in the first sense, whereas in the sense of the second definition it consists of the two isolated vertices a, c.

Next consider the graphs as given in Figs. 7.20–7.22.

As per the first definition, the complement of the graph of Fig. 7.20 is that given in Fig. 7.21, whereas it is the graph of Fig. 7.22 in the sense of the second definition. The graph of Fig. 7.21 is a complete graph on $V = \{a, c, d\}$ and so we cannot talk of its complement according to the first

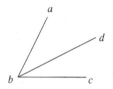

Fig. 7.20.

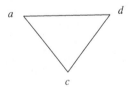

Fig. 7.21.

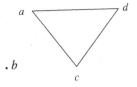

Fig. 7.22.

Fig. 7.23.

Fig. 7.24.

definition, whereas according to the second definition it consists of three isolated vertices a, c, d.

Observe the graph of Fig. 7.24 is the complement of the graph of Fig. 7.23 in the sense of both the definitions.

Since we cannot talk of the complement of a complete n-**graph** in the sense of the first definition, we accept only the second definition of the complement of a graph. If $G' = (V, E')$ is the complement of the graph $G = (V, E)$, then G' is also called the **complementary graph** of the graph G.

Let $G = (V, E)$ be an (n, m) graph and $G' = (V, E')$ be the complement of G. Then $G'' = (V, E \cup E')$ is a complete graph on the vertex set V. In view of the definition of complementary graph, $E \cap E' = \phi$. Then, by the principle of inclusion and exclusion, $o(E \cup E') = o(E) + o(E')$. As $o(V) = n$ and $o(E \cup E') = n(n-1)/2$, $o(E) + o(E') = n(n-1)/2$.

It is fairly easy to see that there is no graph $G = (V, E)$ with complement $G' = (V, E')$ such that the sum of the number of edges in G and in G' equals (i) 18; (ii) 23 or (iii) 24.

Exercise 7.2.

1. If $G = (V, E)$ is a graph with 15 edges and its complement $G' = (V, E')$ has 13 edges, determine the number of vertices of G.
2. If $G = (V, E)$ is a graph with 12 edges and its complement $G' = (V, E')$ has 9 edges, determine the number of vertices of G.

Two graphs $G = (V, E)$ and $G' = (V', E')$ are said to be **isomorphic** if there is a one–one correspondence $\alpha : V \to V'$ and a one–one correspondence $\beta : E \to E'$ such that $\{x', y'\} = \beta(\{x, y\})$ if and only if $x' = \alpha(x), y' = \alpha(y)$ or $x' = \alpha(y)$, $y' = \alpha(x)$. This is equivalent to saying that there is an edge between two vertices in G if and only if there is an edge between the corresponding vertices in G'. In case G, G' are directed graphs, the conditions on α, β take the form: $(x', y') = \beta(x, y)$ if and only if $x' = \alpha(x)$, $y' = \alpha(y)$.

For the graphs given by Figs. 7.25 and 7.26, $1 \to b$, $2 \to a$, $3 \to c$ and $4 \to d$ gives a one–one correspondence between the sets of vertices of the two graphs which induces a one–one correspondence between the corresponding edges. Therefore, the two graphs are isomorphic. Observe that the mapping $1 \to c$, $2 \to a$, $3 \to b$, $4 \to d$ also gives an isomorphism between the two graphs. However, the mapping $1 \to a$, $2 \to d$, $3 \to b$, $4 \to c$ does not give an isomorphism between the two graphs.

Next, consider the directed graphs given by Figs. 7.27 and 7.28. Consider the mapping $1 \to b$, $2 \to d$, $3 \to c$ and $4 \to a$, which gives an isomorphism between the two graphs. The map $1 \to c$, $2 \to a$, $3 \to b$,

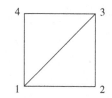

Fig. 7.25.

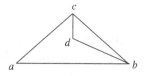

Fig. 7.26.

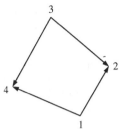

Fig. 7.27.

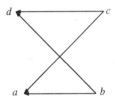

Fig. 7.28.

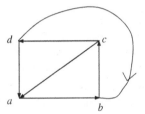

Fig. 7.29.

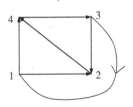

Fig. 7.30.

$4 \to d$ also gives an isomorphism between the two digraphs. However, the map $1 \to d$, $2 \to a$, $3 \to b$, $4 \to c$ does not lead to an isomorphism between the two graphs.

For the graphs given in Figs. 7.29 and 7.30, the mapping $1 \to d$, $2 \to a$, $3 \to c$ and $4 \to b$ gives an isomorphism between the two graphs.

A graph $G = (V, E)$ where V is a non-empty set and E is a multiset of multisets of two elements each from V, is called a **multigraph**. In other words, a graph in which there are some pairs of vertices which are incident with more than one edge is called a multigraph. Similarly, a directed graph, in which there is at least one pair of nodes connected by (or incident with) at least two arrows (or edges) is called a **directed multigraph**. In the language of sets, $G = (V, E)$, where V is a non-empty set and E is a multiset of ordered pairs from $V \times V$ is called a directed multigraph. An example of a multigraph is a graphical representation of a highway map in which an edge between two cities corresponds to a lane in a highway between the cities. The road map of Chandigarh with traffic islands as vertices is another example of a multigraph. As opposed to a multigraph, a graph in which there are no loops and no pair of vertices is incident with more than one edge is called a **simple** or **linear graph**.

The computer system of the branches of the Reserve Bank of India, located in different cities, are connected to each other. If there were only one line connecting pairs of branches and for some technical fault one of the channels failed, work in two or more branches may be held up. To avoid such a breakdown or to minimize the risk of break down of work in branches, a computer network with multiple lines is installed. This gives another example of a multigraph. Such a network may be shown as in Fig. 7.31.

If each edge of a graph is assigned a certain number, then it is called a **weighted graph**. Also, the number assigned to an edge is called the **weight of that edge**. However, there may be weights assigned to vertices instead of edges, or there may be weights assigned to both vertices as well as edges. More formally, a **weighted graph** is either an ordered quadruple (V, E, f, g) or an ordered triple (V, E, f) or an ordered triple (V, E, g) where

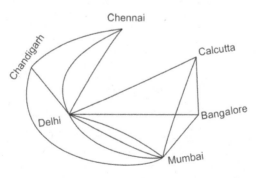

Fig. 7.31.   A complete network with multiple lines.

V is a non-empty set of vertices, E is a set (or multiset) of edges, f is a function with domain V and g is a map with domain E, both f and g taking numerical values.

In the road map of Chandigarh (or any city), if each road or street is assigned a number which gives the density of traffic on that street, we get a weighted graph. We could attach in this graph a number to each traffic island, which represents the number of roads/streets passing through the island. As another example, in the graph of highway connections between cities, an edge may be assigned a weight as the distance between the two cities it connects and a vertex, i.e. a city may be assigned the weight as the population of that city.

As another example of a weighted graph we may consider a computer network in which to each edge in the graph we may assign the response time between the two vertices (i.e. the two stations or the two computer systems) or we may assign the physical distance between the two stations. Two typical graphs may be described by Figs. 7.32(a) and (b).

We end this section by introducing yet another category of graphs. Consider the set V of all married people in a certain town. Two elements a, b from V are said to be incident with each other or that there is an edge connecting a, b if a and b are husband and wife. The set of all such relationships constitutes the edge set. The set V can be partitioned as $V_1 \cup V_2$ where V_1, V_2 are non-empty subsets disjoint from each other. We may take V_1 as the set of all male married persons in the town and V_2 as the set of all female married persons in the town. Then, by the definition of an edge, no two vertices in V_1 are incident with each other and no two vertices in V_2 are incident with each other. As we have taken only the married people in the town, there are no isolated vertices (of course under the presumption that everybody has his or her spouse in the same town or that married people in the town are not living alone or separate from her or his spouse). We could have taken V to be the set of all people in the town, V_1 to be the subset consisting of all male people in the town and V_2 the subset consisting of all female persons in the town. Again $a, b \in V$ are incident with each other if a, b are husband and wife, which means that there is an edge between a and b if they are husband and wife. In this case again, no two elements of V_1 are incident with each other and no two vertices in V_2 are incident with each other. However, every vertex in V_1 may not be incident with a vertex in V_2 and vice versa. So there may be some isolated vertices in this graph. Motivated by such a situation, we give the following definition.

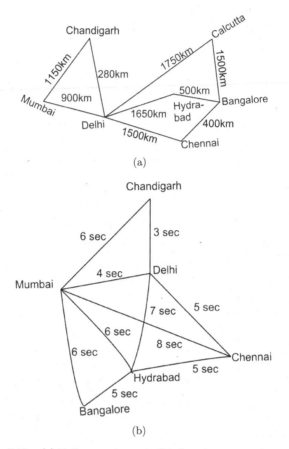

Fig. 7.32.   (a) Rail network graph. (b) Computer network graph.

Definition. A simple graph $G = (V, E)$ is called a **bipartite graph** if its vertex set V is the union of two disjoint non-empty subsets V_1 and V_2 such that every edge in E connects a vertex in V_1 to a vertex in V_2, which is equivalent to saying that no edge in G connects either two vertices in V_1 or two vertices in V_2.

For a bipartite graph $G = (V, E)$ with V partitioned as a union of two non-empty subsets V_1 and V_2, it is not essential that every vertex in V_1 should be adjacent to every vertex in V_2 and conversely. For example, the graph of Fig. 7.33 is a bipartite graph with the vertex set $V = \{a, b, c, d, e, f, g\}$ partitioned as $V_1 \cup V_2$ where $V_1 = \{a, b, g\}$ and $V_2 = \{c, d, e, f\}$. As is clear, no two vertices in V_1 are adjacent and no two vertices in V_2 are adjacent, so the graph is indeed bipartite. Also observe

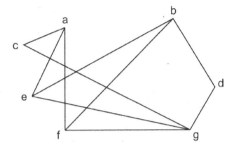

Fig. 7.33.

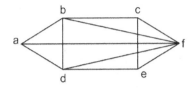

Fig. 7.34.

that the vertex a in V_1 is not adjacent to the vertex d in V_2. Also b in V_1 is not adjacent to c in V_2.

Consider the graph as in Fig. 7.34.

If this graph were bipartite and a belongs to a subset V_1 of the vertex set $V = \{a, b, c, d, e, f\}$ where $V = V_1 \cup V_2, V_1 \cap V_2 = \emptyset$ and $V_1 \neq \emptyset$, $V_2 \neq \emptyset$, then a being adjacent to b, d, f, none of these three vertices is in V_1. Therefore, b, d, f must belong to V_2. As b, d are adjacent, both of these could not belong to V_2. Hence we cannot partition V as required with no two vertices in V_1 being adjacent and no two vertices in V_2 adjacent. Hence the graph under consideration is not bipartite.

Definition. A bipartite graph $G = (V, E)$ is called a **complete bipartite graph** if $V = V_1 \cup V_2, V_1 \cap V_2 = \emptyset$ and $V_1 \neq \emptyset$, $V_2 \neq \emptyset$ and every vertex in V_1 is adjacent to every vertex in V_2 and every vertex in V_2 is adjacent to every vertex in V_1. If the order of V_1 is m and that of V_2 is n, then G is called a complete (m, n) bipartite graph and is denoted by $K_{m,n}$. The complete bipartite graphs $K_{2,2}$, $K_{2,3}$, $K_{3,3}$, $K_{3,4}$, $K_{3,5}$, $K_{2,5}$, $K_{2,4}$ are displayed in Fig. 7.35.

As seen already, the number of edges in a complete graph K_n on n vertices is $n(n-1)/2$. However, in the case of a complete bipartite graph $K_{m,n}$, the number of edges in mn.

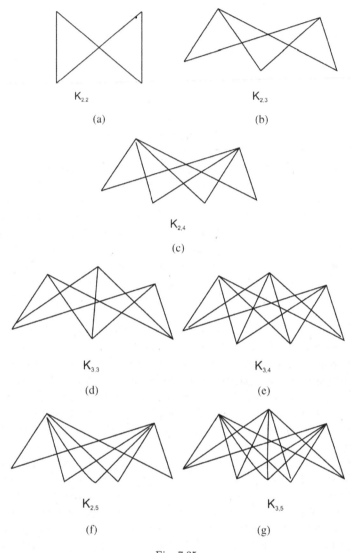

Fig. 7.35.

In a bipartite graph $G = (V, E)$ with V the union of disjoint non-empty subsets V_1 and V_2, we may colour every vertex in V_1 red and every vertex in V_2 green. Then no two red vertices are adjacent and no two green vertices are adjacent. Also, every red vertex need not be adjacent to a green vertex. However, in the case of a complete bipartite graph, every red vertex is adjacent to every green vertex and every green vertex is adjacent to every red vertex.

A simple graph with two vertices and one edge is always bipartite. Therefore, a simple graph that is not bipartite must have at least three vertices. All possible simple graphs with three vertices are (i) the trivial graph, i.e. a graph with three isolated vertices; (ii) a graph having only one edge; (iii) a graph with two edges and (iv) a graph with three edges. A simple graph with three vertices and one or two edges is always bipartite (cf. Fig. 7.36).

In the first case, we can partition $V = \{a, b, c\}$ as $V_1 \cup V_2$ with $V_1 = \{a\}$, $V_2 = \{b, c\}$ while in the second we can take $V_1 = \{b\}$, $V_2 = \{a, c\}$. In the first case we could also take $V_1 = \{b\}, V_2 = \{a, c\}$. There is only one simple graph with three vertices and three edges and it is the complete graph K_3 and is not bipartite. Thus the smallest simple graph which is not bipartite is the complete three-graph K_3.

Example 7.2. We are given four cubes and the six faces of every cube are coloured using four colours, say blue, green, red and white (with exactly one colour on a face and all four colours are used on every cube). Is it possible to stack the cubes one on top of another to form a column such that no colour appears twice on any of the four sides of the column?

Solution. Hit and trial is one way to solve this problem. However, the number of trials involved is huge — in fact it is $3 \times 24 \times 24 \times 24 = 41{,}472$. Therefore, the use of this method is not practical. We will try to solve this problem using graph theory.

Let B, G, R and W denote respectively the colours black, green, red and white. Draw a graph with B, G, R and W as four vertices. Let the four cubes unfolded (say) be as given in Fig. 7.37.

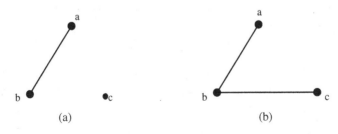

(a)   (b)

Fig. 7.36.

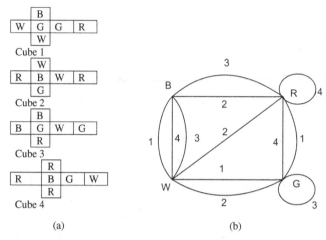

(a)   (b)

Fig. 7.37.   (a) Four cubes unfolded and (b) a graph representing the colours and cubes.

The edges in the graph are given by the colours on the opposite faces of the cubes. Consider cube 1. The three pairs of opposite faces carry the colours G, R; W, G; B, W. Corresponding to these pairs, draw three edges; one between G and R, the second between W and G, and the third between B and W. Also, to indicate that these edges are given by cube 1, each of these edges is marked 1. Draw three edges each for every one of the cubes 2, 3, 4 and the respective edges labelled by the numbers 2, 3, 4 respectively. Then the graph has 12 edges as given in Fig. 7.37(b).

Consider now the graph as given in Fig. 7.37(b). The degree of each vertex represents the total number of faces with the corresponding colour. Thus, the degree of B, G, R and W are 5, 6, 7 and 6 respectively, which match with the number of faces with that colour of the four cubes put together.

Consider now the two opposite vertical sides of the column made by the four cubes which are facing east and west, say. Since all the four cubes are involved in this pair of opposite vertical sides, we get a subgraph containing four edges. These edges carry the labels 1, 2, 3 and 4. The colours on the east-side of the column are all distinct and the colours on the west-side faces of the cubes are all distinct if and only if the degree of every one of the four vertices in this subgraph is 2. Similarly, the colours on the north-side faces of the cubes are all distinct and so are the colours on the south-side faces of the cubes forming the column if and only if the subgraph corresponding to the north- and south-side faces of the column has four edges and four

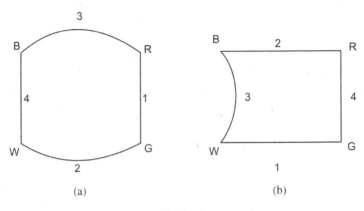

Fig. 7.38.

vertices, each vertex of degree 2. Hence a desired column can be made if and only if we can find two subgraphs of the graph, each having four edges and four vertices and every vertex is of degree 2. Observe that the two subgraphs in Fig. 7.38(a) and (b) satisfy all the requirements and, hence, a required column can be made.

Exercise 7.3. This will give us one way of making the column. Is there another way of arranging the cubes to make a desired column?

7.1.1. *Some typical examples*

Example 7.3. Five married couples on a journey come to a river where they find a boat which can not carry more than three persons at a time. The men do not permit their wives to be left without them in a company where there are other men present. Construct a graph to show how the couples can cross the river to go to the other side.

Solution. Let all allowable configurations be represented as vertices. Let the five couples be $a, a'; b, b'; c, c'; d, d'; e, e'$; where a', b', c', d', e' represent the wives of a, b, c, d, e respectively. Then the initial configuration is A:$\{a, a', b, b', c, c', d, d', e, e' \mid -\}$ i.e. all on this side of the river. Let $\{u, v, x, y, z \mid p, q, r, s, t\}$ denote the configuration where u, v, x, y, z are on this side of the river and p, q, r, s, t have crossed over to the other side of the river. Some other allowable configurations are: B:$\{a, a', b, b', c, d, e \mid c', d', e'\}$, C:$\{a, a', b, b', c, d, e, e' \mid c', d'\}$, E:$\{a, a', b, b', c, c', \mid d, d', e, e'\}$, F:$\{a', b', c' \mid a, b, c, d, d', e, e'\}$, H:$\{a' \mid a, b, b', c, c', d, d', e, e'\}$, I:$\{a, a' \mid b, b', c, c', d,$

$d', e, e'\}$, $J{:}\{-\,|\,a, a', b, b', c, c', d, d', e, e'\}$. There are some intermediary, non-allowable configurations we need: $D{:}\{a, a', b, b', c\,|\,c', d, d', e, e'\}$, $G{:}\{a, a', b', c'\,|\,b, c, d, d', e, e'\}$, $H{:}\{a\,|\,a', b, b', c, c', d, d', e, e'\}$.

Let there be an edge between two vertices if by performing journeys by some one person or a couple or three ladies or three gents or a couple and a gent, the two configurations can be transformed to each other. Let the journeys from an allowable to non-allowable or vice versa or from a non-allowable to non-allowable configuration be represented by dotted edges. Then the transfer made can be represented by the graph of Fig. 7.39.

Example 7.4. Five married couples come to a river where they find a boat which can not carry more than four persons at a time. The men do not permit their wives to be left without them in a company where there are other men present. Construct a graph to show how the five couples can be moved to the other side of the river.

Solution. Let a, b, c, d, e be five men with a', b', c', d', e' respectively their wives. Let all allowable configurations be represented as vertices of a graph. Then the initial configuration is

$$A = \{a, a', b, b', c, c', d, d', e, e'\,|\,-\}$$

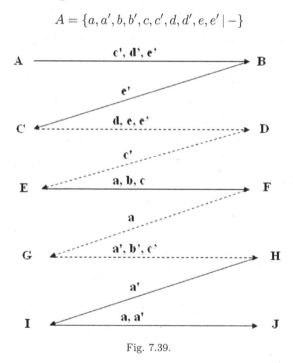

Fig. 7.39.

where, as in the last example, $\{x, y, z, t \mid p, q, r, \ldots\}$ represents a configuration in which the people x, y, z, t are on this side of the river and $p, q, r, \ldots$ are the ones who have crossed over to the other side of the river. Some of the other configurations we come across are

$$B = \{a, a', b, c, d, e \mid b', c', d', e'\}; \quad C = \{a, a', b, b', c, d, e \mid c', d', e'\}$$

$$D = \{a, a', c \mid b, b', c', d, d', e, e'\}; \quad E = \{a, a', c, c' \mid b, b', d, d', e, e'\}$$

$$F = \{- \mid a, a', b, b', c, c', d, d', e, e'\}.$$

Here B, C, E, F are allowable configurations but D is not an allowable configuration. Also F is the final configuration we want to arrive at. Let there be an edge between two configurations (allowable) if by performing journey by some allowable group of persons, the two configurations can be transformed to each other. Let the journey from allowable to non-allowable configuration or vice versa be indicated by a dotted edge. Then the people can be moved from this side of the river to the other as represented by Fig. 7.40.

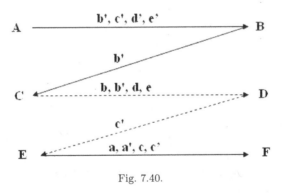

Fig. 7.40.

Example 7.5. A man has with him a dog, a sheep and a basket of cabbages. He has to take all the three across a river using a boat on which he can carry with him only one of the three at a time. For obvious reasons, he cannot leave the dog and the sheep together and he also cannot leave the sheep with the basket of cabbages. Use a graph to show a way that he can finally take all the three across the river.

Solution. Let d, s and c denote respectively the dog, the sheep and the basket of cabbages. Let m represent the man. Let us draw a graph with allowable configurations as vertices. The initial and final configurations are

$$A = \{d, s, c, m \mid -\} \quad \text{and} \quad H = \{- \mid d, s, c, m\}$$

where $\{x, y, \ldots \mid a, b, \ldots\}$ represents a configuration where $x, y, \ldots$ are on this side of the river and $a, b, \ldots$ are on the other side. We need the following allowable configurations:

$$B = \{d, c \mid s, m\}; \quad C = \{d, c, m \mid s\}; \quad D = \{c \mid d, s, m\}$$

$$E = \{c, s, m \mid d\}; \quad F = \{s \mid c, d, m\}; \quad G = \{s, m \mid d, c\}.$$

Let there be an edge between two vertices if the man can travel by boat either alone or with one of the three things so that one configuration can be transformed into the other. Then transfer of the three from one side of the river to the other can be represented graphically by Fig. 7.41.

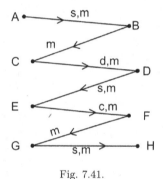

Fig. 7.41.

Example 7.6. Let $G = (V, E)$ be an undirected graph with k components and $o(V) = n, o(E) = m$. Prove that $m \geq n - k$.

Solution. Consider an undirected connected graph $G' = (V', E')$ where $o(V') = s \geq 2$. Consider a vertex v_1. Since G' is a connected graph, there is a vertex adjacent to v_1; let this vertex be v_2. Then v_2 is adjacent to another vertex v_3, v_3 is adjacent to v_4 and so on. Continuing this way, we get a chain, $v_1 \to v_2, v_2 \to v_3, \ldots, v_{s-1} \to v_s$. Therefore, there are at least $s - 1$ edges in G' i.e. $o(E') \geq o(V') - 1$. This result is true even when $o(V') = 1$. Thus, for any connected undirected graph (V', E'), $o(E') \geq o(V') - 1$.

The given graph has k components. Let these be $(E_1, V_1), (E_2, V_2), \ldots,$ (E_k, V_k). Then $E_i \cap E_j = \emptyset$ for $i \neq j$ and $V_i \cap V_j = \emptyset$ for $i \neq j$. Now, as

$$E = E_1 \cup E_2 \cup \ldots E_k \text{ and } V = V_1 \cup V_2 \cup \ldots \cup V_k,$$

$$o(E) = o(E_1) + o(E_2) + \cdots + o(E_k)$$

$$\geq (o(V_1) - 1) + (o(V_2) - 1) + \cdots + (o(V_k) - 1)$$

$$= o(V_1) + o(V_2) + \cdots + o(V_k) - k$$

$$= o(V) - k$$

or $m \geq n - k$.

7.2. Paths and Circuits

In a directed graph, a **path** is a sequence of edges $(e_1, e_2, \ldots, e_k)$ such that the terminal vertex of e_i is the initial vertex of e_{i+1} for every $i, 1 \leq i \leq k-1$. A path is called **simple** if it does not include the same edge more than once. Also a path is said to be **elementary** if it does not meet the same vertex more than once, i.e. no two edges in the sequence have the same terminal vertex. For example, in the graph of Fig. 7.42 $(e_1, e_2, e_3, e_4, e_5)$ is a simple path as well as elementary but the path $(e_1, e_2, e_3, e_4, e_5, e_7)$ is a simple path which is not elementary. The path $(e_1, e_2, e_3, e_4, e_5, e_7, e_{13}, e_5, e_6)$ is neither simple nor elementary.

A **circuit** in a directed graph is a path $(e_1, e_2, \ldots, e_k)$ in which the terminal vertex of e_k is the initial vertex of e_1. A circuit $(e_1, e_2, \ldots, e_k)$ is called (a) a **simple circuit** if the path $(e_1, e_2, \ldots, e_k)$ is simple and (b) an

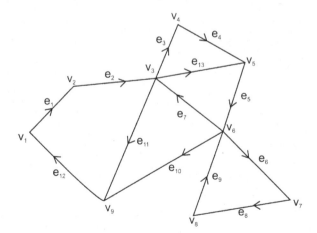

Fig. 7.42.

elementary circuit if the path $(e_1, e_2, \ldots, e_k)$ is elementary. For example, in the graph of Fig. 7.42, $(e_1, e_2, e_3, e_4, e_5, e_{10}, e_{12})$ is a circuit which is simple as well as elementary. The circuit $(e_1, e_2, e_3, e_4, e_5, e_7, e_{11}, e_{12})$ is a simple circuit but is not an elementary circuit as the vertex v_3 is met twice in the circuit or the edges e_2 and e_7 in the graph both have the terminal vertex v_3. However, the circuit $(e_1, e_2, e_3, e_4, e_5, e_7, e_{13}, e_5, e_{10}, e_{12})$ is neither a simple circuit nor an elementary circuit.

Observe that a path/circuit which is not simple cannot be elementary. For example, a path/circuit which is not simple looks like $(\ldots, e_{i-1}, e_i, e_{i+1}, \ldots, e_j, e_i, \ldots)$ then the initial vertex of e_i is the terminal vertex of e_{i-1} as well as e_j. Consider a path/circuit C which is elementary. If C were not simple, then the previous observation shows that C is not elementary, which is a contradiction. We have thus proved the following theorem.

Theorem 7.4. An elementary path/circuit is always simple.

Consider a path $(e_1, e_2, \ldots, e_k)$. Let v_i be the initial vertex of e_i and v_{i+1} be the terminal vertex of e_i for every $i, 1 \leq i \leq k$. If we are discussing simple directed graphs, (v_i, v_{i+1}) is an unambiguous representation of the edge e_i. Therefore, the path $(e_1, e_2, \ldots, e_k)$ could equally well be represented as $(v_1, v_2, \ldots, v_k, v_{k+1})$. Similarly, a circuit $(e_1, e_2, \ldots, e_k)$ with terminal vertex of e_k being $v_{k+1} = v_1$, the circuit may be represented as $(v_1, v_2, \ldots, v_k, v_{k+1} = v_1)$. In the graph of Fig. 7.42, the path (e_1, e_2, e_3, e_4) and the circuit $(e_1, e_2, e_{11}, e_{12})$ may be represented in terms of vertices as $(v_1, v_2, v_3, v_4, v_5)$ and $(v_1, v_2, v_3, v_9, v_1)$ respectively.

We can define a path or a circuit in an undirected graph in a similar fashion. These are defined in terms of edges, say $(e_1, e_2, \ldots, e_k)$ where for any $i, 1 \leq i \leq k, e_i = \{v_i, v_{i+1}\}$. If the graph under consideration is simple, this path shall be represented as $(v_1, v_2, \ldots, v_{k+1})$. However, if the graph under consideration is a multigraph or a graph where multiple edges are allowed between some two vertices, the path cannot be represented as $(v_1, v_2, \ldots, v_k, v_{k+1})$. For example, in Fig. 7.43(a) the path (e_2, e_3, e_4) cannot be represented as (v_2, v_3, v_4, v_1) as the path (e_2, e_3, e_5) could as well be represented as (v_2, v_3, v_4, v_1), the same sequence of vertices.

For a path $(e_1, e_2, \ldots, e_k)$ to be a circuit, we need to have $e_k = \{v_1, v_k\}$ when $e_1 = \{v_1, v_2\}$. We can define simple and elementary paths and circuits exactly as in the case of directed graphs and a path/circuit $(e_1, e_2, \ldots, e_k)$ is called **simple** if no edge occurs more than once in the path/circuit and it is called **elementary** if no vertex is hit more than once in the path/circuit. In the graph of Fig. 7.43(b) the path (q, p, v) is simple but

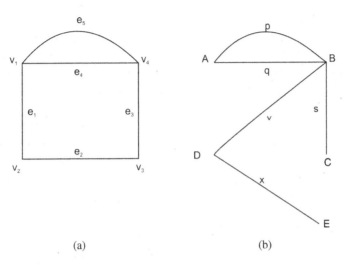

(a)   (b)

Fig. 7.43.

is not elementary, but the path (q, v, x) is simple as well as elementary. Obviously, these paths cannot be represented in terms of vertices. Consider the undirected graph obtained from the graph of Fig. 7.42 by removing the arrowheads from the edges. In this graph, the path $(e_4, e_3, e_{13}, e_7, e_5)$ is simple but not elementary, while the circuit (e_3, e_4, e_5, e_7) is simple as well as elementary.

The number of edges in a path/circuit $(e_1, e_2, \ldots, e_k)$ is called the **length of the path/circuit**. For any path of length k, the number of vertices involved is $k + 1$. However, a circuit of length k involves only k vertices and not $k + 1$.

Theorem 7.5. In a directed or undirected graph with n vertices, if there is a path from vertex v_1 to vertex v_2, then there is a path of no more than $n - 1$ edges from vertex v_1 to vertex v_2.

Proof. Suppose that there is a path from vertex v_1 to vertex v_2. Let $(v_1, \ldots, v_i, \ldots, v_2)$ be a path from v_1 to v_2. If there are l edges in this path, then there are $l + 1$ vertices in the path. If $l \leq n - 1$, we are through. For l larger than $n - 1$, there must be a vertex, say v_k, that appears more than once in this path. Then the path looks like $(v_1, \ldots, v_k, \ldots, v_k, \ldots, v_2)$. Deleting all the edges in this path that lead from v_k back to v_k, we get a path with less than l edges. We repeat this process until we get a path from vertex v_1 to vertex v_2 with, at most, $n - 1$ edges.

An undirected graph is said to be **connected** if there is a path between every two vertices and is said to be **disconnected** otherwise. A graph that is not connected is the union of two or more connected subgraphs, each pair of which has no vertex in common. These connected but disjoint subgraphs are called the **connected components** of the graph. A directed graph is said to be **connected** if the undirected graph obtained from it, by ignoring the directions of edges, is connected; it is said to be **disconnected** otherwise. A directed graph is said to be **strongly connected** if, for every pair of vertices a, b, there is a path from a to b as well as a path from b to a. For example, the directed graphs of Fig. 7.44(a) and (b) are connected graphs, but are not strongly connected. However the graph of Fig. 7.44(c) is a strongly connected directed graph. The graph of Fig. 7.44(d) is disconnected, with two connected components.

Theorem 7.6. There is a simple path between every pair of distinct vertices in a connected undirected graph.

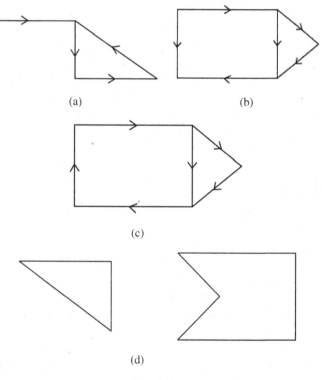

Fig. 7.44.

Proof. Let $G = (V, E)$ be an undirected graph and let u, v be two distinct vertices in G. The graph G being connected, there is a path between u and v. Let $(v_0, v_1, \ldots, v_n)$, with $v_0 = u$ and $v_n = v$ be a path between u and v. If an edge $e = (v_i, v_{i+1})$ appears at least twice in this path, the path looks like $(v_0, \ldots, v_{i-1}, v_i, v_{i+1}, \ldots, v_j, v_i, v_{i+1}, v_{j+1}, \ldots, v_n)$. By removing the vertices occurring after v_{i+1} on the left up to the vertex v_{i+1} on the right, we get a path between u and v in which the repetition of e is reduced by one. We continue this process until all the repetitions of edges are removed and we finally get a path between u and v which is simple.

The existence of simple circuits of length ≥ 3 in graphs is quite useful in deciding the isomorphism or otherwise of two given graphs.

Theorem 7.7. Show that the existence of a simple circuit of length k, where k is a positive integer greater than 2, is an isomorphism invariant.

Proof. Let $G = (V, E)$ and $H = (V', E')$ be two isomorphic graphs. We need to prove that if G has a simple circuit of length $k \geq 3$, then H also has a simple circuit of length k. By the definition of isomorphic graphs, there exist maps $f : V \to V', g : E \to E'$ which are one–one and onto and if $e = \{u, v\} \in E$, then $g(e) = \{f(u), f(v)\}$. Let $(e_1, e_2, \ldots, e_k)$ be a simple circuit in G. If $e_i = \{v_i, v_{i+1}\}, 1 \leq i \leq k$, then the given simple circuit may be taken as $(v_1, v_2, \ldots, v_{k+1})$ with $v_{k+1} = v_1$. The maps f, g being one–one, $g(e_1), g(e_2), \ldots, g(e_k)$ are all distinct. Also $g(e_i) = \{f(v_i), f(v_{i+1})\}$ for $1 \leq i \leq k$ and $f(v_{k+1}) = f(v_1)$. Therefore, $(g(e_1), g(e_2), \ldots, g(e_k))$ or $(f(v_1), f(v_2), \ldots, f(v_{k+1}))$ is a simple circuit of length k in H.

Example 7.7. As an application of this result, consider the graphs (Fig. 7.45) both with six vertices and eight edges. Each of the two graphs has four vertices of degree 3 and two vertices of degree 2. The graph of Fig. 7.45(b) has a simple circuit (u_1, u_2, u_6, u_1) as also (u_3, u_4, u_5, u_3) of length three but as is clear from the construction, the graph in Fig. 7.45(a) does not have a circuit of length three. Hence the two graphs are not isomorphic.

Example 7.8. As another application, consider the graphs of Fig. 7.46.

Either of the two graphs has five vertices and six edges. There are two vertices of degree 3 and three vertices of degree 2 each in G and the same is true about H. Now (u_1, u_3, u_4, u_1) is the only simple circuit of length three in G and (v_3, v_4, v_5, v_3) is the only simple circuit of length three in H.

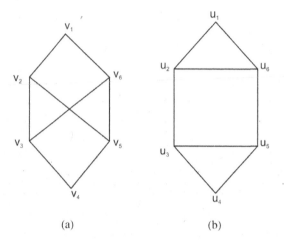

(a)   (b)

Fig. 7.45.

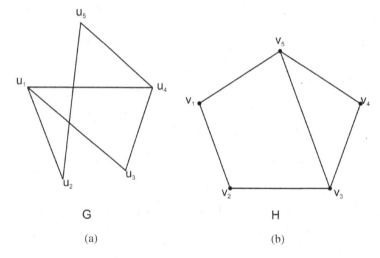

G   H

(a)   (b)

Fig. 7.46.

Hence, if G and H are isomorphic, then under any isomorphism from G to H, the vertex u_3 corresponds to vertex v_4. In G, $(u_1, u_2, u_5, u_4, u_1)$ is the only simple circuit of length four and in H, $(v_1, v_2, v_3, v_5, v_1)$ is the only simple circuit of length four. The circuit $(v_1, v_2, v_3, v_5, v_1)$ is the same as the circuit $(v_3, v_2, v_1, v_5, v_3)$. The vertices v_1, v_2 being vertices of degree 2 each in this circuit and the vertices u_2, u_5 being vertices of degree 2 each in the circuit $(u_1, u_2, u_5, u_4, u_1)$ of length four in G, $\{u_2, u_5\}$ and $\{v_1, v_2\}$ must be the corresponding sets of vertices under any isomorphism, if it exists from

G to H. Taking into consideration the above information, we may consider a map f from the vertex set of G to the vertex set of H given by

$$f(u_1) = v_3, \quad f(u_2) = v_2, \quad f(u_3) = v_4, \quad f(u_4) = v_5, \quad f(u_5) = v_1.$$

The map g from the edge set of G to the edge set of H may be defined by

$$g(\{u_1, u_2\}) = \{v_3, v_2\}, g(\{u_1, u_3\}) = \{v_3, v_4\}, g(\{u_1, u_4\}) = \{v_3, v_5\},$$

$$g(\{u_2, u_5\}) = \{v_1, v_2\}, g(\{u_3, u_4\}) = \{v_4, v_5\} \quad \text{and} \quad g(\{u_4, u_5\}) = \{v_1, v_5\}.$$

Thus f, g establish one–one correspondences between the vertex sets and edge sets of G and H of the desired type.

7.3. Shortest Path in Weighted Graphs

Let $G = (V, E, w)$ be a weighted graph where w is a function from the edge set E to the set of positive real numbers. For every $x \in E$, $w(x)$ is called the **weight** of the edge x. For example, in the graph of highways connecting various cities, with the edge $\{i, j\}$ we may assign the weight $w\{i, j\}$ to be the distance between the cities connected by the edge $\{i, j\}$ or it may be taken as the cost per year for repairing the highway. It could also be taken as the average number of vehicles plying the highway in a year. In the graph modelling a computer network, we may take $w(i, j)$ as the monthly cost of leasing a telephone line between the computers stationed in the two centres connected by $\{i, j\}$ or it may be taken as the response time of the computers over the line, etc. The **length l(p) of a path** p in G is defined to be the sum of weights of the edges in the path (this does not contradict our earlier definition of path length defined when the graph is not a weighted graph as we could assign weight 1 to every edge of the graph which is not a weighted graph). Then the **shortest path** between two vertices has the usual meaning, i.e. among all the paths connecting two vertices, it is that path which has the shortest length. Also the length of shortest path between two vertices is called the **shortest distance** between the two vertices. For example, consider the graph of Fig. 7.47 with the weight attached to each edge being given by the number over that edge. Recall that if two vertices are not adjacent, the weight attached to (the supposed) edge between the two vertices is taken to be infinity (∞). Consider the paths from a to f in this graph.

(i) (a, b, f);   (ii) (a, b, c, f);   (iii) (a, b, c, e, f);

(iv) (a, b, c, d, e, f);   (v) (a, b, c, d, e, c, f).

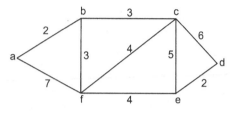

Fig. 7.47.

Observe that for these paths, we have

$$l(a, b, f) = 5; \quad l(a, b, c, f) = 9; \quad l(a, b, c, e, f) = 14;$$
$$l(a, b, c, d, e, f) = 17; \quad l(a, b, c, d, e, c, f) = 22.$$

Among these paths (a, b, f) is the shortest. As there are some other paths still from a to f, it does not prove that the path (a, b, f) is the shortest (although the graph being such a small one, (a, b, f) is indeed the shortest path). Since in a graph there are, in general, only a finite number of vertices and a finite number of edges, there shall be only a finite number of paths from one vertex to another if repetitions of paths are not allowed. Hence it is always possible to find a shortest path connecting two vertices. If there is no path connecting two vertices, the length of a path between these vertices is taken to be ∞.

We now describe a systematic procedure or an algorithm for finding a shortest path from one vertex to another vertex in a weighed connected graph. The procedure we describe is attributed to a Dutch mathematician, E.W. Dijkstra. It is an iterative procedure. If we are to find a shortest path from vertex a to vertex z, we first find a shortest path from a to a first vertex, then a shortest path from a to a second vertex and so on, until we find a shortest path from a to z.

Consider a weighted graph $G = (V, E, w)$ and let a, z be two vertices in G. Let T be a subset of V with $a \notin T$ and let $P = V - T$. For each vertex t in T, let $l(t)$ denote the length of a shortest path among all paths from a to t that do not include any other vertex in T. Then $l(t)$ is called the **index** of t w.r.t. P. Observe that $l(t)$ need not be the length of a shortest path from a to t. Among all the vertices in T, let t_1 be a vertex that has the smallest index.

Theorem 7.8. The shortest distance between a and t_1 is equal to $l(t)$.

Proof. Observe that the shortest distance between a and t_1 is $\leq l(t_1)$. If $l(t_1)$ is not the shortest distance between a and t_1, there exists a path from a to t_1 the length L of which is less than $l(t_1)$. In view of the definition of the index $l(t_1)$ it follows that this path includes one or more of the vertices in $T - \{t_1\}$. Let t_2 be the first vertex in $T - \{t_1\}$ we encounter when we trace this path from a to t_1. Since t_1 and t_2 are connected, $t_1 \neq t_2$ and the edges have positive weights, the length of the portion of the path under consideration from t_2 to t_1 is positive, say it is l'. Then $l(t_2) + l' = L < l(t_1)$ so that $l(t_2) < l(t_1)$. This contradicts the choice of t_1. Hence $l(t_1)$ is the shortest distance between a and t_1.

From the above, it also follows that while computing $l(t)$ when we record the sequences of vertices in the path that yield $l(t)$ for each t in T, we would also have determined a shortest path from a to t_1.

Example 7.9. Consider the graph of Fig. 7.48. To find a shortest path from a to z and the shortest distance, we start with $P = \{a\}$ and $T = V - \{a\}$. Then $l(b) = 2, l(c) = 4$ and the rest of the vertices in T not being adjacent to a, we have $l(x) = \infty$ for every $x \in T - \{b, c\}$. Since $l(b) = 2 < 4 = l(c)$, the choice for enlarging P falls on b. That $l(b)$ is the shortest distance from a is clear. When $P = \{a, b\}, T = V - P = \{c, d, e, z\}$. The vertices in T that are adjacent to b are c, d and e. We calculate $l(t)$ for each one of these. $l(c) = 4, l(d) = 2 + 5 = 7$ and $l(e) = 2 + 6 = 8$. The least among these is $l(c) = 4$. We next adjoin c to P to get $P = \{a, b, c\}$. Now e is the only vertex adjacent to c and also to b. With the present choice of P, $l(e) = \mathrm{Min}\{4 + 2, 2 + 3 + 2, 2 + 6\} = 6$ and among all the paths from a to e, the one which gives 6 is indeed a shortest path from a to e. For the next step, we adjoin e to P to get the new $P = \{a, b, c, e\}$, then $T = \{d, z\}$ and both the vertices in T are adjacent to e. Since $l(d)$ calculated in the second step is 7, which is larger than $l(e)$ calculated in the previous step, $l(d)$ in the present step also is 7. Also $l(z) = l(e)$ for the previous step $+ w(e, z) = 6 + 3 = 9$. The smaller of the two numbers is $l(d) = 7$.

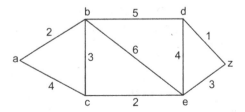

Fig. 7.48.

So we enlarge P by joining d. Thus $P = \{a, b, c, d, e\}$ and $T = \{z\}$. The vertex z is adjacent to d and $l(z)$ equals $\text{Min}\{l(z)$ calculated in the previous step, $l(d)$ calculated at the previous step $+ w(d, z)\} = \text{Min}\{9, 7 + 1\} = 8$. Hence $l(z) = 8$ which is the shortest distance from a to z. Also a shortest path is given by $d \rightarrow z$, preceded by a path leading to $l(d) = 7$ which is $d \leftarrow b \leftarrow a$. Hence a shortest path from a to z yielding the shortest distance is $a \rightarrow b \rightarrow d \rightarrow z$.

Algorithm 7.1. The procedure for computing the shortest distance from a to any vertex in $G = (V, E, w)$:

1. Initially, let $P = \{a\}$ and $T = V - P = V - \{a\}$. Then for every $t \in T, l(t) = w(a, t)$ where it is understood that $w(a, t) = \infty$ if there is no edge between a and t.
2. Select the vertex in T that has the smallest index w.r.t. P. Let x denote this vertex.
3. If x is the vertex we wish to reach from a, stop. If not, let $P' = P \cup \{x\}$ and $T' = T - \{x\}$. For any t in T', compute its index w.r.t. P'.
4. Repeat steps 2 and 3 using P' as P and T' as T.

As seen in Example 7.9, some simplifications can be made in the iterative steps. Initially, $P = \{a\}$ and $T = V - \{a\}$. Instead of calculating $l(t)$ for every $t \in T$, we need to calculate $l(t)$ for those $t \in T$ which are adjacent to a. The second step is the same as above, i.e. choose $x \in T$ for which $l(x)$ is the least among all the indices calculated. If x is the vertex z we wanted to reach, stop. Otherwise adjoin this x to P.

Suppose we are at an intermediate step for which an x with least $l(x)$ has been obtained. We consider the new P by adjoining this x to the previous P. In the new $T = V - P$, let T' be the set of vertices which are adjacent to x together with those t in T for which $l(t)$ was calculated at the previous step. If $y \in T'$ is not adjacent to x, we have $l(y)$ the same as $l(y)$ calculated at the previous step. On the other hand if $y \in T'$ is adjacent to x, then $l(y) = \text{Min}\{w(x, y) + l(x)$ calculated at the previous step, $l(y)$ calculated at the previous step$\}$, where it is understood that if for $y \in T', l(y)$ was not calculated at the previous step, it is taken to be ∞.

Example 7.10. Use Algorithm 7.1 explained in this section and determine a shortest path between a and z in the following graphs where the numbers associated with the edges are the distances between vertices.

Solution. 1. Since we have to find a shortest path from a to z, we take $P = \{a\}$ and then $T = V \setminus P = \{b, c, d, e, f, z\}$. We find the index $l(t)$ for every $t \in T$. Recall that $l(t)$ is the length of a shortest path from a to t that does not pass through any vertex in T except t. Thus, we have $l(b) = 2, l(c) = 1, l(d) = 4$ and any path from a to e or f or z contains at least one other vertex from T. Therefore e, f, z have no index w.r.t. P. The least among the above indices is $l(c) = 1$. We, therefore, next consider $P = \{a, c\}$ and then $T = \{b, d, e, f, z\}$. Indices of vertices in T w.r.t. present P are $l(b) = 2, l(d) = 3, l(e) = 6, l(f) = 8$ and $l(z)$ is not defined or we may say that $l(z) = \infty$. The least among these indices is $l(b) = 2$. Therefore, we next consider $P = \{a, b, c\}$ and then $T = \{d, e, f, z\}$. The indices of d, e, f, z w.r.t. the present P are $l(d) = 3, l(e) = 5, l(f) = 1 + 7 = 8$ and $l(z) = \infty$. The least among these is $l(d) = 3$. Therefore, we consider $P = \{a, b, c, d\}$; and then $T = \{e, f, z\}$. The indices of e, f, z w.r.t. the present P are $l(e) = 2 + 3 = 5, l(f) = 1 + 2 + 4 = 7, l(z) = \infty$. The least among these is $l(e) = 5$. We therefore take $P = \{a, b, c, d, e\}$ and $T = \{f, z\}$. The indices of f, z w.r.t. the P under consideration are $l(f) = 1 + 7 = 8, l(z) = 2 + 3 + 1 = 6$. The least among these is $l(z) = 6$ and we have come to the vertex z we

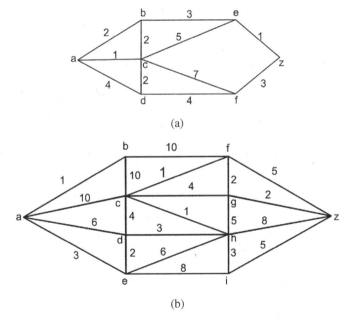

(a)

(b)

Fig. 7.49.

wanted to arrive at. Hence the shortest path from a to z is a to b, b to e and e to z and the length of this path is $2 + 3 + 1 = 6$.

    2. As we want to find a shortest path from a to z, we take

Step 1. $P = \{a\}; T = V \backslash P$.

    Then $l(b) = 1, l(c) = 10, l(d) = 6, l(e) = 3$ and $l(t) = \infty$ for every other $t \in T$. The least among these is $l(b) = 1$.

Step 2. $P = \{a, b\}, T = V \backslash P$.

    $l(c) = 10, l(d) = 6, l(e) = 3, l(f) = 11$ and $l(t) = \infty$ for every other $t \in T$. The least among these is $l(e) = 3$.

Step 3. $P = \{a, b, e\}, T = V \backslash P$.

    $l(c) = 10, l(d) = 5, l(f) = 11, l(h) = 9, l(i) = 11$ and $l(t) = \infty$ for every other t. The least among these is $l(d) = 5$.

Step 4. $P = \{a, b, d, e\}, T = V \backslash P$. $l(c) = 9, l(f) = 11, l(h) = 8, l(i) = 11$, and $l(t) = \infty$ for $t = g$ and z. The least among these is $l(h) = 8$.

Step 5. $P = \{a, b, d, e, h\}, T = \{c, f, g, i, z\}, l(c) = 9, l(f) = 11, l(g) = 13$, $l(i) = 11, l(z) = 16$. The least among these is $l(c) = 9$.

Step 6. $P = \{a, b, c, d, e, h\}, T = \{f, g, i, z\}. l(f) = 10, l(g) = 13, l(i) = 11$, $l(z) = 15$. Since $l(f) = 10$ we may include f in P.

Step 7. $P = \{a, b, c, d, e, f, h\}, T = \{g, i, z\}. l(g) = 13, l(i) = 11, l(z) = 15$. The least among these is $l(i) = 11$.

Step 8. $P = \{a, b, c, d, e, f, h, i\}, T = \{g, z\}. l(g) = 13, l(z) = 15$. The least among these is $l(g) = 13$.

Step 9. $P = \{a, b, c, d, e, f, g, h, i\}, T = \{z\}$ and $l(z) = 14$. The path of shortest distance is a to e, e to d, d to c, c to f, and g to z i.e. $a \rightarrow e \rightarrow d \rightarrow c \rightarrow f \rightarrow g \rightarrow z$ and the distance is $3 + 2 + 4 + 1 + 2 + 2 = 14$.

Example 7.11. Eight communication centres are connected by communication channels as given in the graph below where the centres are represented as vertices and the channels are represented as edges. The communication time delays in minutes are given as the weights of the edges. Suppose that at 1 p.m. the communication centre a broadcasts the news that the Nag surface to air missile has been successfully tested. The other communication centres then broadcast this news through all their channels as soon as they receive it. For the communication centres b, c, d, e, f, g and h, determine the earliest time each receives the news.

Solution. In the language of graphs we need to find the shortest paths from a to the remaining vertices. For this we shall use the algorithm discussed

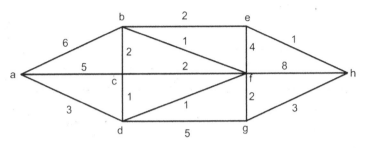

Fig. 7.50.

in this section. We now find shortest paths from a to b, c, d, e, f, g, h. For this we take $P = \{a\}$ and then $T = \{b, c, d, e, f, g, h\}$. The indices of the vertices in T relative to P are $l(b) = 6, l(c) = 5, l(d) = 3, l(t) = \infty$ for other $t \in T$. The least among these is $l(d) = 3$. So, d is the first to receive the news at 3 minutes past 1 p.m.

We next take $P = \{a, d\}, T = \{b, c, e, f, g, h\}$. The indices of elements of T w.r.t. the present P are $l(b) = 6, l(c) = 4, l(f) = 4, l(g) = 8, l(e) = l(h) = \infty$. The least among these is $l(c) = l(f) = 4$. Since the news going to c must pass through b or d or directly along the edge ac, the least time taken for the news to reach c is 4 minutes. Therefore both c and f receive the news simultaneously at 4 minutes past 1 p.m.

We next consider $P = \{a, c, d, f\}$ and then $T = \{b, e, g, h\}$. The indices of b, e, g, h relative to this P are $l(b) = 5, l(g) = 6, l(e) = 8$ and $l(h) = 9$. The least among these is $l(b) = 5$. Therefore, b receives the news at 5 minutes past 1 p.m.

We now take $P = \{a, b, c, d, f\}, T = \{e, g, h\}$. The indices of e, g, h are $l(g) = 6, l(e) = 7$ and $l(h) = 8$. The least being $l(g) = 6, g$ receives the news at 6 minutes past 1 p.m.

Next, consider $P = \{a, b, c, d, f, g\}, T = \{e, h\}$. Then $l(e) = 7, l(h) = 8$. The centre e receives the information at 7 minutes past 1 p.m., and finally h receives the information at 8 minutes past 1 p.m.

The paths followed by the earliest news to reach the centre are:

$$a \xrightarrow{3} d, a \xrightarrow{3} d \xrightarrow{1} c, a \xrightarrow{3} d \xrightarrow{1} f, a \xrightarrow{3} d \xrightarrow{1} f \xrightarrow{1} b, a \xrightarrow{3} d \xrightarrow{1} f \xrightarrow{2} g,$$

$$a \xrightarrow{3} d \xrightarrow{1} f \xrightarrow{1} b \xrightarrow{2} e \quad \text{and} \quad a \xrightarrow{3} d \xrightarrow{1} f \xrightarrow{1} b \xrightarrow{2} e \xrightarrow{1} h.$$

Example 7.12. Determine a shortest path between a and z in the graph of Fig. 7.51 where the numbers associated with the edges are the distances between the vertices concerned.

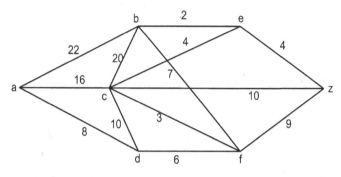

Fig. 7.51.

Solution. To apply the algorithm discussed in this section we first take $P = \{a\}, T\{b, c, d, e, f, z\}$. The indices of the vertices in T are $l(b) = 22, l(c) = 16, l(d) = 8$ and $l(t) = \infty$ for every other $t \in T$. The least among these is $l(d) = 8$.

For the second step, we take $P = \{a, d\}; T = \{b, c, e, f, z\}$. Then $l(b) = 22, l(c) = 16, l(f) = 14$ and $l(t) = \infty$ for $t = e, z$. The least is $l(f) = 14$. So, for the next step, we take $P = \{a, d, f\}, T = \{b, c, e, z\}$. Then $l(b) = 21, l(c) = 16, l(e) = \infty, l(z) = 23$. The least among these is $l(c) = 16$. Take $P = \{a, c, d, f\}, T = \{b, e, z\}$. Again $l(b) = 21, l(e) = 20, l(z) = 23$. The least among these is $l(e) = 20$ and, so, we take $P = \{a, c, d, e, f\}, T = \{b, z\}$. Then $l(b) = 21, l(z) = 23$. We are then required to take $P = \{a, b, c, d, e, f\}, T = \{z\}$ and $l(z) = 23$. The shortest path from a to z is $a \to d \to f \to z$ and the length is 23.

Example 7.13. Determine a shortest path between a and z for the graph given below where the numbers associated with the edges indicate the distance between the vertices concerned.

Solution.

Step 1. $P = \{a\}, T = \{b, c, d, e, f, g, h, i, j, k, l, m, z\}$. Then $l(b) = 2$, $l(c) = 2, l(d) = 2, l(e) = 2$ and $l(t) = \infty$ for other $t \in T$.

Step 2. $P = \{a, b\}, T = \{c, d, \ldots, m, z\}$. Then $l(c) = 2, l(d) = 2, l(e) = 2$, $l(g) = 4, l(f) = 6, l(t) = \infty$ for other $t \in T$.

Step 3. $P = \{a, b, c\}, T = \{d, \ldots, m, z\}$. Then $l(d) = 2, l(e) = 2, l(g) = 4$, $l(f) = 6, l(t) = \infty$ for other $t \in T$.

Step 4. $P = \{a, b, c, d\}, T = \{e, f, \ldots, m, z\}$. Then $l(e) = 2, l(g) = 4$, $l(f) = 6, l(h) = 6, l(i) = 4, l(t) = \infty$ for other $t \in T$.

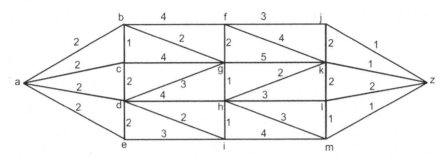

Fig. 7.52.

Step 5. $P = \{a, b, c, d, e\}, T = \{f, \ldots, m, z\}$. Then $l(g) = 4, l(f) = 6$,
$l(i) = 4, l(h) = 6$, $l(t) = \infty$ for other $t \in T$.

Step 6. $P = \{a, b, c, d, e, g\}, T = \{f, h, i, j, k, l, m, z\}$. Then $l(f) = 6$,
$l(i) = 4, l(h) = 6$, $l(k) = 9, l(t) = \infty$ for other $t \in T$.

Step 7. $P = \{a, b, c, d, e, g, i\}, T = \{f, h, j, k, l, m, z\}$. Then $l(f) = 6$,
$l(h) = 5, l(m) = 8, l(k) = 7, l(t) = \infty$ for $t = l, j, z$.

Step 8. $P = \{a, b, c, d, e, g, h, i\}, T = \{f, j, k, l, m, z\}$. Then $l(f) = 6$,
$l(j) = \infty, l(k) = 7, l(l) = 8, l(m) = 8, l(z) = \infty$.

Step 9. $P = \{a, b, c, d, e, f, g, h, i\}, T = \{j, k, l, m, z\}$. Then $l(j) = 9$,
$l(k) = 7, l(l) = 8, l(m) = 8, l(z) = \infty$.

Step 10. $P = \{a, b, c, d, e, f, g, h, i, k\}$, $T = \{j, l, m, z\}$. Then $l(j) = 9$,
$l(l) = 8, l(m) = 8, l(z) = 8$.

Step 11. $P = \{a, b, c, d, e, f, g, h, i, k, l\}, T = \{j, m, z\}$. Then $l(j) = 9$,
$l(m) = 8, l(z) = 8$.

Thus, the shortest path from a to z is $a \xrightarrow{2} d \xrightarrow{2} i \xrightarrow{1} h \xrightarrow{2} k \xrightarrow{1} z$ and its length is 8.

Example 7.14. Determine a shortest path between a and z in the graph of Fig. 7.53, where the numbers associated with the edges are the distances between the vertices concerned.

Solution. For this we use the algorithm described in this section.

Step 1. $P = \{a\}, T = \{b, c, \ldots, o, z\}$. Then $l(b) = 4, l(e) = 3, l(e) = 3$
being lower, we consider

Step 2. $P = \{a, e\}, T = \{b, \ldots, o, z\} \backslash \{e\}$. Then $l(b) = 4, l(f) = 5, l(i) = 6$.
The lowest being $l(b) = 4$, we consider

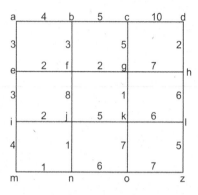

Fig. 7.53.

Step 3. $P = \{a, b, e\}, T = \{c, \ldots, o, z\} \backslash \{e\}$. Then $l(c) = 9, l(f) = 5$, $l(i) = 6$. The lowest among these being $l(f) = 5$, we take

Step 4. $P = \{a, b, e, f\}, T = \{a, \ldots, o, z\} \backslash P$. Then $l(c) = 9, l(i) = 6$, $l(g) = 7, l(j) = 13$. The lowest among these being $l(i) = 6$, we next consider

Step 5. $P = \{a, b, e, f, i\}, T = \{a, \ldots, o, z\} \backslash P$. Then $l(c) = 9, l(g) = 7$, $l(j) = 8, l(m) = 10$. The least among these being $l(g) = 7$, we need to consider

Step 6. $P = \{a, b, e, f, g, i\}, T = \{a, \ldots, o, z\} \backslash P$. The indices of the relevant vertices are $l(j) = 8, l(m) = 10, l(c) = 9, l(k) = 8$, $l(h) = 14$. Since $l(k) = 8$ is the least among these, we consider

Step 7. $P = \{a, b, e, f, g, i, k\}, T = \{a, \ldots, o, z\} \backslash P$. Then $l(j) = 8$, $l(l) = 14, l(o) = 15, l(c) = 9, l(m) = 10, l(h) = 14$. The least among these is $l(j) = 8$. Therefore, we next consider

Step 8. $P = \{a, b, e, f, g, i, j, k\}; T = \{a, \ldots, o, z\} \backslash P$. Then, relevant to this P, we have $l(n) = 9, l(m) = 10, l(o) = 15, l(l) = 14, l(h) = 14, l(c) = 9$. The least among these is $l(n) = 9$. We, therefore, consider

Step 9. $P = \{a, b, e, f, g, i, j, k, n\}, T = \{c, d, h, l, m, o, z\}$. Then $l(h) = 14, l(l) = 14, l(m) = 10, l(o) = 15, l(c) = 9$. The least among these is $l(c) = 9$ and we consider

Step 10. $P = \{a, b, c, e, f, g, i, j, k, n\}, T = \{d, h, l, m, o, z\}$. Then the indices relative to the present P are $l(d) = 19, l(h) = 14, l(l) = 14, l(m) = 10, l(o) = 15$. The least among these indices is $l(m) = 10$. Therefore, we take

Step 11. $P = \{a, b, c, e, f, g, i, j, k, m, n\}, T = \{d, h, l, o, z\}$. Then $l(d) = 16, l(h) = 14, l(l) = 14, l(o) = 15$. The least among these indices is $l(l) = 14$ and we take

Step 12. $P = \{a, b, c, e, f, g, i, j, k, l, m, n\}, T = \{d, h, o, z\}$. Then $l(d) = 16, l(h) = 14, l(o) = 15, l(z) = 19$.

Step 13. $P = \{a, b, c, e, f, g, h, i, j, k, l, m, n\}, T = \{d, o, z\}$. Then $l(d) = 16, l(o) = 15, l(z) = 19$.

Step 14. $P = \{a, b, c, e, f, g, h, i, j, k, l, m, n, o\}, T = \{d, z\}$. Then $l(d) = 16, l(z) = 19$.

Step 15. $P = \{a, b, c, d, e, f, g, h, i, j, k, l, m, n, o\}, T = \{z\}$ and $l(z) = 19$.

A shortest path from a to z is $a \xrightarrow{3} e \xrightarrow{2} f \xrightarrow{2} g \xrightarrow{1} k \xrightarrow{6} l \xrightarrow{5} z$ and the total length of this path is 19.

7.4. Eulerian Paths and Circuits

The River Pregel flows through the city of Königsberg in Germany and land area of Königsberg is divided into four pieces of land, two of these being islands and the land area on the two sides of the river. The four pieces of land are connected to each other by seven bridges. One island is connected to the two sides of the city by two bridges each while the other island is connected to the sides of the city by one bridge each. Also the two islands are connected to each other by a single bridge. The four pieces of land and the bridges over the river are depicted by Fig. 7.54.

In the years 1730–1736, people conducted tours starting from a point on a piece of land in the hope to go over every bridge exactly once and come back to the starting point. None of their efforts was successful. In the year 1736, Leonard Euler, a Swiss mathematician, discovered a general criterion for determining whether there is path in a graph that traverses each of the edges in the graph once and only once. This criterion led him to prove that it was not possible to arrange for a walking tour starting from a point and coming back to it after walking over the seven bridges once and exactly once. Representing each piece of land as a vertex and each bridge between two pieces of land as an edge between the corresponding vertices, we get a model of the land and bridges. The Königsberg bridge problem then translates into the following.

Is it possible to find a path in the graph model starting from any vertex and coming back to it by traversing each and every edge exactly once?

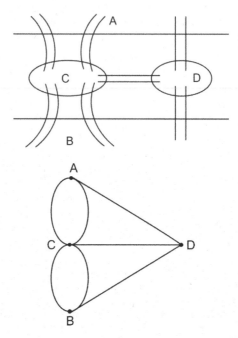

Fig. 7.54.

A path in a graph that traverses each edge in the graph once and only once is called a Eulerian path. In a similar fashion, a circuit in a graph is called a Eulerian circuit if it traverses each edge in the graph once and only once. Observe that a Eulerian path is always a simple path and a Eulerian circuit is a simple circuit but a simple circuit or a simple path need not be Eulerian. For example, the graph of Fig. 7.55(a) does not have a Eulerian path nor a Eulerian circuit but the path (a, e, b, c, e, d) is a simple path and (a, e, b, c, e, d, a) is a simple circuit. The graph (b) has a Eulerian path (a, b, c, e, d, c, a, d) but does not have a Eulerian circuit but (c, d, e, c, b, a, c) is a simple circuit. There do exist some simple paths in this graph. The graph (c) has (b, c, a, d, c, e, d) as a Eulerian path but it does not have a Eulerian circuit (prove it). However, the graph (d) has a Eulerian circuit (c, b, a, c, e, d, c). Observe that a disconnected graph can neither have a Eulerian path nor a Eulerian circuit.

A Eulerian path/circuit is also termed **Euler path/circuit**.

In the case of directed graphs of Fig. 7.56, observe that H_1 has a Euler circuit whereas H_2 does not have a Euler circuit. Also H_3 has a Euler circuit

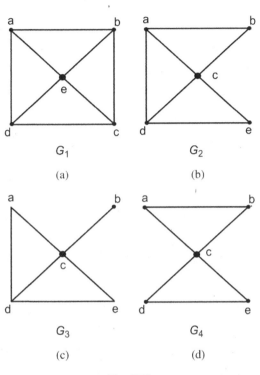

G_1 G_2

(a)   (b)

G_3 G_4

(c)   (d)

Fig. 7.55.

but H_4 does not have a Euler circuit. Similarly H_5 has a Euler circuit but H_6 does not have a Euler circuit (prove it).

In the case of graphs H_1, H_3, H_5 the Euler circuits are (a, b, c, a), (a, b, c, d, a) and $(a, c, e, d, c, f, g, h, f, b, a)$ respectively.

As the number of vertices and edges in a graph (directed or undirected) goes on increasing, the number of possible paths and circuits increases very rapidly and it becomes more and more difficult to decide whether the graph has a Euler path or Euler circuit. Fortunately for us, Euler, while solving the famous Königsberg bridge problem, discovered simple criteria for determining whether a multigraph/graph has a Euler path or a Euler circuit. We now prove these criteria discovered by Euler.

Theorem (Euler) 7.9. An undirected graph possess a Eulerian circuit if and only if it is connected and each of its vertices has even degree.

Proof. There being no edges involving isolated vertices, we may assume that the graph has no isolated vertices. Let G be the given graph with no

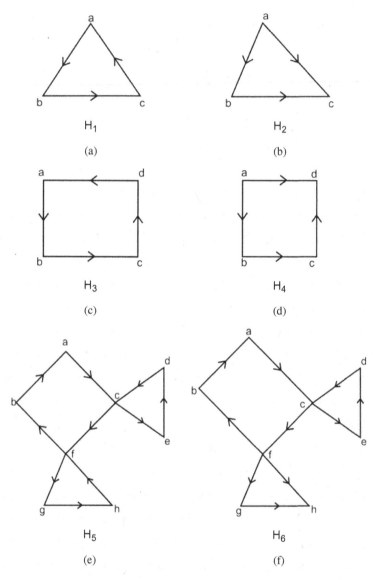

Fig. 7.56.

isolated vertices. Suppose that the graph G has a Eulerian circuit. Since on starting at some vertex, we can have a path which traverses each and every edge in the graph and come back to the starting vertex, G must be a connected graph. Suppose that a Euler circuit begins with a vertex a (say) and that it continues along the edge $\{a, b\}$. The edge $\{a, b\}$ contributes one to the degree of a. When the circuit is traced, every time it crosses a vertex, it adds two to the degree of the vertex since it enters the vertex along one edge and leaves the vertex along another edge that has not been traced earlier. When finally the simple circuit being constructed comes back to the starting vertex a, the edge along which it enters adds one to the degree of a. Thus the degree of a becomes $2 + 2 \times$ (the number of times the circuit has crossed a), which is even. As seen above, the degree of every intermediary vertex is twice the number of times the circuit crosses that vertex. Thus the degree of every vertex in G is even.

Conversely, suppose that the graph G is connected and every vertex of G has even degree. Consider an arbitrary vertex a of G. Let $x_o = a$. We construct a simple path that begins at a. Choose an edge $\{x_o, x_1\}$ incident with a arbitrarily. Then choose an edge $\{x_1, x_2\}$ incident with x_1 and continue building a simple path $\{x_0, x_1\}, \{x_1, x_2\}, \ldots, \{x_{n-1}, x_n\}$ as long as possible. Since the graph G has only a finite number of edges, the simple path being constructed terminates. Each time the path goes through a vertex with even degree, it uses only one edge to enter this vertex, so that at least one edge remains for the path to leave the vertex. Therefore, the path that begins at a along the edge $\{a, x_1\}$, terminates at a with an edge of the form $\{y, a\}$, i.e. we must have $x_n = a$. The path traced may use all the edges or it may not.

If all the edges have been used in the path constructed, we have obtained a Eulerian circuit. Otherwise, consider the subgraph H of G obtained by deleting the edges already used and vertices that are not incident with any remaining edges. Since the graph G is connected, H has at least one vertex in common with the circuit that has been deleted. Let w be one such vertex.

Since all vertices in G have even degree and for each vertex, pairs of edges incident with it have been deleted to form H, every vertex in H has even degree. The subgraph H obtained may or may not be connected (we shall soon see it through an example). Beginning at the vertex w, we can, as for G, obtain a simple circuit in H and this circuit terminates at w. If all the edges of H have been used, we get a Eulerian circuit. Otherwise, we continue the above process of obtaining a subgraph of H, and then construct a simple circuit. Continue this process until all the edges of G have been used. Finally, all the circuits so obtained can be spliced together and we obtain a Eulerian circuit in G.

Example 7.15. Consider the graph of Fig. 7.57(a) which is connected. We begin at a and choose in succession the edges $\{a, b\}, \{b, e\}, \{e, h\}$ and $\{h, a\}$.

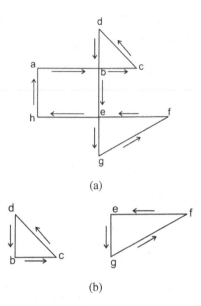

(a)

(b)

Fig. 7.57.

We thus obtain a simple circuit which does not use all the edges of G. By deleting the above chosen edges of G, we obtain a subgraph H (Fig. 7.57(b)) which is not connected but has two connected components. These two connected components of H have vertices b and e in common with the simple circuit traced. Begin at the vertex b for one of the two components and by taking in succession the edges $\{b, c\}, \{c, d\}, \{d, b\}$ we obtained a simple circuit which uses all the edges of one of the connected components of H. Then we begin at the vertex e and take in succession the edges $\{e, g\}, \{g, f\}, \{f, e\}$ to obtain a simple circuit which uses all the edges of the second component of H. We can then splice the three simple circuits as $\{a, b\}, \{b, c\}, \{c, d\}, \{d, b\}, \{b, e\}, \{e, g\}, \{g, f\}, \{f, e\}, \{e, h\}, \{h, a\}$ to obtain a Eulerian circuit which in short be written as $(a, b, c, d, b, e, g, f, e, h, a)$.

Theorem 7.10. A graph has a Eulerian path but not a Euler circuit if and only if it is connected having exactly two vertices of odd degree.

Proof. Let G be the given graph. As in the previous theorem, we can assume that G has no isolated vertices. Suppose that G has a Euler path but not a Euler circuit from a vertex a to a vertex b. Then it is clear that G is connected. The first edge of the Euler path from a to b contributes one to the degree of a. Every other time the path passes through a, it contributes two to the degree of a. The last edge of the Euler path from a to b contributes one to the degree of b. Every other time the path passes through b, it contributes two to the degree of b. Hence both a and b are of odd degree. Every time the Euler path passes through a vertex x other than a and b, it contributes two to the degree of x. Hence every vertex different from a and b has even degree.

Conversely, suppose that the graph G is connected and has exactly two vertices a and b (say) of odd degree. Enlarge the graph G to a graph G' by introducing an edge $\{a, b\}$ to the graph G. Then G' is a connected graph and every vertex of G' is of even degree. By Theorem 7.9, the graph G' has a Euler circuit. Now remove the edge $\{a, b\}$ from the Euler circuit and we obtain a Euler path in G.

Example 7.16. Consider the graphs as given in Fig. 7.58

In the graph G_1, the vertices b and d have degree 3 each while a, c have degree 2 each. Therefore, G_1 has a Euler path. Observe that (b, c, d, b, a, d) is one such path. There are other such paths possible. For example, $(b, a, d, c, b, d), (b, d, a, b, c, d,)$ etc. Every such path begins either at b or d and ends at d or b respectively. There is no such path that begins at a or c and ends at c or a.

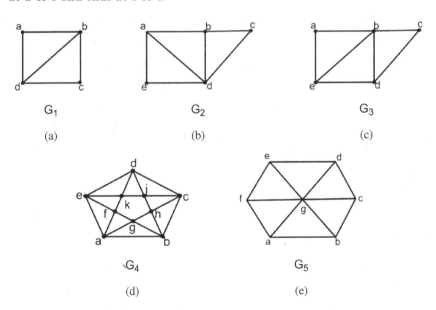

Fig. 7.58.

A Course in Discrete Mathematical Structures

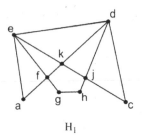

H_1

Fig. 7.59.

In graph G_2, the vertices a and b have degree 3 each while the vertices c, e have degree 2 each and d has degree 4. Therefore, the graph G_2 has a Euler path. One such path is (b, c, d, b, a, e, d, a).

The graph G_3 also has two vertices of degree 3 each, two vertices of degree 2 each and one vertex of degree 4. Therefore, G_3 has a Euler path.

Every vertex in G_4 has degree 4 and, as such, G_4 has a Euler circuit. The graph G_4 looks a little complicated and there are chances that while tracing a Euler circuit some edge may be omitted. To avoid this possibility, we can adopt the following systematic procedure. We first construct a simple circuit (a, b, h, c, b, g, a). By deleting these edges, we get a subgraph as in Fig. 7.59.

The circuit traced has four vertices a, g, h, c in common with the subgraph H_1 obtained. Therefore, we start tracing another simple circuit starting at one of these four vertices. Starting at g we may trace a simple circuit $(g, h, j, c, d, j, k, f, g)$. Now, deleting the edges in this simple circuit, we obtain a subgraph H_2.

The circuit traced has the vertices f, d, k in common with the subgraph H_2.

Then we trace in H_2 a simple circuit beginning at one of these three vertices, say at f. We get a simple circuit (f, e, k, d, e, a, f). We have thus obtained three simple circuits (a, b, h, c, b, g, a), $(g, h, j, c, d, j, k, f, g)$ and (f, e, k, d, e, a, f). Splicing the circuits together, we get $(a, b, h, c, b, g, h, j, c, d, j, k, f, e, k, d, e, a, f, g, a)$ as a Eulerian circuit in G.

The graph G_5 has six vertices each of degree 3 and one vertex of degree 6. Therefore, the graph G_5 does not have a Euler path.

As already seen, the Königsberg bridge problem asks for a Euler path or circuit for the graph as in Fig. 7.54. Since this graph has three vertices A, B, D each of degree 3 and the remaining vertex C of degree 5, i.e. all vertices of odd degree, it follows from Theorem 7.10 that the graph has no Euler path. Hence it is not possible to arrange for a tour starting from one of the land pieces and passing over each and every bridge exactly once.

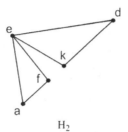

H_2

Fig. 7.60.

There is a fairly simple algorithm attributed to Fleury for finding a Euler circuit in a connected linear graph with no vertices of odd degree. Recall that an edge in a connected graph G is called a **bridge** if deleting this edge results in a disconnected subgraph of G.

Algorithm (Fleury's Algorithm) 7.2.

Let $G = (V, E)$ be a connected graph with each vertex of even degree.

Step 1. Select an edge e_1 that is not a bridge in G. Let the vertices incident with e_1 be v_1, v_2. Let a path π be specified by the subsets $V_\pi = \{v_1, v_2\}$ and $E_\pi = \{e_1\}$. Remove the edge e_1 from E and let G_1 be the resulting connected subgraph of G.

Step 2. Suppose that a simple path π given by $V_\pi = \{v_1, \ldots, v_k\}$ and $E_\pi = \{e_1, e_2, \ldots, e_{k-1}\}$ has been constructed and that all of the edges in E_π and resulting isolated vertices, if any, have been removed from V and E to form a subgraph G_{k-1}. Since v_k has even degree and e_{k-1} ends there, there must be an edge e_k in G_{k-1} that is incident with v_k. If there is more than one such edge, choose one that is not a bridge in G_{k-1}. Let the other vertex incident with e_k be denoted by v_{k+1}. Enlarge V_π and E_π by adding v_{k+1} and e_k to these subsets to obtain an extended path π given by $V_\pi = \{v_1, v_2, \ldots, v_k, v_{k+1}\}$ and $E_\pi = \{e_1, e_2, \ldots, e_k\}$. Then delete e_k and isolated vertices, if any, from G_{k-1} to form a subgraph G_k.

Step 3. Repeat step 2 until no edges remain in E.

We can modify the above algorithm slightly and get an algorithm for finding a Euler path in a connected graph G that has exactly two vertices of odd degree. Instead of starting with any edge e_1 as in step 1, we start with e_1 that is incident with one of the two vertices of odd degree.

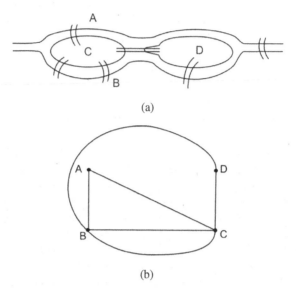

(a)

(b)

Fig. 7.61.

Exercise 7.4. Prove Theorem 7.10 without using Theorem 7.9.

Example 7.17. Decide if it is possible for someone to cross all the bridges shown in Fig. 7.61(a) exactly once and return to the starting point.

Solution. The bridges in the city may be represented in the graph which has four vertices and six edges as in Fig. 7.61(b): Vertices A, D have degree 2 each and both the vertices B, C have degree 4. Therefore, by Theorem 7.9 the graph has a Euler circuit. Therefore, in the present case, it is possible to arrange a walking tour starting at any point and come back to it after crossing over every bridge exactly once. For example, if we start from land A, we go to C, then to B; then to C, then to D, then to B and finally to A, i.e. the Euler circuit is (A, C, B, C, D, B, A). Starting at B, we can construct a circuit (B, D, C, A, B, C, B) or (B, C, A, B, D, C, B).

Example 7.18. In Königsberg, there are two additional bridges, besides the seven that were present in the eighteenth century. These new bridges connect regions A and B and regions A and D. Decide if it is possible to cross all the nine bridges in Königsberg exactly once and return to the starting point.

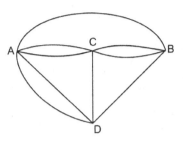

Fig. 7.62.

Solution. Adding the two additional edges in the graph representing the Königsberg bridges, we have the new graph as in Fig. 7.62.

In this graph the vertices A and C have degree 5 each and the vertices B, D have degree 4 each. Therefore, by Theorem 7.10, the graph has a Euler path but not a Euler circuit. Thus, it is not possible to cross all the nine bridges exactly once and return to the starting point. However, it is possible to cross all the nine bridges exactly once by starting at A and ending at C or start at C and end at A.

Example 7.19. In magazines there are sometimes pictures and the reader is asked whether it is possible to draw the picture with a pencil starting from one point, without lifting it and not going over any arm or portion of the picture more than once. These puzzles are solved using Euler's theorem for the construction of a Euler path or Euler circuit. We consider three examples and consider the following graphs.

We will decide if tracing these pictures with pencil starting at a point as described above is possible. Let us name the vertices in the graphs.

(a) In the graph G_2, there are three vertices a, b, f which are of degree 3 each and g is of degree 1. The other three vertices c, d, e are of degree 4 each. In view of Euler's Theorems 7.9 and 7.10 the graph does not have a Euler circuit nor a Euler path.

(b) In the graph G_1, six vertices a, b, c, d, e, f are of degree 2 each, and the other six vertices, namely g, h, i, j, k, l are of degree 4 each. Therefore, by Theorem 7.9, the graph G_1 has a Euler circuit. To trace this figure without lifting the pencil, we may move from

$$a \to g \to l \to k \to j \to i \to h \to g \to e \to l \to b \to k \to d$$
$$\to j \to c \to i \to f \to h \to a$$

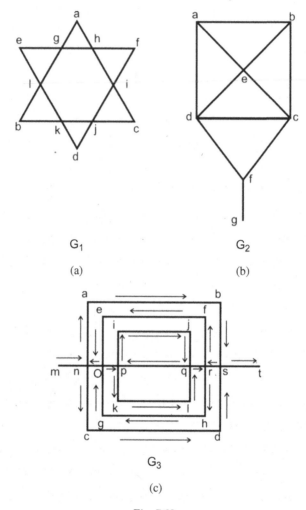

G_1

(a)

G_2

(b)

G_3

(c)

Fig. 7.63.

By giving a cyclic shift to this tracing and then to the resultant circuit and repeating the process, we find that we could start tracing the figure at any vertex. Care needs to be applied for giving a cyclic or an anticyclic shift to the circuit the shift is to be given to the edges rather than to just the vertices. For example, a cyclic shift to the above circuit shall result in the circuit

$$g \to l \to k \to j \to i \to h \to g \to e \to l \to b \to k \to d$$

$$\to j \to c \to i \to f \to h \to a \to g$$

(c) The graph G_3 has two vertices m and t of degree 1 each while every other vertex is of even degree. Therefore, by Theorem 7.10, the graph G_3 has a Euler path. Any such path has to begin at either m or t and terminate at t or m respectively. One such tracing of the figure without lifting the pencil may be done as indicated in the graph G_3 itself. There do exist other ways of tracing the graph. This Euler path is

$$m \to n \to c \to d \to s \to r \to h \to g \to o \to p \to k \to l \to q \to p$$

$$\to i \to j \to q \to r \to f \to e \to o \to n \to a \to b \to s \to t$$

We are moving as in Fig. 7.64(a).

Example 7.20. Consider the graphs G_1, G_2 as in Fig. 7.64(b)–(c).

Decide if these graphs have Eulerian circuits. If not, do they have Euler paths? If such paths/circuits exist, determine one for each graph.

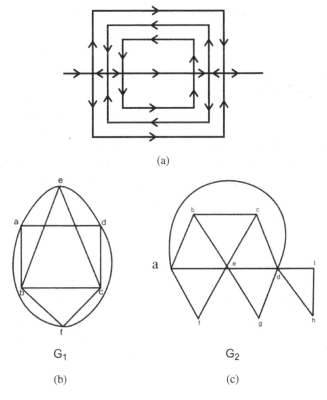

(a)

G_1

(b)

G_2

(c)

Fig. 7.64.

Solution.

(a) Every vertex in the graph G_1 has degree 4. Therefore, the graph G_1 has a Euler circuit. One such circuit is

$$a \to b \to c \to f \to b \to e \to a \to f \to d \to c \to e \to d \to a$$

i.e. $(a, b, c, f, b, e, a, f, d, c, e, d, a)$.

(b) In the graph G_2, vertices b and c are of degree 3 each, a is of degree $4, d, e$, are of degree 6 each while f, g, h, i are of degree 2 each. All the vertices not being of even degree, the graph does not have a Euler circuit. There being only two vertices of odd degree, the graph G_2 has a Euler path. Any such path has to begin at one of the two vertices of odd degree and has to terminate at the other vertex of odd degree. We will consider a Euler path beginning at b. We find that

$$b \to a \to d \to i \to h \to d \to g \to e \to d \to c$$

$$\to e \to f \to a \to e \to b \to c$$

is a Euler path beginning at b and terminating at c.

Example 7.21. n cities are connected by a network of k highways, where a highway is defined to be a road between two cities that does not pass through any other city. Prove that if $k > \frac{1}{2}(n-1)(n-2)$, then one can always travel between any two cities through connecting highways.

Solution. We prove this result by induction on n, the number of cities. If $n = 1$ or 2, then $k > 0$ so that there is at least one road from the city to itself or from one city to the other, as the case may be. Thus one can always travel from one city to the other.

Suppose that $n \geq 2$, there are $n + 1$ cities and k highways with $k > \frac{1}{2}n(n-1)$. Let the $n+1$ cities be denoted by $A_1, A_2, \ldots, A_{n+1}$.

If there is a highway between every two cities we have nothing to prove. Therefore, there are at least two cities between which there is no highway. Let one of these two cities be A_{n+1}. But there being no isolated cities, A_{n+1} is connected to at least one of the cities $A_1, \ldots, A_n$. There can be a maximum of $n - 1$ highways connecting A_{n+1} with the rest of the cities. From the graph of the $n + 1$ cities, ignore A_{n+1} and all the highways connecting A_{n+1} with the rest of the cities. We then get a graph of n cities connected by k' highways where

$$k' \geq k - (n-1) > \frac{1}{2}(n-1)n - (n-1) = \frac{1}{2}(n-1)(n-2).$$

Thus, by induction hypothesis, one can travel between any two cities from $A_1, A_2, \ldots, A_n$ through connecting highways. Now, suppose that we want to travel from A_{n+1} to any of the cities $A_1, A_2, \ldots, A_n$. There is a highway from A_{n+1} to some $A_j, 1 \leq j \leq n$. Since one can travel from A_j to A_i through connecting highways, combining this path from A_j to A_i with the highway from A_{n+1} to A_j, we get a path from A_{n+1} to A_i. This proves that one can travel from A_{n+1} to A_i through connecting highways. Thus one can travel between any two cities from $\{A_1, \ldots, A_{n+1}\}$ through connecting highways. This completes induction. Hence, if $n \geq 2$ cities are connected by k highways with

$$k > \frac{1}{2}(n-1)(n-2),$$

then one can travel between any two cities through connecting highways.

Example 7.22. Let G be a connected graph with no self loops where the edges represent the streets in a city. A police officer wants to make a round-trip to patrol each side of each street exactly once. Furthermore, the officer wants to patrol the two sides of a street in opposite directions. Show that such a trip can always be designed.

Solution. Since both sides of a street are to be patrolled, each street gives rise to two edges between the end points of the street. Also, as one side of the street is to be patrolled in one direction and the other side in the opposite direction, these two edges contribute one to the indegree and one to the outdegree of the vertex (i.e. the endpoint of the street) under consideration. In other words, each street contributes two to the degree of each vertex with which it is incident. This proves that each vertex in the graph is of even degree (or the sum of the indegrees of a vertex is equal to the sum of the outdegree of the vertex). We know that a Euler path (or circuit) is possible in a graph if and only if there is either none or two vertices of odd degree and it is a Euler circuit in the first case. Hence a required trip can always be designed.

We end this section by stating two theorems corresponding to Theorems 7.9 and 7.10 for the existence of Euler circuits and Euler paths in directed graphs. The proofs of these results are on the lines of the corresponding results for undirected graphs. The statement that a vertex is of even degree in the proof of Theorem 7.10 or Theorem 7.9 is replaced by the statement that outdegree of a vertex being equal to indegree of the vertex.

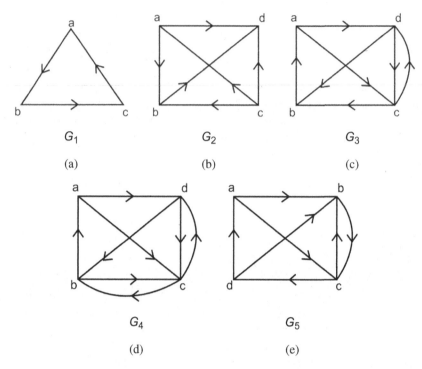

Fig. 7.65.

Theorem 7.11. A directed graph having no isolated vertices has a Euler circuit if and only if it is connected and the indegree and the outdegree of every vertex are equal.

Theorem 7.12. A directed graph having no isolated vertices has a Euler path but not a Euler circuit if and only if it is connected and the indegree and outdegree of each vertex are equal except for two vertices for one of which the indegree is one more than its outdegree while for the other the outdegree is one more than its indegree.

Example 7.23. Consider the graphs as in Fig. 7.65.

(a) In the graph G_1, for each of the vertices a, b, c the indegree = outdegree = 1. Hence the graph has a Eulerian circuit, an obvious one namely $a \to b \to c \to a$ or (a, b, c, a).

(b) In the graph G_2,

$$\text{indegree of } a = 1, \quad \text{outdegree} = 2,$$

for b,   indegree $= 2$,   outdegree $= 1$,

for c,   indegree $= 0$,   outdegree $= 3$,

for d,   indegree $= 3$,   outdegree $= 0$.

The graph does not have a Euler circuit nor does it have a Euler path.

In the case of G_2 we may change the directions of the arrows in whatever manner we like, the graph will never have a Euler path nor a Euler circuit. This is so because for each of the four vertices, we have indegree $+$ outdegree $= 3$.

(c) In the graph G_3

for the vertex a,   indegree $= 1$,   outdegree $= 2$,

for the vertex b,   indegree $= 2$,   outdegree $= 1$,

for the vertex c,   indegree $= 2$,   outdegree $= 2$,

for the vertex d,   indegree $= 2$,   outdegree $= 2$.

The indegree not being equal to the outdegree for all the four vertices, the graph does not have a Euler circuit. However, indegree $=$ outdegree for each of the vertices c, d and

$$\text{indegree} - \text{outdegree} = 1 \quad \text{for } b \quad \text{and}$$

$$\text{indegree} - \text{outdegree} = -1 \quad \text{for } a.$$

Therefore, the graph has a Euler path (a, c, d, c, b, a, d, b).

(d) There are three edges incident with a and five edges incident with c. Therefore, the indegree and outdegree for these two vertices cannot be equal in the graph G_4. Therefore the graph G_4 cannot have a Euler circuit. Also, for the vertex

$$a, \text{indegree} = 1, \quad \text{outdegree} = 2,$$

$$c, \text{indegree} = 3, \quad \text{outdegree} = 2,$$

$$b, \text{indegree} = 2, \quad \text{outdegree} = 2,$$

$$d, \text{indegree} = 2, \quad \text{outdegree} = 2.$$

Hence the graph G_4 has a Euler path, namely $(a, d, c, d, b, c, b, a, c)$

(e) In graph G_5, for vertex

$$a, \text{indegree} = 1, \quad \text{outdegree} = 2,$$

$$c, \text{indegree} = 2, \quad \text{outdegree} = 2,$$

$$b, \text{indegree} = 3, \quad \text{outdegree} = 1,$$

$$d, \text{indegree} = 1, \quad \text{outdegree} = 2.$$

Therefore, this graph neither has a Euler circuit nor a Euler path.

7.5. Hamiltonian Paths and Circuits

We obtained the necessary conditions for the existence of a path/circuit that traverses each and every edge in the graph exactly once. We will now consider the question of the existence of paths/circuits in a graph that pass through each and every vertex of the graph exactly once. A path/circuit that traverses every edge of the graph exactly once will surely pass through every vertex (provided there are no isolated vertices) but it may pass though certain vertices more than once. On the other hand, a path/circuit in a graph that passes through every vertex exactly once may not traverse every edge of the graph. Unlike Euler paths/circuits, there are no simple known necessary and sufficient conditions for the existence of a path/circuit in a graph that passes through every vertex exactly once. However, there do exist some sufficient conditions for the existence of such a path/circuit. In this section we will study this problem.

Definition. A path in a graph that passes through each of the vertices exactly once is called a **Hamiltonian** (or **Hamilton**) **path**. A circuit in a graph that passes through every vertex in the graph exactly once is called a **Hamiltonian (or Hamilton) circuit**. In symbols a path $(x_0, x_1, \ldots, x_n)$ in a graph $G = (V, E)$ is a Hamiltonian path if $x_i \neq x_{i+1}, 0 \leq i \leq n - 1$ and $V = \{x_0, \ldots, x_n\}$ and a Hamiltonian path $(x_0, x_1, \ldots, x_n)$ in G is a Hamiltonian circuit if $x_n = x_0$. Observe that in the graph of Fig. 7.66 the path (a, b, c, a, b) is a Euler path and as this path passes through vertices a and b twice, it is not a Hamiltonian path. On the other hand, the circuit (a, b, c, a) is a Hamiltonian circuit but is not a Eulerian circuit as one of the two edges incident with a and b is not traversed. Observe that in a graph with multiple edges, no Eulerian path or circuit can be Hamiltonian. The above example also shows that although there exists a Hamiltonian circuit

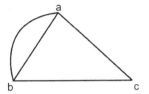

Fig. 7.66.

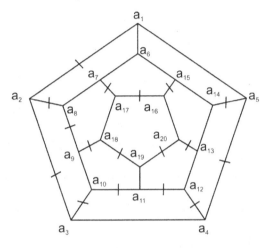

Fig. 7.67.

in a graph, there may not exist a Euler circuit. As another example of this nature, we have the following.

Example 7.24. Consider the graph of Fig. 7.67 which has all the vertices of odd degree. Therefore, this graph does not possess a Euler path or a Euler circuit. However,

$$a_1 \to a_2 \to a_3 \to a_{10} \to a_{11} \to a_{12} \to a_4 \to a_5 \to a_{14} \to a_{13} \to a_{20}$$

$$\to a_{19} \to a_{18} \to a_9 \to a_8 \to a_7 \to a_{17} \to a_{16} \to a_{15} \to a_6 \to a_1$$

is a Hamiltonian circuit. Also

$$a_1 \to a_5 \to a_4 \to a_{12} \to a_{11} \to a_{10} \to a_3 \to a_2 \to a_8 \to a_9 \to a_{18}$$

$$\to a_{19} \to a_{20} \to a_{13} \to a_{14} \to a_{15} \to a_{16} \to a_{17} \to a_7 \to a_6 \to a_1$$

is another Hamiltonian path.

The following gives a sufficient condition for the existence of a Hamiltonian path.

Theorem 7.13. Let G be a linear graph of n vertices. If the sum of the degrees for each pair of vertices in G is at least $n - 1$, then there exists a Hamiltonian path in G.

Proof. We first prove that G is connected. If G is not connected then G has at least two disconnected components. Choose a vertex v_1 from one component having n_1 vertices and choose a vertex v_2 from another disconnected component of G. Suppose that the component of G to which v_2 belongs has n_2 vertices. Since G is a linear graph, multiple edges between any pair of vertices are not allowed. Therefore, $\deg v_1 \leq n_1 - 1, \deg v_2 \leq n_2 - 1$ and $\deg v_1 + \deg v_2 \leq n_1 + n_2 - 2 \leq n - 2 < n - 1$ which is a contradiction. Hence the graph G is connected. Since the case $n = 1$ gives a trivial graph, i.e. it consists of just one vertex and no edges or at best only one edge which is a loop, we assume that $n \geq 2$. The graph G is connected. Therefore, for any two vertices there is a path from one to the other. We pick up two vertices which are adjacent. Thus, we can find a path with just one edge. Let $n > p \geq 2$ and suppose that there exists a path $(v_1, v_2, \ldots, v_p)$ with $p - 1$ edges $(v_1, v_2), (v_2, v_3), \ldots, (v_{p-1}, v_p)$. We want to prove that there exists a path with p edges.

If either v_1 or v_p is adjacent to a vertex v_x which is not in the above path, then by adjoining an edge (v_x, v_1) or (v_p, v_x) as the case may be, we obtain a path $(v_x, v_1, v_2, \ldots, v_p)$ or $(v_1, v_2, \ldots, v_p, v_x)$ with p edges. So, suppose that neither v_1 nor v_p is adjacent to any vertex which is not in the path $(v_1, v_2, \ldots, v_p)$. Thus v_1, v_p are adjacent only to vertices in the path $(v_1, v_2, \ldots, v_p)$. We claim that then there is a circuit which contains exactly the vertices $v_1, v_2, \ldots, v_p$.

If v_1 is adjacent to v_p, then $(v_1, v_2, \ldots, v_p, v_1)$ is such a circuit. Suppose that v_1 is not adjacent to v_p. Let v_1 be adjacent to the vertices $v_{i_1}, v_{i_2}, \ldots, v_{i_k}$ where $2 \leq i_j \leq p - 1, 1 \leq j \leq k$. Then $\deg v_1 = k$. If v_p is not adjacent to any of $v_{i_1-1}, v_{i_2-1}, \ldots, v_{i_k-1}$, then v_p is adjacent to at most $p - k - 1$ vertices and $\deg v_p \leq p - k - 1$. Then $\deg v_1 + \deg v_p \leq k + p - k - 1 = p - 1 < n - 1$, which is a contradiction. Therefore, v_p is adjacent to at least one of the vertices $v_{i_1-1}, v_{i_2-1}, \ldots, v_{i_k-1}$. For simplicity of notation, let this vertex be v_{j-1} i.e. v_p is adjacent to v_{j-1} with v_1 adjacent to v_j and $j \leq p - 1$. Then we have a circuit.

$$(v_1, \ldots, v_{j-1}, v_p, v_{p-1}, \ldots, v_j, v_1). \tag{7.1}$$

This completes the proof of our claim. To include the case of the circuit $(v_1, \ldots, v_p, v_1)$, we may now allow j to take the value p.

Since $p < n$, there exists a vertex not belonging to the set $\{v_1, v_2, \ldots, v_p\}$. Again, since G is connected, this vertex is connected to every vertex in the set $\{v_1, v_2, \ldots, v_p\}$. We can then pick up a vertex $v_x, v_x \notin \{v_1, \ldots, v_p\}$, which is adjacent to some $v_k, 1 \leq k \leq p, k \neq j$ or $j - 1$. Then consider the path

$$(v_x, v_k, v_{k+1}, \ldots, v_{j-1}, v_p, v_{p-1}, \ldots, v_j, v_1)$$

or

$$(v_x, v_k, v_{k+1}, \ldots, v_p, v_{j-1}, \ldots, v_1, v_j, v_{j+1}, \ldots, v_{k-1})$$

which is a path containing p edges. (The two possible paths are according as $k < j - 1$ or $j < k$.)

We can repeat the above construction to obtain a path which contains $n - 1$ edges and, so, all the vertices of G. By the very construction, the path we finally arrive at is a Hamiltonian path.

An argument very similar to the one used in the proof of the above theorem gives the following theorem due to Norwegian mathematician Øystein Ore.

Theorem 7.14. If G is a linear graph with n vertices, $n \geq 3$, such that $\deg(u) + \deg(v) \geq n$ for every pair of non-adjacent vertices u and v in G, then G has a Hamiltonian circuit.

Proof. We prove the result by contradiction. Suppose that G does not have a Hamiltonian circuit. By adding as many edges as possible at each vertex of G without producing a Hamiltonian circuit, obtain a graph H on the same vertices such that addition of one more edge produces a circuit in H. Then surely H possesses a Hamiltonian path, say $(v_1, v_2, \ldots, v_n)$. If v_1, v_n were adjacent, there would result a Hamiltonian circuit in H which is a contradiction. Thus v_1 and v_n are not adjacent. By the given condition $\deg v_1 + \deg v_n \geq n$. Now using an argument similar to the one applied to the path $(v_1, v_2, \ldots, v_p)$ as in the proof of Theorem 7.13, we can get a Hamiltonian circuit $(v_1, \ldots, v_{j-1}, v_n, v_{n-1}, \ldots, v_j, v_1)$ for some $j, 1 < j < n$. This contradicts the construction of H. Hence our assumption that G does not have a Hamiltonian circuit could not be true.

As an immediate corollary of Ore's theorem, we have the following.

Theorem 7.15 (Dirac's theorem). If G is a linear graph with n vertices, $n \geq 3$, such that the degree of every vertex in G is at least $\frac{1}{2}n$, than G has a Hamiltonian circuit.

As another application of Ore's theorem, we have the following.

Theorem 7.16. If G is a linear graph with n vertices, $n \geq 3$, and m edges with $m \geq \frac{1}{2}(n^2 - 3n + 6)$ then the graph G has a Hamiltonian circuit.

Proof. Consider two vertices, u and v of G, that are not adjacent. Let H be the subgraph of G obtained from G by deleting the vertices u, v and all edges which are incident with u and v. Then H is a graph with $n-2$ vertices and the number of edges is $m - \deg u - \deg v$. Since the maximum number of edges that a graph with k vertices can have is $\leq \frac{1}{2}k(k-1)$, we have

$$m - \deg u - \deg v \leq \frac{1}{2}(n-2)(n-3) = \frac{1}{2}(n^2 - 5n + 6).$$

Therefore,

$$\deg u + \deg v \geq m - \frac{1}{2}(n^2 - 5n + 6)$$

$$\geq \frac{1}{2}(n^2 - 3n + 6) - \frac{1}{2}(n^2 - 5n + 6) = n.$$

It then follows from Ore's theorem that G has a Hamiltonian circuit.

The converses of the three Theorems 7.14–7.16 are not true. For example, the graph (Fig. 7.68) has six vertices, six edges and (i) degree of every vertex is $2 < 6/2$; (ii) sum of degrees for every pair non-adjacent vertices is $4 < 6$ and (iii) the number of edges is $6 < \frac{1}{2}(6 \times 6 - 3 \times 6 + 6) = 12$ so that the hypothesis of none of the three theorems is satisfied but the graph has a Hamiltonian circuit.

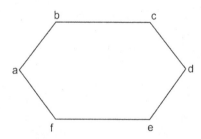

Fig. 7.68.

Fig. 7.69.

There do exist several simple graphs of n vertices having a Hamiltonian path with at least one pair of vertices v_i, v_j such that $\deg v_i + \deg v_j < n - 1$. For example, the graph (Fig. 7.69) is connected, has a Hamiltonian path (a, b, c, d) but $\deg a + \deg d = 1 + 1 = 2 < 4 - 1 = 3$.

We can, in fact, have infinitely many examples. As another example, any polygonal graph with at least six vertices is a connected graph having a Hamiltonian path and every vertex being of degree 2, the sum of degrees of any pair of vertices $= 4 < n - 1$.

Yet another counterexample is provided by the graph of Fig. 7.67 representing Hamilton's, 'A voyage round the world' puzzle which was invented by the Irish mathematician Sir William Rowan Hamilton in 1857. As has already been seen, this graph has a Hamiltonian path, the graph has 20 vertices and every vertex is of degree 3, so that the sum of degrees of any pair of vertices is $6 < 20 - 1 = 19$.

As an application of Theorem 7.13, we will consider the following.

Example 7.25. Consider the problem of scheduling nine examinations in nine days so that two examinations given by the same teacher are not scheduled on consecutive days. If no teacher gives more than five examinations, show that it is possible to schedule the examinations.

Solutions. Construct a graph with nine vertices where each vertex represents an examination (paper) given (set) by some teacher. Let vertices a, b be adjacent if a, b are examinations given by different teachers (i.e. if the examinations, a, b are not given by the same teacher). Let a be an examination given by a teacher A. Since the teacher A does not give more than five examinations, there are at least four examinations which are not given by A. Thus the vertex a is adjacent to at least four vertices. Hence the degree of a is at least 4. Thus, every vertex in the graph has degree ≥ 4. Therefore, for any pair of vertices a, b in the graph

$\deg a + \deg b \geq 4 + 4 = 8 = 9 - 1$. Hence the graph possesses a Hamiltonian path. Since edges in the graph are incident with examinations given by two different teachers, we go on allotting consecutive dates to the vertices as we move along the Hamiltonian path. Hence no two examinations given by the same teacher are scheduled on consecutive days.

Theorem 7.17. There is always a Hamiltonian path in a directed complete graph.

Proof. Let G be a directed complete graph with n vertices, $n \geq 2$. Recall that a directed graph is complete if for every pair of vertices there is exactly one arrow between the two.

Since the number of vertices in G is at least two, there is always a path with one edge. Suppose that $p \geq 2, p < n$ and that there is an elementary path $(v_1, v_2, \ldots, v_p)$ with $p - 1$ edges involving the vertices $v_1, v_2, \ldots, v_p$ all distinct. Let v_x be a vertex in G that does not appear in this path. If there is an edge (v_x, v_1) in the graph G, we can adjoin this edge to the above path and get an elementary path $(v_x, v_1, v_2, \ldots, v_p)$ with p edges. If, on the other hand, there is no edge (v_x, v_1) in G, then there must be an edge (v_1, v_x) in the graph G. If (v_x, v_2) is an edge in the graph, then we can get an elementary path $(v_1, v_x, v_2, v_3, \ldots, v_p)$ in which the vertex v_x has been added between v_1 and v_2. On the other hand, if there is no edge (v_x, v_2) in the graph, then there is an edge (v_2, v_x) in the graph. We can repeat the above argument with v_3 and v_x and, so on. If it is not possible to insert v_x anywhere between v_i and v_{i+1} for $1 \leq i \leq p - 1$, then there is no edge (v_x, v_p) in the graph. Hence there is an edge (v_p, v_x) in the graph and we can consider the path $(v_1, v_2, \ldots, v_p, v_x)$ which has p edges. Thus there is a path which is obtained from the path $(v_1, v_2, \ldots, v_p)$ by adding the vertex v_x to it. We repeat the above argument until all the vertices of the graph G are included in the path. The above construction shows that we get a Hamiltonian path.

Remark.

1. That every undirected complete graph on n vertices with $n \geq 3$, has a Hamiltonian circuit follows immediately from Dirac's theorem (Theorem 7.15).

2. The converse of Theorem 7.17 is not true. There exist infinitely many directed graphs which are not complete but do have Hamiltonian paths. For example, the graphs in Fig. 7.70(a)–(c) are not complete, are directed

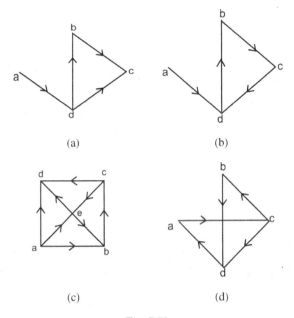

Fig. 7.70.

and have Hamiltonian paths. For (a) and (b), there is one Hamiltonian path, namely (a, d, b, c) whereas in the case of (c) more than one Hamiltonian paths exist. For example, $(a, e, b, c, d), (a, b, c, e, d)$ are two Hamiltonian paths. In case of (d), there is a Hamiltonian circuit, namely (a, c, b, d, a). Apart from this Hamiltonian circuit or cyclic shifts of it, there are no other Hamiltonian paths in this graph.

Hamiltonian paths and circuits can be used to solve several practical problems. Some among these are the following.

Example 7.26. Round-robin tournament. A tournament where each player plays every other player exactly once is called a **round-robin tournament**. To represent this graphically, we represent each player as a vertex and if a, b are two players, we say that there is an edge from a to b if b beats a in the only game played between the two (it is assumed that there is no drawn game between any two players). There then results a directed complete graph. We may consider the problem of ranking players taking part in the tournament. A player a is ranked higher than b if a beats b or if a beats another player c and c beats b or that there is a sequence of players $a, c_1, c_2, \ldots, c_k, b$ where a beats c_1, c_1 beats $c_2, \ldots, c_{k-1}$ beats c_k

and c_k beats b. Since the graph representing the tournament results in a directed complete graph, it follows from Theorem 7.17 that there exists a Hamiltonian path. Hence it follows that it is always possible to rank the players linearly. However, such a ranking may not be unique because there may exist more than one Hamiltonian paths in the graph.

Example 7.27 The travelling salesperson problem. A problem similar to the existence of the Hamiltonian path is the travelling salesperson problem. A salesperson is required to visit the n cities allotted to him by the company in such a way that he visits every city exactly once, subject to the condition that the total distance traveled, or the total travelling expenses involved for the round-trip are the least. Thus he has to plan a Hamiltonian circuit, keeping in mind the restriction imposed. In this problem it is assumed that every two cities are connected to each other and the distance between these or the cost involved in travelling between these is given. A graph model representing this problem is an undirected, weighted complete graph with the n cities to be visited as vertices and there being an edge between every two vertices. Moreover, each edge is assigned a weight, which may either be distance or the cost of travel involved between the two vertices. The problem is of great importance and much effort has been spent by mathematicians in solving this problem but with no result. To date there is no algorithm available to solve this problem in general. There are $(n - 1)!$ (here $m!$ stands for m factorial) Hamiltonian circuits possible to consider. However, a circuit could be traversed in one direction or its opposite. Therefore, there are $(n - 1)!/2$ possible choices available to the salesperson. He can then calculate the cost involved for each circuit. This is easy if n is small, say up to 4, 5 or 6. As n goes on increasing, the amount of work involved in finding the least cost circuit becomes huge and even computers may need years to come to the conclusion. For a near-optimal solution and that too for small values if n, an algorithm known as the **nearest neighbour algorithm**, is available.

Algorithm 7.3. Let $G = (V, E, w)$ be a complete weighted graph with n vertices $v_1, v_2, \ldots, v_n$. The nearest neighbour circuit for G starting at v_1 is obtained as follows:

1. Let $V_1 = V - \{v_1\}$.
2. For $i = 2$ to $n - 1$, do

    (i) choose $v_i =$ the closest vertex in V_{i-1} to v_{i-1} i.e. $w(v_{i-1}, v_i)$
        $= \text{Min} \{w(v_{i-1}, v) | v \in V_{i-1}\}$

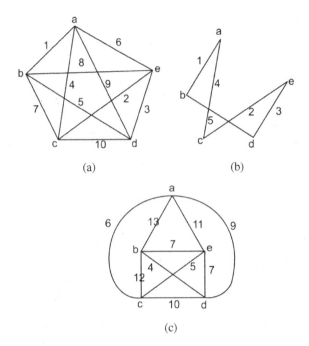

Fig. 7.71.

(ii) $V_i = V_{i-1} - \{v_i\}$.

3. Stop when $V_n = \{v_n\}$.

We consider two examples as applications of this algorithm.

Example 7.28. Solve the salesperson problem, using nearest neighbour algorithm, represented by the weighted graph (Fig. 7.71(a) and (c)).

Let us start with the vertex a in graph (a). The vertex b is closest to a, the distance being 1. Then d is closest to b, e is closest to d, then c is closest to e and finally a. Thus, the circuit obtained is $a \xrightarrow{1} b \xrightarrow{5} d \xrightarrow{3} e \xrightarrow{2} c \xrightarrow{4} a$ and the total distance travelled is $1 + 5 + 3 + 2 + 4 = 15$.

A closer look at the graph (b) suggests that we may start the trip either from c or d or b or e. Let us start with b (say). Then the circuit obtained by using the algorithm shall be $b \xrightarrow{4} d \xrightarrow{7} e \xrightarrow{5} c \xrightarrow{6} a \xrightarrow{13} b$ and the total distance travelled is $4 + 7 + 5 + 6 + 13 = 35$. If we start the trip from a, the circuit is $a \xrightarrow{6} c \xrightarrow{5} e \xrightarrow{7} b \xrightarrow{4} d \xrightarrow{9} a$ and the total distance travelled

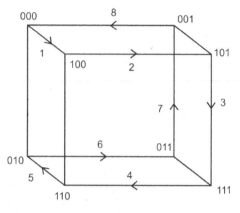

Fig. 7.72.

is 31. There being another choice for going from e, the circuit obtained is $e \xrightarrow{3} c \xrightarrow{6} a \xrightarrow{9} d \xrightarrow{4} b \xrightarrow{7} e$ giving a total distance of 31 again.

Example 7.29 Grey codes. A grey code of length n is a labelling of the 2^n equal segments of a circle with binary strings of length n in such a way that the labels of adjacent segments differ in exactly one digit. Construction of a grey code may be viewed as a graph theoretic problem. The graph model for this problem may be taken on the set V of all binary strings or sequences of length n and for two strings u, v of length n we say that u, v are adjacent to each other if u and v differ exactly in one digit. To find a labelling of segments of the circle is equivalent to finding a Hamiltonian circuit in the above graph. Thus a grey code of length n is nothing but a Hamiltonian circuit in the graph.

For $n = 3$, the three **strings** are $\{000, 001, 010, 011, 100, 101, 110, 111\}$. As per the definition of adjacent strings, we can take one Hamiltonian circuit or grey code of length 3 as $000, 100, 101, 111, 110, 010, 011, 001$. For finding a grey code of length 3, we need to find a Hamiltonian circuit in the graph. The above circuit is as indicated in Fig. 7.72.

Observe that in the above graph every vertex is of odd degree, yet the graph has a Hamiltonian circuit.

Example 7.30 The seating problem. Nine members of a new club meet each day for lunch at a round table. They decide to sit such that every member has different neighbours at each lunch. How many days can this arrangement last?

We will construct a graph $G = (V, E)$, where the set V of vertices is the set of nine persons. For two persons x and y, we say that x, y are adjacent if they sit next to each other. Since every member is allowed to sit next to any other member, G is a complete graph on nine vertices. Then to find the number of days the lunch arrangement lasts is equal to the number of all possible arrangements of seating the nine people so that no two arrangements have a common edge. This is the same as finding all possible distinct Hamiltonian circuits such that no two Hamiltonian circuits have a common edge. In this direction, we can use the following (recorded without proof).

Theorem 7.18. In a complete graph with n vertices, there are $(n-1)/2$ edge-disjoint Hamiltonian circuits, if n is an odd number ≥ 3.

Hence the number of different seating arrangements is $(9-1)/2 = 4$ and, so the lunch arrangements last for four days.

7.6. Planar Graphs

Three houses are to be given connection for water, electricity and gas. The pipes and cables supplying these are underground. Is it possible that the three pipes and cables can be laid at the same depth below earth without any two of these crossing each other? This problem is equivalent to connecting each of the three points, say a, b, c with three points d, e, f by lines straight or curved all lying in a plane and without any two lines cutting each other. This problem gives rise to what are called planar and non-planar graphs.

A graph that can be drawn on a plane in such a way that no edges cross one another except at common vertices is called a **planer graph**. A graph that cannot be drawn in this manner is called a **non-planar graph**. Observe that complete graph K_3 of degree 3 i.e. with three vertices is always a planar graph. The complete graph K_4 as given in Fig. 7.73(a) does not appear to be a planar graph. However, it may be redrawn as in Fig. 7.73(b) and so is a planar graph. The graphs given in Fig. 7.73(c) and (d) are also planar although neither of these is a complete graph.

A **region** of a planar graph is defined to be an area of the plane that is bounded by edges and which cannot be further divided into subareas. Thus graphs in Fig. 7.73(b)–(d) have four, four and five regions as indicated. Consider next the graphs as given in Fig. 7.74(a)–(e) below. Each one of these is a planar graph, two of which have only one vertex and the

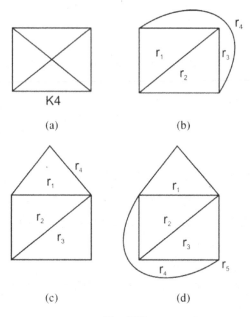

(a)   (b)

(c)   (d)

Fig. 7.73.

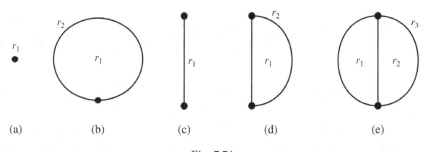

(a)   (b)   (c)   (d)   (e)

Fig. 7.74.

other three have two vertices each. While the graphs in Fig. 7.74(a) and
7.74(c) have one region each, the graphs in Fig. 7.74(b) and (d) have two
regions each but the one in (e) has three regions. Again the graphs given in
Fig. 7.75(a) and (b) below are planar graphs each having only two regions.

A region is said to be **finite** if its area is finite and is called **infinite**
if its area is infinite. It is clear that every planar graph has exactly one
infinite region. An edge, in general, appears in the boundary of at most two
regions. For example, edge a in Fig. 7.75(a) and (b) is not in the boundary

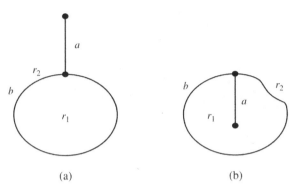

Fig. 7.75.

of any region. However, if an edge appears in the boundary of one region, it will surely appear in the boundary of another region.

Theorem 7.19 (Euler's formula for planar graphs). In any connected planar graph $m - n + k = 2$ where m, n, k are respectively, the number of vertices, edges and regions of the graph.

Proof. We prove the result by induction on the number of edges n in the graph.

If the number of edges in a connected graph is exactly one, the graph is one of the following two types (Fig. 7.76(a) and (b)).

The first of these has two vertices but only one region while the second has only one vertex but two regions. Thus the result $m - n + k = 2$ holds for each one of these graphs.

Next, suppose that G is a graph with $n \geq 2$ edges and that the result holds for all graphs with $\leq n - 1$ edges.

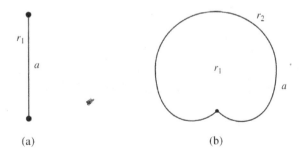

Fig. 7.76.

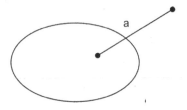

Fig. 7.77.

Case (i) G has a vertex of degree 1. In this case remove this particular vertex and the one edge incident with it. There then results a graph G' with $m-1$ vertices, $n-1$ edges and k regions as removal of an edge of the type a does not decrease the number of regions in the graph. The graph G' is planar and connected. Therefore, by induction hypothesis, $(m-1)-(n-1)+k=2$ which gives $m-n+k=2$.

Case (ii) The graph G has no vertex of degree 1. Then the graph G has at least one finite region. Remove one edge from a finite region of G. This removal then results in decreasing the number of regions by 1. For the subgraph G' obtained from G by removing one edge which is in the boundary of a finite region, the number of vertices is m, the number of edges is $n-1$ and the number of regions is $k-1$. Every subgraph of a planar graph being planar, the induction hypothesis gives $m-(n-1)+(k-1)=2$ or that $m-n+k=2$. This is precisely the result we wanted to prove for G. This completes induction and the proof of the Euler's formula for planar graph.

As an application of Euler's formula, we will prove the following.

Theorem 7.20. Let G be a linear planar connected graph with m vertices and $n \geq 3$ edges. Then $n \leq 3m - 6$.

Proof. Since G is a linear graph, there are no loops and also every region is bounded by at least three edges. If k is the number of regions in G, then $3k \leq$ the total count of edges that appear in the boundaries of the regions. Every edge appears in the boundary of at most two regions. Therefore, the total count of the edges is $\leq 2n$. Combining these two, we get $3k \leq 2n$ or $k \leq \frac{2}{3}n$. Substituting this in Euler's formula $m - n + k = 2$, we get $2 \leq m - n + \frac{2}{3}n = m - \frac{n}{3}$ or $6 \leq 3m - n$ which gives $n \leq 3m - 6$.

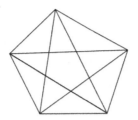

Fig. 7.78.

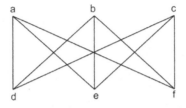

Fig. 7.79.

7.6.1. *Applications*

We next consider some applications of the Euler's formula and Theorem 7.20.

Example 7.31. The complete graph K_5 on five vertices has ten edges and $n = 10 \nleq 3 \times 5 - 6 = 9$. Also the graph is connected. Hence the complete graph K_5 on five vertices is not planar.

Example 7.32. Consider the utilities graph $K_{3,3}$ where three houses are to be connected to three utilities such as water, gas and electricity. Here each of the vertices a, b, c is connected to the vertices d, e, f. Thus the total number of edges is $n = 9$ and the total number of vertices is $m = 6$. Obviously $n = 9 \leq 3 \times 6 - 6 = 3m - 6$ and a direct application of the last theorem does not yield any information regarding the graph being planar or otherwise.

All sets of three vertices each of this graph are

$$\{a, b, c\}, \{a, b, d\}, \{a, b, e\}, \{a, b, f\}, \{b, c, d\}, \{b, c, e\}, \{b, c, f\}, \{b, d, e\},$$
$$\{b, e, f\}, \{a, d, e\}, \{a, d, f\}, \{a, e, f\}, \{b, d, f\}, \{a, c, d\}, \{a, c, e\}, \{a, c, f\},$$
$$\{c, d, e\}, \{c, d, f\}, \{c, e, f\}, \{d, e, f\}$$

and we find that no three vertices are connected to each other.

Therefore, there is no region bounded by three edges and as such every region is bounded by ≥ 4 edges. Let k be the number of regions in this graph. Therefore, the total count of the edges in the boundary of these k region is $\geq 4k$. Every edge appears at most twice in the boundary of regions. Therefore, the total count of edges $\leq 2n$. Therefore, $4k \leq 2n$ or $k \leq n/2$. By Euler's formula $m - n + k = 2$. Therefore, $2 \leq m - n + n/2 = m - n/2$ or $n \leq 2m - 4$. However, here $m = 6, n = 9$ and $9 \nleq 2 \times 6 - 4 = 8$. Therefore, the graph $K_{3,3}$ is not planar.

The above example also proves that Theorem 7.20 is necessary but not sufficient to say that a given connected graph is planar.

Example 7.33. Prove that every finite connected planar graph with at least three vertices has at least one vertex of degree ≤ 5.

Solution. Let G be a finite connected graph with $m \geq 3$ vertices and n edges. If possible, suppose that every vertex of G of degree ≥ 6. Then the sum of the degrees of the vertices is $\geq 6m$. But the sum of the degrees of the vertices is $2n$. Therefore, $2n \geq 6m$ or $n \geq 3m > 3m - 6$ which is a contradiction as in a connected planar graph $n \leq 3m - 6$. Hence G must have at least one vertex of degree ≤ 5.

Example 7.34. Suppose that a connected planar linear graph has 16 vertices, each of degree 3. Determine the number of regions into which a planar graph representation splits the plane.

Solution. Let m be the number of vertices, n the number of edges and k the number of regions into which the plane is split by a planar representation of the group. Then $m = 16$. As each vertex is of degree 3, the sum of the degrees of the vertices is $3 \times m = 48$. Also the sum of the degrees of the vertices is $2n$. Therefore, $2n = 48$ so that $n = 24$. By Euler's theorem $2 = m - n + k = 16 - 24 + k = k - 8$ which gives $k = 8 + 2 = 10$. Hence the plane is split into ten regions by the planar representation of the graph.

Let G be a graph. If an edge is divided into two edges by inserting a new vertex of degree 2 as in Fig. 7.80(a) then the graph G' obtained from G remains planar if G is planar. The same remains true if two edges that are incident with a vertex of degree 2 are combined as a single edge by the removal of this vertex of degree 2. Each of the two transformations is called an **elementary transformation**.

Definition. Two graphs G_1 and G_2 are said to be **isomorphic to within vertices of degree** 2 if either G_1 and G_2 are isomorphic or they can be

(a)   (b)

Fig. 7.80.

transformed into isomorphic graphs by repeated insertions and/or removals of vertices of degree 2.

Consider, for example, the graphs (Fig. 7.81(a) and (b)) which are clearly isomorphic to within vertices of degree 2.

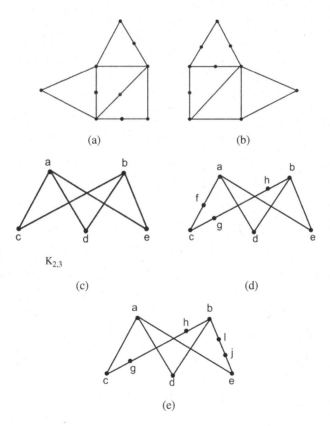

Fig. 7.81.

The three graphs of Fig. 7.81(c)–7.81(e) are isomorphic to within vertices of degree 2.

It follows immediately from the remark just above the definition that if two graphs are isomorphic to within vertices of degree 2 and one of the graphs is planar, then so is the other. We record without proof the following.

Theorem 7.21 (Kuratowski's therorem). A graph is planar if and only if it does not have any subgraph that is isomorphic to within vertices of degree 2 to either the complete graph K_5 or the utility graph $K_{3,3}$.

The complete graph K_5 and the utility graph $K_{3,3}$ are called **Kuratowski graphs.**

7.6.2. Some further examples

Example 7.35. The graph Fig. 7.82(a) may be redrawn as in Fig. 7.82(b) and so is a planar graph. It splits the plane into five regions as indicated.

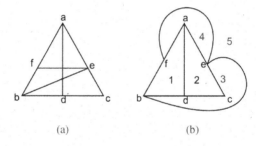

(a)   (b)

Fig. 7.82.

Example 7.36. That the complete graph K_3 is obviously planar and the complete graph K_4 is shown to be planar. We now show that the complete bipartite graphs $K_{2,3}$ and $K_{2,4}$ are planar.

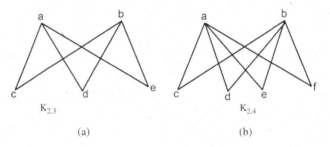

(a)   (b)

Fig. 7.83.

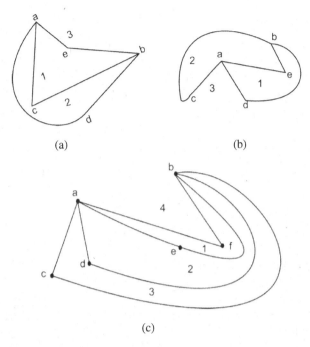

(a)   (b)

(c)

Fig. 7.84.

The graph $K_{2,3}$ may be redrawn as in Fig. 7.84(a) or 7.84(b) whereas $K_{2,4}$ may be redrawn as in Fig. 7.84(c).

The way $K_{2,4}$ has been redrawn also suggests that $K_{2,n}$ for every $n \geq 3$ is planar. The above graph for $K_{2,4}$ splits the plane into four regions and in general $K_{2,n}$ is a planar graph and splits the plane into n regions.

Example 7.37. The graph Fig. 7.85(a) below may be redrawn as in (b) and is clearly a planar graph. Observe that it has only three regions as indicated and Euler's theorem is verified.

Example 7.38. Which of these non-planar graphs have these property that the removal of any vertex and all edges incident with that vertex produces a planar graph?

    (a) K_5;   (b) K_6;   (c) K_7 (d) $K_{3,3}$;
    (e) $K_{3,4}$;   (f) $K_{3,5}$.

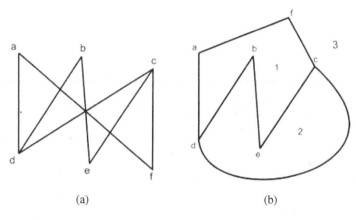

Fig. 7.85.

Solution. (a) Removal of any vertex along with edges incident with that vertex of K_5 results in the complete graph K_4 which has been seen to be a planar graph.

(b) and (c) Removal of any vertex along with edges incident with it from a complete graph K_n produces a complete graph K_{n-1}. For $n = 6$, the resulting graph is K_5 which is non-planar. We can thus use a simple inductive argument to say that for every $n \geq 6$, removal of any one vertex together with all edges incident with it from K_n produces a non-planar graph.

(d) Removal of a vertex together with edges incident with it from the graph $K_{3,3}$ produces a subgraph $K_{2,3}$ or $K_{3,2}$ both of which are planar. Observe that the graphs $K_{m,n}$ and $K_{n,m}$ are the same. As drawn, one is normally upside-down to the other.

(e) Depending on the choice of the vertex removed, the removal from $K_{3,4}$ will produce either $K_{2,4}$ or $K_{3,3}$. If the subgraph that results is $K_{2,4}$, it is planar as already seen in Example 7.36 above. On the other hand, if the subgraph obtained is $K_{3,3}$, it is non-planar.

(f) The removal of a vertex together with edges incident with it from $K_{3,5}$ produces a graph which is either $K_{2,5}$ which is planar or $K_{3,4}$ which has the utility graph $K_{3,3}$ as a subgraph and, hence, $K_{3,4}$ is non-planar.

It is, in fact, immediate from Kuratowski's theorem that the graph $K_{m,n}$ with $m, n \geq 4$ is always non-planar.

Example 7.39. Show that a linear planar graph with less than 12 edges has a vertex of degree 3 or less.

Solution. Let m be the number of vertices and n the number of edges in the given linear planar graph. Then $n < 12$. Suppose that every vertex is of degree ≥ 4. Then total sum of the degrees of the vertices is $2n$ and is $\geq 4m$. Thus $4m \leq 2n \leq 2(3m - 6)$ which gives $m \geq 6$. But then $4m \geq 24$ whereas $2n < 24$ which contradicts $4m \leq 2n$. Hence there is at least one vertex in the graph which is of degree ≤ 3.

7.6.3. *Graph colouring*

Let G be a graph. By **colouring** of G we mean an assignment of colours to the vertices of G such that adjacent vertices have different colours. We say that the graph G is **n-colourable** if there exists a colouring of G that uses n colours. The minimum number of colours needed to paint G is called the **chromatic number** of G and is denoted by $\chi(G)$. Welch and Powell gave an algorithm for colouring of a graph G which ensures colouring of G but does not always yield a minimal colouring of G.

Algorithm (Welch–Powell) 7.3.
The input is a graph G.

Step 1. Order the vertices of G according to decreasing degrees.
Step 2. Assign the first colour C_1 to the first vertex and then in sequential order, assign C_1 to each vertex which is not adjacent to a previous vertex x which was assigned the colour C_1.
Step 3. Repeat step 2 with a second colour C_2 and the subsequence of non coloured vertices.
Step 4. Repeat step 3 with a third colour C_3, then a fourth colour C_4, and so on until all vertices are assigned a colour.
Step 5. Exit.

In a complete graph every two vertices are adjacent. Therefore, the minimum number of colours needed for colouring a complete graph of n vertices is n and, so, chromatic number of K_n is $\chi(K_n) = n$.

Example 7.40. Show that in a connected planar linear graph with 7 vertices and 15 edges, each of the regions is bounded by three edges.

Solution. Since the graph is linear, each region is bounded by at least three edges. If k is the number of regions in the graph, then the total count of edges in the graph is $\geq 3k$. If even one region is bounded by four or more edges, then the total count of edges for the graph is $>3k$. In the present case, we have $7 - 15 + k = 2$ which gives $k = 10$. Therefore the total count of edges for the graph is >30. On the other hand, the total count of edges for the graph is $\leq 2 \times 15 = 30$. Thus, on the one hand total count of edges for the graph is >30 while on the other it is ≤ 30. This is a contradiction. Hence every region must be bounded by exactly three edges.

Remark. The result of the above example is true if we have $e < 3v - 6$, where e is the number of edges and v is the number of vertices.

Example 7.41. Find the chromatic number of the graph of Fig. 7.86.

Solution. Consider the graph G in Fig. 7.86. We use the Welch–Powell Algorithm to obtain a colouring of G. Ordering the vertices according to decreasing degrees, we get the following sequence:

$$A_5, A_1, A_2, A_6, A_7, A_3, A_4, A_8, A_9.$$

The first colour is assigned to A_5, A_3, A_9. The second colour is assigned to A_1, A_6, A_8. The third colour is assigned to A_2, A_7 and finally a fourth colour is assigned to A_4. All the vertices of G have been assigned a colour and so G is 4-colourable. Since every two of A_1, A_2, A_4, A_5 are adjacent, these four vertices must be assigned different colours. Hence G is not 3-colourable. Hence the chromatic number of this graph is four.

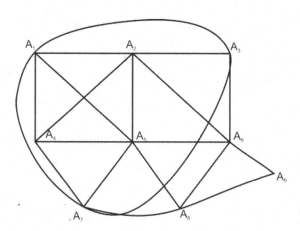

Fig. 7.86.

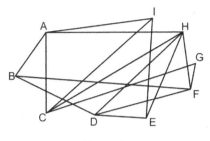

Fig. 7.87.

Example 7.42. Find the chromatic number of the graph of Fig. 7.87.

Solution. The degrees of the vertices are

$$A - 4, B - 3, C - 4, D - 4, E - 3, F - 4, G - 2, H - 5, I - 3.$$

We order these vertices in the decreasing order of degrees as in:

$$H\ A\ C\ D\ F\ B\ E\ I\ G$$

$$1\ 2\ 3\ 2\ 3\ 1\ 3\ 1\ 1.$$

Proceeding sequentially, we use the first colour to paint the vertices H, B, I, G. Then we use the second colour to paint A, D and use the third colour to paint C, E, F. Thus three colours are enough to paint the vertices of the graph. Every two of the vertices A, C, I are adjacent and, so, no two of these can be painted with the same colour. Therefore, the number of colours needed is ≥ 3. Hence the least number of colours for the colouring of the graph is 3, i.e. $\chi(G) = 3$.

Since the complete graph K_n requires n colours for its colouring, there is no limit on the number of colours that may be required for colouring of an arbitrary graph. However, in the case of planar graphs five colours are enough to colour a planar graph. Mathematicians conjectured that only four colours are enough to paint a planar graph. This remained a challenging problem for almost 125 years, when finally in 1976 Apple and Haken proved this conjecture to be true.

Theorem 7.22 (Apple and Haken). Any planar graph is 4-colourable. Proof of this result is beyond the scope of the present text. We will continue with examples on graph colouring.

Example 7.43. Consider a bipartite graph $K_{m,n}$. As it appears there are $m + n$ vertices and mn edges and it may appear that many colours may be needed for colouring this graph. The subsets of vertices of this graph be $V_1 = \{a_1, a_2, \ldots, a_m\}$ and $V_2 = \{b_1, b_2, \ldots, b_n\}$ where no two a_i, a_j are adjacent and no two b_i, b_j are adjacent. Every a_i is of degree n and every b_j is of degree m. Since no two of a_i's are adjacent, we may assign one colour to every a_i. Similarly we can assign the same colour to every b_j. Therefore, $K_{m,n}$ is 2-colourable and two is the least number of colours required. Hence the chromatic number of $K_{m,n}$ is 2.

Example 7.44. Let C_n be a cycle with n vertices. Suppose that $n = 2m$. Then we take the vertices of C_n as $a_1, b_1, a_2, b_2, \ldots, a_m, b_m$ in that order so that no two a's are adjacent and no two b's are adjacent. We may thus assign one colour to all the a's and another colour to the b's. Therefore, C_n for n even is 2-colourable.

Next, consider the case of n odd, say $n = 2m + 1$. We may then take the vertices as $a_1, b_1, a_2, b_2, \ldots, a_m, b_m, c$ where a's and b's alternate and c is the vertex between b_m and a_1. Every a_i is assigned one colour, every b_j is assigned a second colour and c is adjacent to a_1 and also to b_m, c cannot be assigned either of the two colours. Thus only two colours are not enough for the colouring of C_n in this case. Hence the chromatic number of the cycle graph C_n is 2 if n is even and is 3 if n is odd.

Example 7.45. Schedule the final examinations for mathematics papers $M - 1, M - 2, M - 3, M - 4$ and computer science papers $CS - 1, CS - 2, CS - 3, CS - 4$, using the least number of different time slots if there are no students taking both $M - 1$ and $CS - 4$, both $M - 2$ and $CS - 4$, both $M - 4$ and $CS - 1$, both $M - 4$ and $CS - 2$, both $M - 1$ and $M - 2$, both $M - 1$ and $M - 3$, and both $M - 3$ and $M - 4$, but there are students in every other combination of papers.

Solution. Consider a graph having eight vertices A, B, C, D and X, Y, Z, T, where A, B, C, D correspond respectively to the four courses $M - 1, M - 2, M - 3, M - 4$ and X, Y, Z, T correspond respectively to the four courses $CS - 1, CS - 2, CS - 3, CS - 4$. Two vertices are adjacent if the two courses represented by these vertices have common students. In view of the given hypothesis, vertices A and T, B and T, D and X, D and Y, A and B, A and C, C and D are not adjacent whereas every other two vertices are adjacent. The graph is as given in Fig. 7.88.

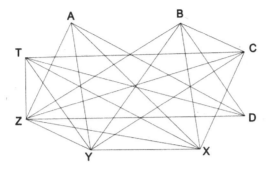

Fig. 7.88.

The vertices are of degrees as under:

$\deg(A) = 4$, $\deg(B) = 5$, $\deg(C) = 5$, $\deg(D) = 4$, $\deg(X) = 6$,
$\deg(Y) = 6$, $\deg(Z) = 7$, $\deg(T) = 5$.

Since Z is adjacent to every other vertex, one colour is needed for Z alone. The vertex X is adjacent to every vertex except D. So another colour is needed to colour D and X. The vertex Y is also adjacent to every vertex except D but D has already been grouped with X. Therefore, a third colour is needed to colour Y. The vertex T is adjacent to every vertex except A, B and so a fourth colour is needed to colour the vertices A, B and T. Only one vertex namely C is left. Although the vertices A, C, D could be grouped together, A and D have already been grouped with T and X respectively. Therefore, a fifth colour is needed for C. So, five colours are needed for the colouring of this graph. Hence five different time slots are needed for fixing the examinations. The time slots needed are

one for $M - 1, M - 2$ and $C - 4$,
one for $M - 4$ and $C - 1$,
one for $M - 3$,
one for $C - 2$,
and one for $C - 3$.

The colouring problem considered arises from assigning colours to different pieces of land or different states in a country or different districts in a state in India so that pieces of land or states or districts that share a common border are assigned different colours. Alternatively, we may not assign any colours to land pieces but only to the boundaries of the land pieces or states or districts. The restriction here is that adjacent boundary

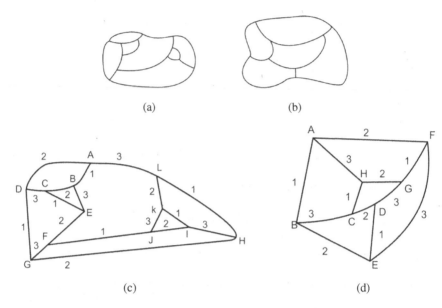

Fig. 7.89.

lines of a state are assigned different colours. The number of different colours used in the bounding lines of a state then indicates the maximum number of states with which the given state shares a common boundary. The resulting colouring is called **edge colouring**. In this case, the graph that represents this colouring problem has the intersection point of two or more land pieces/states as vertices and two vertices are adjacent if these two vertices are immediate neighbours on the boundary of two or more states. For example, the graphs corresponding to the maps Fig. 7.89(a) and (b) are respectively given by Fig. 7.89(c) and 7.89(d).

In the case of the state bounded by the lines AB, BC, CD and DA in the graph only two colours shall be enough to colour the borders while for the region bounded by IJ, JK and KI exactly three colours are needed.

Formally, we may define **edge colouring of a graph** as an assignment of colours to the edges so that edges incident with a common vertex are assigned different colours. The smallest number of colours that can be used in an edge colouring of a graph is called the **edge chromatic number** of the graph.

In the case of graph of Fig. 7.89(c), three colours are enough for the edge colouring of this graph. Since every two edges of the circuit $BCEB$

have to have different colours, no two colours can do the job for this circuit and, therefore, the line chromatic number of this graph is 3.

As indicated by the assignment of different colours to the edges of the graph of Fig. 7.89(d), three colours are enough for the edge colouring of this graph. As there are several edges of degree 3 in this graph, no two colours are enough. Hence the edge chromatic number of this graph is also 3.

For any graph, the edge chromatic number is $\geq$ the maximum number of the degrees of the vertices of the graph. It will be interesting to know whether the edge chromatic number of a graph is exactly equal to the maximum of the degrees of the vertices of the graph. However, if we consider a cycle C_n of length n, we find that the edge chromatic number of C_n is 2 if n is even and it is 3 if n is odd. However, there is no edge of degree 3 in C_n. We have seen that the chromatic number of a bipartite graph, including the complete bipartite graph $K_{m,n}$ is 2. But the edge chromatic number of the complete bipartite graph $K_{m,n}$ is maximum of $m-1$ and $n-1$. The edge chromatic number of the complete graph K_n is $\geq n-1$. For $n=3$, the chromatic number of K_3 is 3.

Exercise 7.5. Find the edge chromatic number of K_n for $n > 3$.

7.7. Matrix Representations of Graphs

We have so far considered diagrammatic representation of a graph and it is very convenient for visual study provided the number of vertices and edges is small. However, it may be more convenient sometimes to represent graphs using associated matrices. This is particularly so for studying structural properties of graphs using computers which can handle matrices and can use many results from linear algebra quite conveniently. In several applications of graph theory, such as in electrical network analysis and operations research (e.g. the transportation problem, assignment problem, travelling salesperson problem, to name a few), matrix theory proves to be the most powerful tool in representing and studying such problems.

There are several matrices that can be associated with any graph, but we study only two of these here.

7.7.1. *Adjacency matrix*

It is sometimes quite convenient to represent a graph by what is called its **adjacency matrix** or **connection matrix**. Let $G = (V, E)$ be a linear graph with n vertices $v_1, v_2, \ldots, v_n$ ordered as listed. The adjacency matrix

A (or A_G or $A(G)$) to indicate the graph G) of G, with the above listing of the vertices, is the square matrix (a_{ij}) of order n where

$$a_{ij} = \begin{cases} 1 & \text{if the vertices } v_i, v_j \text{ adjacent} \\ 0 & \text{otherwise} \end{cases}.$$

We are defining adjacency matrix for an undirected graph and, so, if there is an edge between v_i and v_j, there is an edge between v_j and v_i. Therefore, $a_{ij} = a_{ji}$ for all $i, j, 1, \leq i, j \leq n$. Thus the adjacency matrix of G w.r.t. a given listing of the vertices is always a symmetric matrix. Since there are $n!$ different ordering of the n vertices, there are $n!$ different adjacency matrices of G (one for every ordering of the n vertices). Also there being no loops in a linear graph, $a_{ii} = 0$ for every $i, 1 \leq i \leq n$. If A is adjacency matrix of G corresponding to one ordering of the vertices, then the adjacency matrix of G corresponding to another ordering of the vertices is obtained from A by applying a permutation to the rows of A and by applying the same permutation also to the columns of A.

Let G and H be two graphs. Then G and H are isomorphic if and only if there is a one–one mapping from the vertices of G to the vertices of H which induces a one–one correspondence between the edge sets of G and of H such that if e and e' are corresponding edges in G, H respectively, then the vertices of e' correspond to the vertices of e in G under the given mapping of vertex sets. We may thus regard the set of vertices of H as the same as the set of vertices of G but the ordering of the vertices for H is different from the ordering of the vertices for G. In view of the observation made in the previous paragraph, we may say that the graphs G and H are isomorphic if and only if there exists a non-singular matrix X such that $A(H) = X^{-1}A(G)X$, where $A(G), A(H)$ are the adjacency matrices representing G and H respectively.

In case the graph G is not linear but is allowed to have some loops, but no parallel edges, the adjacency matrix defined above has one change, namely entry a_{ii} becomes 1 in place of 0 if there is a loop around the vertex v_i.

Remarks.

1. In the case of adjacency matrix of a linear graph the number of $1's$ in the ith row or in the ith column of the adjacency matrix A represents the degree of the vertex v_i.

2. If $G = (V, E)$ has two disconnected components, say $G_1 = (V_1, E_1)$ and $G_2 = (V_2, E_2)$, then no vertex in V_1 is adjacent to any vertex in V_2 and

if v_i is in V_1 and v_j is in V_2 then $a_{ij} = a_{ji} = 0$. Therefore, the adjacency matrix A_G can be partitioned in the form

$$A(G) = \left(\frac{A(G_1):0}{0:A(G_2)} \right)$$

where A_{G_1} is the adjacency matrix of G_1 and A_{G_2} is the adjacency matrix of G_2. In case G has k disconnected components, then A_G can be partitioned accordingly.

3. Given any square, symmetric matrix B of order n with entries in the set $\{0,1\}$ of two elements, we can always find a graph G on n vertices such that B is the adjacency matrix of G.

So far, we have considered the adjacency matrix of a linear graph. We may define the adjacency matrix of any graph including graphs having multiple edges as also self loops. For a multigraph $G = (V, E)$, we can define the adjacency matrix $A(G) = (a_{ij})$, corresponding to the listing $v_1, v_2, \ldots, v_n$ of vertices, by taking a_{ij} to be the number of edges between the vertices v_i and v_j and $a_{ii} = 1$ for a loop around a_i. This matrix is also symmetric but all the diagonal entries need not be zero. Also the entries of $A(G)$ are no longer 0 and 1 as in the earlier case.

So far we have considered undirected graphs. We can associate an adjacency matrix with a directed graph which may also have self loops and/or multiple edges between some pairs of vertices. We consider a directed graph G with vertex set V and the vertices in V are ordered as $v_1, v_2, \ldots, v_n$, if there are n vertices. We define the adjacency matrix $A = (a_{ij})$ to represent a directed graph where a_{ij} is the number of edges with initial vertex v_i and terminal vertex v_j. In a directed graph, if there is an edge from v_i to v_j, the same cannot be regarded as an edge from v_j to v_z. Therefore, the adjacency matrix representing a directed graph is, in general, not symmetric. To illustrate this, we consider the adjacency matrices representing the graphs of Fig. 7.90.

Each of the graphs in Fig. 7.90 has four vertices and we order these vertices as a, b, c, d. With this ordering of the vertices the matrices representing these graphs are

$$
\text{(a)} \quad
\begin{array}{c}
\\ a \\ b \\ c \\ d
\end{array}
\begin{array}{c}
a \quad b \quad c \quad d \\
\begin{pmatrix}
0 & 1 & 0 & 0 \\
0 & 0 & 0 & 0 \\
0 & 1 & 0 & 1 \\
1 & 0 & 0 & 0
\end{pmatrix}
\end{array} ; \quad
\text{(b)} \quad
\begin{pmatrix}
0 & 1 & 0 & 0 \\
0 & 0 & 1 & 0 \\
0 & 1 & 0 & 1 \\
1 & 0 & 0 & 0
\end{pmatrix} ;
$$

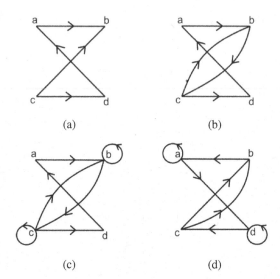

(a)   (b)

(c)   (d)

Fig. 7.90.

$$\text{(c)} \begin{pmatrix} 0 & 1 & 0 & 0 \\ 0 & 1 & 1 & 0 \\ 0 & 1 & 1 & 1 \\ 1 & 0 & 0 & 0 \end{pmatrix} ; \quad \text{(d)} \begin{pmatrix} 1 & 0 & 0 & 1 \\ 1 & 0 & 0 & 0 \\ 0 & 2 & 0 & 0 \\ 0 & 0 & 1 & 1 \end{pmatrix} .$$

Observe that in each of the above four examples the sum of all the entries of the representing matrix is equal to the number of edges in the graph.

Example 7.46. We consider some examples.

1. Use an adjacent matrix to represent each of the graphs in Fig. 7.91.

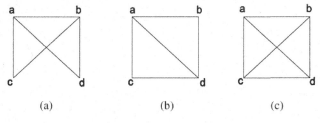

(a)   (b)   (c)

Fig. 7.91.

Solution. We order the vertices as a, b, c, d. Then the matrix representing the graph is

(a)
$$\begin{pmatrix} 0 & 1 & 1 & 1 \\ 1 & 0 & 1 & 1 \\ 1 & 1 & 0 & 0 \\ 1 & 1 & 0 & 0 \end{pmatrix};$$

(b)
$$\begin{pmatrix} 0 & 1 & 1 & 1 \\ 1 & 0 & 0 & 1 \\ 1 & 0 & 0 & 1 \\ 1 & 1 & 1 & 0 \end{pmatrix};$$

(c)
$$\begin{pmatrix} 0 & 1 & 1 & 1 \\ 1 & 0 & 1 & 1 \\ 1 & 1 & 0 & 1 \\ 1 & 1 & 1 & 0 \end{pmatrix}.$$

Example 7.47. Use an adjacency matrix to represent each of the graphs of Fig. 7.92.

Solution. We order the vertices as a, b, c, d, e for the graph (a) and as a, b, c, d for the graphs (b) and (c). Then the matrix representing the graph is

(a)
$$\begin{pmatrix} 0 & 1 & 0 & 1 & 0 \\ 1 & 0 & 0 & 1 & 1 \\ 0 & 0 & 0 & 1 & 1 \\ 1 & 1 & 1 & 0 & 0 \\ 0 & 1 & 1 & 0 & 0 \end{pmatrix};$$

(b)
$$\begin{pmatrix} 0 & 0 & 1 & 0 \\ 0 & 0 & 1 & 2 \\ 1 & 1 & 0 & 1 \\ 0 & 2 & 1 & 0 \end{pmatrix};$$

(c)
$$\begin{pmatrix} 1 & 0 & 2 & 1 \\ 0 & 1 & 1 & 2 \\ 2 & 1 & 1 & 0 \\ 1 & 2 & 0 & 1 \end{pmatrix}.$$

Example 7.48. In (a) to (d) given below draw a graph with the given adjacency matrix.

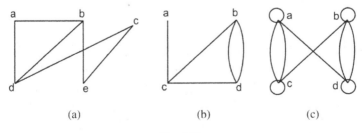

(a)   (b)   (c)

Fig. 7.92.

(a)
$$\begin{pmatrix} 1 & 1 & 1 & 0 \\ 0 & 0 & 1 & 0 \\ 1 & 0 & 1 & 0 \\ 1 & 1 & 1 & 0 \end{pmatrix};$$

(b)
$$\begin{pmatrix} 0 & 0 & 1 & 1 \\ 0 & 0 & 1 & 0 \\ 1 & 1 & 0 & 1 \\ 1 & 1 & 1 & 0 \end{pmatrix};$$

(c)
$$\begin{pmatrix} 1 & 2 & 0 & 1 \\ 2 & 0 & 3 & 0 \\ 0 & 3 & 1 & 1 \\ 1 & 0 & 1 & 0 \end{pmatrix};$$

(d)
$$\begin{pmatrix} 1 & 3 & 2 \\ 3 & 0 & 4 \\ 2 & 4 & 0 \end{pmatrix}.$$

Solution. In case of (a), (b) and (c), the matrices are of order 4 and so, there are four vertices in each of the graphs. Let the vertices be a, b, c and d. In (d), the given matrix is a square matrix of order 3 and, so, the corresponding graph has three vertices. Let the vertices be a, b, c. The fourth column of the matrix (a) is zero showing that the vertex d is not adjacent to any of a, b, c, d. Also this matrix is not symmetric. Therefore, the graph with adjacency matrix (a) is not undirected. Similar is the case with the graph whose adjacency matrix is as in (b). The matrices in (c) and (d) being symmetric, the graphs corresponding to these matrices are undirected. The graphs which are represented by the given matrices are given in Fig. 7.93.

As already observed, in the case of a directed graph, the sum of all the entries representing the graph equals the number of all the edges in the graph. In the case of an undirected graph, the number of edges in the graph equals the sum of all the entries in the upper (or lower) triangular matrix of the representing matrix. In the case of both directed as well as undirected graphs, the sum of the diagonal entries is the number of self loops present in the graph.

Consider an undirected graph G not necessarily a linear graph and $A(G) = (a_{ij})$ the adjacency matrix representing G. By definition of adjacency matrix $a_{ij} =$ the number of edges in G between the vertices v_i and v_j. Therefore, $\sum_{j=1}^{n} a_{ij}$ is the number all edges incident with v_i and, hence, $\sum_{j=1}^{n} a_{ij}$ is the degree of the vertex v_i provided $a_{ii} = 0$. However, if $a_{ii} = 1$, then $\sum_{i=1}^{n} a_{ij} =$ degree of $v_i - 1$ (this is so for if $a_{ii} = 1$, then there is a loop around v_i but a loop contributes 2 to the degree of v_i). In case of a directed graph, a_{ij} is the number of edges with initial vertex v_i and terminal vertex v_j. Therefore, $\sum_{j=1}^{n} a_{ij}, a_{ii} = 0$, represents the outdegree of the vertex v_i. However, when $a_{ii} = 1$, there is a loop around v_i and it contributes one to the indegree of v_i and one to the outdegree of v_i. Hence

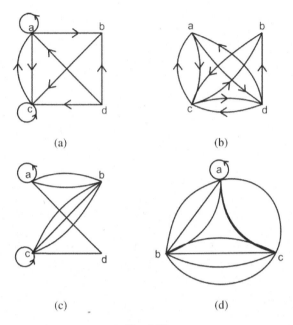

(a)   (b)

(c)   (d)

Fig. 7.93.

$\sum_{j=1}^{n} a_{ij}$ is the outdegree of v_i in this case as well. On the other hand, the sum of the entries in the jth column of $A(G)$ is the indegree of the vertex v_j.

In the case of an undirected graph, the sum of the entries in the jth column of the adjacency matrix is the degree of v_j if $a_{jj} = 0$ and is degree of $v_j - 1$ if $a_{jj} = 1$.

An **edge sequence** is a sequence $e_1, e_2, \ldots, e_k$ of edges in which each edge e_i has one vertex in common with the previous edge e_{i-1} and has another vertex in common with the edge e_{i+1} following it. An edge may appear more than once in an edge sequence.

Theorem 7.23. Let G be a linear graph with adjacency matrix A corresponding to an ordering $v_1, v_2, \ldots, v_n$ of the vertices. Then, for any k, (i, j)th entry of A^k gives the number of distinct k-edge sequences from the vertex v_i to the vertex v_j.

The above theorem does not yield any information about the k-edge sequences but only gives the number of such sequences.

Incidence matrix. Adjacency matrix representing a graph is a square matrix of order n where n is the number of vertices of the graph and the ith vertex being adjacent to the jth vertex contributes one to the (i, j)th

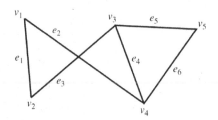

Fig. 7.94.

entry of the adjacency matrix which fully justifies the nomenclature of the representing graph. Another matrix that can be used to represent a graph $G = (V, E)$ on n vertices and m edges is called incidence matrix and this name is also fully justified by the very definition. Let G be an undirected graph with no self loops. Let the vertices be ordered as $v_1, v_2, \ldots, v_n$ and the edges be ordered as $e_1, e_2, \ldots, e_m$. Then the **incidence matrix** w.r.t. the above ordering of V and E representing the graph G is an n by m matrix $A = (a_{ij})$ where a_{ij} are defined by

$$a_{ij} = \begin{cases} 1 & \text{if the } j\text{th edge } e_j \text{ is incident with the } i\text{th vertex } v_i \\ 0 & \text{otherwise} \end{cases}.$$

For example, the incidence matrix representing the graph of Fig. 7.94 is

$$
A = \begin{array}{c}
\begin{array}{cccccc} & e_1 & e_2 & e_3 & e_4 & e_5 & e_6 \end{array} \\
\begin{array}{c} v_1 \\ v_2 \\ v_3 \\ v_4 \\ v_5 \end{array}
\begin{pmatrix}
1 & 1 & 0 & 0 & 0 & 0 \\
1 & 0 & 1 & 0 & 0 & 0 \\
0 & 0 & 1 & 1 & 1 & 0 \\
0 & 1 & 0 & 1 & 0 & 1 \\
0 & 0 & 0 & 0 & 1 & 1
\end{pmatrix}
\end{array}.
$$

Since every edge is incident with two vertices every column of the incidence matrix has two 1s in it.

In the case of multiple edges allowed in a graph, the incidence matrix representing the graph has two (or more) identical columns. The number of 1s in each row equals the degree of the corresponding vertex. A row of all zeros indicates that the corresponding vertex is not an end vertex of any edges, as such, this corresponding vertex is an isolated vertex.

If the graph G consists of two components G_1 and G_2, no vertex in G_1 is adjacent to any vertex in G_2 and conversely. Therefore, as in the case of adjacent matrix, the incident matrix representing G can be partitioned as

$$A(G) = \left(\begin{array}{c|c} A(G_1) & 0 \\ \hline 0 & A(G_2) \end{array} \right)$$

where $A(G_1), A(G_2)$ denote the incident matrices of the subgraphs G_1 and G_2 respectively. We have the following.

Theorem 7.24. Two graphs G and H are isomorphic if and only if one of their incidence matrices $A(G), A(H)$ can be obtained from the other by applying permutations of rows and columns.

Theorem 7.25. The rank of the incidence matrix $A(G)$ of a graph G with n vertices and with no self loops is $n - 1$.

Proof. Let m be the number of edges in G. Then the incident matrix A representing G is n by m matrix. Let $A_1, A_2, \ldots, A_n$ denote the rows of A. Then each A_i is a $1 \times m$ matrix and, so, is a row vector and

$$A = \begin{pmatrix} A_1 \\ A_2 \\ \vdots \\ A_n \end{pmatrix}.$$

The matrix A may be regarded as a binary matrix, i.e. a matrix over the field $B = \{0, 1\}$ of two elements. Since $1 + 1 = 0$ in B and since, every column of A has exactly two 1s, the sum $A_1 + A_2 + \cdots + A_n = 0$ – the 0 row vector. Therefore, $A_1, A_2, \ldots, A_n$ are linearly dependent over B. The rank of a matrix being the maximum number of linearly independent rows or the maximum number of linearly independent columns, the rank of A is $\leq n - 1$. If we ignore one row, say A_n, (no row being a 0 row, it is assumed that there are no isolated vertices in G), and consider the $(n - 1) \times m$ matrix.

$$B = \begin{pmatrix} A_1 \\ A_2 \\ \vdots \\ A_{n-1} \end{pmatrix}.$$

Then at least one column of B has exactly one 1 and, therefore, the rows $A_1, A_2, \ldots, A_{n-1}$ are linearly independent over any field. Hence the rank of

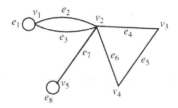

Fig. 7.95.

A is at least $n-1$ and, therefore, rank of A is $n-1$. (Refer to Exercise 7.6 at the end of the chapter).

Example 7.49. We may allow self loops to appear in a graph and the incidence matrix representing this graph may be defined as above with the exception that if there is a loop around a vertex v_i and this loop is marked as edge e_j then we introduce the jth column with exactly one 1 in the ith row. For example, the incidence matrix representing the graph of Fig. 7.95 is

$$
A(G) = \begin{array}{c} \\ v_1 \\ v_2 \\ v_3 \\ v_4 \\ v_5 \end{array}
\begin{array}{c} e_1\ e_2\ e_3\ e_4\ e_5\ e_6\ e_7\ e_8 \\
\begin{pmatrix}
1 & 1 & 1 & 0 & 0 & 0 & 0 & 0 \\
0 & 1 & 1 & 1 & 0 & 1 & 1 & 0 \\
0 & 0 & 0 & 1 & 1 & 0 & 0 & 0 \\
0 & 0 & 0 & 0 & 1 & 1 & 0 & 0 \\
0 & 0 & 0 & 0 & 0 & 0 & 1 & 1
\end{pmatrix}
\end{array}.
$$

When certain loops are present in the graph G, the rank of the incident matrix $A(G)$ is not $n-1$. For the above example let $\alpha_1, \alpha_2, \ldots, \alpha_5$ be scalars such that $\alpha_1 A_1 + \alpha_2 A_2 + \alpha_3 A_3 + \alpha_4 A_4 + \alpha_5 A_5 = 0$. Then $(\alpha_1, \alpha_1 + \alpha_2, \alpha_1 + \alpha_2, \alpha_2 + \alpha_3, \alpha_3 + \alpha_4, \alpha_2 + \alpha_4, \alpha_2 + \alpha_5, \alpha_5) = 0$ which implies that $\alpha_1 = \alpha_2 = \alpha_3 = \alpha_4 = \alpha_5 = 0$. Thus, all the five rows of $A(G)$ in this case are linearly independent and, hence, the rank of $A(G)$ is 5. Next, consider the graph G of Fig. 7.96 which is not connected but consists of two components. The incident matrix $A(G)$ for this graph is the 6×7 matrix

$$
A(G) = \begin{array}{c} \\ v_1 \\ v_2 \\ v_3 \\ v_4 \\ v_5 \\ v_6 \end{array}
\begin{array}{c} e_1\ e_2\ e_3\ e_4\ e_5\ e_6\ e_7 \\
\begin{pmatrix}
1 & 0 & 0 & 0 & 0 & 0 & 0 \\
0 & 1 & 1 & 0 & 0 & 0 & 0 \\
0 & 1 & 1 & 1 & 0 & 1 & 1 \\
0 & 0 & 0 & 1 & 1 & 0 & 0 \\
0 & 0 & 0 & 0 & 1 & 1 & 0 \\
0 & 0 & 0 & 0 & 0 & 0 & 1
\end{pmatrix}
\end{array}.
$$

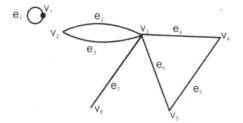

Fig. 7.96.

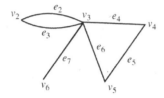

Fig. 7.97.

The rank of a partitioned matrix $\begin{pmatrix} A & O \\ O & B \end{pmatrix}$ being equal to the sum of the ranks of A and B, the rank of $A(G)$ is $1+$ rank of the matrix B, where

$$B = \begin{pmatrix} 1 & 1 & 0 & 0 & 0 & 0 \\ 1 & 1 & 1 & 0 & 1 & 1 \\ 0 & 0 & 1 & 1 & 0 & 0 \\ 0 & 0 & 0 & 1 & 1 & 0 \\ 0 & 0 & 0 & 0 & 0 & 1 \end{pmatrix}.$$

The matrix B being the incidence matrix of the component H (Fig. 7.97) of G, and H being connected with no self loops, the rank of $A(H)$ is 4. Therefore, the rank of $A(G)$ is $1 + 4 = 5 < 6 =$ the number of vertices in G. This leads to the following:

Exercise 7.6. If G is a connected graph with n vertices with at least one loop, is the rank of the incidence matrix $A(G)$ representing G equal to $n-1$?

Given an $n \times m$ matrix with $m \geq n-1$ and every column having either one or two 1s and the rest 0s, we can draw a graph with n vertices and m edges such that the incidence matrix representing the graph is the given matrix.

Example 7.50. Consider the matrices

(a)
$$\begin{pmatrix} 1 & 1 & 0 & 0 & 0 & 1 & 0 & 0 & 0 \\ 1 & 0 & 1 & 1 & 0 & 0 & 1 & 0 & 0 \\ 0 & 1 & 1 & 0 & 0 & 0 & 0 & 1 & 0 \\ 0 & 0 & 0 & 1 & 1 & 0 & 0 & 1 & 1 \end{pmatrix};$$

(b)
$$\begin{pmatrix} 1 & 1 & 0 & 0 & 0 & 1 & 0 & 0 & 0 \\ 1 & 0 & 1 & 1 & 0 & 0 & 1 & 0 & 0 \\ 0 & 1 & 1 & 0 & 0 & 0 & 0 & 1 & 0 \\ 0 & 0 & 0 & 1 & 1 & 0 & 0 & 1 & 1 \\ 0 & 0 & 0 & 0 & 1 & 1 & 1 & 0 & 1 \end{pmatrix};$$

(c)
$$\begin{pmatrix} 0 & 0 & 0 & 1 & 0 & 1 & 0 & 0 \\ 0 & 0 & 0 & 0 & 1 & 1 & 1 & 1 \\ 0 & 0 & 0 & 0 & 0 & 0 & 0 & 1 \\ 1 & 1 & 0 & 0 & 1 & 0 & 0 & 0 \\ 0 & 0 & 1 & 1 & 0 & 0 & 1 & 0 \\ 1 & 1 & 1 & 0 & 0 & 0 & 0 & 0 \end{pmatrix}.$$

(a) The matrix is 4×9 and, so the graph which can have this matrix as its incidence matrix shall have four vertices, say v_1, v_2, v_3, v_4 and nine edges, say $e_1, e_2, \ldots, e_9$ corresponding to the columns 1 to 9 respectively. Looking at each column, we find that the graph is as in Fig. 7.98.

(b) This is a 5×9 matrix and, therefore, the graph has five vertices and nine edges. Let the vertices representing the rows be v_1, v_2, v_3, v_4, v_5 and the edges representing the columns be $e_1, e_2, \ldots, e_9$ in order. Then the graph is as in as in Fig. 7.99.

(c) This is a 6×8 matrix and the graph which has this matrix as its incidence matrix shall have six vertices $v_1, v_2, v_3, v_4, v_5, v_6$ and eight edges representing the eight columns in order as $e_1, e_2, \ldots, e_8$. Then the graph is as in Fig. 7.100.

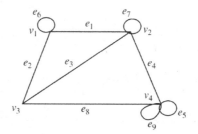

Fig. 7.98.

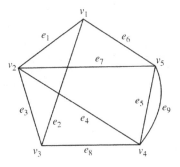

Fig. 7.99.

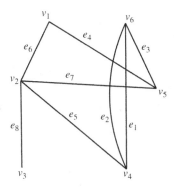

Fig. 7.100.

Please note that Example 7.2 is taken from Narsingh Deo (1974). Also, please refer to Narsingh Deo (1974) for a detailed account of other matrices associated with graphs. For a more detailed treatment of graphs refer to Kolman, Busby and Ross (2005), Liu (2000) and Rosen (2003/2005), in addition to the books on graph theory mentioned in the bibliography.

Exercise 7.7.

1. Consider the floor-plan of a three room structure shown in Fig. 7.101(a). Each room is connected to every room that it shares a wall with and to the outside along each wall. Is it possible to begin in a room or outside and take a walk that goes through each door exactly once? Support your answer by considering graphical model of the problem.

2. Prove or disprove that for every $n \geq 3$, the complete graph K_n has a Euler circuit.

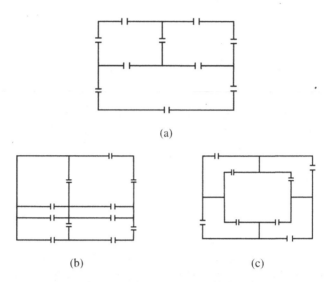

(a)

(b)   (c)

Fig. 7.101.

3. Let G, H be two isomorphic graphs and suppose that G posses a Euler path (circuit). Must H also have a Euler path (circuit)? Justify your answer.

4. The current exhibits of an art museum are arranged in the five rooms as in Fig. 7.101(b). Decide if there is a way to tour the exhibits so that you pass through each door exactly once. If yes, give a sketch of your tour.

5. The floor plan of an historical mansion is as given in Fig. 7.101(c). Is it possible to visit every room in the house by passing through each door exactly once? Give full justification for your answer.

6. In the complete graph K_5 on five vertices, find two Hamiltonian circuits that have no edges in common.

7. Decide if it is possible to find two Hamiltonian paths in the complete graph K_4 on four vertices.

8. Find the number of edges that a Hamiltonian circuit for the complete graph K_n on n vertices with $n \geq 3$ must have.

9. Give an example of a graph with (a) four vertices, (b) five vertices with a circuit that is both a Euler and a Hamiltonian circuit.

10. Give an example of a graph that has a Euler circuit and a Hamiltonian circuit which are not the same.

11. Prove Theorem 7.18.

12. Find a grey code of length 4.

13. Give an example of a graph that has

    (a) a Eulerian circuit and a Hamiltonian circuit;
    (b) a Eulerian circuit but has no Hamiltonian circuit;
    (c) no Eulerian circuit but has a Hamiltonian circuit;
    (d) neither a Eulerian circuit nor a Hamiltonian circuit.

14. Let G be an undirected linear graph with n vertices, $n \geq 3$. Let the vertices of G represent n persons and the edges of G represent a friendship relation among them such that two vertices are connected by an edge if and only if the corresponding persons are friends. Suppose that between any two persons they know all the remaining $n-2$ people. Show that the n people can be stood in line so that everyone stands next to two of his friends, except the two persons at the two ends of the line, each of them stands next to only one of his friends.

15. Eleven students plan to have dinner together for several days. They will be seated at a round table, and each student is planned to have different neighbours at every dinner. For how many days can this be done?

16. Show that the graph in Fig. 7.102(a) has no Hamiltonian circuit.

17. Show that any Hamiltonian circuit in the graph in Fig. 7.102(b) that contains the edge x must also contain the edge y.

18. Show that the sum of the squares of the indegrees over all vertices is equal to the sum of the squares of the outdegrees over all vertices in any directed complete graph.

19. If a graph (connected or disconnected) has exactly two vertices of odd degree, then there exists a path joining these two vertices.

20. A linear graph with n vertices and k components cannot have more than $(n - k)(n - k + 1)/2$ edges.

21. A graph $G = (V, E)$ is disconnected if and only if V can be partitioned into two non-empty disjoint subsets V_1 and V_2 such that there exists

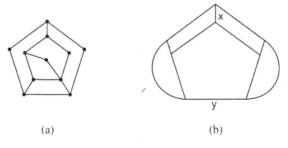

(a)   (b)

Fig. 7.102.

no edge in G that has one end vertex in V_1 and the other end vertex in the subset V_2.

22. Let G be a disconnected graph with n vertices where n is even. If G has two components each of which is complete, prove that G has a minimum of $n(n-1)/4$ edges.

23. Let G be an undirected graph with n vertices and k components. Prove that the number of edges in G is at least $n-k$.

24. Prove Theorem 7.23.

25. Prove Theorem 7.24.

26. Does there exist a simple graph with 15 vertices each of which is of degree 5?

27. A wheel $W_n, n \geq 3$, is a graph obtained from a cycle C_n and an additional vertex which is adjacent to every vertex of C_n. For what values of n is the wheel W_n a complete graph? Give full justification for your answer.

28. Decide if the graphs given in Fig. 7.103 are bipartite.

29. How many edges does a graph have if it has vertices of degree 4, 3, 3, 2, 2, 2? Draw such a graph, if one exists.

30. A graph is called **regular** if every vertex of this graph has the same degree and a regular graph is called n-**regular** if every vertex in this graph has degree n. For which values of m and n is the graph $K_{m,n}$ regular?

31. For which values of n are the graphs (a) C_n; (b) W_n; (c) K_n regular?

32. How many vertices does a regular graph of degree 3 with six edges have? Draw this graph.

33. Find the complementary graphs of the graphs C_n, W_n and K_n.

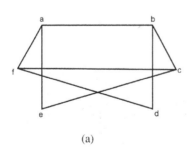 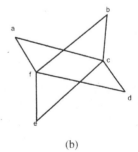

(a)   (b)

Fig. 7.103.

34. The **converse** of a directed graph $G = (V, E)$ is defined to be the directed graph $G^c = (V, F)$, where $(u, v) \in F$ if and only if $(v, u) \in E$ i.e. it is the graph obtained from G by reversing all the arrowheads. It is clear from the alternative definition of converse that $(G^c)^c = G$ for every directed graph. Obtain the necessary and sufficient condition(s) for a directed graph G to be its own converse.

35. Let $G = (V, E)$ be a directed graph. For $u, v \in V$, say that u is related to v (written as uRv) if the edge $(u, v) \in E$. Then R is a binary relation on the non-empty set V of vertices. Is the relation R (i) an equivalence relation? (ii) a partial order relation?

36. Show that an undirected graph can be properly coloured with two colours if and only if it contains no circuit of odd length.

37. Find the chromatic number of the wheel (a) W_6; (b) W_7. Can you generalize this result to W_n for arbitrary n?

38. Find the least number of colours needed for edge colouring the wheel W_6 and W_5.

39. If G is a graph where the longest cycle has odd length $n \geq 3$, prove that $\chi(G) \leq n + 1$.

40. If an undirected graph contains a triangle, prove that the graph is not bipartite.

41. Prove that the complete graph K_n on n vertices with $n \geq 3$ has $(n-1)!$ Hamiltonian circuits.

42. Prove that a graph is a circuit if and only if it is connected and every vertex has degree 2.

43. Give an example of a graph with seven vertices and exactly three components.

44. Prove that the graph of Fig. 7.104 does not have a Euler circuit. Find the minimal number of edges that would have to be travelled twice in order to pass through every edge (vertex) and return to the starting vertex.

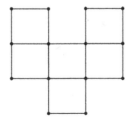

Fig. 7.104.

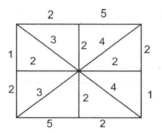

Fig. 7.105.

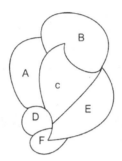

Fig. 7.106.

45. Find a Hamiltonian circuit of minimal weight for the graph (Fig. 7.105).
46. Construct a graph for the map given in Fig. 7.106 and determine the number of colours required for a proper colouring of the map.
47. Construct a graph for a communication network among cities P_1, P_2, P_3, P_4, P_5 with matrix

$$\begin{pmatrix} 0 & 1 & 0 & 0 & 0 \\ 0 & 0 & 1 & 0 & 1 \\ 0 & 0 & 0 & 1 & 0 \\ 0 & 0 & 0 & 0 & 0 \\ 0 & 0 & 1 & 1 & 0 \end{pmatrix}.$$

48. Prove that the existence of a simple circuit of length k, where k is a positive integer greater than two is an isomorphism invariant.
49. Prove that a simple graph is bipartite if and only if it has no circuit with an odd number of edges.
50. If G, H are isomorphic directed graphs and G is self-converse, prove that H is also self-converse.

Chapter 8

TREES

In this chapter we study a special kind of graph called trees. Trees arose in studies of rather unrelated subjects, like projective geometry, electrical networks and differential transformations and are used in a wide variety of applications, such as family trees, organizational charts, computer file structures and a wide range of algorithms in computer science. The British mathematician Arthur Cayley used trees as early as 1857 for counting certain types of chemical compounds. Some of the applications shall be studied in this chapter.

8.1. Introduction and Elementary Properties

A connected undirected graph that contains no simple circuits is called a **tree**. A collection of disjoint trees is called a **forest**. A vertex of degree 1 in a tree is called a **leaf** or a **terminal node** (a vertex of degree 1 in any graph is also called a **pendant vertex**) and a vertex of degree >1 is called a **branch node** or an **internal node**, while an edge is called a **branch**. All the trees (up to isomorphism of graphs) with four, five and six vertices are given in Fig. 8.1.

There are several equivalent conditions for a graph to be a tree. We next consider some of these.

Theorem 8.1. An undirected graph is a tree if and only if there is a unique path between every two of its vertices.

Proof. First, suppose that T is a tree. If possible, suppose that there are two paths between two vertices a and b (say) of the tree. Let the two paths

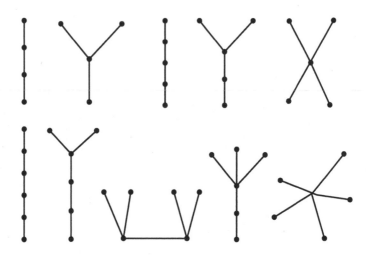

Fig. 8.1.

be $(v_1, v_2, \ldots, v_k)$ and $(u_1, u_2, \ldots, u_l)$ where $v_1 = a = u_1$ and $v_k = b = u_l$. Then $(v_1, v_2, \ldots, v_k, u_{l-1}, \ldots, u_1 = v_1)$ is a circuit in the given tree. If this circuit is not simple, we can remove repetition of edges, if any, and obtain a simple circuit in the tree. This contradicts the definition of a tree. Hence there cannot be two or more paths between any two vertices. A tree being a connected graph, there is always a path between every two vertices. Hence there is a unique path between every two vertices in a tree.

Conversely, assume that there is a unique path between every two vertices in a graph T. Then T is a connected graph. Suppose that the graph T has a simple circuit involving two vertices a and b (say). Since a simple circuit is made up of a simple path from a to b and a simple path from b to a, and the graph being undirected, the simple path from b to a leads to a simple path from a to b, we get two simple paths from a to b. This contradicts our assumption. Therefore, T has no simple circuit and, hence, T is a tree.

The following result, in addition to being needed in the next theorem, is of independent interest as well.

Theorem 8.2. Removal of any edge from a tree leaves a disconnected graph consisting of two components, each of which is a tree.

Proof. Let $T = (V, E)$ be a tree with n vertices and e edges with $n \geq 2$. For $n = 2$, there is only one edge between the two vertices and its removal leads to two isolated vertices each of which is a degenerate tree. (A tree with a single vertex and, so, no edge is called a **degenerate tree**. Other trees are called **non-degenerate**.)

Let $n \geq 3$ and a, b be two vertices in T which are adjacent. Let G' be the subgraph of T obtained by removing the edge $\{a, b\}$. Let x be a vertex from which there is a path to a in G'. Then there is no path from x to b in G' for otherwise there will be a circuit in the tree T. Similarly, if y is a vertex from which there is a path to b in G', then there is no path from y to a in G'. It also follows that x and y are not connected in G'. Take

$$V' = \{x \in V | \text{there is a path in } G' \text{ from } x \text{ to } a\} \cup \{a\},$$

$$V'' = \{y \in V | \text{there is a path in } G' \text{ from } y \text{ to } b\} \cup \{b\}.$$

Then $V' \cap V'' = \varnothing$, and no element of V' is connected to any element of V'' and no element of V'' is connected to any element of V'. If $x \in V$, then T being a connected graph, there is a path from x to b in T. If this path does not involve the edge $\{a, b\}$ then $x \in V''$. If, on the other hand, this path involves the edge $\{a, b\}$ removing the edge $\{a, b\}$, we obtain a path from x to a not involving the edge $\{a, b\}$ and $x \in V'$. Hence $V = V' \cup V''$ Let E' and E'' be the edges in G' which are incident with the vertices in V' and V'' respectively. Then $E = E' \cup E'' \cup \{a, b\}$ and $T' = (V', E'), T'' = (V'', E'')$ are the subgraphs of T obtained by omitting the edge $\{a, b\}$. Since any circuit in T' or in T'' will be a circuit in T which is a tree, there is no circuit in T' and no circuit in T''. Since every two elements of V' are connected in T and no vertex in V' is connected to any vertex in V'', the two vertices in V' must be connected in T'. Similarly any two vertices in T'' are connected. Hence T' and T'' are trees.

Theorem 8.3. A connected graph G with n vertices and e edges is a tree if and only if $e = n - 1$.

Proof. Suppose that $G = (V, E)$ is a tree. We prove by induction on n that $e = n - 1$.

If $n = 2$, obviously there is only one edge in the tree. There has to be at least one edge between the two vertices as there are no isolated vertices.

Suppose that $n \geq 3$. Let a, b be two adjacent vertices in G. Removal of the edge $\{a, b\}$ leads to two components $T' = (V', E'), T'' = (V'', E'')$ of G such that both T', T'' are trees, $V = V' \cup V''$ and $E' \cup E''$ together with the edge $\{a, b\}$ give the edge set E of G. Since neither of V', V'' is empty, $|V'| \leq n - 1$ and $|V''| \leq n - 1$. Therefore, by induction (both T', T'' being trees), $|E'| = |V'| - 1, |E''| = |V''| - 1$. Therefore, $e = |E| = |E'| + |E''| + 1 = |V'| + |V''| - 1 = |V' \cup V''| - 1 = |V| - 1 = n - 1$.

Conversely, suppose G is a connected graph and $e = n - 1$. In order to prove that G is a tree, we need to prove that there is no simple circuit in G. Let C be a simple circuit in G. Suppose that the circuit C has m vertices. The circuit C being simple, the number of edges in C is also m. The graph G being connected, every vertex of G which is not in C must be connected to the vertices in C. However, every edge in G which is not in C can connect at most one additional vertex to C. The number of vertices outside C being $n - m$ there are at least $n - m$ edges of G outside C. Therefore, the total number of edges in G is $\geq m + n - m = n$, which contradicts the assumption that $e = n - 1$. This proves that there is no simple circuit in G and G is a tree.

Theorem 8.4. A tree with two or more vertices has at least two leaves.

Proof. Let $T = (V, E)$ be a tree with n vertices and $n \geq 2$. Then $n = |V| = |E| + 1$. Also $2|E| = \sum_{a \in V} \deg a$ so that $\sum_{a \in V} \deg a = 2(n - 1) = 2n - 2$. If $\deg a \geq 2$ for every $a \in V$ then $\sum \deg a \geq 2n$ or that $2n - 2 \geq 2n$ which is not possible, n being a positive integer. Hence there is at least one $a \in V$ with $\deg a = 1$. If exactly one vertex b in V is of degree 1, then $2n - 2 = \sum_{a \in V} \deg a = \sum_{a \in V, a \neq b} \deg a + 1 \geq 2(n - 1) + 1 = 2n - 1$. Again, this is a contradiction. Hence there are at least two vertices in T which are of degree 1 each, i.e. there are at least two leaves in the tree T.

Theorem 8.5. A graph G with $e = n - 1$ edges, where n is the number of vertices in G, that has no circuit is a tree.

Proof. Let G be a graph with n vertices, $e = n - 1$ edges and having no circuit. To prove that G is a tree, we only need to prove that G is connected. Suppose to the contrary that G is not connected. Let $G', G'', \ldots, G^{(k)}, k \geq 2$, be the connected components of G. Then none of $G', G'', \ldots, G^{(k)}$ has a circuit. Therefore, $G', G'', \ldots, G^{(k)}$ are all trees. By Theorem 8.3, we have $e' = n' - 1, e'' = n'' - 1, \ldots, e^{(k)} = n^{(k)} - 1$ where $e', e'', \ldots, e^{(k)}$ are the number of edges in $G', G'', \ldots, G^{(k)}$ respectively and $n', n'', \ldots, n^{(k)}$ are the number of vertices in $G', G'', \ldots, G^{(k)}$ respectively. Adding all these we get

$$e' + e'' + \cdots + e^{(k)} = n' + n'' + \cdots + n^{(k)} - k$$

or $e = n - k < n - 1$, as $k \geq 2$. This contradicts the hypothesis. Hence G is connected and is, therefore, a tree.

We have defined shortest path and the shortest distance in weighted connected graphs. If we have a connected graph, not necessarily weighted, we may assign weight 1 to every edge and can talk of shortest path and shortest distance between vertices. As already done, we denote the shortest distance between vertices v_i and v_j by $d(v_i, v_j)$. It is fairly obvious that $d(v_i, v_j) = d(v_j, v_i), d(v_i, v_j) = 0$ if and only if $v_j = v_i$ and that $d(v_i, v_k) \leq d(v_i, v_j) + d(v_j, v_k)$.

Let $G = (V, E)$ be a connected graph. For any vertex v_i, we define the eccentricity $\varepsilon(v_i)$ by $\varepsilon(v_i) = \max_{v \in V} d(v_i, v)$. A vertex v in V is called a **centre** of the graph G if $\varepsilon(v) = \min\{\varepsilon(v_i) | v_i \in V\}$. Consider the graph (tree) of Fig. 8.2

$$\varepsilon(a) = 6, \quad \varepsilon(b) = 6, \quad \varepsilon(c) = 5, \quad \varepsilon(d) = 5, \quad \varepsilon(e) = 4, \quad \varepsilon(f) = 5,$$

$$\varepsilon(g) = 6, \quad \varepsilon(h) = \varepsilon(i) = \varepsilon(j) = 7, \quad \varepsilon(n) = 4, \quad \varepsilon(m) = \varepsilon(0) = 5,$$

$$\varepsilon(k) = \varepsilon(l) = 6, \quad \varepsilon(p) = 6, \quad \varepsilon(q) = \varepsilon(r) = 7.$$

In this tree, the minimum of the eccentricities is four and occurs for the vertices e and n. Hence e and n are centres of this tree. The proof of the following theorem suggests an easy way of finding the centre(s) of a tree.

Theorem 8.6. Every tree has one or two centres.

Proof. Let $T = (V, E)$ be a tree. We prove by induction on the number of vertices in T that T has one or two centres. Let $|V| = n$. If $n = 1$ or 2, the

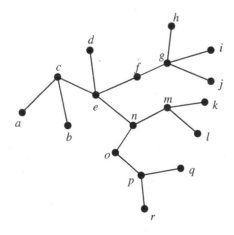

Fig. 8.2.

result is trivial for in the first case T has one centre while in the second case it has two centres. Suppose that $n > 2$. By Theorem 8.4, T has at least two leaves. If v_i is a leaf and $\{v_i, v_j\}$ is an edge in T, then $\varepsilon(v_i) = \varepsilon(v_j) + 1$. This is so, because $d(v_i, v_k) = d(v_k, v_j) + 1$, as every path from v_i to v_k has to pass through v_j. Thus v_i is not a centre of T, i.e. none of the leaves in T is a centre of T. Let $T' = (V', E')$ be the subtree of T obtained by deleting all the leaves of T along with the edges terminating on them. Then every leaf in T' is a branch node of T and is adjacent in T to at least one leaf of T. Also observe that eccentricity of any vertex v of a tree is the length of a path originating from v and terminating at a leaf of the tree. From these observations it follows that if a largest path from v in T terminates at the leaf v_k and v_{k-1} is adjacent to v_k, then the portion of this path terminating at v_{k-1} is a largest path from v in T'. This proves that $\varepsilon(v)$ in T equals $1 + \varepsilon'(v)$, where $\varepsilon'(v)$ denotes the eccentricity of v in T'. If v is a centre of T, then

$$\varepsilon(v) = \mathrm{Min}\{\varepsilon(u)|u \in V\} = \mathrm{Min}\{\varepsilon(u)|u \in V'\} = \mathrm{Min}\{\varepsilon'(u) + 1|u \in V'\}$$

$$= 1 + \mathrm{Min}\{\varepsilon'(u)|u \in V'\} \quad \text{or} \quad \varepsilon(v) - 1 = \mathrm{Min}\{\varepsilon'(u)|u \in V'\} \quad \text{or}$$

$$\varepsilon'(v) = \mathrm{Min}\{\varepsilon'(u)|u \in V'\}$$

which proves that v is a centre of T'. The reverse of the above argument also holds and it follows that centres of T are precisely the centres of T'. As the number of vertices in T' is $\leq n - 2$ (Theorem 8.4), induction hypothesis shows that T' has one or two vertices. Hence T has one or two vertices.

The proof of the above theorem gives a useful technique for finding the centres of a tree with at least two vertices.

The Procedure

Let $T = (V, E)$ be the given tree.

1. Omit all the leaves together with all the edges that terminate at these leaves to obtain a subtree T'.
2. If T' has only two/one vertices, stop. The one/two vertices of T' are the centres of T.
   If T' has more than two vertices, repeat step 1. Continue until we arrive at a subtree having one/two vertices.

The one/two vertices of the subtree finally arrived at are the centre of T.

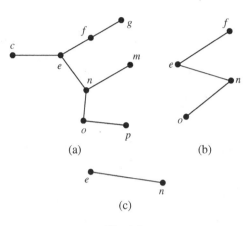

(a)   (b)

(c)

Fig. 8.3.

As an application of this procedure, we consider the tree of Fig. 8.2. Omitting the leaves and the edges terminating there, we get a subtree of Fig. 8.3(a).

Omitting the leaves and edges terminating there of this subtree, we get the subtree of Fig. 8.3(b).

Repeating the above step (step 1) for this tree, we get the subtree of Fig. 8.3(c) which has only two vertices. Hence e and n are the centres of the tree we started with. This also confirms the result already obtained (Theorem 8.6).

Example 8.1. A tree has $3n$ vertices of degree 1, $4n$ vertices of degree 2 and n vertices of degree 3. Determine the number of vertices and edges in the tree.

Solution. Total number of vertices in the graph is $3n + 4n + n = 8n$. Let the number of edges in the graph be e. Then $e = 8n - 1$. Also $2e =$ sum of degrees of all the vertices in the tree $= 3n \times 1 + 4n \times 2 + n \times 3 = 3n + 8n + 3n = 14n$. Using $e = 8n - 1$, we get $2(8n - 1) = 14n$ which gives $n = 1$. Hence the total number of vertices in the tree is $8n = 8$ and the total number of edges is $8n - 1 = 7$.

Example 8.2. A tree has three vertices of degree 2, two vertices of degree 3 and four vertices of degree 4. Determine the number of vertices of degree 1 in the tree.

Solution. Let the number of vertices of degree 1 in the tree be n. Let e be the number of edges in the tree. Then $e + 1 =$ total number of vertices in the tree $= n + 3 + 2 + 4 = n + 9$ or $e = n + 8$. Also $2e =$ sum of degrees of all vertices in the graph $= n \times 1 + 3 \times 2 + 2 \times 3 + 4 \times 4 = n + 6 + 6 + 16 = n + 28$. Therefore, $2(n + 8) = n + 28$ which gives $n = 12$ and, so, there are twelve vertices of degree 1 in the tree.

Example 8.3. A tree has n_2 vertices of degree $2, n_3$ vertices of degree $3, \ldots$, and n_k vertices of degree k. Determine the number of vertices of degree 1 in the tree.

Solution. Let n_1 be the number of vertices of degree 1 and e be the total number of edges in the tree. Then $e = n_1 + n_2 + \cdots + n_k - 1$ and $2e =$ sum of degrees of all the vertices in the tree $= n_1 + 2n_2 + 3n_3 + \cdots + kn_k$ or $2(n_1 + n_2 + \cdots + n_k - 1) = n_1 + 2n_2 + 3n_3 + \cdots + kn_k$ which gives $n_1 = n_3 + 2n_4 + \cdots + (k - 2)n_k + 2$.

Hence the number of vertices of degree 1 in the tree is $n_3 + 2n_4 + \cdots + (k - 2)n_k + 2$.

Example 8.4. Let T be a tree with 40 edges. The removal of a certain edge from T yields two disjoint trees T_1 and T_2. Given that the number of vertices in T_1 equals two more than the number of edges in T_2, determine the number of vertices and the number of edges in T_1 and T_2.

Solution. Let v_1, v_2 be the number of vertices and e_1, e_2 be the number of edges in T_1 and T_2 respectively. Then, by hypothesis,

$$v_1 = e_2 + 2. \tag{8.1}$$

Also we known that $v_1 = e_1 + 1$ and $v_2 = e_2 + 1$. Moreover $e_1 + e_2 = 40 - 1 = 39$ or $e_1 + v_1 - 2 = 39$ (by (8.1)). Then $e_1 + e_1 + 1 - 2 = 39$ which gives $e_1 = 20$. and then, $v_1 = 21$. Also, then $e_2 = 19$ and $v_2 = 19 + 1 = 20$. Hence there are 21 vertices and 20 edges in T_1 while in T_2, the number of vertices is 20 and the number of edges is 19.

8.2. Rooted Trees

A directed graph which becomes a tree when the directions of the edges are ignored is called a **directed tree**. If a directed tree has exactly one vertex whose incoming degree is 0 and the incoming degrees of all other vertices are 1, then it is called a **rooted tree**. The vertex with incoming degree 0 in a rooted tree is called the **root** of the rooted tree. In a rooted tree, a vertex outgoing degree of which is 0 is called a **leaf** or a **terminal node** and a vertex whose outgoing degree is non-zero is called a **branch node** or an **internal node**. A rooted tree could be defined without reference to a directed tree. It is just a tree with a distinguished vertex, called the root. However, directions are assigned to edges by convention while drawing a rooted tree. Structures that can be represented as rooted trees are encountered on many occasions. For example, the organization chart of a company or a university or the family tree of a caste may be represented as rooted trees. A rooted tree representing a university administration is given in Fig. 8.4 (this only gives a broad description of some of the offices/ functionaries and is not a true picture of all the offices/functionaries). For facility, we have used the following abbreviations.

V.C: Vice-Chancellor, Reg: Registrar, COE: Controller of Exams, DR: Deputy Registrar, S: Office Superintendent, Dean: Dean of a Faculty, CP: Chair Person of Teaching Department, DR(C): DR Conduct, DR(S): DR Secrecy, DR(R): DR Results, S(U): S Under graduate, S(P): S Post graduate, S(TE): S Technical education, S(T): S Theory, S(P): S Practicals, S(Pl): S Planning, S(BS): S Bills and Salaries, S(Pf): S Provident fund, S(Pn): S Pension Cell, Dean Sc: Dean Faculty of Science, Dean AL: Dean Faculty of Arts & Languages, Dean En: Dean Faculty of Education, Dean LM: Dean Faculty of Law & Management, CPSc: Chair Persons of Science Departments, CPAL: Chair Persons of Departments of Arts & Languages, CHEn: Chair Persons of the Departments of Education, CPLM: Chair Persons of the Departments of Law & Management.

Let a be a branch node in a rooted tree. If there is an edge from a to a vertex b, then b is called a **son** of a. Also a is called **father** of b. Two vertices which are sons of the same vertex are called **brothers** of each other. If there is a directed path from a to a vertex c, then c is called a **descendent** of a. Also a is then called an **ancestor** of c.

Let a be a branch node in a rooted tree T. The subgraph $T' = (V', E')$ of T where V' contains a and all its descendents and E' the edges in all directed paths emanating from a is called the **subtree with** a as the root. A subtree that has a son of a as a root is called a subtree of a. Observe that the tree in Fig. 8.5(d) is a subtree of the tree in Fig. 8.5(b) with a as the root while that in Fig. 8.5(e) is a subtree of the tree of Fig. 8.5(d).

While drawing a rooted tree we adopt the convention: always place the sons of a branch node below it. If this is done, we may omit the arrowheads on the edges. For example, the tree in Fig. 8.5(c) is a representation of the rooted tree of Fig. 8.5(b).

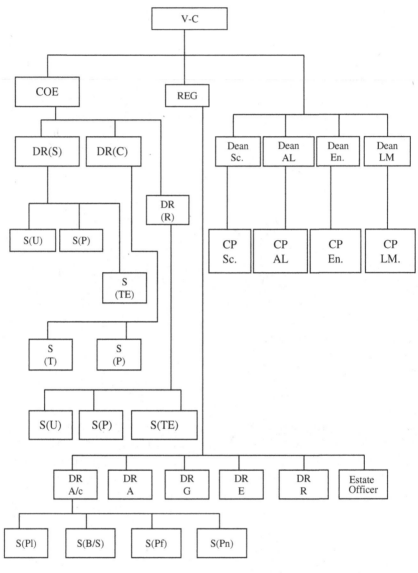

Fig. 8.4.

If we draw a rooted tree with root vertex v_0 (say), the terminal vertices of the edges beginning at v_0 are called level 1 vertices, while v_0 is called a level 0 vertex. The sons of v_0, i.e. the level 1 vertices which have been called sons of v_0 earlier may also be called **offsprings** of v_0. The edges leaving a vertex of level 1 which are normally drawn downwards terminate

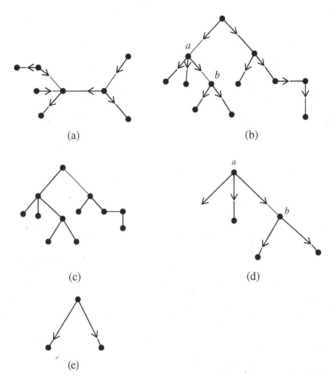

Fig. 8.5. (a) is a directed tree while (b) gives a rooted tree and (c) is a tree.

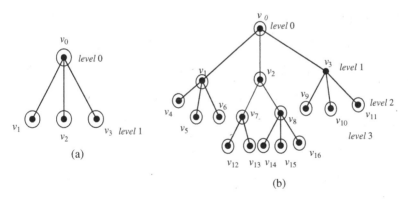

Fig. 8.6.

at various vertices. All these vertices are called level 2 vertices. We may similarly define level 3 vertices, then level 4 vertices and so on. The largest level number is called the **height** of the tree. Consider, for example, the trees in Fig. 8.6.

In the tree in Fig. 8.6 (a) above v_1, v_2, v_3 are level 1 vertices and the height of this tree is 1. In the case of tree (b) v_1, v_2, v_3 are level 1 vertices, $v_4, v_5, v_6, v_7, v_8, v_9, v_{10}, v_{11}$ are level 2 vertices and $v_{12}, v_{13}, v_{14}, v_{15}, v_{16}$ are level 3 vertices. Also the height of this tree is three.

There is an alternative way to define the height of a rooted tree. We define the **path length** of a vertex in a rooted tree to be the distance of the vertex from the root, which is the same as the number of edges in the path from the root to the vertex. Then the maximum of the path lengths of all the vertices is the height of the rooted tree. Observe that this maximum occurs for one of the leaves of the tree.

Consider the family tree of a man who has two sons, the elder son also has two sons, while the younger one has three sons. This may be represented by the rooted tree Fig. 8.7(a).

(a) The graph of Fig. 8.7(b) is isomorphic to the tree as at (a) but (b) may be the family tree of the man who has two sons and the elder son has three sons while younger one has two. This suggests introducing the concept of ordered trees which enables us to refer to a subtree of a branch node in an unambiguous manner. An **ordered tree** is a rooted tree with the edges incident from each branch node labelled with integers $1, 2, \ldots, i, \ldots$ the first, the second, the third, etc. The rooted trees as in Fig. 8.7(a) and (b) can be labelled as in Fig. 8.7(c) and (d).

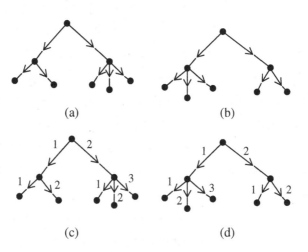

Fig. 8.7.

Two ordered trees are said to be **isomorphic** if there is a one–one correspondence between their vertices and edges that preserves the incident relations and if the labels of the corresponding edges match. It follows from the above definition that trees given in Fig. 8.7(c) and 8.7(d) above are not isomorphic. An ordered tree in which every branch node has at most m sons is called an m-**ary tree** or an m-**tree**. Also, an m-ary tree is said to be **regular** or **complete** or **full** m-**ary** if every one of the branch nodes has exactly m sons. A tree in which every branch node has at most two sons is called a **binary tree**. In the case of a binary tree, its subtrees are referred to as **left subtree** and the **right subtree** instead of the first, second subtree, etc. In a binary tree one node (the root) has degree 2 and all other vertices have degree 1 or 3. An m-tree of height k is called **balanced** if all the leaves are *at* level k or $k - 1$.

Labelled binary trees and/or ordered binary trees also arise while making computations (on computers). A tree is called **labelled** if its vertices and/or edges are assigned certain labels. Consider, for example, the algebraic expression

$$(2 - (3 \times x)) + ((x - 3) - (2 + x)). \tag{8.2}$$

We know that this expression cannot be simplified until we have simplified the expressions $(2 - (3 \times x))$ and $((x - 3) - (2 + x))$. Thus the operation '+' appearing between these two expressions is a central one and cannot be performed without simplifying the two expressions as above. So far as the simplification of the expression $((x - 3) - (2 + x))$ is concerned, the operation '−' appearing between the expressions $x - 3$ and $2 + x$ is central and cannot be performed until $x - 3$ and $2 + x$ are simplified. Similarly, the sign '−' appearing in the expression $2 - (3 \times x)$ is central and cannot be performed unless $3 \times x$ has been computed. Observe that we can represent the expressions $x - 3$ and $2 + x$ as the trees in Fig. 8.8(a) and (b), the first one indicating that the number 3 appearing on the right has to be subtracted from x, which is appearing on the left. Since $x - 3$ is not the same as $3 - x$, we cannot represent $x - 3$ as in Fig. 8.8(c), clearly indicating that the order in which the vertices are labelled is important and we cannot change these at will. We need also to remember the laws followed in mathematics, in which the operations $+, -, \times$ and $\div$ are to be performed. As we know that $a \div b$ is not the same as $b \div a$, we can represent $a \div b$ as in Fig. 8.8(d) and we cannot interchange the labels a, b on the vertices in the tree. We can

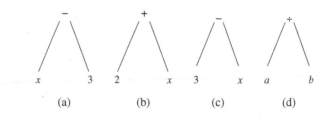

(a)   (b)   (c)   (d)

Fig. 8.8.

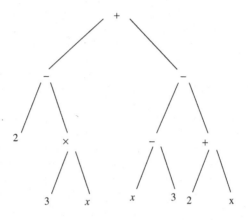

Fig. 8.9.

represent the full expression (8.2) as in the graph in Fig. 8.9, which is a
level 3 rooted tree (ordered and labelled).

We will now express some simple arithmetic expressions as ordered
trees. Consider the following expressions.

$$(a+b)*c, \quad c*d*e, \quad a+b*c, \quad (a+b*c)*d-e,$$

$$((a-b*c)*d+e)/(f+g), (((a-b*c)*d+e)/(f+g))+h*i*j.$$

The ordered trees corresponding to these expressions are, respectively, the
trees given in Fig. 8.10(a)–(f).

We next construct rooted trees in Fig. 8.11 corresponding to the
algebraic expressions

$$(3-(2\times x))+((x-2)-(3+x));$$

$$((3\times(1-x))\div((4+(7-(y+2))))\times(7+(x\div y))).$$

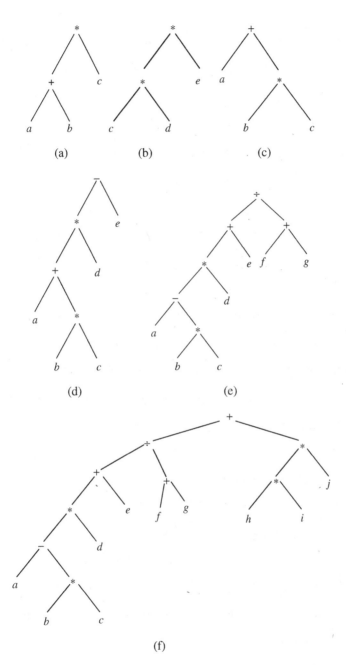

Fig. 8.10.

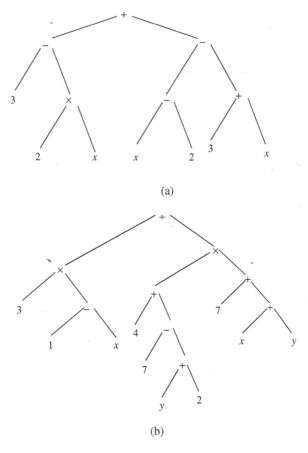

Fig. 8.11.

Remark. Given any tree T, we may choose any one of its vertices as a root and then obtain a rooted tree out of it. For example, consider the tree of Fig. 8.2 and choose e or n as a root. The corresponding rooted trees are given in Fig. 8.12.

Theorem 8.7. If I denotes the sum of the path lengths of all the branch nodes and E denotes the sum of the path lengths of the leaves in a regular m-trees T, then $E = (m - 1)I + mi$, where i denotes the number of branch nodes in the tree.

Proof. Let k be the height of the regular m-tree T. Then, for any $j, i \leq j \leq k$ the number of vertices at level j is m^j. In particular the number of leaves in

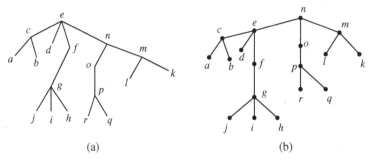

Fig. 8.12.

T is m^k and the number of all branch nodes is $i = 1+m+m^2+\cdots+m^{k-1} = (m^k - 1)/(m - 1)$.

For any vertex at level j, the path length is j. Therefore, the sum of the path lengths of the leaves is $E = k \cdot m^k$. The sum of the path lengths of branch nodes is

$$I = 1.0 + m \cdot 1 + m^2 \cdot 2 + \cdots + m^{k-1}(k - 1)$$

$$= 1 \cdot m + 2 \cdot m^2 + \cdots + (k - 1)m^{k-1}.$$

Then

$$mI = 1 \cdot m^2 + \cdots + (k - 2)m^{k-1} + (k - 1)m^k.$$

Subtracting I from mI, we get

$$(m - 1)I = (k - 1)m^k - (m + m^2 + \cdots + m^{k-1})$$

$$= (k - 1)m^k - (m(m^{k-1} - 1))/(m - 1)$$

$$= ((km^k - m^k)(m - 1) - m^k + m)/(m - 1)$$

$$= km^k - (m(m^k - 1))/(m - 1)$$

$$= E - (m(m^k - 1))/(m - 1) = E - mi$$

or $E = (m - 1)I + mi$.

Corollary. With E, I and i as above for a binary regular tree, $E = I + 2i$.

Example 8.5. Let $V = \{t, u, v, w, x, y, z\}$, $T = \{(t, u), (u, w), (u, x), (u, v), (v, z), (v, y)\}$. Is the graph (V, T) a tree? If yes, find the root of this tree.

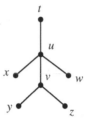

Fig. 8.13.

Solution. There are seven vertices and six edges in the graph. Therefore, this graph can be a tree. If it is indeed a tree, we have to show that there is no circuit. Plotting all the vertices and edges as given, we find that it is a connected graph with no circuits. Hence the graph is indeed a tree. Since there are no paths from x, y, z, u, v, w to t, none of x, y, z, u, v, w is a root. Hence t is the root of this tree. The tree may be represented as in Fig. 8.13. Observe that u is the only vertex of level 1, x, v, w are the vertices of level 2 and y, z are the vertices of level 3. Also, the tree is of height 3.

Example 8.6. If T is a complete n-tree with exactly three levels, prove that the number of vertices of T must be $1 + kn$, where $2 \le k \le n + 1$.

Solution. There is exactly one vertex at level 0, which is the root. The tree being a complete n-tree, there are exactly n vertices at level 1. However, the tree has three levels in all. Therefore, there are some vertices at level 2. The minimum number of vertices at level 2 is n and the maximum number of such vertices is obtained when every vertex at level 1 has offsprings. Each one of these will have exactly n offsprings. Therefore, the maximum number of vertices at level 2 is n^2. If k vertices at level 1 have offsprings, the number of vertices at level 2 shall be kn. Therefore, the total number of vertices that the tree has is equal to $1 + n + kn = 1 + (k + 1)n$, where $1 \le k \le n$. This number is the same as $1 + kn$, where $2 \le k \le n + 1$.

Example 8.7. If T is a regular (complete) n-tree having height $k \ge 1$, prove that the number of vertices of T must be $1 + rn$, where $k \le r \le 1 + n + n^2 + \cdots + n^{k-1}$.

Solution. The proof follows on the same lines as above or we may proceed by induction on k. If $k = 1$, then the number of vertices is clearly $1 + n$. Suppose that the result is true if the height of the tree is $k \ge 1$. Suppose that the height is $k + 1$. At minimum, one vertex at level k has n offsprings.

Therefore, the number of vertices will be $\geq 1 + kn + n = 1 + (k + 1)n$. If l vertices at level k have n offsprings, then the vertices at level $k + 1$ shall be ln. The number of vertices up to level k are $1 + rn$, where $k \leq r \leq 1 + n + n^2 + \cdots + n^{k-1}$. The maximum number of vertices at level k is n^k. Therefore, the number of vertices in the tree is $\geq 1 + (k + 1)n$ and $\leq 1 + rn + n^{k+1}$ or that the number of vertices is $1 + rn$, where $(k + 1) \leq r \leq 1 + n + n^2 + \cdots + n^k$. This completes induction. Hence the number of vertices of a regular n-tree having height $k \geq 1$ is $1 + rn$, where $k \leq r \leq 1 + n + n^2 + \cdots + n^{k-1}$.

Example 8.8. An inter-university basketball tournament begins with 32 teams. One loss eliminates a team. Suppose that the results of the tournament are represented in a binary tree whose leaves are labelled with the original teams and the interior vertices are labelled with the winner of the game between the children of the vertex. Determine the height of the tree created.

Solution. The 32 teams can be divided into 16 pairs of teams and there being a match between each pair, there are 16 winners. These 16 winning teams can be divided into eight pairs of teams and there being a match between teams of each pair, there result eight winners. These eight winning teams are further subdivided into four pairs of teams. There being a match between the two teams of each pair, we get four winners. The four winning teams are divided into two pairs of teams and there being a match between teams of each pair we get two winners. The two winning teams play the final match and the winning team (i.e. the champion) is obtained. Observe that the tree created is a regular (i.e. complete) binary tree. In a complete binary tree of height k, the number of leaves is 2^k. Here the leaves are the

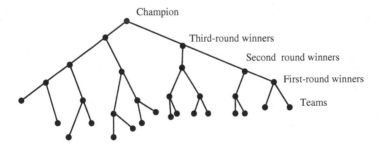

Fig. 8.14.

32 teams taking part in the tournament. Hence $2^k = 32 = 2^5$, which gives $k = 5$. Thus the tree created is of height 5.

The number 32 being large, we draw a binary tree in Fig. 8.14 for the above example with 16 players.

Example 8.9. In a regular n-ary tree with k branch nodes and t leaves, $(n-1)k = t - 1$.

Solution. The tree being regular n-ary, each branch node has n sons. Therefore, the total number of vertices in the tree is $nk + 1$ (one is added for the root). But the total number of vertices is $k+t$. Therefore, $nk+1 = k+t$, which gives $(n-1)k = t - 1$.

Example 8.10. The number of vertices in a binary tree is always odd.

Solution. Let n be the number of vertices in a binary tree T. Then exactly one of the vertices is of degree 2 and all other vertices (i.e. $n-1$) are of odd degree. But the number of vertices of odd degree in any connected graph is even. Therefore, $n-1$ is even and, hence, n is odd.

Example 8.11. The number of pendant vertices p in a binary tree with n vertices is $(n+1)/2$ and the number of internal vertices is $p - 1$.

Solution. There being one vertex of degree 2 and all other branch nodes being of degree 3, the number of edges in the tree is equal to

$$\frac{1}{2}\{p + 3(n - p - 1) + 2\} = n - 1$$

or $(3n - 2p - 1) = 2n - 2$ or $2p = n + 1$ giving $p = (n+1)/2$. The number of internal vertices $= n - p = n - (n+1)/2 = p - 1$.

8.3. Tree Searching or Traversing a Tree

We have considered representing an arithmetical expression as a binary rooted labelled tree. The representation of the expression as a tree will be unambiguous if there is a way to go over or visit the vertices of the tree along the edges in such a way that we get back the given mathematical expression. In general, given a binary rooted labelled tree, we may apply the procedure of visiting the vertices going along the edges to get some simple expression. This simple expression may be represented as a tree. The problem is that the tree we get should be the one we started with. We shall now discuss three procedures or methods of traversing a tree which have the desired property that the expression obtained when represented as a tree gives the tree we start with and conversely.

Let T be a binary rooted labelled tree with vertex v (say) as a root. The root v has two offsprings, say v_L and v_R, where v_L is the left offspring and v_R the right offspring. Let $(T(v_L), v_L)$ be the subtree of T with v_L as a root and let $(T(v_R), v_R)$ be the subtree of T with v_R as a root. We call $(T(v_L), v_L)$ as the left subtree of T and $(T(v_R), v_R)$ as the right subtree of T. We can then talk of the left and the right subtrees of $(T(v_L), v_L)$ as well as that of $(T(v_R), v_R)$, and so on. The three traversing procedures that we would like to discuss are called the **preorder traversal, inorder traversal** and the **postorder traversal.**

Algorithm: preorder

Step 1: Visit the vertex v.
Step 2: If v_L exists, apply the algorithm to $(T(v_L), v_L)$.
Step 3: If v_R exists, apply the algorithm to $(T(v_R), v_R)$. Stop.

The algorithm says the following:

1. Visit the root.
2. Traverse the left subtree of T (by using the procedure) if it exists.
3. Traverse the right subtree of T (by using the procedure) if it exists.

The term 'by using the procedure' given in parenthesis in 2 and 3 is meaningful and it will be made clear in the examples below.

Example 8.12. Let T be the tree given in Fig. 8.15(a). The root of the tree is the vertex labelled A. While preparing an expression on traversing a tree, when a vertex v is visited, the same is printed. In preorder traversing the root is the first vertex to be visited. So A is printed. Then we go to the left subtree which is again a rooted labelled tree with B as its root. For traversing this subtree, again, the root is the first vertex to be visited. Thus B is the next symbol to be printed. The printing so far is AB. The offsprings of B are C and D and it is the subtree with C as its vertex that is to be traversed first. But C being a leaf we need to visit C and, so, C is printed. The printing so far being ABC. For traversing the subtree rooted at B, its left subtree has been taken care of. So, we need to go to the right subtree of B. This is a rooted tree with D as root and this root is to be visited first. So, D is printed next. The printing thus far becomes $ABCD$.

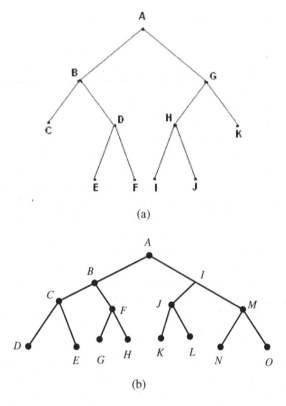

(a)

(b)

Fig. 8.15.

The left subtree of D consists of a single vertex E, which should be printed next. The right subtree of D also consists of a single vertex F which is the next one to be visited and printed. Thus the left subtree of A has been completely traversed and the printing so far is $ABCDEF$. Next, it is the right subtree of A that is to be traversed. It is a rooted tree with G as a root. For this traversing, G is to be printed first of all. Then left subtree of G is a rooted tree with root H which is to be visited first of all and so H is printed next. So the traversing of the right subtree of A yields the printing GH. The left subtree of H has only one leaf, namely I which is to be visited and so, printed next. Then the right subtree of H consisting of J alone is to be visited. Thus J is the next letter to be printed. Thus the traversing of the left subtree of G is complete and the printing so far is $GHIJ$. It is the right subtree of G that is to be traversed. It consists of a single vertex K which is visited and, so, printed. Thus, traversal of the right subtree of

A is complete and the printing so far is *GHIJK*. Combining the two lists obtained we get the string *ABCDEFGHIJK*, which gives the order in which the tree is to be traversed in the preorder procedure.

Example 8.13. Consider the rooted labelled tree *T* as in Fig. 8.15(b). The root of this tree is *A*. After *A* is visited, it is the left subtree of *A* that is to be traversed. This subtree is again rooted with root *B*. Thus *B* is the next vertex to be visited. After visiting *B*, it is the left subtree of *B* that is to be traversed. The left subtree of *B* is a rooted tree with *C* as root. So, *C* is visited next. After this, again it is the left subtree of *C*, which is to be traversed. But this subtree consists of a single vertex, *D* which is to be visited. After visiting *D*, the right subtree of *C* which consists of a single vertex *E* needs to be visited. Thus the traversing of the left subtree of *B* is complete and the printing so far gives the string *ABCDE*. We next need to traverse the right subtree of *B*. This is a rooted tree with *F* as its root, the left subtree of *F* is a single vertex *G* and then the right subtree of *F* is the tree with a single vertex *H*. Thus traversing the right subtree of *B* leads to a string *FGH*. Combine this with the string obtained by traversing the left subtree of *B* and, thus, the left subtree of *A* leads to the string *ABCDEFGH*. We next obtain the string obtained on traversing the right subtree of *A*.

The right subtree of *A* is a rooted tree with root *I*. After *I* is visited, we have to go to the left subtree of *I* which is a rooted subtree with *J* as root. After visiting *J* and, so listing *J*, we have to go to the left subtree of *J* and it consists of a single vertex *K*. After *K* is visited and listed, we have to traverse the right subtree of *J* which also consists of a single vertex *L*. On visiting and listing *L*, the left subtree of *I* leads to the string *IJKL*.

We next need to traverse the right subtree of *I*. This is a rooted tree with root *M*. Visit the root vertex *M* and then go to the left subtree of *M*. This subtree consists of a single vertex *N*, which is then visited and listed. Also after visiting and listing *N*, we need to traverse the right subtree of *M*. This subtree also consists of a single vertex *O* which is then visited and listed. Thus traversing the right subtree of *A* gives a sequence *IJKLMNO*. Combining this with the string obtained on traversing the left subtree of *A*, we get the string *ABCDEFGHIJKLMNO*.

Example 8.14. Let *T* be the tree as given in Fig. 8.11(a). It is a rooted tree with root '+', then the left subtree has root '−', the left subtree of which is vertex with label 3, then we have the right subtree with root '×', the left subtree of which is just a vertex labelled 2 and the right subtree is a vertex labelled x. Thus the left subtree of T leads to a string '$+-3 \times 2x$'. The right subtree of T leads to a string $--x2 \div 3x$. Combining the two strings, the string obtained on traversing the tree in preorder process is $+-3 \times 2x--x2 \div 3x$. This is called the **prefix** or **Polish form** of the algebraic expression represented by the given tree. In this string observe that $+,-,\times,-,-,\div$ are operators or operations.

Example 8.15. Let T be the tree as in Fig. 8.11(b). It is rooted with root '$\div$', the left subtree has root '$\times$', its left subtree has entry '3' etc. The left subtree leads to the string $\div \times 3-1x$. The right subtree gives the string $\times+4-7+y2+7 \div xy$. Combining the two, we get the final string represented by the tree as $\div x3-1x \times +4-7+y2+7 \div xy$ which is the prefix or the Polish form of the string.

Before we consider the inorder and postorder traversing of tree, we consider the method of obtaining an algebraic expression from a given prefix or Polish form of a string. When two operators come together, it is understood that the first operator is the root of the rooted (sub) tree and the second operator is the root of the left subtree of that (sub) tree. Examine the string and replace every triplet consisting of an operator $*$ (say), an entry or variable x (say), an entry or variable y (say), by $x*y$. Continue doing so until we get an algebraic expression. Consider, for example the string

(a) $+-3 \times 2x--x2+3x.$

This gets replaced by the string

$$+-3(2 \times x)-(x-2)(3+x).$$

which in turn gets replaced by

$$+(3-(2 \times x))((x-2)-(3+x))$$

and finally we get

$$(3-(2-x))+((x-2)-(3+x))$$

as the algebraic expression represented by the string. This is the same algebraic expression which was represented by the tree as in Fig. 8.11(a).

(b) $\qquad \div \times 3 - 1x \times +4 - 7 + y2 + 7 \div xy.$

Replacing the triplets as stated, this string gets replaced by

$$\div \times 3(1 - x) \times +4 - 7(y + 2) + 7(x \div y).$$

The above string gets replaced by

$$\div(3 \times (1 - x)) \times +4(7 - (y + 2))(7 + (x \div y)).$$

This in turn gets replaced by

$$\div(3 \times (1 - x)) \times (4 + (7 - (y + 2)))(7 + (x \div y)).$$

This leads to the string

$$\div(3 \times (1 - x))((4 + (7 - (y + 2))) \times (7 + (x \div y)))$$

which finally leads to the algebraic expression

$$(3 \times (1 - x)) \div ((4 + (7 - (y + 2))) \times (7 + (x \div y))).$$

This is the algebraic expression which led to the graph of Fig. 8.11(b).

Given an algebraic expression, we constructed a rooted tree. Traversing this rooted tree by the preorder procedure, we obtain a prefix or Polish form of the corresponding string from which on using the above procedure, we get back the algebraic expression we started with. On the other hand, given a rooted tree in which the branch nodes are labelled with the mathematical operations $+, -, \times, \div$ and the leaves are either numbers or variables, we can get a string involving these operations and numbers/variables using preorder procedure. Then convert it to an algebraic expression using the procedure detailed above. We then construct a rooted tree corresponding to this algebraic expression — in this way we get back the tree we started with.

We next describe the inorder and postorder traversal.

Let T be a rooted labelled tree. For a root vertex v, let v_L and v_R denote the left and right offsprings respectively of v. Let $(T(v_L), v_L)$ and $(T(v_R), v_R)$ denote, as before, the left and right subtrees of T.

Algorithm: inorder

Step 1: Traverse the left subtree $(T(v_L), v_L)$, if it exists.
Step 2: Visit the vertex v.
Step 3: Traverse the right subtree $(T(v_R), v_R)$, if it exists.

Algorithm: postorder

Step 1: Traverse the left subtree $(T(v_L), v_L)$, if it exists.
Step 2: Traverse the right subtree $(T(v_R), v_R)$, if it exists.
Step 3: Visit the vertex v.

The two algorithms may be written in short as:

Left subtree, vertex, right subtree and left subtree, right subtree, vertex. In this vein, the preorder system may be expressed as vertex, left subtree, right subtree.

We now traverse the same trees for which we considered using the preorder process by using the inorder and postorder processes.

Example 8.16. Let T be the tree as in Fig. 8.15(a) (cf. Fig. 8.16 for notations). We first traverse this tree using the inorder process.

To traverse the tree T, we have first to traverse the left subtree T_1 with B as root. Then, we first traverse the subtree T_2, then visit the vertex B, then traverse the right subtree T_3, then the left subtree T_4, then visit the vertex D and finally the right subtree T_5. This traversal gives the string $CBEDF$. After this we need to visit the vertex A followed by the traversal of the right subtree T_6. As in the case of the left subtree T_1, the traversal of the right subtree T_6 leads to the string $IHJGK$. Therefore, the string on traversing the tree T is $CBEDFAIHJGK$.

We next obtain the string on traversing this tree using postorder procedure. For this we search the left subtree T_1 as follows $T_2 \to T_3 \to (T_4 \to T_5) \to D \to B$ which leads to the string $CEFDB$. Traversing the right subtree T_6 leads to the string $IJHKG$ combining the two strings followed by visiting the root A, we get the string $CEFDBIJHKGA$.

Example 8.17. Consider the tree T as in Fig. 8.15(b). Postorder traversals of the left and right subtrees of T give the strings $DECGHFB$ and $KLJNOMI$ respectively. Then the traversal of the tree in the postorder process gives the string $DECGHFBKLJNOMIA$. In the inorder process, the traversal of the left and right subtrees leads to the strings $DCEBGFH$, $KJLINMO$ and so, the inorder process of traversing the tree T gives the string $DCEBGFHAKJLINMO$.

Example 8.18. Consider the tree T as given in Fig. 8.11(a). The inorder process of traversing this tree gives the string $(3 - (2 \times x)) + ((x - 2) - (3 + x))$ which is the algebraic expression to represent which the tree T was constructed. The postorder process of traversing this tree leads to the

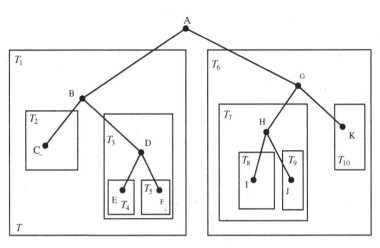

Fig. 8.16.

string $32x \times -x2 - 3x + - +$. The sequence or string obtained from a tree representing an algebraic expression by postorder process is called the **postfix** or the **reverse Polish form** of the expression. To obtain the algebraic expression from its postfix or reverse Polish form, we need to look for triplets of the form $xy*$ where x, y are numbers or variables and $*$ is one of the mathematical operations $+, -, \times, \div$ and replace it by $x * y$. For applying this procedure, we rewrite the string as $32x \times -x2 - 3x + - +$ and can get the new string as $3(2 \times x) - (x - 2)(3 + x) - +$. This string leads to $(3 - (2 \times x))((x - 2) - (3 + x))+$ which finally gives the algebraic expression $(3 - (2 \times x)) + ((x - 2) - (3 + x))$.

Given a string in postfix or reverse Polish form, we can construct a tree as follows: either convert it into an algebraic expression using the procedure given above or look for a portion of the string which ends in two consecutive mathematical operations. The first such substring gives a left subtree, the right most symbol is the root of the tree represented by the given string. The string normally ends in three mathematical operations symbols and the substring starting after the first substring giving the left subtree leading up to the last but one symbol gives the right subtree.

Given a rooted tree in which the branch nodes are labelled by the mathematical operations and the leaves are labelled by numbers or variables it is always a tree of an algebraic expression which can be obtained by traversing the tree by inorder process.

Example 8.19. Finally, we consider the tree given in Fig. 8.11(b). The inorder traversing process leads to the string $(3 \times (1 - x)) \div ((4 + (7 - (y + 2))) \times (7 + (x \div y)))$ which is the algebraic expression using which the tree as in Fig. 8.11(b) was constructed. The postorder traversing process leads to the string $31x - \times 47y2 + - + 7xy \div + \times \div$.

As seen above, given a string involving numbers and/or variables and the mathematical operations $+, -, \times, \div$ in the prefix (Polish) form or postfix (reverse Polish) form, we can always convert it into an algebraic expression which is then used to construct a rooted tree, traversing of which leads to the given string. In the general case, if we are given a listing on the traversing of a tree in one of the three processes namely, preorder, inorder or postorder, is it always possible to construct the tree uniquely using the given listing? Consider the trees of Fig. 8.17.

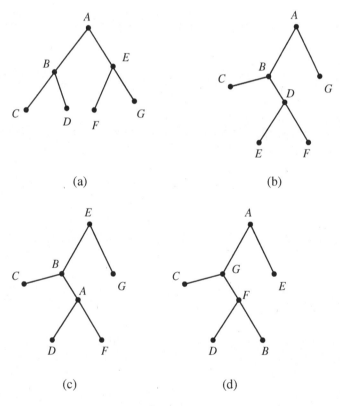

Fig. 8.17.

Then the listings given on traversing these trees in the three processing orders are

(a) $ABCDEFG$ (preorder traversing)
    $CBDAFEG$ (inorder traversing)
    $CDBFGEA$ (postorder traversing).
(b) $ABCDEFG$ (preorder traversing)
    $CBEDFAG$ (inorder traversing)
    $CEFDBGA$ (postorder traversing).
(c) $EBCADFG$ (preorder traversing)
    $CBDAFEG$ (inorder traversing)
    $CDFABGE$ (postorder traversing).
(d) $AGCFDBE$ (preorder traversing)
    $CGDFBAE$ (inorder traversing)
    $CDBFGEA$ (postorder traversing).

Observe that all these are distinct labelled trees but (a) and (b) have the same preorder listing, (a) and (c) have the same inorder listing and (a) and (d) have identical postorder listing. Thus given a listing in any of the three procedures, we cannot construct from it a unique tree. However, if we are given two traversals, one of which is inorder traversal and the other is either preorder traversal or postorder traversal, then we can construct a unique binary tree.

Procedure 8.1

Let preorder and inorder traversal sequences/strings be given.

1. The first entry in the preorder string is the root of the tree.
2. Locate the root in the inorder string.
3. Then, the substring to the left of the root in the inorder string is the inorder traversal string of the left subtree and the substring to the right of the root in the inorder string is the inorder string of the right subtree of the tree.
4. Use the inorder strings of the left and the right subtrees to locate the preorder strings of the left and the right subtrees from the given preorder string of the tree.
5. The preorder and inorder strings of both the left subtree and the right subtree are now known. Use steps 1 to 4 to obtain the preorder and inorder strings of the left and the right subtrees of the left subtree as also the right subtree.

Continue the process until no further substrings are possible.

Procedure 8.2

Let inorder and postorder traversal sequences/strings be given.

1. The last entry in the postorder string is the root of the tree.
2. Locate the root in the inorder string.
3. Then, the substring to the left of the root in the inorder string is the inorder string of the left subtree of the tree and the substring to the right of the root in the inorder string is the inorder string of the right subtree of the tree.
4. Use the inorder strings of the left and right subtrees to locate the postorder strings of the left and the right subtrees from the given postorder string of the tree.
5. The inorder and postorder strings of both the left subtree and the right subtree are now known. Use steps 1 to 4 to obtain the inorder and postorder strings of the left and the right subtrees of the left subtree as also the right subtree obtained in steps 3 and 4.

Continue the process until no further substrings are possible.

The above procedures are best explained through examples.

Example 8.20. Given the preorder and inorder traversal of a binary tree, construct the tree.

$$\text{Preorder}: 6\ 2\ b\ 1\ 3\ 7\ a\ 8\ 4\ 5\ c$$
$$\text{Inorder}\ \ :b\ 2\ 3\ 1\ 7\ 6\ 8\ a\ 5\ 4\ c.$$

Solution. The first entry in the preorder string being 6, 6 is the root of the tree. Then the inorder strings of the left and the right subtree are

$$b\ 2\ 3\ 1\ 7 \qquad 8\ a\ 5\ 4\ c.$$

Comparing the entries in these strings, we find that the preorder strings of the left and the right subtree are

$$2\ b\ 1\ 3\ 7 \qquad a\ 8\ 4\ 5\ c.$$

Thus, for the left and the right subtrees, the preorder and inorder strings are:

$$\text{Preorder :} \quad 2\ b\ 1\ 3\ 7 \mid a\ 8\ 4\ 5\ c$$
$$\text{Inorder \ \ :} \quad b\ 2\ 3\ 1\ 7 \mid 8\ a\ 5\ 4\ c.$$

For the left subtree, two is the root whereas a is the root of the right subtree. Left subtrees of both of these are one element trees and the right subtrees have the preorder and inorder strings as below

$$\text{Preorder :} \quad 1\ 3\ 7 \mid 4\ 5\ c$$
$$\text{Inorder \ \ :} \quad 3\ 1\ 7 \mid 5\ 4\ c.$$

These subtrees are, respectively given by Figs. 8.18(a) and 8.18(b). Collecting the above information we find the tree to be as in Fig. 8.18(c).

Example 8.21. Given the inorder and postorder traversal of a binary tree, construct the tree.

$$\text{Inorder \ \ \ :} \quad p\ n\ q\ m\ r\ \boxed{l}\ t\ s\ v\ u\ w$$
$$\text{Postorder :} \quad p\ q\ n\ r\ m\ t\ v\ w\ u\ s\ \boxed{l}.$$

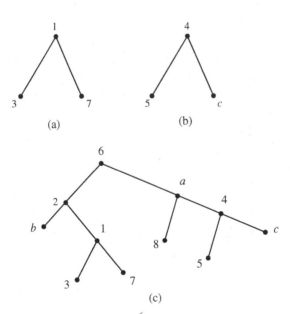

(a) (b)

(c)

Fig. 8.18.

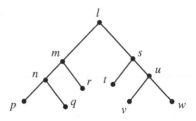

Fig. 8.19.

Solution. The last entry in the postorder string being the root, the root of the tree is l. Then inorder and postorder strings of the left and the right subtrees are

$$\text{Inorder} \quad : \quad p\ n\ q\ \boxed{m}\ r\ \big|\ t\ \boxed{s}\ v\ u\ w$$
$$\text{Postorder} : \quad p\ q\ n\ r\ \boxed{m}\ \big|\ t\ v\ w\ u\ \boxed{s}.$$

The left subtree of the left subtree and the right subtree of the right subtree have inorder and postorder strings as

$$\text{Inorder} \quad : \quad p\ \boxed{n}\ q\ \big|\ v\ \boxed{u}\ w$$
$$\text{Postorder} : \quad p\ q\ \boxed{n}\ \big|\ v\ w\ \boxed{u}.$$

The roots of the subtrees have been enclosed in squares. The required tree that can be read from the above information is constructed in Fig. 8.19.

In order to avoid any confusion between the variable x and the multiplication symbol $\times$, we write $*$ for the multiplication operation. Also the operation giving powers is denoted by $\uparrow$, for example, $2^3, 3^2$ etc. are represented as $2 \uparrow 3, 3 \uparrow 2$ etc. With these notations we now calculate the values of some prefix and postfix expressions.

Example 8.22. Find the value of each of the following

(a) Prefix expressions

   (i) $\uparrow - * 33 * 425$;

   (ii) $+ - \uparrow 32 \uparrow 23/6 - 42$;

   (iii) $* + 3 + 3 \uparrow 3(3+3)3$.

(b) Postfix expressions

   (i) $521 - -314 + +*$;

   (ii) $93/5 + 72 - *$;

   (iii) $32 * 2 \uparrow 53 - 84/ * -$.

Solution.

(a) (i)
$$\uparrow - * 33 * 425 = \uparrow -(3*3)(4*2)5$$
$$= \uparrow -985$$
$$= \uparrow (9-8)5 = \uparrow 15 = 1 \uparrow 5 = 1^5 = 1.$$

(ii)
$$+ - \uparrow 32 \uparrow 23/6 - 42 = + - (3 \uparrow 2)(2 \uparrow 3)/6(4-2)$$
$$= + - (3^2)(2^3)/6 \ 2$$
$$= + - 98(6/2)$$
$$= +(9-8)3 = +13 = 1 + 3 = 4.$$

(iii)
$$* + 3 + 3 \uparrow 3 + 333 = * + 3 + 3 \uparrow 3(3+3)3$$
$$= * + 3 + 3 \uparrow 363$$
$$= * + 3 + 3(3 \uparrow 6)3$$
$$= * + 3 + 3(3^6)3$$
$$= * + 3 + 3(729)3$$
$$= * + 3(3 + 729)3$$
$$= * + 3(732)3$$
$$= *(3 + 732)3 = *(735)3 = 735 * 3 = 2205.$$

(b) (i)
$$521 - -314 + + * = 5(2-1) - 3(1+4) + *$$
$$= 51 - 35 + *$$
$$= (5-1)(3+5) *$$
$$= 48 * = 4 * 8 = 32.$$

(ii)
$$93/5 + 72 - * = (9/3)5 + (7-2) *$$
$$= 35 + 5 *$$
$$= (3+5)5 * = 85 * = 8 * 5 = 40.$$

(iii)
$$32 * 2 \uparrow 53 - 84/ * - = (3 * 2)2 \uparrow (5-3)(8/4) * -$$
$$= 62 \uparrow 22 * -$$
$$= (6 \uparrow 2)(2 * 2) -$$
$$= (6^2)(4) - = (36)4 - = 36 - 4 = 32.$$

We have considered obtaining an expression given its prefix or postfix form. Given arithmetical expression, how to find its prefix or postfix form? For this, given an expression, construct a binary tree representing the expression. Then obtain the prefix or the postfix form from the tree

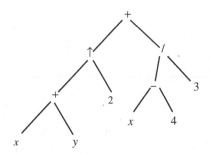

Fig. 8.20.

constructed. For example, consider an expression

$$((x + y) \uparrow 2) + ((x - 4)/3).$$

The binary tree representing this expression is as in Fig. 8.20.
Then the preorder and postorder traversals of this tree give

Prefix form : $+ \uparrow + xy2/ - x43$
Postfix form : $xy + 2 \uparrow x4 - 3/ + $.

On the other hand, given an expression in the prefix or postfix form, we
can get the corresponding arithmetical expression. Once the arithmetical
expression is obtained, we can construct the tree representing this
expression and then using the preorder traversal or the postorder traversal,
we can obtain the required prefix or postfix expression, as the case may be
for the given postfix or prefix expression.

8.4. Applications of Trees

Trees have applications in various disciplines. We consider only some of
these without going into greater detail.

8.4.1. *Prefix codes*

Consider a problem in telecommunication, where messages are to be
transferred from one station to another. For this purpose messages are first
to be converted into strings or sequences of 0s and 1s because these are the
two symbols recognized by the machine (computer). For this conversion of
message each character of the English alphabet is to be associated with
a string of 0s and 1s. If strings of the same length are associated with
every alphabet, there being 26 letters and $2^4 < 2\ 6 < 2^5$, each letter may

be represented as a binary sequence of length 5. There are two problems associated with communication and these are the storage capacity and transmission time. The goal is to minimize the requirement of storage capacity and also to minimize the transmission time. We are not considering here the problem of minimizing error in transmission. Among the English alphabet there are some letters which occur more frequently than the others. For example, a and e appear much more frequently than q and z. To achieve minimization of storage space and transmission time, the letters which occur more frequently are represented by binary strings of smaller length (less than 5) and ones which occur less frequently by larger binary strings.

In the case of fixed length code (i.e. the code where every alphabet is associated with a length 5 binary sequence), on the receiving end, the string received is divided into substrings each of length 5 and then the message received in the form of binary string is converted into a message in the English language. However, in the case of variable length code (i.e. where different alphabets are represented as binary strings of not necessarily the same length), conversion of the received binary sequence into English text is not that easy. Moreover, this conversion may not be unique. For example, if we represent the letters a, e, d, m by the sequences 00, 1, 01 and 000 respectively and the sequence received is 0001, then it could be translated as ad and also as me. This problem of decoding can be overcome by using what are called prefix codes.

Definition. A set of binary sequences is called a **prefix code** if no sequence in the set is the initial part (prefix) of another sequence in the set. For example $\{1, 00, 01, 000, 001\}$ is not a prefix code where as $\{000, 001, 01, 1\}$ and $\{000, 001, 01, 10, 11\}$ are prefix codes.

Two problems arise. First, is there an easy way to get a prefix code? And second, given a binary sequence, is it possible to divide it into sequences representing the letters in a message unambiguously if the letters are associated with sequences in a prefix code?

Let a binary tree T be given. Label the edges in the tree as follows: if a vertex a in T has two children, label the edge leading to the left child of a with 0 and the edge leading to the right child of a with 1. Label each leaf of the tree with the binary sequence that is sequence of labels of the edges in the path from the root to the respective leaf. Since no leaf is an ancestor of another leaf, no sequence which is the label of a leaf is a prefix of another sequence which is the label of another leaf. Thus, the set of all sequences which are labels of the leafs of the binary tree is a prefix code.

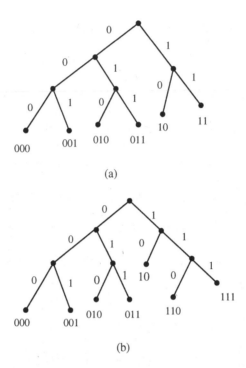

(a)

(b)

Fig. 8.21.

For example, consider the binary trees as in Fig. 8.21. The prefix codes given by the trees of Fig. 8.21(a) and (b) are respectively $\{000, 001, 010, 011, 10, 11\}$ and $\{000, 001, 010, 011, 10, 110, 111\}$. Let a binary prefix code be given. Let h be the length of a largest possible sequence in the code. Construct a full regular binary tree of height h. Observe that the number of sequences in the given prefix code is at most $2 + 2^2 + \cdots + 2^h$ and the number of vertices in a full regular binary tree of height h is $1 + 2 + 2^2 + \cdots + 2^h$. Label the edges of the tree as done earlier, i.e. the edge leading to a left child of a vertex a (say) is labelled 0 and the edge leading to a right child of the vertex a is labelled 1. Also assign to each vertex a of the tree a binary sequence that is the sequence of labels of the edges in the path from the root to that vertex. This way all the binary sequences of length 1 to h appear as labels of vertices in this tree. The sequences in the given prefix code being of length $\leq h$ appear in the labelling of the vertices of the tree. Identify the sequences in the prefix code and encircle the corresponding vertices of the tree. Prune the tree by deleting all the descendents of the encircled vertices along with the edges/paths leading to the descendents. We are then left with a binary tree with the originally encircled vertices as leaves. This is the binary tree, the labels on the leaves of which are the sequences of the given prefix code.

The number of leaves in any binary tree of height h being $\leq$ the number of leaves in a full regular binary tree of height h, it follows that the number of sequences in a binary prefix code is $\leq 2^h$, where h is the length of a largest binary sequence in the prefix code.

We now address the question of decoding a binary sequence. Let a string of 0s and 1s that is received be given. Let also the prefix code used be given. First construct a binary tree the leaves of which are labelled by the sequences in the given prefix code in the manner described. Now, look at the left entry in the given string. Starting from the root of the tree, go along the left edge (or the edge labelled 0) if the entry is 0 and along the other edge if the entry is 1. Then look up the second entry and go further down the tree from the vertex we had arrived at (after the first step). Again go to the left if the second entry is 0 and along the right edge if this entry is 1. Continue moving down in the tree in the above fashion until we arrive at a leaf in the tree. Once a leaf is reached, we have obtained a sequence in the prefix code. Then we look up the rest of the given sequences and proceed exactly the way we have done in obtaining a sequence in the prefix code but starting once again from the root of the tree. On getting to a leaf in the tree, we get another sequence (repetitions are allowed) in the prefix code. Then, again go back to the root of the tree and adopt the procedure for the sub sequence obtained from the given sequence by deleting (or separating) the two sequences in the prefix code already arrived at. Since the string we start with is finite, the process must terminate after a finite number of steps.

Example 8.23. Let a prefix code $\{000, 001, 010, 011, 10, 11\}$ be given. The length of a largest sequence here is three. So we construct a full regular binary tree of height 3 and label the edges by 0 and 1, 0 on an edge to the left child from a vertex and 1 on an edge to the right child from a vertex. We also label the vertices by sequences of 0s and 1s, the sequences given by the paths from the root to the vertices. Encircle the vertices which are labelled by the sequences in the given code. We obtain the tree of Fig. 8.22.

Now, pruning the tree by deleting the descendents of the marked vertices and the edges leading to the descendents, we get the tree (as in Fig. 8.23(a)) representing the given prefix code.

A prefix code may be represented using a binary tree, where the characters are the labels of leaves in the tree. The edges in the tree are

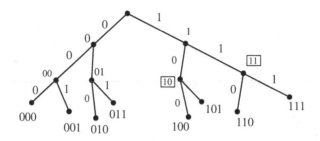

Fig. 8.22.

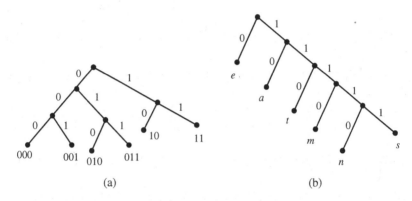

(a)   (b)

Fig. 8.23.

labelled so that an edge leading to the left child of a vertex is labelled 0 and an edge leading to the right child of a vertex is labelled 1. The bit string that is used to encode a character is the sequence of labels of the edges in the path from the root to the leaf. For example, the tree in Fig. 8.23(b) represents the encoding of e by 0, a by 10, t by 110, m by 1110, n by 11110 and s by 11111.

Example 8.24. Consider the prefix code as described in the paragraph above. We like to divide the strings 11111, 110, 10, 0, 1110, 0, 11110, 10, 11111 and 11010111111100 into subsequences which are sequences in the prefix code.

Solution. (a) In the first string we have seven 1s initially. If we follow the path with 1s all through, we find that after five 1s we come to leaf (labelled s). Then going back to the root, we trace 110100111.... We find that two 1s followed by a 0 lead to another leaf (labelled t). Then, going back to the tree we follow the next entries. But the two entries 10 lead to the leaf (labelled a). The next 0 leads us to another leaf (labelled e). The next three 1s followed by 0 lead to a leaf (labelled m). The next 0 leads to the leaf (labelled e). The four 1s and a 0 lead to the leaf (labelled n). The next 10 leads to a leaf (labelled a) and finally the five 1s lead to the leaf labelled s. Thus, the sequences breaks up as $11111, 110, 10, 0, 1110, 0, 11110, 10, 11111$ which can be read as 'staemens'.

(b) The string 110 10 11111 11 0 0 breaks up as $110, 10, 11111, 110, 0$ and reads 'taste'.

We have not so far taken into account the frequencies of occurrence of letters in the text. We now consider the construction of binary trees for the coding of letters along with the frequencies of their occurrence. Consider n symbols (letters) $a_i, 1 \leq i \leq n$, with the frequency of occurrence of a_i being $w_i, 1 \leq i \leq n$. We call w_i the weight of a_i. In order to give a code (known as **Huffman coding**) for these symbols, we start with a forest of n single vertex trees consisting of $a_1, a_2, \ldots, a_n$ along with their weights indicated $w_1, w_2, \ldots, w_n$ respectively. We assume that no two weights are equal. Among these trees, choose vertices a_i, a_j which have the least possible weights with $w_i < w_j$. In the forest, replace these vertices with a rooted tree with w_j as the weight of the left child and w_i of the right child. The root is assigned the weight $w_i + w_j$. The forest is thus reduced to one consisting of $n-2$ isolated vertices and a rooted tree. Among the $n-1$ weights assigned to the $n-2$ isolated vertices and the root of the rooted tree, again pick up two lowest weights and replace these two trees by a single rooted tree, with the tree with the larger weight as the left subtree and the one with smaller weight as its right subtree and assign the sum of weights of these two trees as the weight to the root of the resultant rooted tree. Continue the process until we finally get a single rooted tree in which the given symbols $a_i, a_2, \ldots, a_n$ appear as leaves. Label all the edges of the tree in the usual way namely assign 0 to the edge leaving the root going to the left and 1 to the edge going to the right from the root. In the final tree, find the bit strings which are given by the paths from the root to the leaves of the tree. These strings then give the coding of the symbols $a_1, a_2, \ldots, a_n$.

We can also describe ternary and, in general, *m-ary* Huffman coding by using the symbols 0, 1, 2 for ternary and $0, 1, 2, \ldots, m-1$ for *m-ary* Huffman coding to label the edges. We consider a couple of examples.

Example 8.25. Consider the symbols A, B, C and D with frequencies

$$A\!:\!0.70; \quad B\!:\!0.16; \quad C\!:\!0.09 \quad \text{and} \quad D\!:\!0.05.$$

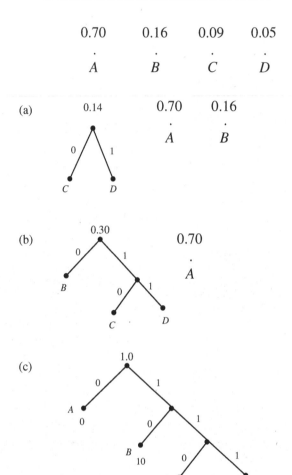

Fig. 8.24.

The various steps needed for the encoding are given by the following

$$
\begin{array}{cccc}
0.70 & 0.16 & 0.09 & 0.05 \\
A & B & C & D
\end{array}
$$

The coding given is $A : 0; B : 10; C : 110; D : 111$. The average number of bits used to encode a symbol using this encoding is $1 \times 0.70 + 2 \times 0.16 + 3 \times 0.09 + 3 \times 0.05 = 0.70 + 0.32 + 0.27 + 0.15 = 1.44$.

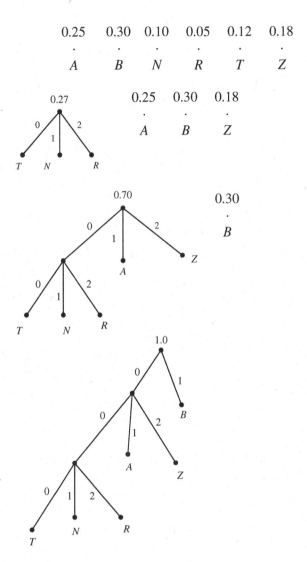

Fig. 8.25.

Example 8.26. Consider forming a ternary Huffman coding to encode the letters with the given frequencies:

$$A : 0.25, \quad B : 0.30, \quad N : 0.10, \quad R : 0.05, \quad T : 0.12, \quad Z : 0.18.$$

The various steps involved for the encoding are given in 8.2.5.

The Huffman Code gives

$$A : 01; \quad B : 1; \quad N : 001; \quad R : 002; \quad T : 000; \quad Z : 02.$$

8.4.2. Binary search trees

A person asks a friend to pick a card from a pack of playing cards. He likes to know or guess the card that has been picked. A straightforward way of asking would be: is the card an ace of spades or a king of spades or a queen of spades and so on. Once the spade cards are exhausted and he has not arrived at the card picked, he may go on asking similar questions with spades replaced by clubs, then diamonds and finally hearts. The number of questions asked can be anything up to 52, a rather tedious way of knowing the card picked up. There is a much more efficient way of knowing the card. The first question he asks could be: Is the card black? to which the friend replies by saying yes or no. If the answer is yes, the person need not worry about the cards of diamond and hearts and the search has been reduced to only 26 cards. If the answer is no, the 26 cards of spades and clubs are ignored. Supposing the answer is yes. The next question shall be: Is the card a spade? If the answer is yes, the 13 cards of clubs are to be ignored and the 13 cards of spades are to be ignored otherwise. Suppose that the answer is yes. Now the strategy changes slightly. Answer to question from now on has to be either 'yes' or 'less than' or 'greater than'. He asks, is the card a '7'? If the answer is yes the card is determined. Otherwise, the choice has been narrowed down to 1 to 6 or 8 to 13. Suppose that the answer is 'less than'. The next question shall be, is it a '3'? If the answer is yes, the card is determined. If the answer is 'less than', the next question shall be is it a '2'? Whatever the answer to this question is, the card is determined. On the other hand, if the answer is 'greater than', the next question shall be, is it a '5'? The answer to this also determines the card for either it is a '5' or less than '5' and so, it is a '4' whereas, if it is 'more than', the only choice is a '6'. Thus the total number of questions to be asked is reduced to 5.

The problem discussed above is known as searching an item from among a given list of items. The given items are assumed to be linearly ordered. For example, these may be numbers, or these may be words in which the ordering is alphabetical or alphanumeric (for example PAN numbers issued by the income tax authority to various taxpayers) and so on. The goal is to implement a searching algorithm that finds items efficiently. This is achieved through the use of a **binary search tree**. It is a binary tree in which each child of a vertex is either called a right or left child, and no vertex has more than one right or left child. Each vertex of this tree is labelled with a key, which is one of the given items. The vertices are assigned keys so that the key of a vertex is larger than the keys of all vertices in its left subtree and is smaller than the keys of all vertices in its right subtree.

For constructing a binary search tree for a list of items, a recursive procedure is used. The first item in the list is assigned as key of the root. To add a new item, (when some have already been added), first compare it with the keys of the vertices already in the tree, starting at the root and moving to the left if the item is less than the key of the respective vertex, if this vertex has a left child, or moving to the right if the item is greater than the key of the respective vertex, if this vertex has a right child. When the item is less than the respective vertex and this vertex has no left child, then a new vertex with this item as its key is inserted as a new left child. Similar step is taken when the item is greater than the respective vertex and this vertex has no right child.

The above construction of a binary search tree is best explained through a couple of examples.

Example 8.27. Construct a binary search tree for the words mathematics, physics, geography, zoology, geology, psychology, Sanskrit, chemistry and botany using alphabetical order.

Solution. Steps used to form this binary search tree are explained in Fig. 8.26 below.

The word mathematics (maths) is the key of the root. As physics (phys) comes after mathematics we go to the right from the root and add the vertex with key physics. The word geography (geog) comes before mathematics and so, go to left from the root and add a vertex with geography (geog) as its key. The word zoology (zool) comes after physics, so we go to right from the vertex with key physics and add a new vertex with key zoology. The word geology (geol) comes before mathematics but after geography. Therefore, we move left from the vertex with key geography to the right and add a new vertex with key geology. The next word in the list is psychology (psy) which comes after mathematics but before zoology. So, from the vertex with key zoology we move to the left and add a vertex with key psychology. The word Sanskrit comes after psychology but before zoology. So we go to right from the vertex with psychology as key and add a new vertex with Sanskrit (skt) as its key. The next word chemistry (chem) comes before mathematics and also before geography. Therefore, we move left from the vertex with key geography and add a new vertex with chemistry as key. Finally, the

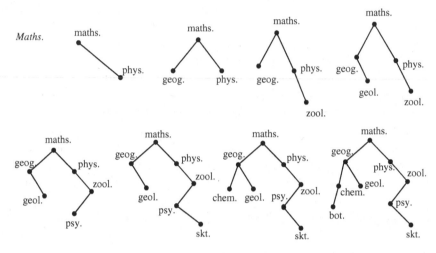

Fig. 8.26.

word botany (bot) comes before chemistry and we move to the left from the vertex with key chemistry and add a new vertex with botany as its key.

Example 8.28. Form a search tree for the words mathematics, physics, geography, zoology, meteorology, geology, Sanskrit and chemistry.

Solution. The steps involved in the construction of the binary search tree are explained in Fig. 8.27 below. Observe that mathematics <

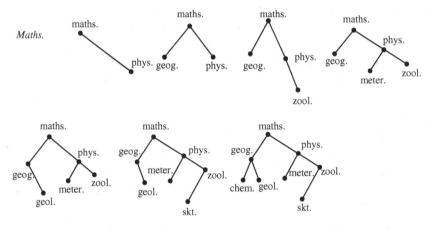

Fig. 8.27.

physics, geography < mathematics, physics < zoology, mathematics < meteorology< physics, geology < mathematics, geography < geology, physics < psychology, psychology < zoology, psychology < Sanskrit, Sanskrit < zoology, chemistry < geography < mathematics, chemistry < geography < geology < mathematics

We use the above information in the steps for construction.

Suppose we are to look for the word 'theta' in the list of items of Examples 8.26 and 8.27. Observe that 'theta' does not appear in either of the two lists. The word mathematics < theta and, so, we need to search for it in the right subtree of the binary search tree. The word theta < zoology and, so, we have to search for it in the left subtree below zoology. However, Sanskrit < theta, the word theta must appear as a right child of Sanskrit. Since it is not in either of the two lists, this needs to be inserted. The two binary search trees then appear as in Fig. 8.28.

In general, let $K_1, K_2, \ldots, K_n$ be n items in ordered list, say $K_1 < K_2 < \cdots < K_n$, which are known as keys. Let x be an item that is to be searched in this list. The search has $2n + 1$ possible outcomes, namely x could be equal to any one of the given n items, or it could be $<K_1$, or is could be $>K_n$ or it could be strictly between some K_i and K_{i+1} for $1 \leq i \leq n - 1$. The binary search tree has n branch nodes but $n + 1$ leaves, the leaves being $K_0, K_1, \ldots, K_n$. For the branch node with label K_i, the left subtree contains only vertices with labels $K_j, j < i$ and its right subtree contains only vertices with labels $K_j, j \geq i$. For example, a binary search tree for the keys K_1, K_2, K_3, K_4, K_5 is given by Fig. 8.29.

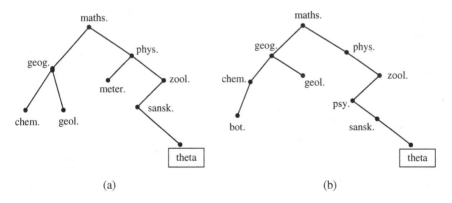

(a)   (b)

Fig. 8.28.

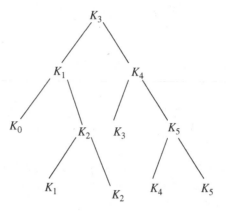

Fig. 8.29.

Observe that there is a variation in this case from the binary search trees of Examples 8.27 and 8.28 above. In the case of Examples 8.27 and 8.28, the nine and eight items are not given in the increasing order although these can be linearly ordered. If we represent these as $k_1, k_2, \ldots$ then the items as given are

(a) $k_5, k_6, k_3, k_9, k_4, k_7, k_8, k_2$ and k_1.
(b) $k_4, k_5, k_2, k_8, k_6, k_3, k_7$ and k_1.

In the sense of the above discussion, binary search tree for (a) need to have nine branch nodes and ten pendant vertices while in the case of (b), the binary search tree need to have eight branch nodes and nine pendant vertices. The two binary search trees are given in Fig. 8.30.

8.4.3. On counting trees

While trying to count the number of structural isomers of the saturated hydrocarbons $C_k H_{2k+2}$, Arthur Cayley, a British mathematician, discovered trees in 1857. He used a connected graph to represent the $C_k H_{2k+2}$ molecules where each carbon atom was represented by a vertex of degree 4 and a hydrogen atom was represented by a vertex of degree 1. The total number of vertices is $k + 2k + 2 = 3k + 2$ and the number of edges is $\frac{1}{2}(4 \times k + 2k + 2) = 3k + 1$. Hence the connected graph is a tree. Thus the problem of counting structural isomers of a given hydrocarbon becomes the problem of counting trees having certain properties. In this direction, Cayley proved the following theorem, which is known as Cayley's theorem.

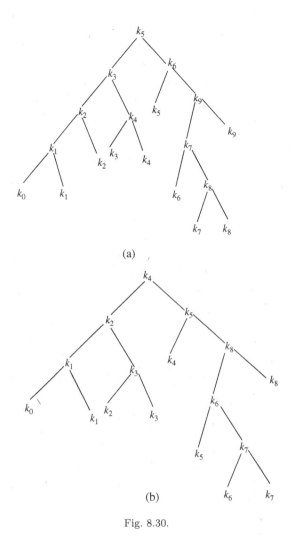

(a)

(b)

Fig. 8.30.

Theorem 8.8. Cayley's theorem. The number of labelled trees with n vertices, $n \geq 2$, is n^{n-2}.

Proof. Let T be a labelled tree with n vertices and suppose that the vertices are labelled with $1, 2, 3, \ldots, n$. Remove the pendant vertex which is labelled with the smallest number along with the edge incident with it. Let a_1 be the label of the removed vertex and b_1 be the vertex adjacent to a_1 so that the edge (a_1, b_1) is removed along with a_1. In the resulting subtree, let

a_2 be the smallest label among the pendant vertices and b_2 be the vertex adjacent to a_2. Remove the vertex with label a_2 along with the edge (a_2, b_2). Repeat this process of removing a pendant vertex (with the smallest label among all the pendant vertices) along with the adjacent edge $n-2$ times so that a subtree with just two vertices remains. This removal process gives a unique sequence $(b_1, b_2, \ldots, b_{n-2})$ of length $n-2$, where $b_1, b_2, \ldots, b_{n-2}$ are positive integers and each one of these is $\leq n$. Observe that $b_1, b_2, \ldots, b_{n-2}$ are not all distinct, for if there are two pendant vertices with labels $1, 2$ and both these are adjacent with the same vertex, say b_1, then $b_2 = b_1$. Conversely, let a sequence

$$(b_1, b_2, \ldots, b_{n-2}) \tag{8.3}$$

of positive integers with every $b_i \leq n$, be given. We construct a labelled tree with labels $1, 2, \ldots, n$ as follows. Determine the first number in the sequence

$$1, 2, 3, \ldots, n \tag{8.4}$$

that does not appear in the sequence (8.3). Let this number be a_1. Then the edge (a_1, b_1) is determined. Remove b_1 from sequence (8.3) and a_1 from the sequence (8.4). Among the resultant subsequence of (8.4), determine the first number that does not appear in the subsequence $(b_2, b_3, \ldots, b_{n-2})$ of (8.3). Let this number be a_2. Then an edge (a_2, b_2) is determined. Remove a_2 from the resultant subsequence of (8.4) and b_2 from the subsequence $(b_2, b_3, \ldots, b_{n-2})$. Continue this process until all the entries of the sequence (8.3) have been used and the edges $(a_1, b_1), (a_2, b_2), \ldots, (a_{n-2}, b_{n-2})$ have been obtained. The two remaining terms of the sequence (8.4), say a_{n-1}, a_n determine the remaining edge (a_{n-1}, a_n). By our choice it follows that for any $i \geq 1, a_i \neq b_j$ for any $j, i \leq j \leq n-2$. Also $a_i \neq a_j$ for any $i \neq j$. Moreover, the two elements a_{n-1}, a_n that remain uncovered, both these cannot be in the multiset $\{b_1, b_2, \ldots, b_{n-2}\}$.

If possible, let there be a circuit $(a_{i_1}, b_{i_1}), \ldots, (a_{i_k}, b_{i_k})$ in the graph constructed with $i_1 < i_2 < \cdots < i_k$. For this to be a cycle $b_{i_k} = a_{i_1}$ which is not possible, as a_{i_1} is different from the elements in the subsequence $b_{i_1}, b_{i_1+1}, \ldots, b_{n-2}$ and b_{i_k} is in this subsequence. If the circuit terminates in (a_{n-1}, a_n), then $a_{n-1} = b_{i_k}$ so that a_n is not a b of the given sequence of $n-2$ numbers. But then $a_n \neq a_{i_1}$. Hence there is no circuit in the graph constructed. Hence the graph constructed has n vertices, $n-1$ edges and has no circuits. Therefore, the graph is a tree which is labelled with labels $1, 2, \ldots, n$.

We have thus proved that there is a one–one correspondence between the set of labelled trees with n vertices and the set of sequences $b_1, b_2, \ldots, b_{n-2}$ of positive integers with every $b_i \leq n$. But the number of such sequences is n^{n-2}. Hence there are n^{n-2} labelled trees with n vertices.

We consider a couple of examples on the use of the procedure adopted in the proof of Theorem 8.8 for constructing a tree (labelled) from a given sequence $(b_1, b_2, \ldots, b_{n-2})$ where b_i are all positive and every $b_i \leq n$. We also consider an example on the construction of such a sequence given a labelled tree on n vertices.

Example 8.29. Let $n = 7$ and the given sequence be $(4, 3, 2, 1, 1)$. The first number in the sequence

$$1, 2, 3, 4, 5, 6, 7 \qquad (8.5)$$

that does not appear in the given sequence

$$4, 3, 2, 1, 1 \qquad (8.6)$$

is 5. Thus, the first edge of the required tree is $(5, 4)$. On deleting 5 from the sequence (8.5) that does not appear in the sequence (8.6) and when 4 is omitted from the sequence (8.6), the first number that appears in the subsequence of (8.5) that does not appear in the subsequence of (8.6) is 4. So the next edge is $(4, 3)$. Now delete both 4 and 5 from the sequence (8.5). The first number in the resulting subsequence which does not appear in the sequence (8.6) when 4, 3 are omitted from it is 3 so that the next edge is $(3, 2)$. Once 5, 4, 3 are deleted from (8.5) and 4, 3, 2 are deleted from (8.6), the choice of the next a_i gives the edge $(2, 1)$ and, then finally, using the above procedure gives the edge $(6, 1)$. All the entries of the sequence (8.6) are exhausted and the two entries left from (8.5) are 1, 7. Thus, the last edge is then $(1, 7)$. Therefore, the six edges of the required tree are $(5, 4)$, •

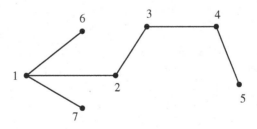

Fig. 8.31.

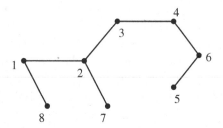

Fig. 8.32.

$(4, 3)$, $(3, 2)$, $(2, 1)$, $(6, 1)$ and $(1, 7)$. Hence the required labelled tree is as in Fig. 8.31.

Example 8.30. Let $n = 8$ and the sequence

$$6, 4, 3, 2, 2, 1 \qquad (8.7)$$

be given. The first number (term) in the sequence

$$1, 2, 3, 4, 5, 6, 7, 8 \qquad (8.8)$$

that does not appear in (8.7) is 5 so that the first edge of the required tree is $(5, 6)$. Omit 5 from the sequence (8.8) and 6 from the sequence (8.7). The first term of the subsequence of (8.8) that does not appear in the subsequence of (8.7) is 6 so that the next edge is $(6, 4)$. Next omit both 5, 6 from the sequence (8.8) and 6, 4 from the sequence (8.7). The choice of the next term of the subsequence of (8.8) leads to the next edge as $(4, 3)$. The procedure applied three more times leads to the edges $(3, 2)$, $(7, 2)$, $(2, 1)$. The remaining two terms of the sequence 1, 8 yield an edge $(1, 8)$. Thus, all the seven edges are: $(5, 6)$, $(6, 4)$, $(4, 3)$, $(3, 2)$, $(7, 2)$, $(2,1)$ and $(1, 8)$. To get these edges quickly, write the two sequences

$$1, \ 2(6), \ 3(4), \ 4(3), \ 5(1), \ 6(2), \ 7(5), \ 8$$
$$6, \ 4, \ 3, \ 2, \ 2, \ 1$$

and the numbers written in parentheses adjacent to the given numbers give the order in which those numbers are to be associated with the numbers in the sequence (8.8). Finally, the two remaining entries of (8.8) lead to the edge $(1,8)$. The required tree is as in Fig. 8.32.

Example 8.31. Construct a labelled tree with nine vertices given a sequence $b = (9, 7, 6, 4, 2, 2, 1)$.

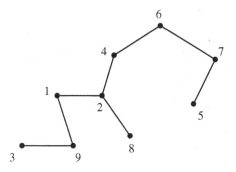

Fig. 8.33.

Solution. Consider the two sequences

$$1, \ 2(7), \ 3(1), \ 4(5), \ 5(2), \ 6(4), \ 7(3), \ 8(6), \ 9$$
$$9, \ 7, \ 6, \ 4, \ 2, \ 2, \ 1.$$

The seven edges obtained by the procedure are:

$$(3,9), (5,7), (7,6), (6,4), (4,2), (8,2), (2,1).$$

Also the remaining two terms of the top sequence yield an edge (1, 9). The required tree is as in Fig. 8.33.

Example 8.32. Adopt the procedure as adopted in the proof of Theorem 8.8 and obtain a sequence of length 6 from a given tree as in Fig. 8.34.

Solution. We encircle the pendant vertices and the resultant pendant vertices. We get $b_1 = 2, b_2 = 1, b_3 = 1, b_4 = 6, b_5 = 1, b_6 = 2, b_7 = 2$.

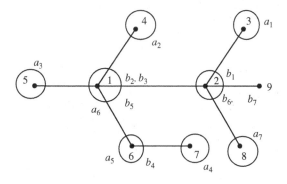

Fig. 8.34.

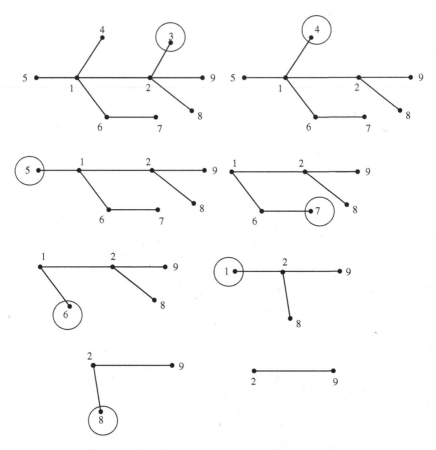

Fig. 8.35.

Hence the required sequence of length 7 is: (2, 1, 1, 6, 1, 2, 2). The steps involved in obtaining this 7 sequence are as given in Fig. 8.35.

So the required sequence is $2, 1, 1, 6, 1, 2, 2$.

Example 8.33. Consider the tree as given in Fig. 8.36.

Pendant Vertex	Corresponding b_i
4	2
5	2
2	1
1	3
6	3
7	3
3	8

Hence the sequence of $b's$, is 2, 2, 1, 3, 3, 3, 8.

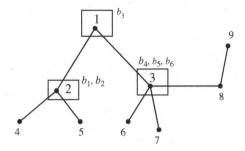

Fig. 8.36.

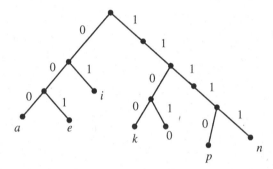

Fig. 8.37.

8.4.4. Some further examples

Example 8.34. Find the codes for a, e, i, k, o, p and n if the encoding scheme is represented by the tree as in Fig. 8.37.

Solution. Following the paths from the root of the tree to the pendant vertices, we find that codes for the given alphabet are:

$a : 000$, $e : 001$, $i : 01$, $k : 1100$, $o : 1101$, $p : 11110$ and $n : 11111$.

Example 8.35. Construct the binary tree with prefix codes representing the coding schemes:

(a) a: 11, e: 0, c: 101, s: 100.
(b) a: 1, e: 01, c: 001, d: 0001, n: 00001.
(c) a: 1010, e: 0, t: 11, s: 1011, u: 1001, v: 10001.

Solution. The trees for the three coding schemes are as in Fig. 8.38.

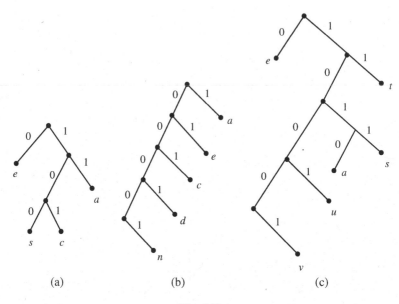

Fig. 8.38.

Example 8.36. Which of the following codes are prefix codes?

(a) a: 11, b: 00, c: 10, d: 01.
(b) a: 0, b: 1, c: 01, d: 001.
(c) a: 101, b: 11, c: 001, d: 011, n: 010.
(d) a: 1010, b: 0, c: 11, d: 1011, e: 1001, f: 10001.

Solution. We draw rooted trees (Fig. 8.39) corresponding to the codes.

In the rooted trees (a), (c), (d), none of a, b, c, d; a, b, c, d, n and a, b, c, d, e, f is a descendent of the other characters. Therefore, each one of these three codes is a prefix code. In the case of tree (b), we find that both c, d are descendent of a. Therefore, the code as given in (b) is not a prefix code.

Example 8.37. Given the coding schemes $a : 001$, $b : 0001$, $e : 1$, $r : 0000$, $s : 0100$, $t : 011$, $x : 01010$, find the words represented by the sequences
(a) 01110100011, (b) 0001110000, (c) 0100101010, (d) 01100101010.

Solution. The sequences can be partitioned as

(a) 011,1,0100,011 which translates as *test*.

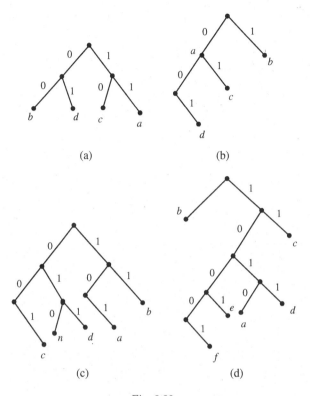

(a)   (b)

(c)   (d)

Fig. 8.39.

(b) 0001,1,1,0000 which translates as *beer*.
(c) 0100,1,01010 which translates as *sex*.
(d) 011,001,01010 which translates as *tax*.

Example 8.38. Consider the Huffman code tree as in Fig. 8.40.

(a) Find the string that represents the word
   (i) care; (ii) sea; (iii) ace; (iv) case; (v) ear; (vi) scar.
(b) Decode each of the following messages
   (i) 1111101110 (ii) 1100101110 (iii) 11101011110 (iv) 1101111101110
   (v) 11101011110110 (vi) 11011111011100.

Solution. The coding given by the given tree is:

   e: 0; *a*: 10; *s*: 110; *r*: 1110; *c*: 1111.

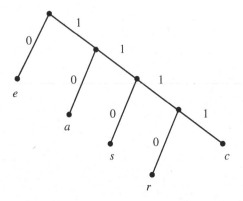

Fig. 8.40.

(a) The words given are represented by the strings:
    (i) 11111011100; (ii) 110010; (iii)1011110; (iv) 1111101100; (v)0101110;
    (vi) 1101111101110 respectively.
(b) We subdivide the strings as follows:
    (i) 1111,10,1110; (ii) 110,0,10,1110; (iii) 1110,10,1111,0
    (iv) 110,1111,10,1110; (v) 1110,10,1111,0,110 (vi) 110,1111,10,1110,0
    and find that the strings represent the words (i) car; (ii) sear; (iii) race;
    (iv) scar; (v) races and (vi) scare, respectively.

Example 8.39. Consider the tree as given in Fig. 8.41 and the
accompanying list of words:

1. one
2. cow

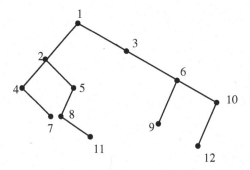

Fig. 8.41.

3. see
4. never
5. purple
6. never
7. I
8. A
9. I
10. I
11. saw
12. Hope.

Suppose that visiting a node means printing out the word corresponding to the number that labels the node. Give the sentence that results from doing a

(a) Preorder search.
(b) Inorder search.
(c) Postorder search of the tree.

Solution. The preorder, inorder and postorder traversal of the tree give the sequences:

(a) 1, 2, 4, 7, 5, 8, 11, 3, 6, 9, 10, 12.
(b) 4, 7, 2, 8, 11, 5, 1, 3, 9, 6, 12, 10.
(c) 7, 4, 11, 8, 5, 2, 9, 12, 10, 6, 3, 1.

Substituting for the numbers from the given list, the three messages are:

(a) One cow never I purple a saw see never II hope.
(b) Never I cow a saw purple one see I never hope I.
(c) I never saw a purple cow I hope I never see one.

Example 8.40. Suppose we are given a sequence of numbers and we are required to arrange these either in increasing order or in decreasing order. For this purpose a sorting algorithm known as **tournament sort** is applied. This procedure/algorithm works exactly the way a champion among various players in a round-robin tournament is obtained/determined. In this procedure, we list the numbers to be sorted in a row and we construct a rooted tree with the given numbers as pendant vertices. We compare the given numbers in pairs starting from left and going to right. When two adjacent numbers i, j are compared, a node is introduced with i, j as its sons and the node introduced is labelled with the larger of the two numbers.

Having done this, for all the numbers, we apply the same procedure to the nodes introduced. If we continue, a rooted tree is formed. The label of the root of the rooted tree is the largest number among the given numbers. The procedure is then repeated, replacing the largest number arrived at by $-\infty$. Continue the process until all the numbers have been arranged in the decreasing order. In case the numbers are to be arranged in increasing order, when the adjacent pair of numbers i, j are compared, the node introduced is labelled with the lower of the two numbers i, j. When finally a rooted tree is obtained, the label of the root is the smallest of the given set of integers.

For example, let us arrange the numbers 21, 7, 13, 16, 2, 8, 26, 11 in decreasing order. We construct the rooted tree for these numbers as in Fig. 8.42(a).

Thus, the largest number is 26. We again consider the given numbers with 26 replaced by $-\infty$ (meaning thereby that 26 is replaced by a number which is lower than the lowest). We draw the tree for resulting numbers as in Fig. 8.42(b).

The root is labelled 21 so that the next number in decreasing order is 21. Now we replace 21 by $-\infty$ and redraw the rooted tree which gives the next number in the decreasing order.

The next number in order is 16. Replace 16 also by $-\infty$ and draw the resulting rooted tree.

The next number in order is 13. Replace 13 also by $-\infty$ and draw the graph for the resulting numbers.

The next number in order is 11. Replace 11 also by $-\infty$ and construct the next tree.

The next number in order is determined as 8. Draw the next tree.

The next number in order is 7. Only one more number namely 2 is left and is less than 7. Hence the numbers are arranged as 26, 21, 16, 13, 11, 8, 7, 2.

8.5. Spanning Trees and Cut-Sets

Let G be a connected graph where the vertices represent the cities in India and the edges represent connecting highways between the cities. We wish to determine a subset of highways which should be kept open all the time so that we can reach one city from another through these highways. We would also like to determine the subset of highways whose blockage will separate some cities from the others. We may consider similar problem of all airports

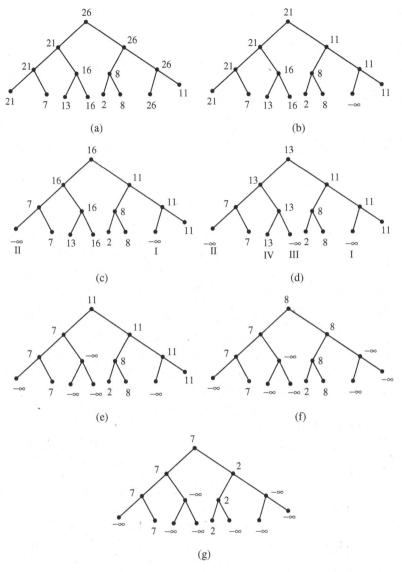

Fig. 8.42.

in the country with the edges as air roots connecting the various airports. This study leads to what are called spanning trees of connected graphs.

By a **tree of a graph** we mean a subgraph of the graph which is a tree. Recall that a spanning subgraph of a graph G is a subgraph of G which

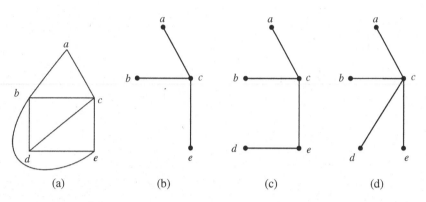

(a)   (b)   (c)   (d)

Fig. 8.43.

contains all the vertices of G. A **spanning tree** of a connected graph is a spanning subgraph of the graph which is a tree. For example,

Figure 8.43(b) is a tree of the graph Fig. 8.43(a) but not a spanning tree, whereas (c) and (d) are spanning trees of the graph. An edge of the graph which is in a tree is called a **branch** of the tree whereas an edge of the graph which is not in a tree is called a **chord** or a **link** of the tree. For example, for the tree 8.43(b) $(a,c),(b,c),(c,e)$ are branches where as $(a,b),(b,d),(c,d),(d,e),(b,e)$ are chords or links of this tree. For the spanning tree $(c),(a,c),(b,c),(c,e),(d,e)$ are branches of this tree but $(a,b),(b,d),(b,e),(c,d)$ are chords or links of this tree. Again, for the spanning tree (d), (a,c), (b,c), (c,d), (c,e) are branches but (a,b), (b,d), (d,e), (b,e) are chords or links of this tree.

The set of chords of a tree is called the **complement of the tree**. Thus $\{(a,b),(b,d),(b,e),(c,d)\}$ is the complement of the tree (c), $\{(a,b),(b,d),(c,d),(d,e),(b,e)\}$ is the complement of the tree (b) and $\{(a,b),(b,d),(b,e),(d,e)\}$ is the complement of the tree (d).

Observe that if $T = (V,E')$ is a spanning tree of a graph $G = (V,E)$ then T has $o(V)-1$ branches and has $o(E)-o(V)+1$ chords. Spanning trees could also be defined through spanning sets. If $G = (V,E)$ is a connected undirected graph, then a **spanning set** for G is a subset E' of E such that (V,E') is connected. Therefore, if E' is a spanning set for $G = (V,E)$, then $T = (V,E')$ is a spanning tree.

Theorem 8.9. A connected graph always contains a spanning tree.

Proof. Let $G = (V,E)$ be a connected graph. If G has no circuit, then G is a tree and, therefore, G is its own spanning tree. Suppose that G has one or more circuits. We first consider one of the circuits in G and remove edges from it one by one without loosing connectivity of the graph. We continue doing so until there is no circuit in the resulting graph. Since no vertices are deleted in this process, we get a spanning subgraph of G which is connected and has no circuits. We thus arrive at a spanning tree.

Definition. A **connecting subgraph** of a graph is a spanning subgraph that is connected. A connecting subgraph H of a graph G is called **minimal**, if H has no proper connecting subgraph. It is then clear that a minimal connecting subgraph of a connected graph is a spanning tree of the graph. A **cut-set** is a set of edges in a graph such that removal of the edges in the set increases the number of connected components of the resulting subgraph by one whereas the removal of edges in any proper subset of it does not. Thus, a cut-set is a minimal set of edges with this property. Since in a connected graph there is one connected component, removal of a cut-set from a connected graph will separate the graph into two connected parts.

Let $G = (V, E)$ be a connected graph and E' be a cut-set of G. Then G is partitioned into two connected components, say G_1 and G_2. Let V_1, V_2 be the set of all the vertices of G_1, G_2 respectively. The subgraphs G_1, G_2 being connected, every two vertices of G_1 are connected by a path lying wholly in G_1, any two vertices in G_2 are connected by a path lying wholly in G_2 and no vertex of G_1 is connected to any vertex of G_2 and vice versa. Thus, all the edges in G which connect vertices in G_1 with those in G_2 and vice versa have been removed. These edges of G are, therefore, contained in the cut-set E'. If this set of edges is denoted by E'', then $E'' \subseteq E'$. If $E'' \neq E'$, then removal of the edges in E'' breaks the graph G into two connected components. Since E' is the minimal set of edges that does this job, we have a contradiction. Therefore $E'' = E'$. This leads us to the following alternative definition of a cut-set.

If the graph G is divided into two parts G_1, G_2 such that any two vertices in G_1 are connected to each other by a path lying wholly in G_1, any two vertices of G_2 are connected by a path lying wholly in G_2, then the set of edges which connect vertices in G_1 and G_2 is a cut-set.

Example 8.41. Consider the graph of Fig. 8.44.

If we consider possible pairs of components of this graphs as:

(a) $G_1 = \{\{v_1, v_2\}, e_1\}$, $\quad G_2 = \{\{v_3, v_4, v_5\}, e_3, e_7, e_8\}$;

(b) $G_1 = \{\{v_2, v_3\}, e_2\}$, $\quad G_2 = \{\{v_1, v_4, v_5\}, e_4, e_5, e_8\}$;

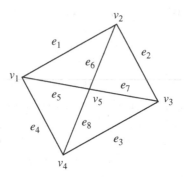

Fig. 8.44.

(c) $G_1 = \{\{v_3, v_4\}, e_3\}, \quad G_2 = \{\{v_1, v_2, v_5\}, e_1, e_5, e_6\}$;

(d) $G_1 = \{\{v_1, v_4\}, e_4\}, \quad G_2 = \{\{v_2, v_3, v_5\}, e_2, e_6, e_7\}$;

(e) $G_1 = \{\{v_1, v_5\}, e_5\}, \quad G_2 = \{\{v_2, v_3, v_4\}, e_2, e_3\}$.

then the corresponding cut-sets are respectively $\{e_2, e_4, e_5, e_6\}, \{e_1, e_3, e_6, e_7\}, \{e_2, e_4, e_7, e_8\}, \{e_1, e_3, e_5, e_8\}$ and $\{e_1, e_4, e_6, e_7, e_8\}$.

We can also get three other cut-sets if we consider the following partitioning of the set of vertices:

$V_1 = \{v_2, v_5\}, \quad V_2 = \{v_1, v_3, v_4\}$;

$V_1 = \{v_3, v_5\}, \quad V_2 = \{v_1, v_2, v_4\}$;

$V_1 = \{v_4, v_5\}, \quad V_2 = \{v_1, v_2, v_2\}$.

Exercise 8.1. Find all other possible cut-sets for the graph considered above.

It is now clear from the above discussion that if the highways or the airways in a cut-set are blocked then it is not always possible to go from one city to another or from one airport to another. Therefore, if the cities or airports are to remain connected (as is necessary in the case of cantonments at the time of war) then at least one highway falling in every cut-set must always remain operational.

The concepts of spanning trees, circuits and cut-sets are closely related. We explore this relationship here.

Remark 1. Addition of a chord to the spanning tree of a graph yields a subgraph that contains exactly one circuit.

Proof. Suppose that a chord $\{v_1, v_2\}$ is added to a spanning tree. As there is already a path from v_1 to v_2, this path together with the chord $\{v_1, v_2\}$

yields a circuit in the graph. If the addition of this chord $\{v_1, v_2\}$ results in yielding two or more circuits in the graph, there must be two or more paths from v_1 to v_2 in the spanning tree. Therefore, there must be a circuit in the spanning tree — this circuit is obtained by combining the two paths from v_1 to v_2. This is a contradiction as in a tree there are no circuits. This proves the assertion.

Remark 2. For a spanning tree there are $o(E) - o(V) + 1$ chords and addition of each chord results in exactly one circuit in the graph. Thus, we obtain, for a given spanning tree, $o(E) - o(V) + 1$ circuits in the graph. This system of circuits in the graph is called the fundamental system of circuits relative to the spanning tree. Each circuit in the fundamental system is called a **fundamental circuit**.

Each fundamental circuit contains exactly one chord of the spanning tree and, as such, this circuit is also called the **fundamental circuit corresponding to the chord**.

For the graph considered in Fig. 8.44 above, a spanning tree is as in Fig. 8.45 and its chords are e_2, e_6, e_3 and e_8. Then the fundamental circuit corresponding to the chord e_2 is $\{e_1, e_2, e_7, e_5\}$; that corresponding to e_3 is $\{e_5, e_7, e_3, e_4\}$; that corresponding to e_6 is $\{e_1, e_6, e_5\}$; and that corresponding to e_8 is $\{e_5, e_8, e_4\}$.

Remark 3. The removal of a branch from a spanning tree breaks the spanning tree up into two trees. One or both of these parts may consist of a single vertex. Thus corresponding to a branch in a spanning tree there is a division of vertices in the graph into two subsets corresponding to the vertices in the two subtrees. Thus for every branch in a spanning tree there is a corresponding cut-set (consisting of this branch together with other edges of the graph which connect vertices in the two subsets but have been removed to obtain the spanning tree).

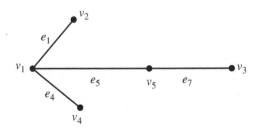

Fig. 8.45.

Consider the graph of Fig. 8.44 and the spanning tree as in Fig. 8.45. If the branch e_5 of the spanning tree is removed, we get the two sets of vertices as $\{v_1, v_2, v_4\}$ and $\{v_5, v_3\}$. The corresponding cut-set is $\{e_2, e_3, e_6, e_5, e_8\}$. If on the other hand the branch e_1 is removed, the set of vertices is subdivided into the subsets $\{v_2\}$, $\{v_1, v_3, v_4, v_5\}$ and the corresponding cut-set is $\{e_1, e_2, e_6\}$. Since there are four branches in the spanning tree, we can get two more cut-sets in this way.

For a given spanning tree, the set of $o(V) - 1$ cut-sets corresponding to the $o(V) - 1$ branches of the spanning tree is called the **fundamental system of cut-sets relative to the spanning tree**. A cut-set in the fundamental system of cut-sets is called a **fundamental cut-set**. In each fundamental cut-set there is one tree branch and it is called the **fundamental cut-set corresponding to this branch**.

The fundamental system of cut-sets for the example investigated above is:

$$\{e_2, e_3, e_5, e_6, e_8\}, \quad \{e_1, e_2, e_6\}, \quad \{e_2, e_3, e_7\} \quad \text{and} \quad \{e_3, e_4, e_8\}.$$

We next investigate some properties of circuits and cut-sets. We investigate these for connected graphs.

Theorem 8.10. A circuit and the complement of any spanning tree must have at least one edge in common.

Proof. Recall that the complement of a tree is the set of all chords of the tree i.e. it is the set of those edges of the graph which are not in the tree. Therefore, if a circuit has no edge in common with the complement of a spanning tree, every edge in the circuit is contained in the spanning tree. But a tree has no circuit. Hence we arrive at a contradiction. Therefore every circuit and the complement of a spanning tree must have at least one edge in common.

Theorem 8.11. A cut-set and any spanning tree must have at least one edge in common.

Proof. Suppose that a cut-set has no common edge with a spanning tree. Then removal of all the edges in the cut-set from the graph will leave the tree intact. Since it is a spanning tree, it contains all the vertices of the graph and every two of these are connected. Therefore, with this removal of edges, the graph is not subdivided into two components. This contradicts the definition of a cut-set. Hence a cut-set and any spanning tree must have at least one edge in common.

Theorem 8.12. Every circuit has an even number of edges in common with every cut-set.

Proof. Consider a graph $G = (V, E)$ with a cut-set E' (say). When the edges in E' are removed from the graph, then the set V of vertices is partitioned into two subsets A, B say such that $V = A \cup B$ and $A \cap B = \emptyset$ with neither $A = \emptyset$ nor $B = \emptyset$. Consider a circuit which we may assume a path from vertex v to itself. Suppose that v is in A. If the circuit under consideration does not involve any edge lying in E', then the circuit has 0 edges in common with E' and we are through. If the circuit involves an edge lying in E', then the circuit contains an edge joining a vertex v in A with a vertex in B. Since the circuit has to come back to v which is in A, this circuit must contain another edge going from a vertex in B to a vertex in A i.e. it must involve another edge lying in E'. Thus for every edge of the circuit which is in common with the edges in E', there is another edge common to the circuit and the cut-set. Thus the number of common edges between the circuit and the cut-set must be even.

A close relationship between the fundamental system of circuits and the fundamental system of cut-sets relative to a spanning tree is revealed by the following results.

Theorem 8.13. For a given spanning tree, let $D = \{e_1, e_2, \ldots, e_k\}$ be a fundamental cut-set in which e_1 is a branch and $e_2, e_3, \ldots, e_k$ are chords of the spanning tree. Then e_1 is contained in the fundamental circuits corresponding to e_i for $i = 2, 3, \ldots, k$. Moreover e_1 is not contained in any other fundamental circuits.

Proof. Let $G = (V, E)$ be a connected graph and T a spanning tree of G with e_1 as a branch. Let C be the fundamental circuit corresponding to the chord e_2. Then every element of C except e_2 is a branch of T. Also $e_2 \in C \cap D$. In view of Theorem 8.12 $o(C \cap D)$ is even and, so, at least 2. Since e_1 is the only branch of T in D, $\mathrm{O}(C \cap D) = 2$ and $C \cap D = \{e_1, e_2\}$. Thus e_1 is in the fundamental circuit of e_2. Since the above argument could have been given for $e_i, i \neq 1$, instead of e_2, it follows that e_1 is contained in the fundamental circuits corresponding to e_i for $i = 2, 3, \ldots, k$. Let e be a chord of T such that $e \neq e_i$, $i = 2, \ldots, k$. Let C' be the fundamental circuit of T relative to e. Since all the elements of C' except e are branches of T, if

$C' \cap D \neq \emptyset$, then C' and D have an odd number of common edges. Hence $C' \cap D = \emptyset$ and e_1 is not in C'.

Theorem 8.14. For a given spanning tree, let $C = \{e_1, e_2, \ldots, e_k\}$ be a fundamental circuit in which e_1 is a chord and $e_2, \ldots, e_k$ are branches of the spanning tree. Then, e_1 is contained in the fundamental cut-sets corresponding to e_i for $i = 2, 3, \ldots, k$. Moreover, e_1 is not contained in any other fundamental cut-set.

Proof. Let G be the given connected graph and T a spanning tree. Let D be the fundamental cut-set corresponding to the branch e_2 of the spanning tree T. Then D contains only the branch e_2 of T. The rest of the elements of D are chords of T. Thus e_2 is in $C \cap D$ and it is a branch of T. As the other elements of D are chords of T, e_1 is the only chord in C and $C \cap D$ has an even number of elements, therefore, $e_1 \in D$. We can repeat the above argument with e_2 replaced by e_i for $i = 3, \ldots, k$. Therefore, e_1 is in every fundamental cut-set determined by e_i for $i = 2, 3, \ldots, k$. Suppose that e_1 is in the fundamental cut-set D' corresponding a branch e of T where $e \neq e_i$ for any i, $2 \leq i \leq k$. Since all the elements of D' except e are chords of T, if $c \cap D' \neq \emptyset$, then $C \cap D' = \{e_1\}$ i.e. C and D' have an odd number of common edges. Hence $C \cap D' = \emptyset$ and e_1 is not in D'.

Example 8.42. Prove that the complement of a spanning tree does not contain a cut-set and that the complement of a cut-set does not contain a spanning tree.

Solution. If possible, suppose that a cut-set is contained in the complement of a spanning tree. We know that a cut-set and any spanning tree must have at least one edge in common. Therefore, the given cut-set and the given spanning tree have at least one edge in common. This common edge must then belong to the complement of the spanning tree. Thus, this edge belongs to the spanning tree and also to its complement. This is a contradiction. Hence the complement of a spanning tree does not contain a cut-set. Interchanging cut-set and spanning tree in the above argument proves the second part.

Example 8.43. Prove that a pendant edge (an edge whose one end vertex is of degree 1) in a connected graph G is contained in every spanning tree of G.

Solution. Let $e = (v, v_1)$ be a pendant edge in G with $\deg v = 1$. Let T be any spanning tree of G. Since all the vertices of G are in T, both v, v_1 are vertices in T. Now T being a tree, every two vertices of T are connected in T. In particular, there is a path from v to v_1 in T. But e being the only edge from v to v_1 in G, this edge must be a part of the tree T. Hence e is contained in the spanning tree T.

Example 8.44. If a graph G has n vertices and the subgraph T of G has $n - 1$ edges and no circuits, then G is connected and T is a spanning tree in G.

Solution. We prove the result by induction on the number n of vertices in G. Let H be a graph with m vertices and $n \geq m$ edges. A simple induction argument then shows that H has a circuit. It follows from this that the subgraph T with $n - 1$ edges and no circuits must have $\geq n$ vertices. But G itself being a graph with n vertices, T is a graph with n vertices and $n - 1$ edges. If every vertex in T is of degree at least 2, then $2(n - 1) =$ sum of the degrees of the vertices is $\geq 2n$ which is not possible. Hence T has at least one vertex of degree 1.

Thus G is a graph with n vertices, T a subgraph of G with $n - 1$ edges, n vertices and no circuits. Moreover, T has a vertex of degree 1. If $n = 2$, then G is obviously connected and T is its spanning tree. Indeed, T is G itself.

Suppose that the result holds for an $n \geq 2$. Let G be a graph with $n + 1$ vertices, T is a subgraph of G with $n + 1$ vertices, n edges and no circuits. Let v be a vertex of degree 1 in T, with e edge ending with v. Consider the subgraph G' of G obtained by omitting e and v and T' be the subgraph of G' obtained from T on omitting e and v. Thus G' is a graph with n vertices and T' is a subgraph of G' with $n - 1$ edges and no circuits, so that T' is also having n vertices. By induction hypothesis, T' is a spanning tree of G' and G' is connected. Since v is adjacent to a vertex v_1 of G' and there is a path from v_1 to every other vertex in G', v is connected to every vertex of G'. Hence G is connected. Also T' being a tree is connected. Therefore, T is also connected. Thus T is a subgraph of G with $n + 1$ vertices and n edges and, therefore, T is a spanning tree of G. This completes induction.

Example 8.45. Prove that any given edge of a connected graph G is a branch of some spanning tree of G. Is it also true that any arbitrary edge of G is a chord of some spanning tree of G?

Solution. Let $e = (v_1, v_2)$ be a given edge of the connected graph G. Since every connected graph has a spanning tree, let T be a spanning tree of G and $\overline{T}$ be the complement of T in G so that $G = T \cup \overline{T}$. If e is in T, then we are through. If e is not a branch of T, consider $T_1 = T \cup \{e\}$. Since v_1, v_2 are already connected in T and e being another path from v_1 to v_2, T_1 has a circuit C in which e also appears. Remove an edge e' other than e from C. Then $T' = T_1 - \{e'\}$ is a connected subgraph of G with n vertices and has no circuit. Therefore, T' is a spanning tree of G and T' has e as a branch.

It is not always possible for an edge of a connected graph to be a chord of some spanning tree of G. For example, G itself may be a tree, so that G is a spanning tree of itself. None of the edges of G can be deleted without making the resulting subgraph (to have all the vertices of G) of G disconnected. Thus, any edge of G may not always be a chord of some spanning tree of G.

Alternatively, if e is a pendant edge of G (provided G has one) then e is contained in every spanning tree of G and hence e cannot be a chord of any spanning tree of G.

Example 8.46. Consider the connected graph G as given in Fig. 8.46 with the edges as mentioned.

Observe that $T_1 = \{b_1, b_2, b_3, b_4, b_5, b_6\}$ and $T_2 = \{b_1, b_2, b_3, b_4, b_5, c_6\}$ are spanning trees of G. Observe that $b_6 \in T_1$ but $b_6 \notin T_2$. Then $c_6 \in T_2$ but $c_6 \notin T_1$ such that $(T_1 - \{b_6\}) \cup \{c_6\} = T_2$ and $(T_2 - \{c_6\}) \cup \{b_6\} =$

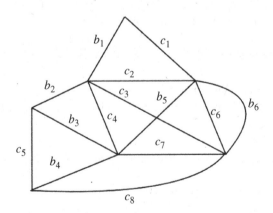

Fig. 8.46.

T_1 which are spanning trees of G. Let us consider another spanning tree $\{c_1, c_2, c_6, c_7, c_8, c_5\} = T_3$.

(a) Consider a branch b_2 of T_1 which is not in T_3. We find that c_2 is a branch of T_3 which is a chord for T_1. Then $(T_1 - \{b_2\}) \cup \{c_2\} = \{b_1, c_2, b_3, b_4, b_5, b_7\}$ and $(T_2 - \{c_2\}) \cup \{b_2\} = \{c_1, b_2, c_6, c_7, c_8, c_5\}$ are again spanning trees.

(b) Let us next consider T_1 and branch b_1 of T_1. We find that c_1 is a branch of T_3 which is a chord of T_1. Observe that $(T_1 - \{b_1\}) \cup \{c_1\} = \{c_1, b_2, b_3, b_4, b_5, b_6\}$ and $(T_3 - \{c_1\}) \cup \{b_1\} = \{b_1, c_2, c_6, c_7, c_8, c_5\}$ are again spanning trees of G.

(c) b_3 is a branch of T_1 which is a chord for T_3. We consider c_5 which is a branch of T_3 but it is a chord of T_1. Then $(T_1 - \{b_3\}) \cup \{c_5\} = \{b_1, b_2, c_5, b_4, b_5, b_6\}$ and $(T_3 - \{c_5\}) \cup \{b_3\} = \{c_1, c_2, c_6, c_7, b_3, c_8\}$ are again spanning trees of G.

(d) Next consider the branch b_4 of T_1. Then $(T_1 - \{b_4\}) \cup \{c_8\} = \{b_1, b_2, b_3, b_5, b_6, c_8\}$ and $(T_3 - \{c_8\}) \cup \{b_4\} = \{c_1, c_2, c_6, c_7, b_4, c_5\}$ are spanning trees of G.

(e) Next consider the branch b_5 of T_1. Then $(T_1 - \{b_5\}) \cup \{c_7\} = \{b_1, b_2, b_3, b_4, c_8, b_6\}$ and $(T_3 - \{c_7\}) \cup \{b_5\} = \{c_1, c_2, c_6, b_5, c_8, c_5\}$ are spanning trees of G.

(f) Finally, we consider the branch b_6 of T_1. Corresponding to this, we find that c_6 is a branch of T_3 which is not in T_1 and $(T_1 - \{b_6\}) \cup \{c_6\} = \{b_1, b_2, b_3, b_4, b_5, c_6\}$ and $(T_3 - \{c_6\}) \cup \{b_6\} = \{c_1, c_2, b_6, c_7, c_8, c_5\}$ are again spanning trees of G.

Proof of Theorem 8.9 suggests one way of finding the spanning tree of a connected graph. This method is, as a matter of fact, not so simple as at every step we are required to identify a circuit in the graph. There are two other ways of finding a spanning tree of a connected graph which are to some extent simpler than the method suggested by the proof of Theorem 8.9. We now describe these two methods.

Depth-First Search

This method allows us to form a rooted tree which contains all the vertices of the given connected graph and then the underlying undirected graph is a spanning tree of G.

First, choose a vertex of G arbitrarily and with this vertex as root, form a path by successively adding vertices and edges, where each new edge is

incident with the last vertex in the path and a vertex not already in the
path. Continue doing so as long as it is possible. If this path contains all
the vertices of G, we have arrived at a tree (ignoring the directions), which
is a spanning tree. If all the vertices of G have not been taken care of, we
go to a vertex just above the last vertex in the path and with this vertex as
root, construct a path but adding only edges and vertices which have not
already occurred in the path constructed. If it is not possible to add any
edges and vertices starting with the last but one vertex, we go back in the
first path by another step or two steps backward or three steps backward,
etc. and construct a second path. If, again, all the vertices of G have already
appeared in the tree constructed so far, we are through. On the other hand,
if all the vertices of G do not appear in the tree that is constructed, repeat
the above procedure. Continue until all the vertices of G have been visited.
Since G has only a finite number of vertices, the process terminates in a
finite number of steps and we arrive at a spanning tree.

The depth-first search procedure is also called the **backtracking**
procedure. To illustrate this procedure, we consider a couple of examples.

Example 8.47. Use a depth-first search to find a spanning tree for the
graph G given in Fig. 8.47.

Solution. The steps used in the depth-first search method to produce a
spanning tree of G are given in Fig. 8.48.

We arbitrarily start with the vertex i. A path is built by successively
adding edges incident with vertices not already in the path, as long as this
is possible. The path constructed is i, j, k, l, m and we can go no further.
The last vertex in this fashion is m and the previous one is l. However, no
more edges can be added starting with l and we go to the vertex k previous
to l. Again, no more edges can be added starting with k as well. So, we go
to the vertex j previous to k on this path. Starting with j we can build
the path j, h, g, f, n and cannot go further. On this path built, the vertex

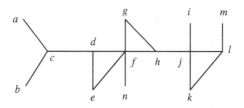

Fig. 8.47.

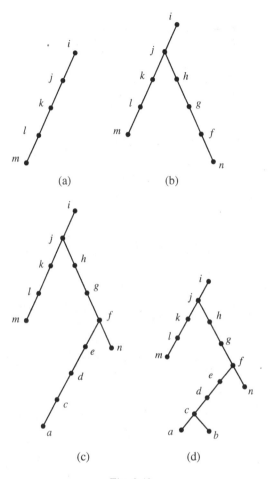

Fig. 8.48.

previous to n is f and starting with f we build the path f, e, d, c, a. We cannot go any further and, so, we go to the last vertex but on this path. This vertex is c and starting with c we build the path c, b consisting of just one edge. Thus the spanning tree is given in the last graph in Fig. 8.48(d).

Example 8.48. Use a depth-first search to find a spanning tree for the graph shown in Fig. 8.49.

Solution. The steps of the depth-first search procedure are shown in Fig. 8.50. We choose the starting vertex e arbitrarily.

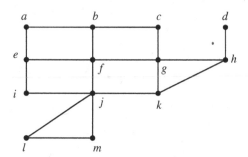

Fig. 8.49.

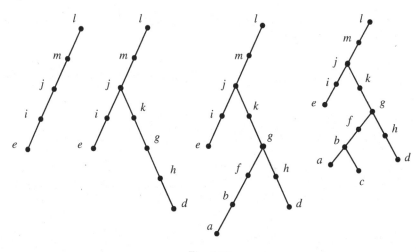

Fig. 8.50.

We can first trace a path e, i, j, m, l in which we cannot go further. In this path the last but one vertex is m and we cannot trace a path starting with m. Therefore, we go to the vertex j, which comes just before m on this path. Starting with j we may trace a path j, k, g, h, d and we cannot go beyond d. On this path the vertex prior to d is h and we cannot trace a path starting with h. So, we go to vertex g which occurs prior to h on this path. Then starting with g we can trace a part g, f, b, a. Again we cannot continue beyond a. The vertex which occurs just before a on this path is b and starting with b we can trace the path b, c. Since all the vertices of G have been taken care of, we get a spanning tree of G as in the last graph of Fig. 8.50.

Breadth-First Search

Another method to produce a spanning tree of a simple graph is known as **breadth-first search**. In this method again a rooted tree is constructed and the underlying undirected graph of this rooted tree forms a spanning tree of the graph we started with. In this method, we choose a vertex arbitrarily and add all edges incident with this vertex. The vertices added in this way become vertices at level 1. Order these level 1 vertices arbitrarily. For each vertex at level 1, visited in order, add all edges incident with the particular level 1 vertex as long as it does not result in a simple circuit. The new vertices added at this stage are level 2 vertices. We continue the process until all the vertices of the graph have been visited. Again the procedure ends in a finite number of steps and there results a rooted tree. The underlying undirected graph of this rooted tree is a spanning tree of the given graph.

We next construct spanning trees of the graphs considered in the two examples above by the method of breadth-first search.

Example 8.49. Use breadth-first search to find a spanning tree of the graph G shown in Fig. 8.47.

Solution. The steps used in the breadth-first search to produce a spanning tree of G are given in Fig. 8.51. This time, we choose the vertex h to be the root.

The spanning tree is then the undirected graph of the graph in Fig. 8.51(d).

Example 8.50. Use a breadth-first search to obtain a spanning tree of the graph G shown in Fig. 8.49.

Solution. The steps involved in the breadth-first search to produce a spanning tree of G are given in Fig. 8.52. As in the case of depth-first search, we obtain a rooted tree with e as a root. The degree of e being three, there are three vertices a, f, i each at level 1. Then there are three vertices, namely b, g, j, which are at level 2. Five vertices, namely l, m, k, h, c, are at level 3 and finally only one vertex d is at level 4.

The underlying undirected graph of the rooted tree as in Fig. 8.52(d) is a spanning graph of G.

In the case of both a depth-first search and breadth-first search, the choice of a different vertex as root will lead to different spanning tree of the graph.

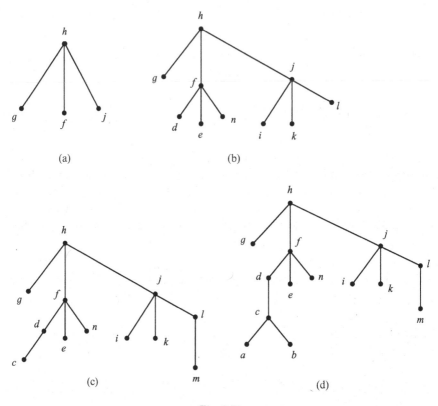

Fig. 8.51.

8.6. Minimal/Minimum/Shortest Spanning Tree

Consider a connected weighted graph where some real numbers are assigned as weights to the edges. Then the **weight of a tree** is defined as the sum of the weights of the branches of the tree. Among all spanning trees, one with the minimum weight is called a **minimal/minimum/shortest spanning tree**. A physical situation where we come across a minimal spanning tree is a network of cities in a country/state connected with roads or communication channels. Each communication channel between two cities is assigned the cost of setting up and maintaining the link as its weight. We may like to set up a communication network connecting all the cities at minimum cost. This problem then asks for finding a minimum spanning tree. As a connected, weighted graph has only a finite number of spanning trees, it always has a minimum spanning tree.

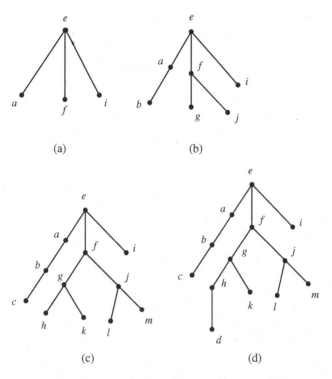

Fig. 8.52.

Lemma. Let G be a connected weighted graph and e be an edge in G of least possible weight. Then there exists a minimum spanning tree of G with e as a branch.

Proof. Let T be a minimum spanning tree of G. If $e \in T$, we have nothing to prove. So, suppose that $e \notin T$. Let G_1 be a subgraph of G obtained by adjoining e to T. Then G_1 has a unique circuit C (say) with e in C. Let e' be an edge in C with $e' \neq e$ and with least weight among all the branches of T in C. Consider T' obtained from G_1 by deleting e'. Then T' is connected and the number of branches in T' equals the number of branches of T. Hence T' is a tree — indeed, a spanning tree. By hypothesis, the weight of e is the least among the weights of all edges of G. Also T' is obtained from T by replacing $e' \in T$ by e. Therefore, $wt(T') \leq wt(T)$. But T is a minimum spanning tree of G. Therefore, $wt(T') = wt(T)$ and T' is a minimum spanning tree.

We consider two algorithms for obtaining a minimum spanning tree.

Kruskal's Algorithm

Let $G \doteq (V, E)$ be a connected weighted graph.

(a) Choose an initial edge $e_1 = \{v_1, v_2\}$ of minimal weight and its vertices v_1, v_2 and let $T_2 = (\{v_1, v_2\}, \{e_1\})$ be the initial subtree.

(b) Assume that for $i \geq 2$, $T_i = (V_i, E_i)$ is the current subtree already constructed. Among all edges $\{v, v_k\}$ with $v \in V - V_i$, $v_k \in V_i$, choose one edge, say $\{v, v_k\}$ of minimal weight and set $v = v_{i+1}$, $e_i = \{v, v_k\} = \{v_{i+1}, v_k\}$ and set $T_{i+1} = (V_{i+1}, E_{i+1})$, where $V_{i+1} = V_i \cup \{v_{i+1}\}$, $E_{i+1} = E_i \cup \{e_i\}$.

(c) Repeat step (b) until a spanning tree $T_n, n = o(V)$, is obtained.

Theorem 8.15. The spanning tree T_n obtained in Kruskal's Algorithm is a minimal spanning tree of $G = (V, E)$.

Proof. Since the edge $e_i = (v, v_k) = (v_{i+1}, v_k)$ as in (b) above has one end v_k in T_i and the other end v_{i+1} is not in T_i, adjoining of edge e_i to T_i does not create a circuit in T_{i+1} if T_i does not have a circuit already. Thus T_{i+1} is a tree if T_i is already one. Since T_2 is trivially a subtree of G, we have an increasing sequence $T_2 \subset T_3 \subset \cdots \subset T_n$ of subtrees of G with T_n as a spanning subtree.

Let $P(i)$ be the proposition 'there is a minimal subtree T of G with $T_i \subseteq T$'. It follows from the lemma above (Kruskal's Algorithm) that $P(i)$ holds for $i = 2$. Assume that $i \geq 2$ and that $P(i)$ holds. Let T be a minimal spanning tree such that $T_i \subseteq T$. By the algorithm (step (b)), $T_{i+1} = (V_{i+1}, E_{i+1})$ where $V_{i+1} = V_i \cup \{v_{i+1}\}$ and $E_{i+1} = E_i \cup \{e_i\}$. If $e_i \in T$, then $T_{i+1} \subset T$. So, suppose that $e_i \notin T$. Then the subgraph G_1 of G obtained by adjoining e_i to T contains a circuit C which is of the form $(e_i, s_1, s_2, \ldots, s_r)$ where the $s_j's$ are in T. The edges $s_1, s_2, \ldots, s_r$ cannot all be from T_i for otherwise C will be a circuit in T_{i+1} which is a contradiction. Let s_j be the edge with smallest index j that is not in T_i. Then s_j has one vertex in T_i and the other vertex not in T_i. This means that s_j was also available when e_i was chosen by Kruskal's Algorithm. Therefore the weight of s_j is greater than or equal to the weight of e_i. Hence, if T' is the tree obtained from G_1 by deleting the edge s_j, then T' is a spanning tree of G with $wt(T') \leq wt(T)$ (as the T' is obtained from T by replacing the edge s_j by e_i). Since T is a minimal spanning tree of G, we must have $wt(T') = wt(T)$ and T' is a minimal spanning tree of G. By the very construction $e_i \in T'$ and, so, $T_{i+1} \subseteq T'$.

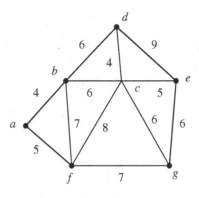

Fig. 8.53.

This completes induction. Hence T_n is contained in a minimal spanning tree T of G. T_n itself being a spanning tree, T_n is a minimal spanning tree.

Example 8.51. Using Kruskal's Algorithm, find a minimal spanning tree of the graph in Fig. 8.53.

Solution. The given graph has seven vertices and 12 edges. There are two edges of least weight 4, namely $\{c, d\}$ and $\{a, b\}$. Let us select the edge $\{c, d\}$ (we could equally well choose the other edge $\{a, b\}$). So, the first subtree T_2 as given by the algorithm is $T_2 = (V_2, E_2)$, with $V_2 = \{c, d\}$, $E_2 = \overline{cd}$. In order to avoid confusion, we write $\overline{xy}$ for the edge $\{x, y\}$. The vertices of $V - V_2$ which are adjacent to c are b, e, f, g and those adjacent to d are b, e. The six edges incident with c and d are $\overline{ce}$ of weight 5, $\overline{cb}$ of weight 6, $\overline{cf}$ of weight 8, $\overline{cg}$ of weight 6, and $\overline{db}$ of weight 6, $\overline{de}$ of weight 9.

The edge of least weight among these is $\overline{ce}$ with weight 5. Therefore $T_3 = (V_3, E_3)$, with $V_3 = \{c, d, e\}$, $E_3 = \{\overline{cd}, \overline{ce}\}$. The vertices of $V - V_3$ that are adjacent to c, d, e are b, f, g; b; g respectively giving edges $\overline{cb}$ of weight 6, $\overline{cf}$ of weight 8, $\overline{cg}$ of weight 6; $\overline{db}$ of weight 6; $\overline{eg}$ of weight 6.

The least among these weights being 6, we may choose one of the edges $\overline{cb}$, $\overline{cg}$, $\overline{db}$ and $\overline{eg}$. Choosing $\overline{cb}$ at this stage gives the subtree $T_4 = (V_4, E_4)$, with $V_4 = \{c, d, e, b\}$, $E_4 = \{\overline{cd}, \overline{ce}, \overline{cb}\}$. The vertices of $V - V_4$ adjacent to vertices in V_4 are a, f, g and the edges incident with these are $\overline{ba}$ of weight 4, $\overline{bf}$ of weight 7, $\overline{cf}$ of weight 8, $\overline{cg}$ of weight 6 and $\overline{eg}$ of weight 6.

The least among these weights being 4, the next subtree is $T_5 = (V_5, E_5)$, with $V_5 = \{c, d, e, b, a\}$, $E_5 = \{\overline{cd}, \overline{ce}, \overline{cb}, \overline{ba}\}$. The vertices of $V - V_5$ that are adjacent to vertices in V_5 are f, g and edges not in E_5

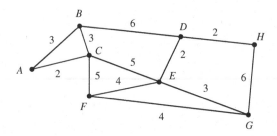

Fig. 8.54.

that are incident with these vertices are $\overline{af}$, $\overline{bf}$, $\overline{cf}$, $\overline{cg}$, $\overline{eg}$ the weights of which are 5, 7, 8, 6, 6 respectively. The least among these is 5 with the corresponding edge $\overline{af}$. Therefore, the next subtree is $T_6 = (V_6, E_6)$, with $V_6 = \{c, d, e, b, a, f\}$, $E_6 = \{\overline{cd}, \overline{ce}, \overline{cb}, \overline{ba}, \overline{af}\}$. There is only one vertex in $V - V_6$ and the edges incident with it are $\overline{fg}$, $\overline{gc}$, $\overline{ge}$ of weights 7, 6, 6, respectively. We may thus choose either $\overline{cg}$ or $\overline{eg}$ each of which is of weight 6. Thus, the final subtree and, so, a minimal spanning trees is $T_7 = (V_7, E_7) = (V, E_7)$ with $E_7 = \{\overline{cd}, \overline{ce}, \overline{cb}, \overline{ba}, \overline{af}, \overline{cg}\}$.

Example 8.52. Using Kruskal's Algorithm, find a minimal spanning tree of the graph in Fig. 8.54.

The edge $\overline{AC}$ being of least weight, the initial subtree is $T_2 = (V_2, E_2)$, with $V_2 = \{A, C\}$; $E_2 = \{\overline{AC}\}$. Using the procedure (and suppressing the detailed discussion), the sequence of subtrees is

$T_3 = (V_3, E_3)$, with $V_3 = \{A, C, B\}$, $E_3 = \{\overline{AC}, \overline{CB}\}$;
$T_4 = (V_4, E_4)$, with $V_4 = \{A, C, B, E\}$, $E_4 = \{\overline{AC}, \overline{CB}, \overline{CE}\}$;
$T_5 = (V_5, E_5)$, with $V_5 = \{A, C, B, E, D\}$, $E_5 = \{\overline{AC}, \overline{CB}, \overline{CE}, \overline{ED}\}$;
$T_6 = (V_6, E_6)$, with $V_6 = \{A, C, B, E, D, H\}$, $E_6 = \{\overline{AC}, \overline{CB}, \overline{CE}, \overline{ED}, \overline{DH}\}$;
$T_7 = (V_7, E_7)$, with $V_7 = \{A, C, B, E, D, H, G\}$, $E_7 = \{\overline{AC}, \overline{CB}, \overline{CE}, \overline{ED}, \overline{DH}, \overline{EG}\}$;
$T_8 = (V_8 = V, E_8)$, with $V_8 = V$, $E_8 = \{\overline{AC}, \overline{CB}, \overline{CE}, \overline{ED}, \overline{DH}, \overline{EG}, \overline{EF}\}$.

Since all the vertices have been taken care of, we have obtained a spanning tree, indeed a minimal spanning tree. Observe that weight of the minimal spanning tree obtained is

$$wt(T_8) = wt(\overline{AC}) + wt(\overline{CB}) + wt(\overline{CE}) + wt(\overline{ED}) + wt(\overline{DH})$$
$$+ wt(\overline{EG}) + wt(\overline{EF})$$
$$= 2 + 3 + 5 + 2 + 2 + 3 + 4 = 21.$$

Lemma. Among all the edges incident with a vertex, an edge with the smallest weight is contained in some minimum spanning tree.

Proof. Let v_1 be a vertex and $\{v_1, v_2\}$ be an edge weight of which is less than or equal to the weights of all edges incident with v_1. Let T be a minimal spanning tree not containing the edge $\{v_1, v_2\}$. Adjoin the edge $\{v_1, v_2\}$ to T and let U be the resulting graph. Then U contains exactly one circuit, namely the fundamental circuit corresponding to the chord $\{v_1, v_2\}$ relative to the tree T. This circuit is made up of the edge $\{v_1, v_2\}$ and the path from v_1 to v_2 in T. Let this path be $\{v_1, v_{i_1}, v_{i_2}, \ldots, v_{i_k}, v_2\}$. Removing the edge $\{v_1, v_{i_1}\}$ from U results in a spanning tree T' in which $\{v_1, v_2\}$ is a branch and the weight of T' is less than or equal to the weight of T. Since T is a minimal spanning tree and T' is a spanning tree with weight less than or equal to the weight of T, T' is a minimal spanning tree.

We use the above observation to give another procedure for obtaining a minimum spanning tree of a connected weighted graph. This procedure uses the method of coalescing of vertices. First we describe the method of coalescing.

Let $G = (V, E)$ be a given graph and v_1, v_2 be two vertices of G. The graph $G' = (V', E')$ obtained on coalescing v_1 and v_2 is such that $V' = (V - \{v_1, v_2\}) \cup \{v^*\}$ i.e. the vertices v_1, v_2 are replaced by a new vertex v^* and E' contains all the edges of E with the change that the edges incident with v_1 and v_2 are now incident with the new vertex v^*. For example, consider the graph as in diagram Fig. 8.55(a) as below.

The graphs on coalescing v_1 and v_2; v_1 and v_3; v_1 and v_4 are the graphs given in Figs. 8.55(b), (c) and (d) respectively.

Theorem 8.16. Let e be an edge with smallest weight that is not a loop in G. Let G' be the graph obtained from $G = (V, E)$ by coalescing the vertices v_1 and v_2 with which the edge e is incident in G, and T' be a minimum spanning tree of G'. Then the subgraph T of G consisting of all the edges in T' together with the edge e is a minimum spanning tree of G.

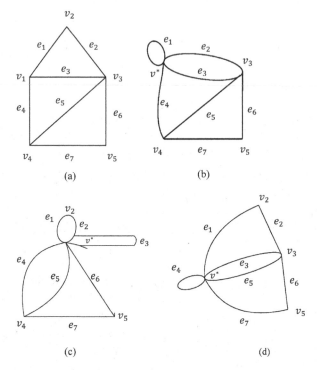

Fig. 8.55.

Proof. Let the vertices v_1, v_2 of G coalesce into v^*. Then T' contains all the vertices of G, the vertex v^* but not the vertices v_1, v_2. When the edge e is added to T', we can suppose the vertex v^* removed and v_1, v_2 introduced. The number of edges in T' is $(o(V)-1)-1$ and, so, the number of edges in T is $0(V)-1$. Hence T is a spanning tree of G. We also have $W(T) = W(T') + w(e)$ where $W(T'), W(T)$ are respectively the sum of weights of the edges of T', T and $w(e)$ is the weight of the edge e.

If T is not a minimal spanning tree of G, there exists a minimum spanning tree $\hat{T}$ of G with $W(\hat{T}) < W(T)$ and since e is an edge of minimum weight in G, we assume that $\hat{T}$ must contain the edge e. Let $\hat{T}'$, denote the tree obtained from $\hat{T}$ on coalescing the vertices with which e is incident and the loop e is removed. Then $\hat{T}'$ is a spanning tree of G'. Then $W(\hat{T}) = W(\hat{T}')+w(e)$. Therefore, $W(\hat{T}')+w(e) < W(T')+w(e)$ or $W(\hat{T}') < W(T')$ which contradicts the assumption that T' is a minimum spanning tree of G'. Hence T is a minimum spanning tree of G.

Procedure (using coalescing).

Let $G = (V, E)$ be a connected weighted graph. A minimum spanning tree is obtained in a step-by-step manner.

1. Find an edge of least possible weight in G. Coalesce the vertices of G with which this edge is incident. Remove the self loops that result and let G' be the subgraph so obtained.
2. Repeat step 1 for G' and continue the process until we arrive at a graph that has a single vertex.
3. Consider the subgraph of G with all the vertices of G with the vertices v_i and v_j joined by the edge e which necessitated their coalescing. It is better to mark the edges $\{v_i, v_j\}$ in the original graph and then to remove the unmarked edges.

There results a minimum spanning tree of G. A slight variation of this procedure is an algorithm which is called Prim's Algorithm.

A vertex u in a weighted graph $G = (V, E)$ is called a **nearest neighbour** of vertex v if u and v are adjacent and no other vertex of G is joined to v by an edge the weight of which is less than the weight of the edge $\{u, v\}$. A vertex v is called a **nearest neighbour of a set of vertices** $V_1 = \{v_1, v_2, \ldots, v_k\}$, in G if v is adjacent to some member v_i of V_1 and no other vertex adjacent to a member of V_1 is joined by an edge the weight of which is less than the weight of the edge $\{v, v_i\}$. The vertex v may or may not belong to V_1.

Prim's Algorithm

Let G be a graph on n vertices.

Step 1. Choose a vertex v_1 in G. Let $V = \{v_1\}$, and $E = \phi = \{\ \}$.

Step 2. Choose a nearest neighbour v_i of V that is adjacent to v_j, $v_j \in V$, and for which the edge $\{v_i, v_j\}$ does not form a cycle with members of E. Add v_i to V and add the edge $\{v_i, v_j\}$ to E.

Step 3. Repeat step 2 until $O(E) = n - 1$. Then V contains all the n vertices of G and E contains the edges of G which form a minimal spanning tree of G.

Using the Lemma proved after Example 8.52 and an argument similar to the one in the proof of Theorem 8.15, we can prove the following.

Theorem 8.17. Prim's algorithm produces a minimum spanning tree of a connected weighted graph.

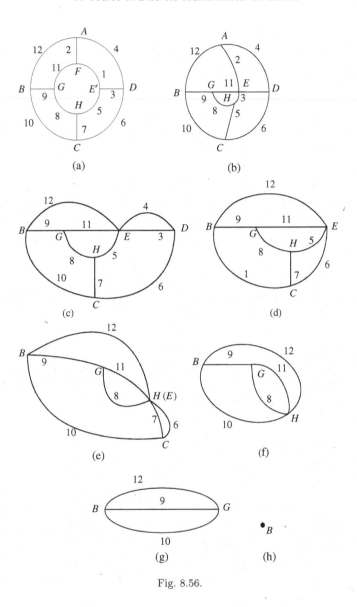

Fig. 8.56.

Example 8.53. We apply the above procedure to obtain a minimum spanning tree of the graph of Fig. 8.56(a).

An edge of minimum weight is $\{E, F\}$. On coalescing the vertices E and F we obtain the graph as in (b). We next coalesce the vertices A and E

and obtain the graph as in (c). The edge $\{D, E\}$ being of minimum weight, coalesce D, E to obtain the graph (d) (the edge of weight 4 becomes a self loop on coalescing and is, therefore, omitted). Next is the turn to coalesce E and H which results in the graph (e). Now coalesce D and C. The edge (C, H) of weight 7 becomes a self loop and, so, is also omitted. The resulting graph is given in (f). Next is the turn to coalesce G and H because of the (G, H) being an edge of weight 8. The edge of weight 11 also becomes a self-loop and is omitted. The resulting graph is as in (g). Now B and G are to the coalesced and there results a graph with single vertex.

The edges removed because of the coalescing of vertices are $\{E, F\}_1$, $\{A, F\}_2$, $\{D, E\}_3$, $\{D, H\}_5$, $\{D, C\}_6$, $\{H, G\}_8$, $\{G, B\}_9$ and the minimal spanning tree obtained is as given in Fig. 8.57.

Example 8.54. Consider the graph of Fig. 8.58.

We use the method of coalescing (i.e. of merger) to obtain a minimal spanning tree. We start with the vertex A. The edge $\{A, C\}$ being of weight 2, we coalesce A and C. Once this is done, we get two edges $\{B, C\}$ of weight 3 each. Let us coalesce one of these two, say the original vertices A and B. Then we cannot coalesce B and C for that will lead to a cycle (A, B, C, A). Now we look at the vertex C again. Let us coalesce C and F. So the edges included in the tree so far are $\{A, C\}$, $\{A, B\}$, $\{C, F\}$. Having come to F, we coalesce F and G, then G and E. Then we coalesce D and E and finally D and H. We thus finally get the spanning tree as in Fig. 8.59 the weight of which is 21. It is a minimal spinning tree.

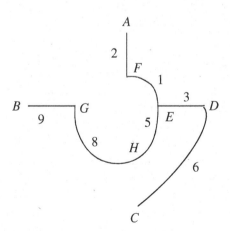

Fig. 8.57.

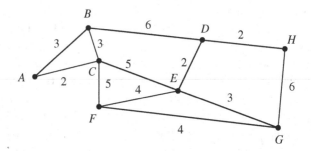

Fig. 8.58.

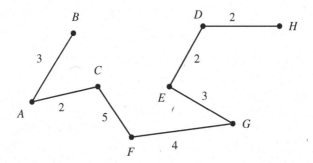

Fig. 8.59.

Let us next examine all the edges of the given graph (Fig. 8.58) in increasing order of their weights. We have $\{A,C\}_2$, $\{D,E\}_2$, $\{D,H\}_2$, $\{A,B\}_3$, $\{B,C\}_3$, $\{E,G\}_3$, $\{E,F\}_4$, $\{F,G\}_4$, $\{C,E\}_5$, $\{C,F\}_5$, $\{B,D\}_6$, $\{G,H\}_6$.

Retaining all the three edges of weight 2 does not lead to any cycle and so, we retain these. The next edge under consideration is $\{A,B\}$ and its inclusion dos not lead to a cycle and we retain it. Inclusion of $\{B,C\}$ leads to a cycle and so this edge is omitted. However, inclusion of the edge $\{E,G\}$ does not lead to a problem and retain it. Adding the edge $\{E,F\}$ does not result in a cycle and we retain it. The edge $\{F,G\}$ if added leads to a cycle and, so, is omitted. We can next add the edge $\{C,E\}$ but will have to omit the remaining edges $\{C,F\}$, $\{B,D\}$ and $\{G,H\}$ as the inclusion of any one of them leads to a cycle. The minimal spanning tree thus obtained is (as in Fig. 8.60). The weight of this spanning tree is also 21.

We next use Prim's algorithm to obtain a minimal spanning tree for the same graph.

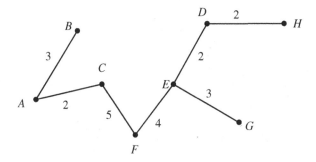

Fig. 8.60.

We choose A as the first vertex. The nearest neighbour to this vertex is the vertex C. So we choose the edge (A, C). We consider the set $S_1 = \{A, C\}$ consisting of two vertices A and C. The nearest neighbour to S_1 is B and we arbitrarily choose the edge (B, C) for inclusion. So, we have the set $S_2 = \{A, B, C\}$ of three vertices and the set $E_2 = \{(A, C), (B, C)\}$ containing two edges. E, F both being at the same distance 5 from C are nearest neighbours of S_2 and we arbitrarily choose the edge (C, E) for inclusion. We then have $S_3 = \{A, B, C, E\}$ and $E_3 = \{(A, C), (B, C), (C, E)\}$. The vertex D is at a distance 2 from E and, therefore, D is the nearest neighbour to S_3. We then consider $S_4 = \{A, B, C, E, D\}$ and $E_4 = \{(A, C), (B, C), (C, E), (E, D)\}$. The nearest neighbour to S_4 is H and we next consider $S_5 = \{A, B, C, E, D, G\}$, $E_5 = \{(A, C), (B, C), (C, E), (E, D), (D, H)\}$. The nearest neighbour to S_5 is H and we, get $S_6 = \{A, B, C, E, D, G, H\}$, $E_6 = \{(A, C), (B, C), (C, E), (E, D), (E, G), (D, H)\}$. The next nearest neighbour to S_6 is F and it is equidistant from E and G. We may choose either of the two edges (E, F) or (F, G). Inclusion of neither of the two edges leads to a cycle. If we include the edge (E, F) respectively (F, G), the corresponding minimal spanning tree is as in Fig. 8.61(a) respectively 8.61(b).

We now apply Kruskal's Algorithm to obtain a minimal tree of the same graph.

We consider an edge (A, C) which is of least weight. Set $E_1 = E \setminus \{(A, C)\}$. In E_1 an edge of least weight we choose as (E, D). Then $S_2 = \{(A, C), (E, D)\}$ and we set $E_2 = E \setminus S_2$. An edge of least weight in E_2 is (D, H) and this together with S_2 does not result in a cycle. So, we set $S_3 = \{(A, C), (E, D), (D, H)\}$ and $E_3 = E \setminus S_3$. In E_3 the edges $(A, B), (B, C)$ and (E, G) are each of weight 3. Choosing any one of these does not result

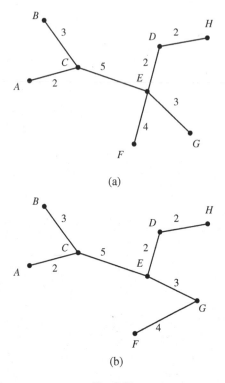

(a)

(b)

Fig. 8.61.

in a cycle in $S_3 \cup$ {this edge}. We choose (E, G) as an edge of least weight in E_3. We get $S_4 = \{(A, C), (E, D), (D, H), (E, G)\}$ and $E_4 = E \backslash S_4$.

Edges of least weight in E_4 are $(A, B), (B, C)$ each one of which is of weight 3. Let us choose (A, B) as an edge for inclusion. We get $S_5 = \{(A, C), (E, D), (D, H), (E, G), (A, B)\}$ and $E_5 = E \backslash S_5$.

An edge of least weight in E_5 is (B, C) but adding that results in a cycle. So, we cannot include (B, C). Edges of next least possible weight are (E, F) and (F, G) each of which is of weight 4. Including either of these two to S_5 does not result in a cycle in the graph constructed so far. We choose (E, F) for inclusion. Then $S_6 = \{(A, C), (E, D), (D, H), (E, G), (A, B), (E, F)\}$ and $E_6 = E \backslash S_6$.

An edge of least possible weight in E_6 is (F, G) but adjoining to the graph constructed so far results in the formation of a cycle. Therefore, (F, G) cannot be added to S_6. The next edges of lowest possible weight – in E_6 are (C, E) and (C, F) and either of these may be added to

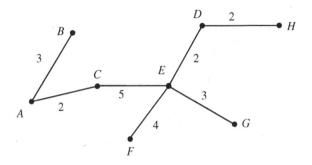

Fig. 8.62.

S_6. We select (C, F). Thus $S_7 = \{(A, C), (E, D), (D, H), (E, G), (A, B), (E, F), (C, F)\}$.

Thus, we have got seven edges as required and a minimal tree obtained is given in Fig. 8.62.

This method is the same as the method of obtaining a minimal spanning tree by listing the edges of the graph in increasing order of their weights, examining the list so obtained from left to right and ignoring step by step, edges which result in yielding a cycle in the graph.

Example 8.55. The distances between eight cities are given by the following table. Use Kruskal's Algorithm to find a minimal spanning tree the vertices of which are these cities. What is the total distance for this tree?

	A	**B**	**C**	**D**	**E**	**F**	**G**	**H**
A	—	69	121	30	113	70	135	63
B	69	—	52	97	170	117	163	16
C	121	52	—	149	222	160	206	59
D	30	97	149	—	92	63	122	93
E	113	170	222	92	—	155	204	174
F	70	117	160	63	155	—	66	101
G	135	163	206	122	204	66	—	147
H	63	16	59	93	174	101	147	—

Solution. We use Kruskal's Algorithm version which lists all the edges in the increasing order of their weights. But first we draw the graph. Observe that every two edges are connected and so it is a complete graph.

We list the edges in the increasing order of their weights.

$(B, H)_{16}, (A, D)_{30}, (B, C)_{52}, (C, H)_{59}, (A, H)_{63}, (D, F)_{63}, (F, G)_{66},$

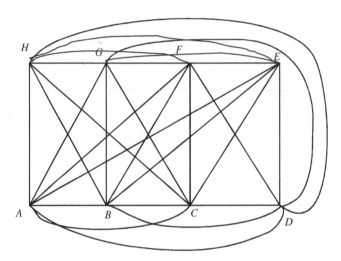

Fig. 8.63.

$(A, B)_{69}$, $(A, F)_{70}$, $(D, E)_{92}$, $(D, H)_{93}$, $(B, D)_{97}$, $(F, H)_{101}$, $(A, E)_{113}$, $(B, F)_{117}$, $(A, C)_{121}$, $(D, G)_{122}$, $(A, G)_{135}$, $(G, H)_{147}$, $(C, D)_{149}$, $(E, F)_{155}$, $(C, F)_{160}$, $(B, G)_{163}$, $(B, E)_{170}$, $(E, H)_{174}$, $(E, G)_{204}$, $(C, G)_{206}$, $(C, E)_{222}$.

Proceeding in order, we find that inclusion of $(B, H), (A, D), (B, C)$ does not create any cycle in the graph being constructed. Adding the next edge (C, H) generates a cycle in the graph and so (C, H) is ignored. Adding (A, H) creates no problem and so is the case with $(D, F), (F, G)$. However, if (A, B) or (A, F) is added next, there results a circuit. Thus $(A, B), (A, F)$ are ignored. Adding (D, E) is no problem but adding (D, H) creates a circuit. So (D, H) is ignored. For giving rise to a circuit, if added, the edges $(B, D), (F, H), (A, E), (B, F)$ are ignored. In fact, since we have already got seven edges as required for a tree on eight vertices, no further edges need to be added and we have already got a minimal spanning tree (Fig. 8.64) of the graph. The weight of this tree is $52 + 16 + 63 + 30 + 63 + 66 + 92 = 382$.

Exercise 8.2. Find a minimal spanning tree for the graph in the last example by using the method of coalescing of vertices by starting with the vertex B.

Example 8.56. Use Prim's Algorithm and find a minimal spanning tree of the connected graph given in Fig. 8.65 beginning with the vertex E.

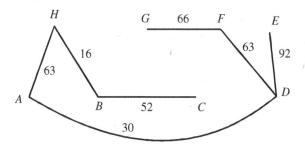

Fig. 8.64.

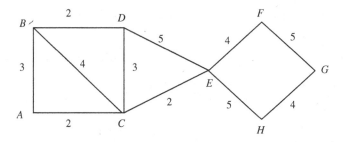

Fig. 8.65.

Solution. The nearest neighbour of E is the vertex C. So, we take $S_1 = \{E, C\}$ and $E_1 = \{(C, E)\}$. A nearest neighbour of S_1 is the vertex A and so, we take $S_2 = \{A, C, E\}$ and $E_2 = \{(C, E), (A, C)\}$. We can take next nearest neighbour of S_2 as either B or D neither of which creates a cycle. So, we take

$$S_3 = \{A, C, E, B\}, \quad E_3 = \{(C, E), (A, C), (A, B)\}.$$

Nearest neighbour of S_3 is D and we take

$$S_4 = \{A, C, E, B, D\}, \quad E_4 = \{(C, E), (A, C), (A, B), (B, D)\}.$$

Next, nearest neighbour of S_4 is C and including the edge (D, C) leads to a circuit. Therefore (D, C) cannot be included. Therefore, next nearest neighbour of S_4 is F and we take

$$S_5 = \{A, C, E, B, D, F\}, \quad E_5 = \{(C, E), (A, C), (A, B), (B, D), (E, F)\}.$$

Both the vertices H and G are at the same distance from S_5 and inclusion of either the edge (F, G) or (E, H) does not lead to a circuit. So, let us take

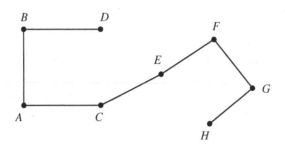

Fig. 8.66.

G as the next nearest neighbour. So, we get

$$S_6 = \{A, C, E, B, D, F, G\},$$

$$E_6 = \{(C, E), (A, C), (A, B), (B, D), (E, F), (F, G)\}.$$

The vertex H is a nearest neighbour of S_6 and we add this vertex and the edge (G, H) and get

$$S_7 = \{A, C, E, B, D, F, G, H\}, E_7 = \{(C, E), (A, C), (A, B), (B, D),$$

$$(E, F), (F, G), (G, H)\}.$$

We have got all the eight vertices and seven edges. Hence, we have arrived at a minimal spanning tree which is given in Fig. 8.66.

We have so far considered the question of minimal spanning tree. Consider a network of pipes through which water/gas/oil is flowing and we want a maximal flow. We may like a system of connections such that we pass through every junction and get a maximal flow. Thus, we want to get a maximal spanning tree of the graph.

Kruskal's Algorithm (for maximal spanning tree).

Let G be a connected graph with n vertices and let $S = \{e_1, e_2, \ldots, e_k\}$ be the set of weighted edges of G.

Step 1. Choose an edge e_i in S of largest weight. Let $E = \{e_i\}$. Replace S with $S \backslash \{e_i\}$.

Step 2. Select an edge e_i in S of largest weight that will not make a circuit with members of E. Replace E with $E \cup \{e_i\}$ and S with $S \backslash \{e_i\}$.

Step 3. Repeat step 2 until $0(E) = n - 1$.

The above algorithm is equivalent to the following procedure.

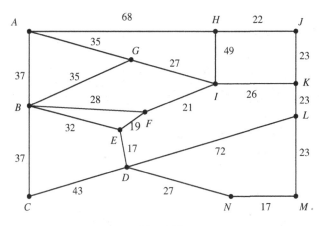

Fig. 8.67.

Arrange all the edges of G in the decreasing order of their weights. Examine these edges in order from left to right step by step and at any given step retain an edge if its inclusion does not result in producing a cycle or a circuit with the edges already chosen and ignore it otherwise. Continue this until either we have already selected $n-1$ edges or we come to the last edge in the list prepared in the beginning.

Example 8.57. Suppose that the graph as given in Fig. 8.67 represents possible flows through a system of pipes. Find a spanning tree that gives the maximum possible flow in this system.

Solution. The edges of the graph in decreasing order of their weights are:

$(D, L)_{72}, (A, H)_{68}, (H, I)_{49}, (C, D)_{43}, (A, B)_{37}, (B, C)_{37}, (B, G)_{35},$
$(A, G)_{35}, (B, E)_{32}, (B, F)_{28},$
$(G, I)_{27}, (D, N)_{27}, (I, K)_{26}, (J, K)_{23},$
$(K, L)_{23}, (L, M)_{23}, (H, J)_{22}, (F, I)_{21},$
$(E, F)_{19}, (D, E)_{17}, (M, N)_{17}.$

Observe that retaining the edges $(D, L), (A, H), (H, I), (C, D), (A, B),$ (B, C) and (B, G), does not create any cycle. However, if we retain the next edge (A, G) in order, there results a circuit and, therefore, (A, G) is ignored. Adding (B, E) and (B, F) in order does not result in creating a cycle and are, therefore, retained. Adding (G, I) to the edges already chosen, results in creating a circuit along with the edges earlier chosen. Therefore, (G, I) is ignored. Next adding the edges $(D, N), (I, K), (J, K)$ to the ones already

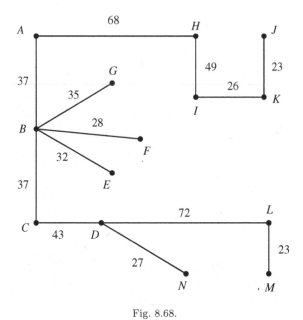

Fig. 8.68.

chosen does not lead to creating a cycle. These edges are therefore retained. Including the next edge (K, L) to the system creates a circuit and, so, the edge (K, L) is ignored. However, adding the edge (L, M) does not create any problem. So, this edge is added. We have already chosen 13 edges. The number of vertices being 14, we have already obtained a tree, a maximal tree (so to say) (Fig. 8.68).

The weight of this maximal spanning tree obtained is

$$72 + 68 + 49 + 43 + 37 + 37 + 35 + 32 + 28 + 27 + 26 + 23 + 23 = 500.$$

So, the maximal flow through this path is 500.

Please note that the study of trees and, in particular, their applications is mainly based on and influenced by the presentation of the subject in Dornhoff and Hohn (1978), Kolman, Busby and Ross (2005), Liu (2000), Deo (2007) and Rosen (2003/2005). The theorem of Cayley on counting of trees is particularly taken from Deo (2007). Study of eccentricity and of centre of a tree is influenced by Dornhoff and Hohn (1978).

Exercise 8.3.

1. Prove that (up to isomorphism) there are only three trees with five vertices (as shown in Fig. 8.1).

2. Prove that the only trees (up to isomorphism) with six vertices are those given in Fig. 8.1.

3. Construct all possible trees (up to isomorphism) with one, two and three vertices.

4. Prove that any tree with n vertices, $n \geq 2$, must have at least two vertices of degree 1.

5. If G is a connected graph and the removal of any edge leaves a disconnected subgraph consisting of two components, each of which is a tree, prove that G itself is a tree.

6. If any two non-adjacent vertices in a tree are joined by an edge, prove that the resultant graph has exactly one circuit.

7. Let G be a graph that has no circuits. When any two non-adjacent vertices in G are joined by an edge, the resultant graph has a single circuit. Prove that G is a tree.

8. Find the centres of the trees of Fig. 8.69.

9. Find all pairwise nonisomorphic rooted trees with six vertices.

10. If n is the number of vertices in a binary tree, prove that n is odd.

11. If the binary tree T has n vertices, find the number of leaves in T and also the number of vertices of degree 3 in T.

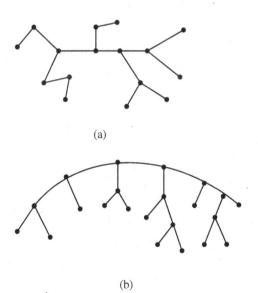

(a)

(b)

Fig. 8.69.

12. If i is the number of branch nodes in a regular m-tree, prove that the number of edges in the tree is mi.

13. Find the value of each of the prefix expressions

 (a) $+ - *235/\uparrow 234$.

 (b) $+ * + - 53214$.

 (c) $*/93 + *24 - 76$.

14. Find the value of each of the postfix expressions

 (a) $723 * -4 \uparrow 93/+$.

 (b) $21 - 324 \div +*$.

 (c) $432 \div -5 * 42 * 5 * \div \div$.

15. Draw the ordered rooted tree corresponding to each of the arithmetic expressions of Question 13 above written in prefix notation.

16. Draw the ordered rooted tree corresponding to each of the arithmetic expressions of Question 14 above written in postfix notation.

17. Construct a binary tree when its inorder and postorder traversal strings are

$$\text{inorder}: \; gdbheiacf$$
$$\text{postorder}: \; gdhiebfca.$$

18. Show the sequential orders in which the vertices of the trees given in Fig. 8.70 are visited in a preorder traversal, an inorder traversal, and a postorder traversal.

 Also, determine the height of each one of these trees. Decide also the centre(s) of each one of these trees. Determine also the diameter in each case and also the path length.

19. Given orders of traversals in (a) preorder and inorder or (b) inorder and postorder, we can construct a binary rooted tree. Given preorder and postorder traversal orders of a rooted tree, is it possible to construct a rooted tree which corresponds to the two traversal orders?

20. Build a binary search tree for the words

 (a) banana, peach, apple, pear, coconut, mango and papaya.

 (b) oenology, phrenology, campanology.

21. Construct the labelled tree that represents the Huffman code:

$$a: 000, \quad b: 01, \quad c: 001, \quad d: 1100, \quad e: 1101.$$

22. In the tree of Fig. 8.71 show the result of performing (a) preorder, (b) inorder and (c) postorder search.

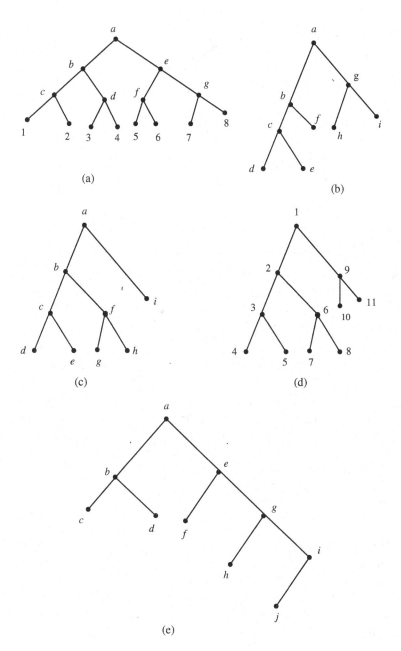

(a)

(b)

(c)

(d)

(e)

Fig. 8.70.

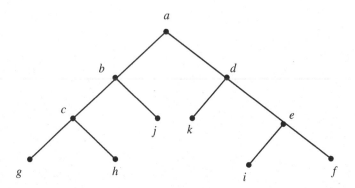

Fig. 8.71.

23. Use the tournament sort to sort the list

(a) 23, 9, 15, 17, 4, 10, 28, 12.
(b) 16, 3, 2, 4, 12, 9, 13, 5.

Show the labels of the vertices at each step.

24. Prove that a minimal connecting subgraph of a connected graph is a spanning tree of the graph.

25. Find the numbers of vertices in a full (a) 5-ary tree (b) 3-ary tree with 100 internal vertices.

26. Using backtracking find five positions on a 5×5 chess board so that no two of these positions are in the same row, same column or the same diagonal. Recall that a diagonal consists of all positions (i, j) with $i + j = m$ for some m or $i - j = m$ for some m.

27. Find the number of different spanning trees for each of the simple graphs

(a) K_3;   (b) K_4;   (c) $K_{2,2}$;   (d) C_5.

28. Use a depth-first search to produce a spanning tree for the simple graphs given in Fig. 8.72. Choose a as the root of this spanning tree, assuming the vertices are ordered alphabetically.

29. Use a breadth-first search to produce a spanning tree for either of the simple graphs in the above exercise. Choose a as the root of either spanning tree.

30. Use backtracking to find a subset, if it exists, of the set $\{28, 25, 20, 15, 12, 9\}$ with sum

(a) 23;   (b) 45;   (c) 65.

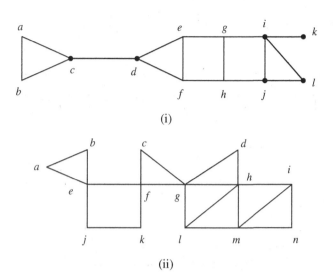

(i)

(ii)

Fig. 8.72.

31. Let G be a connected graph and T be a spanning tree of G constructed using breadth-first search. Show that an edge of G not in T must connect vertices at the same level or at levels that differ by one in the spanning tree.

32. Use Prim's algorithm to find a minimum spanning tree for each of the weighted graphs in Fig. 8.73.

33. Use Kruskal's Algorithm to find a minimum spanning tree for each of the weighted graphs in Question 32 above.

34. Find a maximum spanning tree for each of the weighted graphs in Question 32.

35. Find necessary and sufficient conditions for a connected weighted graph to have a unique minimum spanning tree.

36. Prove Theorem 8.17.

37. Draw a binary tree whose preorder search produces the string.
    (a) *KCBDEJIFHG*; (b) *DBUTBOEEPHT*.

38. Draw a binary tree whose postorder search produces the string.
    (a) *RESEARCHING*; (b) *FARMHOUSE*.

39. If v is a vertex in a connected graph G, show that there is a spanning tree T of G such that for all vertices v_i, the distance $d(v, v_i)$ is the same in G and in T.

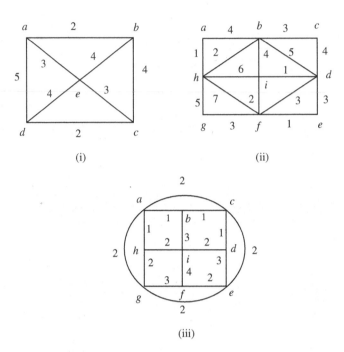

(i)   (ii)

(iii)

Fig. 8.73.

40. If e is an edge in a connected graph G, show that there is a spanning tree T_1 in G which contains e. Also show that if e is in some circuit in G, then some spanning tree T_2 in G does not contain e.

41. Prove that a subgraph of a connected graph is a Hamiltonian path if and only if it is a spanning tree with all vertices of degree at most 2.

42. Let L be a circuit in a graph G and a, b be any two edges in L. Prove that there exists a cut-set C such that $L \cap C = \{a, b\}$.

43. Determine a minimum spanning tree for each of the connected weighted graphs in Fig. 8.74.

44. Draw a tree in which the diameter is not equal to twice its radius. Under what condition(s) does equality hold?

45. Prove that a pendant edge in a connected graph G is contained in every spanning tree of G.

46. Let T_1, T_2 be two spanning trees of a connected graph G. If an edge e is in T_1 but not in T_2, prove that there exists an edge f in T_2 but not in T_1 such that the subgraphs $(T_1 - e) \cup f$ and $(T_2 - f) \cup e$ are also spanning trees of G.

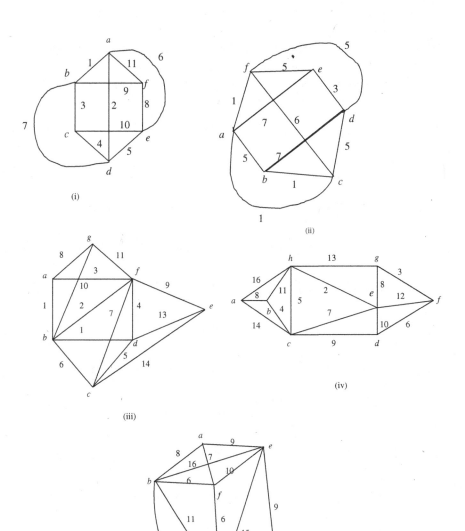

Fig. 8.74.

47. State Prim's Modified Algorithm to find a maximal spanning tree of a connected weighted graph.

48. Use the modified Prim's Algorithm to find a maximal spanning tree of the graph in the Example 8.57.

49. Does a tree exist with $2n$ vertices of degree 1, $3n$ vertices of degree 2 and $4n$ (respectively $3n$) vertices of degree 3? Justify.

Chapter 9

GROUPS

In this chapter we will study some algebraic structures. These structures are built on non-empty sets with some binding conditions/relations on every set. These binding conditions/relations are what we call 'binary composition'. Some very elementary properties of groups, subgroups, quotient groups and symmetric groups are studied.

9.1. Groups: Preliminaries

Let A be a non-empty set. A **binary composition** in A is a map from the Cartesian product $A \times A$ to A itself. Equivalently, a binary composition in (or on) A is a rule or a law that associates to every ordered pair a, b of elements of A a unique element of A again. The unique element of A corresponding to the pair of elements a, b may be denoted by $a * b$ or $a \circ b$ or $a \cdot b$ or simply ab. This unique element may also be written as $a \oplus b$ or just $a+b$. If the element of A associated with a, b is written as ab, the composition in A is called a 'multiplicative composition' while if this element is written as $a + b$ we call the composition an 'additive composition'. We have the following familiar compositions on some known sets:

1. On the set Z of integers we have

   (a) $(m, n) \to m + n$, $m, n \in Z$;
   (b) $(m, n) \to mn$, $m, n \in Z$;
   (c) $(m, n) \to m - n$, $m, n \in Z$.

   i.e. the usual addition, multiplication of the number system and taking difference.

2. The above compositions are also defined in the set Q of all rational numbers, the set R of all real numbers, the set C of all complex numbers and on the set $2Z$ of all even integers.

3. On the set $2Z + 1$ of all odd integers, multiplication of the number system gives a composition while addition and subtraction do not give a composition.

4. On the set N of all natural numbers, although addition and multiplication of the number system give binary compositions, taking difference (i.e. subtraction) does not give a composition.

5. On the sets $Z^* = Z\backslash\{0\}$ of all non-zero integers, $Q^* = Q\backslash\{0\}$ of all non-zero rational numbers, $R^* = R\backslash\{0\}$ of all non-zero real numbers, $C^* = C\backslash\{0\}$ of all non-zero complex numbers and $2Z^* = 2Z\backslash\{0\}$ of all non-zero even integers, multiplication of the number system induces compositions while addition of the number system does not induce a composition.

6. On the sets $Z, 2Z, Q, R$ and C we may also define a composition by $a \cdot b = a + b - ab$, for a, b in the set concerned.

7. Let $M_2(R), M_2(R)^*$ and $M_2(R)^{**}$ denote respectively the set of all square matrices of order 2, the set of all non-singular square matrices of order 2 and the set of all square matrices of order 2 with determinant 1, all matrices with entries in the set R of real numbers. Since the product of non-singular matrices is non-singular and since the determinant of a product of matrices each of determinant 1 is again 1, the product of matrices induces compositions in $M_2(R), M_2(R)^*$ and $M_2(R)^{**}$. Again, the sum of non-singular matrices need not be non-singular and the sum of matrices each of determinant 1 need not have determinant 1. Therefore, although addition of matrices induces a composition in $M_2(R)$ it does not induce a composition in $M_2(R)^*$ or $M_2(R)^{**}$.

8. The vector product of two three-dimensional vectors is again a three-dimensional vector. Therefore, the vector product gives a composition in the set V of all three-dimensional vectors.

9. Let $n \geq 2$ be an integer and G_n be the set of all nth roots of unity. Recall that an nth root of unity is a complex number z such that $z^n = 1$. The product of two nth roots of unity is again an nth root of unity. Therefore, the product of complex numbers induces a composition in the set G_n. The sum of two nth roots of unity need not be an nth root of unity. For example, if $n = 4$, both 1 and $i = \sqrt{-1}$ are fourth roots of unity but

$$(1 + i)^4 = 1 + 4i + 6i^2 + 4i^3 + i^4 = -4 \neq 1$$

and $1 + i$ is not a fourth root of unity. Therefore, the sum of complex numbers does not induce a composition in G_n.

10. Let A be an alphabet set and A^+ be the set of all possible finite strings over the alphabet set A. For two strings s_1 and s_2 in A^+, define $s_1 * s_2$ by concatenation of strings, i.e. $s_1 * s_2 = s_1 s_2$. Then $*$ is a binary composition in A^+.

A binary composition defined on a non-empty set S is said to satisfy the associative law if for all $a, b, c \in S, a(bc) = (ab)c$. A non-empty set S with a binary composition which satisfies the associative law is called a **semigroup**.

Let S be a non-empty set with a binary composition defined on it. Let e, e' be the elements of S such that

$$ae = ea = a \quad \text{for every } a \in S \tag{9.1}$$

and

$$ae' = e'a = a \quad \text{for every } a \in S. \tag{9.2}$$

Then, by (9.1) $ee' = e'$ while, by (9.2) $ee' = e$. Therefore, $e' = e$. The unique element $e \in S$ for which $ea = ea = a$ for every $a \in S$ is called the **identity element** of S.

A semigroup with identity is called a **monoid**. Thus, a monoid M is a non-empty set with a binary composition which satisfies the associative law and has identity element. A non-empty set G with a binary composition defined in it is called a **group** if

(a) the composition satisfies the associative law, i.e. if $(ab)c = a(bc)$ for all $a, b, c \in G$,

(b) G has identity, i.e. there exists an element $e \in G$ such that $ae = ea = a$ for every $a \in G$,

(c) for every $a \in G$ there exists an element $b \in G$ such that $ab = ba = e$.

As observed above, every group has a unique identity. Let G be a group and $a \in G$. Suppose that $b, b' \in G$ are elements such that

$$ab = ba = e \tag{9.3}$$

and

$$ab' = b'a = e \tag{9.4}$$

where e is the identity of G.

Consider the element $b'(ab)$ of G. Now, by (9.3), $b'(ab) = b'e = b'$.
Again, using the associative law in G and (9.4), $b' = b'e = b'(ab) =$
$(b'a)b = eb = b$. Thus, for $a \in G$, there is a unique element $b \in G$ such that
$ab = ba = e$. This unique element $b \in G$ is called the **inverse** of a and we
write $b = a^{-1}$.

Observe that every group is a monoid and every monoid is a semigroup.
A group G in which the composition satisfies the commutative law, i.e.
$ab = ba$ for all $a, b \in G$ is called an **Abelian group**.

Example 9.1.

(a) $Z, 2Z, Q, R, C$ are all groups (in fact Abelian groups) with respect to
   addition of the number system.
(b) While $2Z$ is a semigroup with respect to multiplication, it is not a
   monoid. However, $2Z + 1$ is a monoid with respect to multiplication.
(c) Every one of the sets Z, Q, R, C is a monoid with respect to
   multiplication but none of these is a group with respect to
   multiplication.
(d) While $Z^* = Z\backslash\{0\}$ is a monoid with respect to multiplication but not
   a group, every one of $Q^* = Q\backslash\{0\}, R^* = R\backslash\{0\}, C^* = C\backslash\{0\}$ is a group
   with respect to multiplication.
(e) The set $M_2(R)$ is a group with respect to addition of matrices but
   not a group with respect to multiplication of matrices. With respect to
   multiplication $M_2(R)$ is a monoid.
(f) The sets $M_2(R)^*$ and $M_2(R)^{**}$ are groups with respect to matrix
   multiplication but neither of these is Abelian.
(g) The set $M_2(2Z)$ of all square matrices of order 2, with entries as
   even integers is a semigroup but not a monoid with respect to matrix
   multiplication.
(h) For an integer $n \geq 2$, the set G_n of all nth roots of unity is an Abelian
   group with respect to multiplication.
(i) The set V of all three-dimensional vectors is a group with respect to
   addition of vectors but is not even a semigroup with respect to vector
   multiplication.
(j) For an alphabet A, the set A^+ of all strings over A is a monoid but not
   a group.

In some of the examples, we restricted ourselves to matrices of order 2
over R. We could take these over Q or C. Also, we could have considered
matrices of any order $n \geq 2$.

Let G be a group. For any $a \in G$ and integer n, we define the power a^n as follows:

$$a^0 = e \quad a^n = \underbrace{a.a.....a}_{n \text{ times}}, \quad \text{if } n > 0,$$

and

$$a^n = (a^{-1})^{-n}, \quad \text{if } n < 0.$$

In fact, for $n > 0, a^n$ could also be defined inductively as follows: $a^1 = a$, and if for $n > 0, a^n$ is already defined, then $a^{n+1} = (a^n)a$.

Using a simple induction argument, we can have the following.

Theorem 9.1. Let G be a group. For any integers m, n and $a, b \in G$, $(a^m)^n = a^{mn}$ and $a^m.a^n = a^{m+n}$. If G is Abelian and $a, b \in G$, then $(ab)^n = a^n b^n$.

The **order** of a group G is the number of elements in the set G. If this number is finite, say n, we say G is a **finite group of order** n. Otherwise, we say that G is an **infinite group** or that G is a group of **infinite order**. We write $o(G)$ or also $|G|$ for the order of G.

The **order of an element** a of a group G is the smallest positive integer n, if it exists, such that $a^n = e$. If no such n exists then a is said to be of **infinite order**. We write $o(a)$ or also $|a|$ for order of a. Observe that the order of the identity e of any group G is always 1. Also, for any $a \in G$, the order of a and a^{-1} are equal. All the groups encountered in Example 9.1 except those in (h) and (j) are infinite. The one in Example 9.1(h) is of order n.

A group G is called a **cyclic group** if there exists an element $a \in G$ such that every element of G is a power of a. (In case G is a group with respect to addition, then every element of G needs to be a multiple of a instead of a power of a.) Then a is called a **generator** of G and we say that G is a cyclic group generated by a. Observe that every power of a is also a power of a^{-1} and conversely. Therefore, if G is a cyclic group generated by a, then it is also generated by a^{-1}. If G is a cyclic group generated by a, we write $G = \langle a \rangle$.

Theorem 9.2. Given any integer $n \geq 1$, there exists a group of order n.

Proof. If $n = 1$, the set $\{1\}$ consisting of the integer 1 is a group with respect to multiplication and its order is 1. So, suppose that $n \geq 2$. Let

$$G_n = \text{the set of all nth roots of unity}$$

$$= \{x \in C \mid x^n = 1\}.$$

The product of nth roots of unity is again an nth root of unity. Therefore, multiplication of complex numbers gives a composition in G_n. The integer $1 \in G_n$ and if $x \in G_n$, then $x^n = 1, x \neq 0$ and therefore,

$$\left(\frac{1}{x}\right)^n = \frac{1}{x^n} = 1.$$

Thus the inverse of x is again in G_n. Since multiplication satisfies the associative law for all complex numbers, it does so for any subset of C. Hence G_n is a group, in fact an Abelian group. Every element of G_n is a root of the polynomial $X^n - 1$ of degree n and every root of the polynomial $X^n - 1$ is in G_n. Since a polynomial of degree n has exactly n roots, G_n has n elements.

For any integer $k, 0 \leq k < n, e^{2k\pi i/n} = \cos 2k\pi/n + i \sin 2k\pi/n$ is a complex number with $(e^{2k\pi i/n})^n = e^{2k\pi i} = \cos 2k\pi + i \sin 2k\pi = 1$. Therefore, $e^{2k\pi i/n} \in G_n$. It is not hard to prove that no two of these n elements are equal. Hence $G_n = \{e^{2k\pi i/n} \mid 0 \leq k < n\}$ and, so, every element of G_n is a power of $e^{2\pi i/n}$. Hence G_n is a cyclic group.

Example 9.2. Every integer is a multiple of the integer 1. Therefore, the additive group Z of integers is a cyclic group generated by 1.

Example 9.3. The set $G = \{1, -1\}$ is a group with respect to multiplication and is cyclic generated by -1.

Example 9.4. The set $G = \{1, -1, i, -i\}$ is a group with respect to multiplication and every element of G is a power of i. Therefore, $G = <i>$.

We record without proof the following simple but useful result and the student is encouraged to supply a proof.

Theorem 9.3. Let G be a non-empty set with a binary composition that satisfies the associative law. Then

(a) G is a group if the two cancellation laws hold in G;

(b) G is a group if for any $a, b \in G$, the equations $ax = b$ and $ya = b$ have unique solutions in G.

9.2. Subgroups

Let G be a group. The composition in G induces a composition in a non-empty subset H of G if $a, b \in H$ imply that $ab \in H$. A non-empty subset H of a group G is called a **subgroup** of G if

(i) the composition in G induces a composition in H;
(ii) H is a group with respect to the induced composition.

Theorem 9.4. Let G be a group and H be a subgroup of G. Then identity of the subgroup H is the same as the identity of G. Also, for any $a \in H$, the inverse of a as an element of H is the same as the inverse of a as an element of G.

Proof. Let e be the identity of the group G and e' be the identity of the subgroup H. Then $e'e' = e'$ and $e'e = e'$. Therefore, $e'e' = e'e$ and the left cancellation law in G implies that $e' = e$.

Now, let $a \in H$ and b be the inverse of a as an element of G and c be the inverse of a as an element of H. Therefore, $ab = e = ac$. The left cancellation in G then implies that $b = c$.

Observation. It is easy to observe that if H is a subgroup of K and K is a subgroup of G, then H is a subgroup of G.

Example 9.5.

(a) Additive group Z of integers is a subgroup of the additive group Q of rational numbers, the additive group Q is a subgroup of the additive group R of real numbers, the additive group R is a subgroup of the additive group C of complex numbers and $2Z$ is a subgroup of the group Z.

(b) The set Q^* of non-zero rational numbers is a subgroup of the group R^* and R^* is a subgroup of group C^*.

(c) The group $\{1, -1\}$ is a subgroup of the group $\{1, -1, i, -i\}$.

(d) The group $M_2(R)^{**}$ is a subgroup of the group $M_2(R)^*$.

(e) Let G be the group of all matrices of the form $\begin{pmatrix} a & 0 \\ b & c \end{pmatrix}$, where $a, b, c \in R$ (the set of real numbers) and $ac \neq 0$. Then G is a subgroup of the multiplicative group $M_2(R)^*$.

(f) The group G_n of all nth roots of unity is a subgroup of the multiplicative group C^*.

(g) If a positive integer m is a divisor of a positive integer n, then the group G_m of all mth roots of unity is a subgroup of the group G_n of all nth roots of unity.

A subgroup H of a group G is called a **normal subgroup** of G if for every $x \in G, a \in H$, the element $x^{-1}ax \in H$. We write $H \triangle G$ to express that H is a normal subgroup of G.

Example 9.6.

(a) Every subgroup of an Abelian group G is normal in G.

(b) If B is a non-singular matrix of order 2 with real entries and A is a square matrix of order 2 with determinant 1, then

$$\det(B^{-1}AB) = \det(B^{-1})\det(A)\det(B) = \det(B^{-1})1\det(B)$$

$$= \det(B^{-1})\det(B) = det(B^{-1}B) = \det(I) = 1.$$

Hence $B^{-1}AB \in M_2(R)^{**}$. Thus $M_2(R)^{**}$ is a normal subgroup of $M_2(R)^*$. We will now give some criteria for a non-empty subset of a group to be a subgroup.

Theorem 9.5. A non-empty subset H of a group G is a subgroup of G if and only if $a, b \in H$ imply $a^{-1}b \in H$.

Proof. Let H be a non-empty subset of a group G such that $a, b \in H$ imply $a^{-1}b \in H$. Since $H \neq \phi$, there exists an $a \in H$. Then $a, a \in H$ and by the given condition $a^{-1}a \in H$ i.e. $e \in H$.

Let $b \in H$. Then $b, e \in H$ and so, $b^{-1}e \in H$ i.e. $b^{-1} \in H$. Thus H contains e and also contains the inverse of every element of H. Next, let $a, b \in H$. As $a \in H$, $a^{-1} \in H$. Thus $a^{-1}, b \in H$ and, therefore, $(a^{-1})^{-1}b \in H$, i.e. $ab \in H$. Hence the composition in G induces a composition in H. Since associative law holds for all the elements of G, it holds for the elements of H. This completes the proof that H is a subgroup of G. The converse is easy to prove.

Theorem 9.6. A non-empty subset H of a finite group G is a subgroup of G if and only if $a, b \in H$ imply that $ab \in H$.

Proof. If H is a subgroup of G, then the composition in G induces a composition H. Therefore, $a, b \in H$ imply that $ab \in H$. Conversely, suppose that H is a non-empty subset of G such that $a, b \in H$ imply that $ab \in H$. Since $H \neq \phi$, there exists an element $a \in H$. By the given condition on H, $aa = a^2 \in H$. A repeated application of this criterion shows that the elements

$$a, a^2, a^3, \ldots, a^k, \ldots \qquad (9.5)$$

are all in H. All the elements in the sequence (9.5) cannot be distinct because H is a subset of a finite set G and so has only a finite number of elements. Therefore, there exist positive integers $i, j, i \neq j$, such that $a^i = a^j$. Since $i \neq j$, one of these is smaller than the other. Let us suppose that $i > j$. Then $i - j > 0$ and $i - j - 1 \geq 0$. Therefore, $a^{i-j} \in H$. But $a^{i-j} = a^i.a^{-j} = a^j.a^{-j} = a^{j-j} = a^0 = e$. Therefore, $e \in H$. As $i - j - 1 \geq 0, a^{i-j-1} \in H$ and $aa^{i-j-1} = a^{i-j} = e, a^{i-j-1}a = a^{i-j} = e$.

Therefore, $a^{-1} = a^{i-j-1} \in H$. As $a \in H$ was chosen arbitrarily, the inverse of every element of H is in H. Since the associative law holds for all the elements of G, it holds for the elements of H. Hence H is a subgroup of G.

There is a slight variation of the above result, the proof of which is exactly the same (with a slight change at just one place).

Theorem 9.7. A non-empty finite subset H of a group G is a subgroup of G if and only if $a, b \in H$ imply that $ab \in H$.

Similarly, there is a result corresponding to Theorem 9.5, proof of which is analogous to that of Theorem 9.5.

Theorem 9.8. A non-empty subset H of a group G is a subgroup of G if and only if $a, b \in H$ imply that $ab^{-1} \in H$.

9.2.1. *Lagrange's theorem*

Let G be a group and H be a subgroup of G. For an element $x \in G$, the subset $xH = \{xa | a \in H\}$ of G is called a **left coset** of H in G. Also x is called a **representative** of the left coset xH. Right cosets of H in G are defined similarly, i.e. if $x \in G$, the subset $Hx = \{ax | a \in H\}$ of G is called a **right coset** of H in G. Also, x is called a **representative** of the right coset Hx. A left coset of H in G (and for that matter, a right coset of H in G) is, in general, only a subset of G and not a subgroup of G. For example, if $G = \{1, -1, i, -i\}$ and $H = \{1, -1\}$, then $iH = (i, -i\}$, which is not a subgroup of G. Also observe that $-iH = \{-i, i\} = \{i, -i\} = iH$, which shows that a left coset may have more than one coset representatives.

Theorem 9.9. Left cosets of H in G have the following simple properties.

(a) $xH \neq \phi$, in fact $x \in xH$;
(b) $o(xH) = o(H)$ for every $x \in G$;
(c) For $x, y \in G$, $y \in xH$ if and only if $x^{-1}y \in H$;
(d) For $x, y \in G$, $y \in xH$ if and only if $yH = xH$;
(e) Two left cosets of H in G are either disjoint or are identical.

Proof. Let x, $y \in G$.

(a) Since $e \in H, x = xe \in xH$. In particular, it follows that $xH \neq \phi$.
(b) Define a map $\theta : H \to xH$ by

$$\theta(a) = xa, \quad a \in H.$$

Since any element of xH is of the form xa for some $a \in H$ and $xa = \theta(a)$, the map θ is onto. On the other hand, let $a, b \in H$ and suppose that $\theta(a) = \theta(b)$. Then $xa = xb$ and the left cancellation law which holds in G implies that $a = b$. This proves that θ is one–one. Thus θ gives a one–one correspondence between the elements of H and those of xH. Hence $o(xH) = o(H)$.
(c) Let $y \in xH$. Then there exists an $a \in H$ such that $y = xa$, which implies that $x^{-1}y = x^{-1}(xa) = (x^{-1}x)a = ea = a \in H$. Conversely, suppose that $x^{-1}y \in H$. Let $x^{-1}y = a$. Then

$$y = ey = (xx^{-1})y = x(x^{-1}y) = xa \in xH.$$

(d) Suppose that $yH = xH$. By property (a) above, $y \in yH$ and, therefore, $y \in xH$. Conversely, suppose that $y \in xH$. Let $a \in H$ such that

$$y = xa. \tag{9.6}$$

This also implies that

$$x = ya^{-1}. \tag{9.7}$$

Let $z \in yH$. Then there exists $b \in H$ such that $z = yb$. By (9.6) $z = yb = (xa)b = x(ab) \in xH$, as $a, b \in H$ imply $ab \in H$.
Therefore, $yH \subseteq xH$. Similarly, using (9.7) instead of (9.6), we can prove that $xH \subseteq yH$. Combining the two inclusions, we get $yH = xH$.
(e) Consider the left cosets xH and yH. If $xH \cap yH = \phi$, we have nothing to prove. So, suppose that $xH \cap yH \neq \phi$. Let $z \in xH \cap yH$. Then $z \in xH$

and $z \in yH$. By (d) above, $zH = xH$ and $zH = yH$. Hence $xH = yH$. Right cosets of H in G have similar properties which we state without proof.

Theorem 9.10. For $x, y \in G$

(a$'$) $Hx \neq \phi$; in fact $x \in Hx$;
(b$'$) $o(Hx) = o(H)$;
(c$'$) $y \in Hx$ if and only if $yx^{-1} \in H$;
(d$'$) $y \in Hx$ if and only if $Hy = Hx$;
(e$'$) Two right cosets of H in G are either disjoint or are identical.

Let G be a group and H be a subgroup of G. Let $L_G(H)$ denote the set of all left cosets of H in G and $R_G(H)$ denote the set of all right cosets of H in G. Define a map $\theta : L_G(H) \to R_G(H)$ by

$$\theta(xH) = Hx^{-1}, \quad x \in G.$$

For any $x \in G$, $Hx = H(x^{-1})^{-1} = \theta(x^{-1}H)$ and θ is onto.

Let $x, y \in G$ and suppose that $\theta(xH) = \theta(yH)$. Therefore, $Hx^{-1} = Hy^{-1}$. By property (d$'$) of right cosets $y^{-1} \in Hx^{-1}$ and then property (c$'$) shows that $y^{-1}(x^{-1})^{-1} \in H$ i.e. $y^{-1}x \in H$. Property (c) of left cosets shows that $x \in yH$ and then (d) shows that $xH = yH$. This completes the proof that θ is one–one. Thus θ is a one–one correspondence between the elements of $L_G(H)$ and $R_G(H)$ and, therefore, their orders are equal.

We have thus proved the following.

Theorem 9.11. The number of left cosets of H in G equals the number of right cosets of H in G.

The number of left (or right) cosets of H in G is called the **index** of H in G and is denoted by $[G : H]$.

We now have all the machinery to prove the celebrated theorem attributed to Lagrange.

Theorem 9.12 (Lagrange's theorem). The order of every subgroup of a finite group divides the order of the group.

Proof. Let G be a finite group and H be a subgroup of G. Consider all the possible left cosets of H in G. The set G being finite, the number of all possible subsets of G is finite. Every left coset of H in G being a subset of G, the number of left cosets of H in G is finite, say k. Let $x_1H, x_2H, \ldots, x_kH$

be all the distinct left cosets of H in G. Then, as

$$x_i H \subseteq G, \ \bigcup_{i=1}^{k} x_i H \subseteq G.$$

Let $x \in G$. Then xH is a left coset of H and must equal one of the k left cosets listed above. Let $xH = x_j H$, for some $j, 1 \le j \le k$. Then $x \in x_j H$ and, so, $x \in \bigcup_{i=1}^{k} x_i H$. This proves that $G \subseteq \bigcup_{i=1}^{k} x_i H$ and, hence,

$$G = \bigcup_{i=1}^{k} x_i H.$$

Since, for $i \ne j, x_i H \cap x_j H = \phi$, we have

$$o(G) = o\left(\bigcup_{i=1}^{k} x_i H\right) = \sum_{i=1}^{k} o(x_i H) = \sum_{i=1}^{k} o(H) = k o(H) = [G : H]o(H).$$

Therefore, $o(H)$ divides $o(G)$. The above relation also shows that the index of H in G also divides $o(G)$.

Corollary. The order of every element of a finite group divides the order of the group.

Proof. Let G be a group of finite order $m \ge 2$ (say). Let $a \in G$. If $o(a) = 1$, we have nothing to prove. So, suppose that $o(a) \ge 2$. Consider the subset $H = \{a, a^2, a^3, \ldots\}$ of G consisting of all powers of a. Since $H \subseteq G$ and G is finite, all the elements of H cannot be distinct. Then there exist positive integers $i, j, i \ne j$ such that $a^i = a^j$. Suppose that $i > j$. Then $a^{i-j} = a^i.a^{-j} = a^j.a^{-j} = a^{j-j} = a^0 = e$. As $i - j - 1 \ge 0, a^{i-j-1} \in H$ and $a^{i-j-1}a = a^{i-j} = e$, it follows that $a^{-1} = a^{i-j-1} \in H$. Hence the set H consisting of all the distinct powers of a is a subgroup of G. We have also, in fact, proved that $o(a) < \infty$. Let $o(a) = k$. Then H, which consists of the powers of a contains exactly k elements, namely $a, a^2, \ldots, a^{k-1}, a^k = e$. Thus $H = \langle a \rangle$ is a cyclic group of order k. By Lagrange's theorem $o(H) \mid o(G)$ and so, $k = o(a)$ divides $o(G)$.

Corollary. A group of prime order is cyclic.

Proof. Let G be a group of prime order p. Since $p \ge 2$, there exists an element $a \in G, a \ne e$. Let $H = \langle a \rangle$ be the cyclic subgroup of G generated by a. Then $o(H) = o(a)$ and by Lagrange's theorem $o(H)$ divides $o(G) = p$, i.e. $o(a)$ divides p. Since p is a prime and $o(a) \ne 1, o(a) = p$. Thus $H \subseteq G$

and $o(H) = o(G) = p$. Therefore, $G = H = \langle a \rangle$ i.e. G is cyclic group generated by a.

9.3. Quotient Groups

We will begin with an exercise.

Exercise 9.1. A subgroup H of a group G is a normal subgroup of G if and only if for every $x \in G, xH = Hx$.

Let G be a group and H be a normal subgroup of G. Let G/H denote the set of all left (or right) cosets of H in G. We define a composition in G/H as follows.

For $x, y \in G$, define $xH.yH = xyH$. Since a left coset can have more than one coset representatives, we need to prove that the above composition is independent of the choice of coset representatives. We then say that the composition is well defined. For this, suppose that $xH = x'H$ and $yH = y'H$. Then $x' = xa$ and $y' = yb$ for some $a, b \in H$. Then $x'y'H = xaybH = xy(y^{-1}ay)bH = xyH$, as $y^{-1}ay \in H$, H being normal and, so, $(y^{-1}ay)b \in H$. This proves that $x'H.y'H = xH.yH$, i.e. the composition is well defined.

Since associative law holds in G, for $x, y, z \in G, (xHyH)zH = xyH.zH = (xy)zH = x(yz)H = xH.yzH = xH(yHzH)$, i.e. the composition in G/H satisfies the associative law.

For any $x \in G$,

$$xH.eH = xeH = xH = exH = eHxH$$

and

$$x^{-1}HxH = x^{-1}xH = eH = xx^{-1}H = xHx^{-1}H.$$

Thus $eH = H$ is the identity element of G/H and $(xH)^{-1} = x^{-1}H$. Hence G/H is a group. The group G/H is called the **quotient group** of G modulo, or with respect to, the normal subgroup H.

Example 9.7. Consider the additive group Z of integers and $n \geq 2$ an integer. The group Z of integers being Abelian, $nZ = \{nk \mid k \in Z\}$ is a normal subgroup of Z. We can, therefore, talk of the quotient group $Z/nZ = \{0 + nZ, 1 + nZ, \ldots, n - 1 + nZ\}$. We can write the elements of this group simply as $0, 1, 2, \ldots, n - 1$, i.e. $Z/nZ = \{0, 1, 2, \ldots, n - 1\}_{mod\ n}$, where addition is defined *modulo* n. We write this group as Z_n. It is clearly a cyclic group generated by one.

Exercise 9.2. Prove that every quotient of a cyclic group is cyclic.

Let G, G' be groups. A map $f : G \to G'$ is called a **homomorphism** if $f(xy) = f(x)f(y)$ for all $x, y \in G$.

A homomorphism $f : G \to G'$ is called

(a) a **monomorphism** if the map f is one–one,
(b) an **epimorphism** if the map f is onto,
(c) an **isomorphism** if the map f is both one–one and onto.

Lemma. Let $f : G \to G'$ be a homomorphism of groups. Then

(a) $f(e) = e'$, where e is the identity of G and e' is that of G',
(b) for any $x \in G, f(x^{-1}) = f(x)^{-1}$.

Proof.

(a) $f(e)e' = f(e) = f(ee) = f(e)f(e)$ and the left cancellation law in G' implies that $f(e) = e'$.
(b) Let $x \in G$. Then

$$e' = f(e) = f(xx^{-1}) = f(x)f(x^{-1})$$

and

$$e' = f(e) = f(x^{-1}x) = f(x^{-1})f(x),$$

which together imply that $f(x)^{-1} = f(x^{-1})$.

Definition. Let $f : G \to G'$ be a homomorphism. The subset $\ker f = \{x \in G \mid f(x) = e'\}$ of G is called the **kernel** of f.

Theorem 9.13. Let $f : G \to G'$ be a homomorphism. Then $\ker f$ is a normal subgroup of G.

Proof. As proved above $f(e) = e'$. Therefore, $e \in \ker f$ and $\ker f \neq \phi$. Let $a, b \in \ker f$ and $x \in G$. Then

$$f(a^{-1}b) = f(a^{-1})f(b) = f(a)^{-1}f(b) = (e')^{-1}e' = e'.e' = e'.$$

Therefore, $a^{-1}b \in \ker f$ and, so, $\ker f$ is a subgroup of G. Also

$$f(x^{-1}ax) = f(x^{-1})f(a)f(x) = f(x^{-1})e'f(x) = f(x^{-1})f(x)$$
$$= f(x^{-1}x) = f(e) = e'.$$

Therefore, $x^{-1}ax \in \ker f$ which proves that $\ker f \triangle G$.

9.4. Symmetric Groups

Let M be a finite non-empty set. A map $f : M \to M$ which is both one–one and onto is called a **permutation** of M. If f and g are two permutations of M, then the composition $gf : M \to M$ is the map defined by $(gf)(x) = g(f(x))$, $x \in M$. If $x, y \in M$ and $gf(x) = (gf)(y)$ then $g(f(x)) = g(f(y))$. The map g being one–one, $f(x) = f(y)$. Again f being one–one this shows that $x = y$. Let $x \in M$. The map g being onto there exists an element $y \in M$ such that $x = g(y)$. Also f is onto and, so, there exists $z \in M$ such that $y = f(z)$. Therefore, $x = g(y) = g(f(z)) = (gf)(z)$, which proves that gf is onto. Hence gf is a permutation of the set M. That for permutations $f, g, h, h(gf) = (hg)f$ is easy to see. The identity map $1 : M \to M, 1(x) = x, x \in M$ is one–one and onto and is a permutation of M. It is a trivial thing to check that $f1 = 1f = f$ for every permutation f of M. Let f be a permutation of M. Suppose that $M = \{a_1, a_2, \ldots, a_n\}$ and let $f(a_i) = b_i, 1 \le i \le n$. Then $M = \{b_1, b_2, \ldots, b_n\}$ as well and define $g : M \to M$ by $g(b_i) = a_i$, for $1 \le i \le n$. Since corresponding to every a_i there is exactly one b_i and the map being one–one no two b_i are equal, therefore, the map g is also bijective, i.e. g is a permutation of M. That $gf = 1$ and $fg = 1$ are clear. Hence, every permutation of M is invertible and the set $S(M)$ of all permutations of M is a group.

The group $S(M)$ of permutations of the set M is independent of the choice of the type of elements of M but depends only on the number of elements of M. If, as above, $o(M) = n$, then $S(M)$ the group of permutations of M is called the **symmetric group of degree** n and can be taken as the group of permutations of the set $M = \{1, 2, \ldots, n\}$. This group is denoted by S_n. Any subgroup of S_n is called a **permutation group** (of degree n).

Let f be a permutation of the set $M = \{1, 2, \ldots, n\}$. Suppose that $f(i) = a_i, 1 \le i \le n$. Then f being one–one and onto $\{a_1, a_2, \ldots, a_n\} = M$ and we write

$$f = \begin{pmatrix} 1 & 2 & \ldots & n \\ a_1 & a_2 & \ldots & a_n \end{pmatrix}$$

i.e. we write the elements $1, 2, \ldots, n$ in a row in order and below every number we write the image of that number under f. We next explain how to obtain the composition gof in this notation. Let g be another permutation of M. Let g

$$= \begin{pmatrix} 1 & 2 & 3 & \ldots & n \\ b_1 & b_2 & b_3 & \ldots & b_n \end{pmatrix}.$$

Then

$$gf = \begin{pmatrix} 1 & 2 & 3 & \ldots & n \\ b_1 & b_2 & b_3 & \ldots & b_n \end{pmatrix} \begin{pmatrix} 1 & 2 & 3 & \ldots & n \\ a_1 & a_2 & a_3 & \ldots & a_n \end{pmatrix}$$

$$= \begin{pmatrix} 1 & 2 & 3 & \ldots & n \\ c_1 & c_2 & c_3 & \ldots & c_n \end{pmatrix}$$

where $c_1, c_2, \ldots, c_n$ are determined as follows: The image of i under the permutation f is a_i. Since $\{a_1, \ldots, a_n\} = \{1, 2, \ldots, n\}, a_i = j$ where $1 \leq j \leq n$. Then locate j in the upper row of g. Once this is located, the image of j under g is the element b_j and, so, we take $c_i = b_j$. We do this for every $i, 1 \leq i \leq n$, and the composition gf is computed. We illustrate this with an example.

Example 9.8. Let $M = \{1, 2, \ldots, 6\}$,

$$f = \begin{pmatrix} 1 & 2 & 3 & 4 & 5 & 6 \\ 6 & 1 & 2 & 3 & 5 & 4 \end{pmatrix}, \quad g = \begin{pmatrix} 1 & 2 & 3 & 4 & 5 & 6 \\ 3 & 5 & 4 & 6 & 2 & 1 \end{pmatrix}.$$

Then

$$gf = \begin{pmatrix} 1 & 2 & 3 & 4 & 5 & 6 \\ 3 & 5 & 4 & 6 & 2 & 1 \end{pmatrix} \begin{pmatrix} 1 & 2 & 3 & 4 & 5 & 6 \\ 6 & 1 & 2 & 3 & 5 & 4 \end{pmatrix}$$

$$= \begin{pmatrix} 1 & 2 & 3 & 4 & 5 & 6 \\ 1 & 3 & 5 & 4 & 2 & 6 \end{pmatrix}$$

while

$$fg = \begin{pmatrix} 1 & 2 & 3 & 4 & 5 & 6 \\ 6 & 1 & 2 & 3 & 5 & 4 \end{pmatrix} \begin{pmatrix} 1 & 2 & 3 & 4 & 5 & 6 \\ 3 & 5 & 4 & 6 & 2 & 1 \end{pmatrix}$$

$$= \begin{pmatrix} 1 & 2 & 3 & 4 & 5 & 6 \\ 2 & 5 & 3 & 4 & 1 & 6 \end{pmatrix}.$$

Since, under gf the image of 1 is 1 while under fg the image of 1 is 2, the above example shows that $fg \neq gf$. However, this is so, provided $n \geq 3$.

Theorem 9.14. The order of the symmetric group S_n of degree n is $n!$

Proof. Let f be a permutation of the set $M = \{1, 2, \ldots, n\}$. Then the number of choices for $f(1)$ is n, once the image of 1 is chosen, the number of choices for $f(2)$ is $n - 1$, then the number of choices for $f(3)$ in $n - 2$. Continuing in this fashion, we find that the number of choices for f is

$n(n-1)(n-2)\ldots 2.1 = n!$ Hence the number of elements of $S(M)$ is $n!$ i.e. order of S_n is $n!$

Example 9.9. We consider the group S_n for some small values of n.

(a) $n = 1$. In this case there is only one element of S_n, namely $\begin{pmatrix} 1 \\ 1 \end{pmatrix}$.

(b) $n = 2$.

Here the elements of S_n are $\begin{pmatrix} 1 & 2 \\ 1 & 2 \end{pmatrix}$ and $\begin{pmatrix} 1 & 2 \\ 2 & 1 \end{pmatrix}$ so that the group is of order 2 and, hence, Abelian.

(c) $n = 3$. Order of $S_3 = 3! = 6$ and all the elements of S_3 are:

$$\begin{pmatrix} 1 & 2 & 3 \\ 1 & 2 & 3 \end{pmatrix}, \begin{pmatrix} 1 & 2 & 3 \\ 2 & 1 & 3 \end{pmatrix}, \begin{pmatrix} 1 & 2 & 3 \\ 3 & 2 & 1 \end{pmatrix}, \begin{pmatrix} 1 & 2 & 3 \\ 1 & 3 & 2 \end{pmatrix},$$

$$\begin{pmatrix} 1 & 2 & 3 \\ 2 & 3 & 1 \end{pmatrix}, \begin{pmatrix} 1 & 2 & 3 \\ 3 & 1 & 2 \end{pmatrix}.$$

Observe that

$$\begin{pmatrix} 1 & 2 & 3 \\ 2 & 1 & 3 \end{pmatrix} \begin{pmatrix} 1 & 2 & 3 \\ 3 & 2 & 1 \end{pmatrix} = \begin{pmatrix} 1 & 2 & 3 \\ 3 & 1 & 2 \end{pmatrix},$$

$$\begin{pmatrix} 1 & 2 & 3 \\ 3 & 2 & 1 \end{pmatrix} \begin{pmatrix} 1 & 2 & 3 \\ 2 & 1 & 3 \end{pmatrix} = \begin{pmatrix} 1 & 2 & 3 \\ 2 & 3 & 1 \end{pmatrix}$$

confirming that

$$\begin{pmatrix} 1 & 2 & 3 \\ 2 & 1 & 3 \end{pmatrix} \begin{pmatrix} 1 & 2 & 3 \\ 3 & 2 & 1 \end{pmatrix} \neq \begin{pmatrix} 1 & 2 & 3 \\ 3 & 2 & 1 \end{pmatrix} \begin{pmatrix} 1 & 2 & 3 \\ 2 & 1 & 3 \end{pmatrix}.$$

Other products can be computed similarly.

Again,

$$\begin{pmatrix} 1 & 2 & 3 \\ 2 & 1 & 3 \end{pmatrix} \begin{pmatrix} 1 & 2 & 3 \\ 2 & 1 & 3 \end{pmatrix} = \begin{pmatrix} 1 & 2 & 3 \\ 1 & 2 & 3 \end{pmatrix},$$

$$\begin{pmatrix} 1 & 2 & 3 \\ 3 & 2 & 1 \end{pmatrix} \begin{pmatrix} 1 & 2 & 3 \\ 3 & 2 & 1 \end{pmatrix} = \begin{pmatrix} 1 & 2 & 3 \\ 1 & 2 & 3 \end{pmatrix},$$

$$\begin{pmatrix} 1 & 2 & 3 \\ 1 & 3 & 2 \end{pmatrix} \begin{pmatrix} 1 & 2 & 3 \\ 1 & 3 & 2 \end{pmatrix} = \begin{pmatrix} 1 & 2 & 3 \\ 1 & 2 & 3 \end{pmatrix}.$$

Next,

$$\begin{pmatrix} 1 & 2 & 3 \\ 2 & 3 & 1 \end{pmatrix}^2 = \begin{pmatrix} 1 & 2 & 3 \\ 2 & 3 & 1 \end{pmatrix}\begin{pmatrix} 1 & 2 & 3 \\ 2 & 3 & 1 \end{pmatrix} = \begin{pmatrix} 1 & 2 & 3 \\ 3 & 1 & 2 \end{pmatrix},$$

$$\begin{pmatrix} 1 & 2 & 3 \\ 2 & 3 & 1 \end{pmatrix}^3 = \begin{pmatrix} 1 & 2 & 3 \\ 3 & 1 & 2 \end{pmatrix}\begin{pmatrix} 1 & 2 & 3 \\ 2 & 3 & 1 \end{pmatrix} = \begin{pmatrix} 1 & 2 & 3 \\ 1 & 2 & 3 \end{pmatrix}$$

and

$$\begin{pmatrix} 1 & 2 & 3 \\ 3 & 1 & 2 \end{pmatrix}^2 = \begin{pmatrix} 1 & 2 & 3 \\ 3 & 1 & 2 \end{pmatrix}\begin{pmatrix} 1 & 2 & 3 \\ 3 & 1 & 2 \end{pmatrix} = \begin{pmatrix} 1 & 2 & 3 \\ 2 & 3 & 1 \end{pmatrix},$$

$$\begin{pmatrix} 1 & 2 & 3 \\ 3 & 1 & 2 \end{pmatrix}^3 = \begin{pmatrix} 1 & 2 & 3 \\ 2 & 3 & 1 \end{pmatrix}\begin{pmatrix} 1 & 2 & 3 \\ 3 & 1 & 2 \end{pmatrix} = \begin{pmatrix} 1 & 2 & 3 \\ 1 & 2 & 3 \end{pmatrix}.$$

Thus, in S_3 there are three elements of order 2 and two elements of order 3 besides the identity element, which is always of order 1.

The order of every subgroup of S_3 being a divisor of 6, S_3 has the identity subgroup of order 1 and some subgroups of order 2 and some others of order 3. The subgroups of order 2 are:

$$\left\{ \begin{pmatrix} 1 & 2 & 3 \\ 1 & 2 & 3 \end{pmatrix}, \begin{pmatrix} 1 & 2 & 3 \\ 2 & 1 & 3 \end{pmatrix} \right\}, \quad \left\{ \begin{pmatrix} 1 & 2 & 3 \\ 1 & 2 & 3 \end{pmatrix}, \begin{pmatrix} 1 & 2 & 3 \\ 3 & 2 & 1 \end{pmatrix} \right\},$$

$$\left\{ \begin{pmatrix} 1 & 2 & 3 \\ 1 & 2 & 3 \end{pmatrix}, \begin{pmatrix} 1 & 2 & 3 \\ 1 & 3 & 2 \end{pmatrix} \right\}.$$

There is only one subgroup of order 3 and it is

$$\left\{ \begin{pmatrix} 1 & 2 & 3 \\ 1 & 2 & 3 \end{pmatrix}, \begin{pmatrix} 1 & 2 & 3 \\ 2 & 3 & 1 \end{pmatrix}, \begin{pmatrix} 1 & 2 & 3 \\ 3 & 1 & 2 \end{pmatrix} \right\}.$$

Consider subgroups H and K of a group G. We define

$$HK = \{ab | a \in H, b \in K\},$$

which is in general a subset of G but not a subgroup of G. If we take $a = e$, then $ab = eb \in HK$ for every $b \in K$. Therefore, $K \subseteq HK$. Similarly $H \subseteq HK$. We will now give an example to show that HK need not be a subgroup of G.

Example 9.10. Consider $G = S_3$ and $H = \{1, (12)\}, K = \{1, (13)\}$. (Refer to notation below.) Then

$$HK = \{1, (12), (13), (12)(13)\}.$$

Now

$$(12)(13) = \begin{pmatrix} 1 & 2 & 3 \\ 2 & 1 & 3 \end{pmatrix} \begin{pmatrix} 1 & 2 & 3 \\ 3 & 2 & 1 \end{pmatrix} = \begin{pmatrix} 1 & 2 & 3 \\ 3 & 1 & 2 \end{pmatrix} = (132).$$

Thus $HK = \{1, (12), (13), (132)\}$ the order of which is 4. Since 4 does not divide $6 = 3!$ which is the order of S_3, HK is not a subgroup of S_3 (refer to Lagrange's theorem).

A permutation of a set M is called a **cyclic permutation** if the elements permuted can be written in a row in such a way that the image of the first element in the row is the second element, the image of the second element is the third element, the image of the third element is the fourth element and the image of the last element is the first element in the row. If, in a cyclic permutation, $a_1 \to a_2, a_2 \to a_3, \ldots, a_{k-1} \to a_k$ and $a_k \to a_1$ we write it as $(a_1 a_2 \ldots, a_k)$.

An **element** $i \in M$ is said to be **fixed by a permutation** f if $f(i) = i$. If $f(i) \neq i$, we say that i is **moved** by f. In the notation

$$\begin{pmatrix} 1 & 2 & 3 & \ldots & n \\ a_1 & a_2 & a_3 & \ldots & a_n \end{pmatrix},$$

if an element i is fixed by this permutation, then i is omitted from the first row as well as the second row. For example,

$$\begin{pmatrix} 1 & 2 & 3 & 4 & 5 & 6 \\ 6 & 1 & 2 & 3 & 5 & 4 \end{pmatrix} = \begin{pmatrix} 1 & 2 & 3 & 4 & 6 \\ 6 & 1 & 2 & 3 & 4 \end{pmatrix} \quad \text{which is the cycle } (16432)$$

$$\begin{pmatrix} 1 & 2 & 3 & 4 & 5 & 6 \\ 1 & 3 & 5 & 4 & 2 & 6 \end{pmatrix} = \begin{pmatrix} 2 & 3 & 5 \\ 3 & 5 & 2 \end{pmatrix} \quad \text{which is the cycle } (235).$$

$$\begin{pmatrix} 1 & 2 & 3 & 4 & 5 & 6 \\ 2 & 5 & 3 & 4 & 1 & 6 \end{pmatrix} = \begin{pmatrix} 1 & 2 & 5 \\ 2 & 5 & 1 \end{pmatrix} \quad \text{which is the cycle } (125).$$

We write 1 or I for the identity permutation, i.e. the permutation f in which $f(a_i) = a_i$ for every $i, 1 \leq i \leq n$.

$$\begin{pmatrix} 1 & 2 & 3 \\ 2 & 1 & 3 \end{pmatrix} = \begin{pmatrix} 1 & 2 \\ 2 & 1 \end{pmatrix} = (12), \quad \begin{pmatrix} 1 & 2 & 3 \\ 3 & 2 & 1 \end{pmatrix} = \begin{pmatrix} 1 & 3 \\ 3 & 1 \end{pmatrix} = (13),$$

$$\begin{pmatrix} 1 & 2 & 3 \\ 1 & 3 & 2 \end{pmatrix} = \begin{pmatrix} 2 & 3 \\ 3 & 2 \end{pmatrix} = (23), \quad \begin{pmatrix} 1 & 2 & 3 \\ 2 & 3 & 1 \end{pmatrix} = (123),$$

$$\begin{pmatrix} 1 & 2 & 3 \\ 3 & 1 & 2 \end{pmatrix} = (132).$$

The **length of the cyclic permutation** $(b_1 b_2 \ldots b_k)$ is defined as k and we also call it a **cycle of length** k. A cycle of length 2 is called a **transposition**.

Definition. Two permutations f and g of M are said to be **disjoint** if the elements moved by f are kept fixed by g and those moved by g are kept fixed by f.

Observe that no two non-identity permutations in S_3 are disjoint. Similarly, no two of the permutations (16432), (235), (125) are disjoint. On the other hand, the permutations

$$\begin{pmatrix} 1 & 2 & 3 & 4 & 5 & 6 \\ 3 & 2 & 1 & 4 & 5 & 6 \end{pmatrix} \quad \text{and} \quad \begin{pmatrix} 1 & 2 & 3 & 4 & 5 & 6 \\ 1 & 2 & 3 & 5 & 6 & 4 \end{pmatrix} \quad \text{are disjoint.}$$

We next state without proof a few results on permutations.

Theorem 9.15. Two disjoint permutations commute with each other.

Theorem 9.16. Any permutation is a product of disjoint cycles.

Theorem 9.17. Any cycle is a product of transpositions. Thus, every permutation is a product of transpositions.

Definition. A permutation which is a product of an even number of transpositions is called an **even permutation** and the one which is a product of an odd number of transpositions is called an **odd permutation**. Thus every transposition is an odd permutation.

Definition. We say that a group G can be embedded in a group K if there exists a monomorphism $\theta : G \to K$.

Theorem 9.18. Every finite group can be embedded in a symmetric group S_n for a suitable n.

As already defined, any subgroup of S_n is called a permutation group. The above theorem thus shows that every finite group can be regarded as a permutation group. We illustrate this idea by expressing the groups $\{1, -1\}$ and $\{1, -1, i, -i\}$ as permutation groups.

Example 9.11. (a) Consider the group $G = \{1, -1\}$ and also consider the underlying set $M = \{1, -1\}$. We consider the maps $T_1 : M \to M$ and $T_{-1} : M \to M$, where $T_1(x) = x, x \in M$ and $T_{-1}(x) = -x, x \in M$. The maps T_1 and T_{-1} are clearly one–one and onto and, so, are permutations of M. We can write these permutations as

$$T_1 = \begin{pmatrix} 1 & -1 \\ 1 & -1 \end{pmatrix} = 1; \quad T_{-1} = \begin{pmatrix} 1 & -1 \\ -1 & 1 \end{pmatrix} = (1, -1).$$

We can thus identify G with the group $\{1, (1-1)\}$.

(b) Next, consider the group $G = \{1, -1, i, -i\}$. Consider the permutations of the set G as follows:

$$1, \begin{pmatrix} 1 & -1 & i & -i \\ -1 & 1 & -i & i \end{pmatrix}, \begin{pmatrix} 1 & -1 & i & -i \\ i & -i & -1 & 1 \end{pmatrix}, \begin{pmatrix} 1 & -1 & i & -i \\ -i & i & 1 & -1 \end{pmatrix}.$$

Observe that

$$\begin{pmatrix} 1 & -1 & i & -i \\ i & -i & -1 & 1 \end{pmatrix}^2 = \begin{pmatrix} 1 & -1 & i & -i \\ i & -i & -1 & 1 \end{pmatrix} \begin{pmatrix} 1 & -1 & i & -i \\ i & -i & -1 & 1 \end{pmatrix}$$

$$= \begin{pmatrix} 1 & -1 & i & -i \\ -1 & 1 & -i & i \end{pmatrix}$$

$$\begin{pmatrix} 1 & -1 & i & -i \\ i & -i & -1 & 1 \end{pmatrix}^3 = \begin{pmatrix} 1 & -1 & i & -i \\ -1 & 1 & -i & i \end{pmatrix} \begin{pmatrix} 1 & -1 & i & -i \\ i & -i & -1 & 1 \end{pmatrix}$$

$$= \begin{pmatrix} 1 & -1 & i & -i \\ -i & i & 1 & -1 \end{pmatrix}$$

and

$$\begin{pmatrix} 1 & -1 & i & -i \\ i & -i & -1 & 1 \end{pmatrix}^4 = \begin{pmatrix} 1 & -1 & i & -i \\ -i & i & 1 & -1 \end{pmatrix} \begin{pmatrix} 1 & -1 & i & -i \\ i & -i & -1 & 1 \end{pmatrix}$$

$$= \begin{pmatrix} 1 & -1 & i & -i \\ 1 & -1 & i & -i \end{pmatrix} = 1.$$

Thus the four permutations considered are the powers of the permutation

$$\begin{pmatrix} 1 & -1 & i & -i \\ i & -i & -1 & 1 \end{pmatrix}$$

i.e.

$$\left\{ 1, \begin{pmatrix} 1 & -1 & i & -i \\ -1 & 1 & -i & i \end{pmatrix}, \begin{pmatrix} 1 & -1 & i & -i \\ i & -i & -1 & 1 \end{pmatrix}, \begin{pmatrix} 1 & -1 & i & -i \\ -i & i & 1 & -1 \end{pmatrix} \right\}$$

$$= \left\langle \begin{pmatrix} 1 & -1 & i & -i \\ i & -i & 1 & -1 \end{pmatrix} \right\rangle,$$

the cyclic group generated by

$$\begin{pmatrix} 1 & -1 & i & -i \\ i & -i & -1 & 1 \end{pmatrix} = (1i - 1 - i).$$

Also the given group is cyclic generated by i. Hence both are cyclic groups of order 4. As cyclic groups of equal orders are isomorphic, the given group is the permutation group $\langle (1i - 1 - i) \rangle$.

Theorem 9.19. For $n \geq 2$, the set A_n of all even permutations is a group and its order is $\frac{n!}{2}$.

Proof. Let A_n be the set of all even permutations. As $1 = (12)(12)$ is a product of two transpositions, $1 \in A_n$. Let $f, g \in A_n$. Then f is a product of an even number of transpositions and so is g. Writing the decompositions of f and g in fg, we find that fg is again a product of an even number of transpositions. Therefore, $fg \in A_n$. As S_n is a finite group, it follows that A_n is a subgroup of S_n.

Let $A_n = \{\alpha_1, \alpha_2, \dots, \alpha_m\}$ i.e. let $\alpha_1, \alpha_2, \dots, \alpha_m$ be the even permutations in S_n and $\beta_1, \beta_2, \dots, \beta_k$ be all the odd permutations in S_n. Since the order of S_n is $n!$, $n! = m + k$. Let β be any odd permutation. For any $i, 1 \leq i \leq k, \beta_i$ is a product of an odd number of transpositions and also β is a product of an odd number of transpositions. Therefore, $\beta\beta_i$ is a product of an even number of transpositions and $\beta\beta_i$ is an even permutation. Also, no two elements among $\beta\beta_1, \beta\beta_2, \dots, \beta\beta_k$ are equal. For otherwise, we can apply the left cancellation law, which holds in S_n and the corresponding two $\beta_i's$ will be equal, which is a contradiction. Hence $\beta\beta_1, \beta\beta_2, \dots, \beta\beta_k$ are all distinct and each one is an even permutation. The total number of even permutations being m, we get $k \leq m$. Next we consider the products $\beta\alpha_1, \beta\alpha_2, \dots, \beta\alpha_m$. Each one of these products is an odd permutation and no two of these are equal. Since the total number of odd permutations is k, we get $m \leq k$. Hence $m = k$ and then $m + k = n!$ implies that $m = n!/2$ i.e. $o(A_n) = n!/2$. Then $[S_n : A_n] = 2$ and since every subgroup of index 2 is normal in the group, A_n is a normal subgroup of S_n.

Alternatively, let $\alpha \in A_n$ and $\beta \in S_n$. Suppose that $\beta = (a_1 a_2) \cdots (a_k a_{k+1})$ is a product of k transpositions. Then $\beta^{-1} = (a_k a_{k-1})(a_{k-1} a_k) \cdots (a_1 a_2)$, which is again a product of k transpositions. Let α be a product of $2t$ transpositions, where t is a positive integer. Then $\beta^{-1}\alpha\beta$ is a product of $2t + 2k = 2(t + k)$ transpositions and, therefore, $\beta^{-1}\alpha\beta \in A_n$. Hence A_n is a normal subgroup of S_n.

Exercises 9.1.

1. Find the orders of all the elements of the groups G as in Examples 9.3 and 9.4.
2. Prove that 1 and -1 are the only elements of finite order in Q^* or R^*, which are groups under multiplication.
3. Prove that in the additive groups Z, Q, R and C, the additive identity 0 is the only element of finite order.
4. If $a \in G$ is an element of odd order in the group G, prove that $o(a) = o(a^2)$.
5. If a is an element of even order in a group G, is it essential that $o(a) = o(a^3)$? (In G_6, consider the element $e^{2\pi i/6}$.)
6. Let G be a group. Prove that G is Abelian in each of the following cases.

   (a) If $a^2 = e$ for every $a \in G$;
   (b) If $a = a^{-1}$ for every $a \in G$;
   (c) If $(ab)^2 = a^2 b^2$ for all $a, b \in G$.

7. Prove that every cyclic group is Abelian.
8. In a group G the two cancellation laws hold, i.e. if $a, b, x \in G$, then $ax = bx$ implies that $a = b$ and $xa = xb$ implies that $a = b$.
9. Let G be a group and $a, b \in G$. Then the equations $ax = b$ and $ya = b$ have unique solutions in G.
10. Define a group, order of a group and order of an element of a group. Give an example of

    (a) a finite group,
    (b) an infinite group,
    (c) a non-identity element of finite order in an infinite group.

11. Prove that any subgroup of the additive group Z of integers is of the form nZ, where n is a non-negative integer.
12. Define a subgroup of a group. Is every subgroup of an infinite group infinite? Justify.
13. Prove that every subgroup of a cyclic group is cyclic.
14. Prove that Z_{10} is a group under addition *modulo* 10. Is it a group under multiplication *modulo* 10?
15. Repeat Question 14 with 10 replaced by 6, 8, 12, 14.
16. Construct a subgroup of order 4 in the group Z_{12} of integers with addition *modulo* 12. Also find all cosets of this subgroup in the group.

17. Prove that the map $f : G \to G$ given by $f(x) = x^{-1}, x \in G$, is a homomorphism if and only if G is Abelian.

18. If H is a subgroup and K is a normal subgroup of a group G, prove that HK is a subgroup of G.

19. If $f : G \to G'$ is a homomorphism and H is a subgroup of G, prove that $f(H) = \{x' \in G' \mid x' = f(h)$ for some $h \in H\}$ is a subgroup of G'

20. Find all possible subgroups of Z_{12} — the group of integers with addition *modulo* 12.

21. Prove that a subgroup of index 2 is always a normal subgroup of the group. Is the converse also true? Justify.

22. If G is an Abelian group, prove that the map $f : G \to G$ given by $f(x) = x^2, x \in G$, is a homomorphism.

23. If K is a normal subgroup of a finite group G find the order of the quotient group G/K.

24. Is the additive group Q of rational numbers cyclic? Justify.

25. Is it correct to say that $1, -1, i, -i$ are the only elements of finite order in the multiplicative group C^* of non-zero complex numbers? Justify.

26. Consider the group $G = \{1, 2, 3, 4, 5, 6\}$ under multiplication *modulo* 7. Find

   (a) the multiplication table of G;
   (b) $2^{-1}, 3^{-1}, 6^{-1}$;
   (c) the subgroup generated by 2 and 3 and the order of this subgroup;
   (d) the orders of the elements 2, 3.

27. Consider $G = \{1, 5, 7, 11, 13, 17\}$ under multiplication *modulo* 18.

   (a) Construct the multiplication table for G. Is G a group?
   (b) Find $5^{-1}, 7^{-1}, 17^{-1}$ if these exist.
   (c) Find the subgroup generated by (a) 5 and (b) 7. Also find the orders of these subgroups.
   (d) Is G cyclic? Justify.

28. Let G be a group and $x, y \in G$. Prove that (a) $(xy)^{-1} = y^{-1}x^{-1}$, and (b) $(x^{-1})^{-1} = x$.

Chapter 10

RINGS

In the previous chapter, we studied sets with one binary composition. In this chapter we study rings, which are sets with two binary compositions satisfying certain properties. We discuss some very elementary properties of rings which, we believe, are essential for computer science/engineering students but are in no way adequate for mathematics majors.

10.1. Rings

We study sets with two binary compositions. Such sets do exist. For example, the sets Z, Q, R and C all have two compositions, namely addition and multiplication of the number system. The set $M_n(R)$ of all square matrices of order n with real entries has two compositions, namely addition and multiplication of matrices.

Definition. A non-empty set R with two binary compositions, say addition and multiplication, is called a **ring** if

(a) R is an Abelian group with respect to addition;
(b) multiplication satisfies the association law i.e. $(ab)c = a(bc)$ for all $a, b, c \in R$;
(c) multiplication is distributive over addition, i.e.

$$a(b + c) = ab + ac; \quad \text{and} \quad (a + b)c = ac + bc \text{ for all } a, b, c \in R.$$

A ring R is said to be **commutative** if the multiplication satisfies the commutative law, i.e. $ab = ba$ for $a, b \in R$. A ring R is said to have an

identity, or is called a **ring with identity** if there exists an element $1 \in R$ such that $1.a = a.1 = a$ for every $a \in R$. As in the case of a group, we can prove that the identity element in a ring R, if it exists, is unique. The additive identity of a ring R is always denoted by 0.

Theorem 10.1. Let R be a ring. Then

(a) $a.0 = 0.a = 0$ for every $a \in R$;
(b) $a(-b) = (-a)b = -(ab)$ for all $a, b \in R$;
(c) $(-a)(-b) = ab$ for all $a, b \in R$.

Proof.

(a) For any $a \in R$, $0 + a.0 = a.0 = a(0 + 0) = a.0 + a.0$ which, by the cancellation law that holds in the additive group R, implies that $a.0 = 0$. That $0.a = 0$ follows on similar lines.
(b) Let $a, b \in R$. Then
$ab + a(-b) = a(b + (-b)) = a.0 = 0$ and
$a(-b) + ab = a(-b + b) = a.0 = 0$,
which together show that $a(-b) = -(ab)$.
That $(-a)b = -(ab)$ follows similarly.
(c) For $a, b \in R$,
$(-a)(-b) = -(a(-b)) = -(-(ab)) = ab$.

Example 10.1.

(a) Z, Q, R and C are all commutative rings with identity with respect to the addition and multiplication of the number system.
(b) $2Z$ — the set of all even integers is a commutative ring but does not have identity.
(c) The set $M_2(Z)$ of all square matrices of order 2 with integer entries is a ring with identity but is not commutative because

$$\begin{pmatrix} 1 & -1 \\ 0 & -1 \end{pmatrix} \begin{pmatrix} -1 & 0 \\ 1 & 1 \end{pmatrix} = \begin{pmatrix} -2 & -1 \\ -1 & -1 \end{pmatrix}$$

while

$$\begin{pmatrix} -1 & 0 \\ 1 & 1 \end{pmatrix} \begin{pmatrix} 1 & -1 \\ 0 & -1 \end{pmatrix} = \begin{pmatrix} -1 & 1 \\ 1 & -2 \end{pmatrix}$$

and

$$\begin{pmatrix} 1 & -1 \\ 0 & -1 \end{pmatrix} \begin{pmatrix} -1 & 0 \\ 1 & 1 \end{pmatrix} \neq \begin{pmatrix} -1 & 0 \\ 1 & 1 \end{pmatrix} \begin{pmatrix} 1 & -1 \\ 0 & -1 \end{pmatrix}.$$

(d) The set $M_2(2Z)$ of all square matrices of order 2 with entries as even integers is a ring which is neither commutative nor has identity.

The compositions in the above two examples are addition and multiplication of matrices.

(e) $Z/10Z = \{0, 1, 2, \ldots, 9\}$ with addition and multiplication defined *modulo* 10 is a commutative ring with identity. In fact, for any given positive integer n, we can consider the ring $Z_n = Z/nZ = \{0, 1, 2, \ldots, n - 1\}$ of integers *modulo* n, i.e. with addition and multiplication defined *modulo* n which is a commutative ring with identity.

We have seen that multiplication of any element of a ring with the zero element is always zero. There do exist rings with elements that are both non-zero, but their product is zero. For example, in $M_2(Z)$, $\begin{pmatrix} 1 & 0 \\ 0 & 0 \end{pmatrix}$ and $\begin{pmatrix} 0 & 0 \\ 0 & 2 \end{pmatrix}$ are non zero elements but

$$\begin{pmatrix} 1 & 0 \\ 0 & 0 \end{pmatrix} \begin{pmatrix} 0 & 0 \\ 0 & 2 \end{pmatrix} = \begin{pmatrix} 0 & 0 \\ 0 & 0 \end{pmatrix} \quad \text{and} \quad \begin{pmatrix} 0 & 0 \\ 0 & 2 \end{pmatrix} \begin{pmatrix} 1 & 0 \\ 0 & 0 \end{pmatrix} = \begin{pmatrix} 0 & 0 \\ 0 & 0 \end{pmatrix}.$$

Definition. An element a in a ring R is called a **zero divisor** if there exists an element $b \in R, b \neq 0$ such that $ab = 0$ or $ba = 0$. Observe that the element 0 in a ring R is always a zero divisor.

Definition. A commutative ring R having no non-zero zero divisors is called an **integral domain**. It can be easily proved that R is an integral domain if and only if for $a, b \in R$ with $ab = 0$ implies that either $a = 0$ or $b = 0$.

Since the product of two numbers is zero if and only if one of the two numbers is zero, $Z, 2Z, Q, R, C$ are integral domains.

Definition. A non-empty subset S of a ring R is called a **subring** of R if

(a) the compositions in R induce compositions in S, i.e. if $a, b \in S$ then $a + b \in S$ and $ab \in S$;

(b) with respect to the induced compositions, S is a ring.

Definition. A non-empty subset A of a ring R is called an **ideal** of R if

(a) A is a subgroup of R with respect to addition;
(b) for any $a \in A$, $x \in R$, both ax and xa are in A.

Observe that every ideal in a ring R is a subring of R. The converse is not always true. For example, Z is a subring of Q but is not an ideal in Q; Q is a subring of R but is not an ideal in R; R is a subring of C but is not an ideal in C.

An ideal A in a ring R is called a **principal ideal** if there exists an element $a \in A$ such that A is the smallest ideal in R containing the element a. If R is a commutative ring with identity and a is an element of R, then $aR = \{ax | x \in R\}$ is an ideal of R and is, in fact, the smallest ideal in R containing a. Thus aR is a principal ideal in R and is called the principal ideal generated by a. Also, we write $aR = \langle a \rangle$. The element a is also called a **generator** of the ideal aR.

Definition. An integral domain R with identity is called a **Principal Ideal Domain (PID)** if every ideal in R is principal.

In view of Exercise 10.2(3) below, it follows that the ring Z of integers is an integral domain in which every ideal is of the form nZ for some non-negative integer n. Thus Z is a *PID*.

10.2. Polynomial Rings

Let R be a commutative ring with identity and x be a variable. Let $R[x]$ denote the set of all finite formal expressions of the form $a_0 + a_1 x + a_2 x^2 + \cdots + a_n x^n$, where $a_i \in R$ and $n \geq 0$. A finite formal expression of the form $f(x) = a_0 + a_1 x + a_2 x^2 + \cdots + a_n x^n$ is called a **polynomial** in x over the ring R, i.e. with coefficients in R. If $a_n \neq 0$, then $f(x) = a_0 + a_1 x + \cdots + a_n x^n$ is called a polynomial of **degree** n. Also, if $a_i = 0$ for all $i \geq 1$ in $f(x)$, then $f(x) = a_0$ and is called a **constant polynomial**. Thus R is a subset of $R[x]$. For $f(x) = a_0 + a_1 x + \cdots + a_n x^n, g(x) = b_0 + b_1 x + \cdots + b_m x^m$ two polynomials in x over R, we say that $f(x) = g(x)$ if $b_i = a_i$ for every i. In $R[x]$, we define addition and multiplication as follows:

For $f(x) = a_0 + a_1 x + \cdots + a_n x^n$, $g(x) = b_0 + b_1 x + \cdots + b_m x^m$, define

$$f(x) + g(x) = (a_0 + b_0) + (a_1 + b_1)x + (a_2 + b_2)x^2 + \cdots$$
$$= c_0 + c_1 x + c_2 x^2 + \cdots + c_k x^k,$$

where $k = Max\{m, n\}$ and $c_i = a_i + b_i$ for every $i \leq Min\{m, n\}$ and for $i > Min\{m, n\}$,

$$c_i = \begin{cases} a_i & \text{if } n > m \quad \text{and} \\ b_i & \text{if } m > n; \end{cases}$$

and

$$f(x)g(x) = a_0 + (a_0b_1 + a_1b_0)x + (a_0b_2 + a_1b_1 + a_2b_0)x^2 + \cdots + (a_0b_i$$
$$+ a_1b_{i-1} + a_2b_{i-2} + \cdots + a_ib_0)x^i + \cdots + a_nb_mx^{m+n}.$$

Observe that $deg(f(x) + g(x)) = Max\{deg\, f(x), deg\, g(x)\}$ except in one case when $deg\, f(x) = n = deg\, g(x)$ and $b_n = -a_n$ in which case $deg(f(x) + g(x)) < n$. Also $deg(f(x)g(x)) \leq m + n = deg\, f(x) + deg\, g(x)$. It is not essential that $deg(f(x)g(x)) = deg\, f(x) + deg\, g(x)$.

Example 10.2. Let $R = Z_6$, the ring of integers *modulo* 6. Take $f(x) = 2 + 3x + 2x^2$ and $g(x) = 1 + 2x + x^2 + 3x^3$, so that $deg\, f(x) = 2$, $deg\, g(x) = 3$. Now

$$f(x)g(x) = (2 + 3x + 2x^2)(1 + 2x + x^2 + 3x^3)$$
$$= 2 + (4 + 3)x + (6 + 2 + 2)x^2 + (4 + 3 + 6)x^3 + (2 + 9)x^4 + 6x^5$$
$$= 2 + x + 4x^2 + x^3 + 5x^4, \quad \text{as } 6 = 0.$$

Thus $deg(f(x)g(x)) = 4 < 5 = deg\, f(x) + deg\, g(x)$.

Now, back to the set $R[x]$ of all polynomials in x over a commutative ring R with identity. With respect to addition and multiplication of polynomials as defined above, $R[x]$ becomes a commutative ring which is again with identity. It also follows from the way addition and multiplication of polynomials have been defined, that the ring R is a subring of the ring $R[x]$. The ring $R[x]$ is called the **polynomial ring** in the variable x over R.

In general, there may exist non-zero elements in $R[x]$ the product of which is zero. For this, consider again the ring $R = Z_6$ and in $R[x]$ the elements $f(x) = 2 + 4x$, $g(x) = 3 + 3x^2$, neither of which is zero. But $f(x)g(x) = (2 + 4x)(3 + 3x^2) = 6 + 12x + 6x^2 + 12x^3 = 0$, as $6 = 0$ in R. However, the following is true.

Theorem 10.2. If R is an integral domain with identity, then so is $R[x]$.

Proof. Let R be an integral domain with identity. As observed already, $R[x]$ is a commutative ring with identity. We thus only need to prove that $R[x]$ has no non-zero zero divisors. For this, consider $f(x), g(x) \in R[x]$, $f(x) \neq 0$, $g(x) \neq 0$. Suppose that $f(x) = a_0 + a_1 x + \cdots + a_m x^m$, $g(x) = b_0 + b_1 x + \cdots + b_n x^n$ where $a_m \neq 0, b_n \neq 0$ so that $deg\, f(x) = m$, $deg\, g(x) = n$. It follows from the product of polynomials defined that the coefficient of x^{m+n} in $f(x)g(x) = a_m b_n$. Also, as $a_m \neq 0$, $b_n \neq 0$ and R is an integral domain, $a_m b_n \neq 0$. Hence $f(x)g(x) \neq 0$ and $R[x]$ is an integral domain. Again,

$$f(x)g(x) = c_0 + c_1 x + c_2 x^2 + \cdots + c_{m+n} x^{m+n},$$

where $c_0 = a_0 b_0$, $c_1 = a_0 b_1 + a_1 b_0, \ldots, c_i = a_0 b_i + a_1 b_{i-1} + \cdots + a_i b_o$, and $c_{m+n} = a_m b_n \neq 0$. As coefficients of powers of x higher than $m + n$ are all zero, $deg(f(x)g(x)) = m + n = deg\, f(x) + deg\, g(x)$.

We next show that if R is a *PID*, then $R[x]$ need not always be a *PID*.

Example 10.3. Consider the polynomial ring $Z[x]$, where the ring Z of integers as already observed is a *PID*. Consider the subset

$$I = \{2f(x) + xg(x) | f(x), g(x) \in Z[x]\}$$

of $Z[x]$. It is fairly easy to check that I is an ideal of $Z[x]$. Suppose that I is a principal ideal generated by a polynomial $a(x)$. Thus every element of I is a multiple of $a(x)$, i.e. it is of the form $a(x)h(x)$ for some $h(x) \in Z[x]$. Taking $f(x) = 1$ and $g(x) = 0$ and again $f(x) = 0$, $g(x) = 1$, we find that $2 = 2.1 + x.0 \in I$ and $x = 2.0 + x.1 \in$ I. Therefore,

$$2 = a(x)h(x), \quad x = a(x)k(x)$$

for some $h(x), k(x) \in Z[x]$. As Z is an integral domain,

$$0 = deg\, 2 = deg(a(x)h(x)) = deg\, a(x) + deg\, h(x).$$

As the degree of a polynomial is non-negative, we find that $a(x) = a \in Z$ is a constant polynomial. Then

$$x = ak(x) = a(k_0 + k_1 x + \cdots).$$

Comparing the coefficients of various powers of x on the two sides of this relation, we get $ak_i = 0$ for all $i \neq 1$ and $ak_1 = 1$. As a, k_1 are integers, it follows that $a = 1$ or -1. For any polynomial $\lambda(x) \in Z[x]$, we have $\lambda(x) = 1.\lambda(x) = (-1)(-\lambda(x)) \in I$. This proves that $Z[x] = I$. As $3 \in Z[x]$, $3 = 2f(x) + xg(x)$ for some $f(x), g(x) \in Z[x]$. Observe that the constant term in $xg(x)$ is 0. Therefore, comparing the constant terms on the two sides of the above relation gives $3 = 2a_0$, where a_0 is the constant term of $f(x)$. This is an absurdity, as 3 is odd and the right-hand side is even. Hence $I \neq Z[x]$ and so I is not a principal ideal. Therefore, $Z[x]$ is not a *PID*.

Definition. A non-constant polynomial $f(x) \in R[x]$ is said to be **irreducible** if, whenever $f(x) = g(x)h(x), g(x), h(x) \in R[x]$, then either $deg\, g(x) = 0$ or $deg\, h(x) = 0$. Otherwise, $f(x)$ is said to be **reducible**, i.e. if there exist non-constant polynomials $g(x), h(x) \in R[x]$ such that $f(x) = g(x)h(x)$, then $f(x)$ is reducible.

Observe that every polynomial of degree 1 in $R[x]$, where R is an integral domain, with identity is irreducible. If R is not an integral domain, then there can be polynomials of degree 1 which are reducible. For example, if $R = Z_6$, the ring of integers *modulo* 6, then $x = (2 + 3x)(3 + 2x)$.

Thus there can be a systematic study of reducible or irreducible polynomials in $R[x]$ only when R is an integral domain.

Let R be a commutative ring with identity. An element $u \in R$ is called a **unit** if there exists an element $v \in R$ such that $uv = 1$. An element $a \in R$ is called **irreducible** if, whenever $a = bc, b, c \in R$, then either b is a unit or c is a unit. Otherwise, a is called **reducible**, i.e. if a is reducible, then there exist elements $b, c \in R$, neither of which is a unit and $a = bc$.

An element $a \in R$ is said to **divide an element** $b \in R$ if there exists an element $c \in R$ such that $b = ac$. We express this fact by writing $a|b$. If a does not divide b, we write $a\backslash b$. An element $p \in R$ is called a **prime element** if, whenever p divides $ab, a, b \in R$, then either p divides a or p divides b.

Let R be an integral domain with identity and p be a prime element of R. Suppose that $p = ab$ for some $a, b \in R$. Then p divides ab. The element p being prime, $p|a$ or $p|b$. If $p|a$, let $a = pr$ for some $r \in R$. Then $p = prb$, which gives, R being an integral domain, $1 = rb$, i.e. b is a unit. Similarly, if $p|b$ then a is a unit. This proves that p is irreducible.

Thus we get the following result.

Theorem 10.3. In an integral domain with identity every prime element is irreducible. In the case of a PID, every irreducible element is a prime element.

However, we do not go into its proof here.

10.3. Quotient Rings and Homomorphisms

Let R be a ring and A be an ideal of R. Then A is a subgroup of the additive group of R. Since R is an Abelian group with respect to addition and every subgroup of an Abelian group is normal, A is a normal subgroup of the additive group R. We can thus talk of the quotient group R/A of R with respect to A. Observe that

$$R/A = \{x + A | x \in R\}$$

with addition defined by:

$$(x + A) + (y + A) = x + y + A, \quad x, y \in R.$$

For $x, y \in R$, define

$$(x + A)(y + A) = xy + A.$$

If $x + A = x' + A$ and $y + A = y' + A$, then $x' = x + a$, $y' = y + b$ for some $a, b \in A$. Then

$$\begin{aligned}
(x' + A)(y' + A) &= x'y' + A \\
&= (x + a)(y + b) + A \\
&= xy + (xb + ay + ab) + A.
\end{aligned}$$

Since $a, b \in A$ and A is an ideal of R, $xb + ay + ab \in A$ and, therefore,

$$xy + (xb + ay + ab) + A = xy + A = (x + A)(y + A).$$

Hence

$$(x' + A)(y' + A) = xy + (xb + ay + ab) + A = (x + A)(y + A)$$

and the product defined is independent of the choice of coset representatives. Multiplication in R/A satisfies the associative law and is distributive over addition because the two laws hold in R. Thus R/A becomes a ring. If the ring R is commutative, then so is R/A and if R is a ring with identity 1, then R/A is also a ring with identity with $1 + A$ as its identity.

Definition. An ideal A in a commutative ring R is called a **prime ideal** if for $a, b \in R$, whenever $ab \in A$, then either $a \in A$ or $b \in A$.

Definition. An ideal M in a ring R is called a **maximal ideal** if

(a) $M \neq R$; (b) whenever A is an ideal of R with $M \subseteq A \subseteq R$, then either $A = M$ or $A = R$.

Exercise 10.1. In the ring Z of integers, prove that

(a) nZ is a prime ideal if and only if n is a prime;
(b) every prime ideal is maximal.

We state without proof two theorems.

Theorem 10.4. In a commutative ring R, an ideal P is a prime ideal if and only if R/P is an integral domain.

Theorem 10.5. In a commutative ring R with identity, an ideal M is maximal if and only if R/M is a field.
(Refer to the next chapter for the definition of a field.)

Let R, R' be rings. A map $f \colon R \to R'$ is called a **homomorphism** if

(a) $f(x + y) = f(x) + f(y)$, (b) $f(xy) = f(x)f(y)$ for all $x, y \in R$.

Since R, R' are Abelian groups with respect to addition, condition (a) of the definition of homomorphism shows that f is a homomorphism of groups and, therefore,

(i) $f(0) = 0$;
(ii) $f(-x) = -f(x)$ for every $x \in R$.

If R is a ring with identity, it is not essential that $f(1)$ is identity of R'.

Example 10.4. Let $R = Z, R' = M_2(Z)$, the ring of square matrices of order 2 with integer entries and $f : Z \to M_2(Z)$ be defined by

$$f(m) = \begin{pmatrix} m & 0 \\ 0 & 0 \end{pmatrix}, \quad m \in Z.$$

The map f is clearly a homomorphism. Then

$$f(1) = \begin{pmatrix} 1 & 0 \\ 0 & 0 \end{pmatrix}$$

which is not the identity in $M_2(Z)$.

A ring homomorphism $f : R \to R'$ is called a **monomorphism** if the map f is one–one and it is called an **epimorphism** if the map f is onto. If the map f is both one–one and onto, then the homomorphism f is called an **isomorphism**. Two rings R and R' are called **isomorphic** if there exists an isomorphism $f : R \to R'$ or $R' \to R$. Indeed, if there exists an isomorphism: $R \to R'$, then there exists an isomorphism: $R' \to R$. If R, R' are isomorphic, we express it by $R \cong R'$.

Definition. Let R, R' be rings, and $f : R \to R'$ be a homomorphism. The subset $\ker f = \{a \in R | f(a) = 0\}$ of R is called the **kernel** of f.

Theorem 10.6. Let R, R' be rings and $f : R \to R'$ be a homomorphism. Then $\ker f$ is an ideal of R.

Proof. Since $f(0) = 0, 0 \in \ker f$ and so $\ker f \neq \phi$. Let $a, b \in \ker f$ and $x \in R$. Then $f(a-b) = f(a+(-b)) = f(a)+f(-b) = f(a)-f(b) = 0-0 = 0$ and $a - b \in \ker f$.

Therefore, $\ker f$ is a subgroup of the additive group of R.

Again, $f(xa) = f(x)f(a) = f(x)0 = 0$ and $f(ax) = f(a)f(x) = 0f(x) = 0$.

Therefore, $ax, xa \in \ker f$. Hence $\ker f$ is an ideal of R.

Theorem 10.7. Let R, R' be rings and $f : R \to R'$ be an onto ring homomorphism. Then $R/\ker f \cong R'$.

Proof. $\ker f$ is an ideal of R and, therefore, we can define the quotient ring $R/\ker f$. Define a map $\theta : R/\ker f \to R'$ by

$$\theta(x + \ker f) = f(x), \quad x \in R.$$

If $x + \ker f = y + \ker f$, then $y = x + a$ for some $a \in \ker f$ and

$$\theta(y + \ker f) = f(y) = f(x + a) = f(x) + f(a)$$
$$= f(x) + 0 = f(x) = \theta(x + \ker f).$$

Therefore, θ is well defined. For $x, y \in R$,

$$\theta((x + \ker f) + (y + \ker f)) = \theta(x + y + \ker f)$$
$$= f(x + y)$$
$$= f(x) + f(y)$$
$$= \theta(x + \ker f) + \theta(y + \ker f)$$

and

$$\theta((x + \ker f)(y + \ker f)) = \theta(xy + \ker f) = f(xy) = f(x)f(y)$$

$$= \theta(x + \ker f)\,\theta(y + \ker f).$$

Hence θ is a homomorphism.

Let $x' \in R'$. As f is onto, there exists an $x \in R$ such that $x' = f(x) = \theta(x + \ker f)$ which shows that θ is onto. If $x, y \in R$ and $\theta(x + \ker f) = \theta(y + \ker f)$, then $f(x) = f(y)$. But then $f(y - x) = f(y) - f(x) = 0$, i.e. $y - x \varepsilon \ker f$ which implies that $x + \ker f = y + \ker f$. This completes the proof that θ is an isomorphism. Therefore, $R/\ker f \cong R'$.

Exercises 10.2.

1. Check that the rings Z_5, Z_7, Z_{11} of integers with addition and multiplication defined *modulo* 5, 7, 11 respectively are integral domains.

2. Prove that none of the rings $Z_4, Z_6, Z_8, Z_9, Z_{10}, Z_{12}$ of integers with addition and multiplication defined *modulo* 4, 6, 8, 9, 10, 12 respectively is an integral domain.

3. Prove that every ideal of Z is of the form nZ for some non-negative integer n.

4. Prove that $\{0\}$ and Q are the only ideals of the ring Q of rational numbers.

5. Prove that $\{0\}$ and R are the only ideals of the ring R of real numbers.

6. Prove that $\{0\}$ and C are the only ideals of the ring C of complex numbers.

7. If R is a ring with identity and A is an ideal of R with $1 \in A$, prove that $A = R$.

8. If a non-zero element x of a ring R with identity has a multiplicative inverse, prove that x cannot be a zero divisor. (Recall that an element x of a ring R with identity is called invertible if there exists an element y in R such that $xy = yx = 1$. The element y is unique and is called the inverse of x.)

9. If R is a commutative ring with identity and $a \in R$, prove that the subset $aR = \{ax | x \in R\}$ is an ideal of R and $a \in aR$.

10. Give an example of each of the following:

    (a) an infinite ring;
    (b) a finite ring;
    (c) a commutative ring without identity;

(d) a commutative ring with identity; and,

(e) a non-commutative ring without identity.

11. Give an example of a ring R with a subring which is not an ideal in R.

12. Which of the following rings are integral domains? Explain with full justification.

 (a) the ring Z_{15} of integers *modulo* 15;

 (b) the ring Z_{31} of integers *modulo* 31; and,

 (c) the ring Z_{71} of integers *modulo* 71.

13. Prove that the subsets

 (a) $\{a + \sqrt{3}b | a, b \in Z\}$

 (b) $\{a + \sqrt{5}b | a, b \in Z\}$

 of the ring R of real numbers are subrings of R. Are these ideals in R? Justify.

14. Prove that $\{2a + \sqrt{5}b | a, b \varepsilon Z\}$ is an ideal in the ring $\{a + \sqrt{5}b | a, b \in Z\}$. Is this ideal principal?

15. In the ring $Z[x]$, prove that the ideal $\langle x \rangle$ generated by x is prime. Is it maximal? Justify.

16. In a finite commutative ring, prove that every proper prime ideal is maximal.

17. In the ring Z of integers, prove that

 (a) an ideal nZ is a prime ideal if and only if $n = p$ a prime.

 (b) Is every prime ideal in Z maximal? Justify.

18. Prove that in the ring $2Z$ of even integers, every ideal is of the form $2mZ$ where m is a non-negative integer.

19. Let R be a commutative ring not having identity and $a \in R$. Prove that the subset $\{ra + ma | r \in R, m \in Z\}$ of R is an ideal of R. Prove that if A is an ideal of R with $a \in A$, then $\{ra + ma | r \in R, m \in Z\} \subseteq A$. Thus $\{ra + ma | r \in R, m \in Z\}$ is the principal ideal $\langle a \rangle$ generated by a.

20. Prove that in the ring $2Z$ of even integers $2pZ$, where p is a prime, is a prime ideal. Is every prime ideal in $2Z$ of this form? Justify.

21. Let R be the ring of square matrices of order 2 with integer entries. Prove that the subset

$$S = \left\{ \begin{pmatrix} a & 0 \\ b & 0 \end{pmatrix} | a, b \in Z \right\}$$

is a subring of R. Is S an ideal in R? Justify.

22. Let R be the ring as in Question 21 above. Prove that the sets

$$S = \left\{ \begin{pmatrix} a & 0 \\ b & c \end{pmatrix} \mid a, b, c \in Z \right\}, T = \left\{ \begin{pmatrix} a & b \\ 0 & c \end{pmatrix} \mid a, b, c \in Z \right\}$$

are subrings of R. Are any of these an ideal of R? Justify.

23. Prove that the set

$$S = \left\{ \begin{pmatrix} 1 & 0 \\ a & 0 \end{pmatrix} \mid a \in Z \right\}$$

is a subring of R where R is the ring as in Question 21. Also prove that S is not an ideal of R.

24. Prove that $R = \{a + b\sqrt{13} \mid a, b \in Z\}$ is not an integral domain. Prove that $\pm 1, 18 \pm 5\sqrt{13}$ and $-18 \pm 5\sqrt{13}$ are units in R. Are there any other units in R? Give justification.

25. In the Z_{12} of integers *modulo* 12,

    (a) find the units;
    (b) find the roots of the polynomial $f(x) = x^2 + 4x + 3$; and,
    (c) find $-4, -7$, and 5^{-1}.

26. Prove that a commutative ring R is an integral domain if and only if $ab = ac, a, b, c \in R, a \neq 0$ implies $b = c$.

27. If $a^2 = a$ for every a in a ring R, prove that R is commutative.

    Let R be an integral domain and suppose that with every non-zero $a \in R$ is associated a positive number $d(a)$ such that (i) $d(a) \leq d(ab)$, for all $a, b \in R, ab \neq 0$, (ii) for all $a, b \in R, a \neq 0$, there exist elements q, c in R such that $b = qa + c$, where either $c = 0$ or $d(c) < d(a)$. Then R is called a **Euclidean domain.**

28. Prove that every ideal in a Euclidean domain R is of the form aR, for some $a \in R$. Deduce that a Euclidean domain is a ring with identity.

Chapter 11

FIELDS AND VECTOR SPACES

Fields play a central role in many areas of mathematics. Here we define a field, extension of a field, its characteristic and splitting field of a polynomial. Splitting fields of some polynomials of small degree over some fields of small orders are obtained. We also introduce the concept of vector space over a field. The basis of a finite dimensional vector space is defined and it is proved that every two bases of a finite dimensional vector space have an equal number of elements.

11.1. Fields

Definition. A commutative ring F with identity $1 \neq 0$ is called a **field** if every non-zero element of F is invertible w.r.t. multiplication. Equivalently, a set F having at least two elements and with two binary compositions, say addition and multiplication is called a field if

(a) F is an Abelian group w.r.t. addition;

(b) multiplication in F satisfies associative and commutative laws;

(c) multiplication is distributive over addition, i.e.

$$a(b + c) = ab + bc \quad \text{for all } a, b, c \in F;$$

(d) there exists an element $1 \in F$ such that $1a = a$ for every $a \in F$;

(e) for every $a \in F, a \neq 0$, there exists an element $b \in F$ such that $ab = 1$.

Definition. A subset K of field F having at least two elements is called a **subfield** of F if for $a, b \in K$ (i) $a + b \in K$; (ii) $ab \in K$ and w.r.t. the induced compositions K is a field.

Example 11.1. The sum of two rational numbers and product of two rational numbers are again rational numbers. Also reciprocal of a non-zero rational number is a rational number. The other properties are easy to check and the set Q of all the rational numbers is a field.

Example 11.2. The sum and product of two real numbers are again real numbers. Addition and multiplication of real numbers satisfy both the associative law and the commutative law. The numbers 0 and 1 are, respectively, the additive and multiplicative identities. For a real number $a, -a$ is its additive inverse while if $a \neq 0$, then $\frac{1}{a}$ is its multiplicative inverse. Since the distributive laws are also satisfied, the set R of all real numbers is a field.

Example 11.3. The set C of all complex numbers w.r.t. addition and multiplication of number system is also a field.

Example 11.4. The set Z of all integers is not a field w.r.t. addition and multiplication of the number system. Justify.

All the fields considered above have infinitely many elements. There also exist fields which have a finite number of elements. We consider a couple of examples.

Example 11.5. Consider the set $B = Z_2 = \{0, 1\}$ of two elements in which addition and multiplication are given by

+	0	1		×	0	1
0	0	1		0	0	0
1	1	0		1	0	1

Then Z_2 is a field and has only two elements.

Example 11.6. The set $Z_3 = \{0, 1, 2\}$ in which addition and multiplication are given by the tables

+	0	1	2		×	0	1	2
0	0	1	2		0	0	0	0
1	1	2	0		1	0	1	2
2	2	0	1		2	0	2	1

is a field having three elements.

We will next consider a general case in which Examples 11.5 and 11.6 are particular cases.

Example 11.7. Let p be a prime and $Z_p = \{0, 1, 2, \ldots, p-1\}$. In Z_p define addition and multiplications as follows. For $a, b \in Z_p$ define

$a \oplus b =$ The least non-negative remainder when the sum $a+b$ of the integers a, b is divided by p;

$a \odot b =$ The least non-negative remainder on dividing the product ab of the integers a, b by p.

The numbers 0 and 1 are clearly the additive and multiplicative identities, respectively. As in any ring 0 is its own additive inverse and 1 is its own multiplicative inverse. For $a \in Z_p, a \neq 0, p - a \in Z_p$ and $a \oplus (p - a) = 0$. Since for integers $a, b, a + b = b + a$ and $ab = ba, \oplus$ and $\odot$ in Z_p satisfy commutative law. Consider $a \in Z_p, a \neq 0, 1$. Then $g.c.d.$ $(a, p) = 1$. Therefore, there exist integers x and y such that $xa+py = 1$. Let $x = pr+b$, where $0 \leq b \leq p - 1$. If $b = 0$, then $x = pr$ and we get $p(y + r) = 1$ which cannot happen as p is a prime and $y+r$ is an integer. Therefore, $1 \leq b \leq p-1$ i.e. $b \in Z_p$ and $b \neq 0$. Then $1 = xa+py = (pr+b)a+py = ba+p(ra+y)$ which shows that $b \odot a = a \odot b = 1$. Hence a is invertible w.r.t. multiplication.

Let $a, b, c \in Z_p$. Let $a + b = pk + r, 0 \leq r < p$ and $r + c = pt + s$, $0 \leq s < p$. Then $(a \oplus b) \oplus c = s$. On the other hand, let $b + c = pk' + r'$, $0 \leq r' < p$ and $a + r' = pt' + s', 0 \leq s' < p$. Then $a \oplus (b \oplus c) = s'$. However,

$$(a + b) + c = pk + r + c = pk + pt + s = p(k + t) + s,$$

$$a + (b + c) = a + pk' + r' = pk' + pt' + s' = p(k' + t') + s'$$

showing that

$$p(k + t) + s = p(k' + t') + s', \quad 0 \leq s, s' < p.$$

This implies that $s = s'$ or $(a \oplus b) \oplus c = a \oplus (b \oplus c)$. A similar argument shows that $(a \odot b) \odot c = a \odot (b \odot c)$ and $a \odot (b \oplus c) = a \odot b \oplus a \odot c$. Hence Z_p is a field.

Theorem 11.1. A finite integral domain having at least two elements is a field.

Proof. Let R be a finite integral domain having at least two elements. Let $R = \{a_1, a_2, \ldots, a_n\}$. Choose an element $a \in R, a \neq 0$ and consider the elements $aa_1, aa_2, \ldots, aa_n$ which are again in R. Suppose that $aa_i = aa_j$. Since $a \neq 0$ and R is an integral domain, $a(a_i - a_j) = 0$ implies that $a_i - a_j = 0$ or that $a_i = a_j$. Thus all the elements $aa_1, aa_2, \ldots, aa_n$ which are n in number are distinct. Hence $R = \{aa_1, aa_2, \ldots, aa_n\}$. Again $a \in R$ implies that there exists an $i, 1 \leq i \leq n$, such that $a = aa_i = a_i a$. For any $j, 1 \leq j \leq n$, $(aa_j)a_i = a(a_j a_i) = a(a_i a_j) = (aa_i)a_j = aa_j$ which proves that a_i is identity in R. Let us write $a_i = 1$. Then $1 \in \{aa_1, aa_2, \ldots, aa_n\}$ which shows that there exists $j, 1 \leq j \leq n$ such that $aa_j = a_j a = 1$ i.e. the element $a \in R$ is invertible w.r.t. multiplication. Since $a \in R, a \neq 0$ was chosen arbitrarily, every non-zero element of R is invertible and R is a field.

Theorem 11.2. A commutative ring R with identity $1 \neq 0$ is a field if and only if $\{0\}$ and R are the only ideals of R.

Proof. If R is a field then it follows from Exercise 11.1(5) that $\{0\}$ and R are the only ideals of R.

Conversely, suppose that $\{0\}$ and R are the only ideals of R. Let $a \in R, a \neq 0$. Then aR is an ideal of R and $a = a1 \in aR$. Therefore, $aR \neq \{0\}$. By hypothesis $aR = R$. Therefore, $1 \in aR$ and so there exists an element $b \in R$ such that $ab = ba = 1$. This proves that every non-zero element of R is invertible w.r.t. multiplication and R is a field.

11.1.1. *Field extensions and minimal polynomial*

Let K be a field and F be a subfield of K. Then K is called an **extension** of F. An element $\alpha \in K$ is called **algebraic** over F if there exists a polynomial $f(x) \in F[x]$ such that $f(\alpha) = 0$ i.e. α is a root of $f(x)$. Consider the subset

$$A = \{f(x) \in F[x] | f(\alpha) = 0\}$$

of the polynomial ring $F[x]$. If $f(x), g(x) \in A$ and $h(x) \in F[x]$, then $(f \pm g)(\alpha) = f(\alpha) \pm g(\alpha) = 0 \pm 0 = 0$ and $(hf)(\alpha) = h(\alpha)f(\alpha) = h(\alpha)0 = 0$. Therefore, $f(x) \pm g(x) \in A$ and $h(x)f(x) = f(x)h(x) \in A$. Hence A is an ideal of $F[x]$. But $F[x]$ is a Euclidean domain. Therefore, there exists a polynomial $f(x) \in A$ such that for every $g(x) \in A$ we can find an $h(x) \in F[x]$ satisfying $g(x) = h(x)f(x)$. Then $\deg g(x) = \deg h(x) + \deg f(x)$. Therefore, the degree of $f(x)$ is the smallest among the degrees of all polynomials in $F[x]$ which have α as a root. Suppose that degree of $f(x)$ is m. Then $f(x) = a_0 x^m + a_1 x^{m-1} + \cdots + a_{m-1} x + a_m$, where $a_0, a_1, \cdots, a_m \in F$ and $a_0 \neq 0$. Then $f(x) = a_0 f_1(x)$ where $f_1(x) = x^m + a_0^{-1} a_1 x^{m-1} + \cdots + a_0^{-1} a_{m-1} x + a_0^{-1} a_m \in F[x]$ and degree of $f_1(x)$ is again m. Also $0 = f(\alpha) = a_0 f_1(\alpha)$ shows that $f_1(\alpha) = 0$. The polynomial $f_1(x)$ is called the **minimal polynomial** of α over F. In other words the polynomial of smallest possible degree having α as a root and with the coefficient of highest power of x equal to 1 is called the minimal polynomial of α over F. We have already defined reducible and irreducible polynomials over a commutative ring with identity. It is fairly easy to prove that the minimal polynomial $f(x)$ of an element $\alpha \varepsilon K$ over the subfield F is irreducible over F.

11.1.2. *Characteristic of a field*

Characteristic of a field F is the least positive integer m, if it exists, such that $ma = 0$ for every $a \in F$. If no such m exists, the characteristic of F is defined to be zero. It is fairly easy to see that characteristic of each of the fields Q, R and C is zero. The characteristic of Z_2 is 2, of Z_3 is 3 and in general if p is a prime then the characteristic of Z_p is p. We record the following without proof.

Theorem 11.3. The characteristic of a field is either 0 or a prime.

Theorem 11.4. If the characteristic of a field F is 0, then F contains Q as a subfield, while if the characteristic of F is a prime p, then F contains Z_p as a subfield.

It is immediate from the above theorem that a finite field has always to be of prime characteristic. Moreover, if F is a finite field of characteristic p, then order of F is p^n for some positive integer n. On the other hand given any prime p and positive integer n, these exists a field having order p^n. The proofs of these statements are beyond the scope of the present text.

11.1.3. *Splitting field*

Let K be an extension of a field F. A polynomial $f(x) \in F[x]$ is said to split over K if $f(x)$ can be written as a product of linear factors all belonging to $K[x]$.

Definition. Let F be a field and $f(x) \in F[x]$. A field K is called a **splitting field** of $f(x)$ if

(a) $f(x)$ splits into linear factors over K;
(b) K contains F as a subfield, i.e. K is an extension of F;
(c) K is the smallest field with the properties (a) and (b) above.

Example 11.8. Consider the polynomial $x^2 - 2$ over Q. Observe that the roots of $x^2 - 2$ are $\sqrt{2}, -\sqrt{2}$ which are in R but not in Q. Both these elements are in $Q(\sqrt{2})$, which is a field (refer to Question 1(a) at the end of this section). Therefore, $x^2 - 2$ splits into linear factors over $Q(\sqrt{2})$. If $a \in Q$, we can write $a = a + 0\sqrt{2}$ and so, $a \in Q(\sqrt{2})$. Thus Q is a subfield of $Q(\sqrt{2})$. If F is any field which contains Q as a subfield and also contains $\sqrt{2}$, then it contains every element $a + \sqrt{2}b$ for $a, b \in Q$. Thus $Q(\sqrt{2}) \subseteq F$ and $Q(\sqrt{2})$ is the smallest field which contains Q and over which $x^2 - 2$ splits into linear factors. Hence $Q(\sqrt{2})$ is the splitting field of $x^2 - 2$.

Example 11.9. Consider the polynomial $x^2 + 1 \in Q[x]$. The roots of this polynomial are i and $-i$ and belong to $Q(i)$, which shall be shown to be a field (see (2) in Exercises 11.1). It contains Q as a subfield. If F is a field which contains i and contains Q as a subfield, then F contains $Q(i)$ as a subfield. Hence $Q(i)$ is the splitting field of $x^2 + 1$ over Q.

Example 11.10. If we regard $x^2 + 1$ as a polynomial over R, then the splitting field of this polynomial is C because $R(i) = \{a + ib | a, b \in R\} = C$.

Example 11.11. Consider a polynomial $f(x) = x^2 + x + 1$ over the field Z_2 of two elements. Observe that neither 0 nor 1 is a root of this polynomial. Let α be a root of this polynomial. Then $\alpha^2 + \alpha + 1 = 0$ or $\alpha^2 = \alpha + 1$ as $-1 = 1$ in Z_2. Consider the set $Z_2(\alpha) = \{0, 1, \alpha, 1 + \alpha\}$. As $\alpha^2 = 1 + \alpha$ and $2 = 0$ in Z_2, addition and multiplication tables for $Z_2(\alpha)$ may be taken as:

$+$	0	1	α	$1 + \alpha$	$\times$	0	1	α	$1 + \alpha$
0	0	1	α	$1 + \alpha$	0	0	0	0	0
1	1	0	$1 + \alpha$	α	1	0	1	α	$1 + \alpha$
α	α	$1 + \alpha$	0	1	α	0	α	$1 + \alpha$	1
$1 + \alpha$	$1 + \alpha$	α	1	0	$1 + \alpha$	0	$1 + \alpha$	1	α

With these compositions $Z_2(\alpha)$ becomes a field. It is clear from the tables above that additive inverse of any element a of $Z_2(\alpha)$ is a itself whereas multiplicative inverse of 1 is 1, of α is $1 + \alpha$ and of $1 + \alpha$ is α. Also, $1 + (1 + \alpha) + (1 + \alpha)^2 = 1 + 1 + \alpha + 1 + \alpha + \alpha + \alpha^2 = 0$. Therefore the roots of $f(x)$ are α and $1 + \alpha$ both of which belong to $Z_2(\alpha)$. Clearly Z_2 is a subfield of $Z_2(\alpha)$. If F is a field which contains Z_2 and α it must also contain $1 + \alpha$ and, so, $Z_2(\alpha) \subset F$. Hence $Z_2(\alpha)$ is the splitting field of $f(x)$ over Z_2.

Example 11.12. Consider the polynomial $f(x) = x^3 + x^2 + 1$ over Z_2. Observe that neither 0 nor 1 is a root of $f(x)$ and, therefore, $f(x)$ is irreducible over Z_2. Let α be one of the roots of $f(x)$. Then $\alpha^3 + \alpha^2 + 1 = 0$ and since $2 = 0$, $\alpha^3 = \alpha^2 + 1$. Consider the set

$$Z_2(\alpha) = \{0, 1, \alpha, 1 + \alpha, \alpha^2, 1 + \alpha^2, \alpha^2 + \alpha, 1 + \alpha + \alpha^2\}.$$

In $Z_2(\alpha)$ define addition and multiplication to satisfy the associative laws, the commutative laws and the distributive laws subject to the relation $\alpha^3 = 1 + \alpha^2$. Then 0, 1 are respectively the additive and multiplicative identity. Since $1 = -1$ in Z_2, the additive inverse of $a \in Z_2(\alpha)$ is a itself. Observe that

$$\alpha(\alpha + \alpha^2) = \alpha^2 + \alpha^3 = \alpha^2 + 1 + \alpha^2 = 1$$

$$\alpha^2(1 + \alpha) = \alpha^2 + \alpha^3 = \alpha^2 + 1 + \alpha^2 = 1$$

$$(1 + \alpha^2)(1 + \alpha + \alpha^2) = 1 + \alpha + \alpha^2 + \alpha^2 + \alpha^3 + \alpha^4$$

$$= \alpha + \alpha^2 + \alpha^4 = \alpha + \alpha^2 + \alpha(\alpha^2 + 1)$$

$$= \alpha^3 + \alpha^2 = \alpha^2 + 1 + \alpha^2 = 1.$$

Therefore, $\alpha, \alpha + \alpha^2$ are inverse of each other, $\alpha^2, 1 + \alpha$ are inverse of each other and $1 + \alpha^2, 1 + \alpha + \alpha^2$ are inverse of each other w.r.t. multiplication. Therefore, $Z_2(\alpha)$ is a field and contains α and Z_2. Since

$$1 + (\alpha^2)^2 + (\alpha^2)^3$$

$$= 1 + \alpha^4 + (\alpha^3)^2 = 1 + \alpha^4 + (1 + \alpha^2)^2$$

$$= 1 + \alpha^4 + 1 + \alpha^4 = 0$$

$$1 + (1 + \alpha + \alpha^2)^2 + (1 + \alpha + \alpha^2)^3$$

$$= 1 + 1 + \alpha^2 + \alpha^4 + (1 + \alpha + \alpha^2)(1 + \alpha^2 + \alpha^4)$$

$$= \alpha^2 + \alpha^4 + 1 + \alpha^2 + \alpha^4 + \alpha + \alpha^3 + \alpha^5 + \alpha^2 + \alpha^4 + \alpha^6$$

$$= 1 + \alpha + \alpha^3 + \alpha^2(1 + \alpha^2) + \alpha^2 + \alpha^4 + (1 + \alpha^2)^2$$

$$= 1 + \alpha + \alpha^3 + 1 + \alpha^4 = \alpha(1 + \alpha^2 + \alpha^3) = 0.$$

Thus α^2 and $1 + \alpha + \alpha^2$ are the other roots of $f(x)$ and these also belong to $Z_2(\alpha)$. Therefore, $f(x)$ splits into linear factors over $Z_2(\alpha)$, Z_2 is a subfield of $Z_2(\alpha)$.

Let F be a field which contains Z_2 as a subfield and also contains the roots $\alpha, \alpha^2, 1 + \alpha + \alpha^2$ of $f(x)$. Then F contains

$$\alpha + \alpha^2, \alpha + 1 + \alpha + \alpha^2 = 1 + \alpha^2,$$

$$\alpha^2 + 1 + \alpha + \alpha^2 = 1 + \alpha, \alpha + \alpha^2 + 1 + \alpha + \alpha^2 = 1.$$

Hence F contains all the elements of $Z_2(\alpha)$. This proves that $Z_2(\alpha)$ is the splitting field of $f(x)$ over Z_2.

Example 11.13. Consider the polynomial $x^2 + x + 2 = f(x)$ over the field Z_3 of order 3. None of 0, 1 or 2 is a root of $f(x)$ and $f(x)$ is irreducible over Z_3. Let α be a root of $f(x)$. Then $\alpha^2 + \alpha + 2 = 0$ or $\alpha^2 = -\alpha - 2 = 2\alpha + 1$. Consider the set

$$Z_3(\alpha) = \{0, 1, 2, \alpha, 2\alpha, 1 + \alpha, 1 + 2\alpha, 2 + \alpha, 2 + 2\alpha\}.$$

In $Z_3(\alpha)$ define addition and multiplication to satisfy associative, commutative and distributive laws subject to the relation $\alpha^2 = 2\alpha + 1$. Then $Z_3(\alpha)$ is a commutative ring with identity. We find that

$$\alpha(1 + \alpha) = \alpha + \alpha^2 = \alpha + 2\alpha + 1 = 1, \quad 2.2 = 1,$$

$$2\alpha(2\alpha + 2) = \alpha(\alpha + 1) = 1,$$

$$(1 + 2\alpha)(2 + \alpha) = 2 + \alpha + 4\alpha + 2\alpha^2 = 2 + 2\alpha + 2(2\alpha + 1) = 1.$$

Therefore, $\alpha, 1 + \alpha$ are inverse of each other and so are $2\alpha, 2\alpha + 2$ and $1 + 2\alpha, 2 + \alpha$. Thus every non-zero element of $Z_3(\alpha)$ is invertible w.r.t. multiplication and $Z_3(\alpha)$ is a field. This field contains Z_3 as a subfield. Observe that $(2 + 2\alpha)^2 + (2 + 2\alpha) + 2 = (1 + \alpha)^2 + 2\alpha + 1 = \alpha^2 + \alpha + 2 = 0$. Thus the roots of $f(x)$ are $\alpha, 2\alpha + 2$ both of which belong to $Z_3(\alpha)$. If F is a field which contains Z_3 and the roots $\alpha, 2 + 2\alpha$ of $f(x)$, then $\alpha + \alpha = 2\alpha, 1 + 2\alpha, 1 + 2(2\alpha) = 1 + \alpha, 2 + \alpha$ all belong to F. This proves that $Z_3(\alpha)$ is contained in F. Hence $Z_3(\alpha)$ is the smallest field which contains Z_3 and over which $f(x)$ splits into linear factors. Thus $Z_3(\alpha)$ is the splitting field of $f(x)$.

Example 11.14. Consider the polynomial $f(x) = x^2 + 2$ over Z_5 the field of order 5. None of the elements $0, 1, 2, 3, 4$ of Z_5 is a root of $f(x)$. Therefore, $f(x)$ is irreducible over Z_5. Let α be a root of $f(x)$. Then $\alpha^2 + 2 = 0$ or $\alpha^2 = 3$. Consider

$$Z_5(\alpha) = \{a + b\alpha \mid a, b \in Z_5\}.$$

In $Z_5(\alpha)$ define addition and multiplication to satisfy associative, commutative and distributive laws subject to $\alpha^2 = 3$. Then $Z_5(\alpha)$ becomes a commutative ring with identity having 25 elements. The elements $\alpha, 2\alpha, 3\alpha, 4\alpha$ are invertible with $2\alpha, \alpha, 4\alpha, 3\alpha$ respectively as their inverses. Thus in order to prove that $Z_5(\alpha)$ is a field it only remains to be proved that every element of the form $a + b\alpha, a, b \in Z_5, a \neq 0, b \neq 0$, is invertible. Consider such an element $a + b\alpha$. Let $c + d\alpha \in Z_5(\alpha)$ such that $(a + b\alpha)(c + d\alpha) = 1$. Then $ac + 3bd = 1$ and $ad + bc = 0$. Since $a \neq 0, d = -a^{-1}bc$ and $1 = ac - 3a^{-1}b^2c$. Therefore, $(a^2 - 3b^2)c = a$. In $Z_5, 1^2 = 1, 2^2 = 4 = 3^2, 4^2 = 1$. Thus 3 is not a square in Z_5 and, therefore, $a^2 - 3b^2 \neq 0$ in Z_5 and we get $c = a(a^2 - 3b^2)^{-1}$.

Substituting this value in $d = -a^{-1}bc$, we get $d = -b(a^2 - 3b^2)^{-1}$ in Z_5 which yields $(a + b\alpha)(c + d\alpha) = 1$. This completes the proof that $Z_5(\alpha)$ is a field.

The two roots of $f(x)$ are α and $-\alpha = 4\alpha$ both of which belong to $Z_5(\alpha)$. Therefore, $f(x)$ splits as a product of two linear factors over $Z_5(\alpha)$. If F is a field containing Z_5 as a subfield and both the roots $\alpha, 4\alpha$ of $f(x)$, then F contains all the elements of $Z_5(\alpha)$. Hence $Z_5(\alpha)$ is the splitting field of $f(x) = x^2 + 2$ over Z_5.

Exercise 11.1.

1. Prove that the sets
   (a) $Q(\sqrt{2}) = \{a + \sqrt{2}b | a, b \in Q\}$;
   (b) $Q(\sqrt{3}) = \{a + \sqrt{3}b | a, b \in Q\}$;
   (c) $Q(\sqrt{5}) = \{a + \sqrt{5}b | a, b \in Q\}$
   are fields and each contains the field Q of rational numbers
2. Prove that the set $Q(i) = \{a + bi | a, b \in Q\}$, where $i = \sqrt{-1}$ is a field and Q is a subfield of $Q(i)$.
3. Prove that the set $Z(\sqrt{2}) = \{a + b\sqrt{2} | a, b \in Z\}$ is an integral domain containing Z as a subring but is not a field.
4. Prove that every field is an integral domain.
5. If $A \neq 0$ is an ideal of a field F, prove that $A = F$.
6. Let K be an extension of a field F. Prove that minimal polynomial of an $\alpha \in K$ over F, if it exists, is unique and is irreducible.
7. The minimal polynomial of $\sqrt{2}$ over the field Q of rational number is $x^2 - 2$.
8. The minimal polynomial of $i = \sqrt{-1}$ over Q is $x^2 + 1$. Also $x^2 + 1$ is the minimal polynomial of i over R.

9. For any $a \in Q, x - a$ is the minimal polynomial of a over Q.

10. Let F be a field and $f(x)$ be a polynomial over F. An element $\alpha \in F$ is a root of $f(x)$ if and only if $f(x) = (x - \alpha)g(x)$ for some polynomial $g(x) \in F[x]$.

11. Let $f(x) \in F[x]$, F a field, be a polynomial of degree ≤ 3. Then $f(x)$ is reducible if and only if some element of F is a root of $f(x)$. Equivalently $f(x)$ is irreducible over F if and only is none of the elements of F is a root of $f(x)$.

12. Prove that $x^2 + x + 1$ is irreducible over $Z_2 = B = \{0, 1\}$. Also prove that this is the only irreducible polynomial of degree 2 over B.

13. Prove that the polynomials $x^3 + x + 1$ and $x^3 + x^2 + 1 \in B[x]$ are irreducible over B. If possible, find all other irreducible polynomials of degree 3 over B.

14. Prove that the polynomial $x^2 - 2 \in Z_3[x]$ is irreducible over Z_3.

15. Find all polynomials of degree 2 in $Z_3[x]$ which are irreducible over Z_3.

16. Construct a field of order 4.

17. Find the splitting field of
    (a) $x^2 - 3$ over Q;   (b) $x^2 - 3$ over R;
    (c) $x^2 + 2$ over Q;   (d) $x^2 + 4$ over Q.

18. Find the splitting field of
    (a) $x^3 + 1$ over Z_2;   (b) $x^3 + x + 1$ over Z_2;
    (c) $x^4 + x + 1$ over Z_2;   (d) $x^4 + x^3 + 1$ over Z_2;
    (e) $x^4 + x^3 + x^2 + x + 1$ over Z_2.

19. Prove that the polynomials (a) $x^2 + 1$; (b) $x^2 + 2x + 2$ are irreducible over Z_3. Find the splitting fields of these polynomials over Z_3.

20. Prove that the polynomials
    (a) $x^3 + 2x^2 + 1$;   (b) $x^3 + 2x + 1$;
    (c) $x^3 + 2x^2 + x + 1$; (d) $x^3 + x^2 + 2x + 1$.
    are irreducible over Z_3. Find the splitting field of each of these polynomials over Z_3.

21. Prove that the polynomials
    (a) $x^2 + 3$;   (b) $x^2 + x + 1$;   (c) $x^2 + x + 2$;
    (d) $x^2 + 2x + 3$; (e) $x^2 + 2x + 4$;   (f) $x^2 + 3x + 3$;
    (g) $x^2 + 3x + 4$; (h) $x^2 + 4x + 1$;   (i) $x^2 + 4x + 2$
    are irreducible over Z_5. Find the splitting field of each of these polynomials over Z_5.

22. If F is a field, prove that the polynomial ring $F[x]$ is a Euclidean domain.

11.2. Vector Spaces

We start with some examples.

Example 11.15. Let V be the set of all vectors in space. Recall that if $\hat{i}, \hat{j}, \hat{k}$ denote unit vectors along the positive directions of the x-axis, y-axis and z-axis respectively, then any vector can be expressed in the form $a\hat{i} + b\hat{j} + c\hat{k}$, where $a, b, c \in R$ is the set of real numbers. If $\bar{r} = a\hat{i} + b\hat{j} + c\hat{k}$ and $\bar{s} = a'\hat{i} + b'\hat{j} + c'\hat{k}$ are two vectors, then we know that

$$\bar{r} + \bar{s} = (a + a')\hat{i} + (b + b')\hat{j} + (c + c')\hat{k}$$

which is again a vector. With respect to this addition of vectors, the set V becomes an Abelian group. Also for any real number α and element $\bar{r} = a\hat{i} + b\hat{j} + c\hat{k}$, in V, $\alpha\bar{r} = (\alpha a)\hat{i} + (\alpha b)\hat{j} + (\alpha c)\hat{k}$ is again in V. This scalar product of real numbers and the elements of V satisfies the following properties.

$$(\alpha + \beta)\bar{r} = \alpha\bar{r} + \beta\bar{r} \quad \alpha(\beta\bar{r}) = (\alpha\beta)\bar{r}$$

$$\alpha(\bar{r} + \bar{s}) = \alpha\bar{r} + \alpha\bar{s} \quad 1 \cdot \bar{r} = \bar{r}$$

for all $\alpha, \beta \in R, \bar{r}, \bar{s} \in V$ and where 1 is the real number 1. Moreover, if $a\hat{i} + b\hat{j} + c\hat{k} = 0$, then taking dot product with $\hat{i}, \hat{j}$ and $\hat{k}$ respectively and using the fact that $\hat{i} \cdot \hat{i} = \hat{j} \cdot \hat{j} = \hat{k} \cdot \hat{k} = 1, \hat{i} \cdot \hat{j} = \hat{j} \cdot \hat{i} = \hat{i} \cdot \hat{k} = \hat{k} \cdot \hat{i} = \hat{j} \cdot \hat{k} = \hat{k} \cdot \hat{j} = 0$, we find that $a = b = c = 0$. This proves that every element of V can be uniquely written in the form $a\hat{i} + b\hat{j} + c\hat{k}$.

Example 11.16. Let m, n be positive integers and let $M_{m,n}(R)$ denote the set of all $m \times n$ matrices with real entries. Sum of two matrices in $M_{m,n}(R)$ is again in $M_{m,n}(R)$ and w.r.t. the matrix addition $M_{m,n}(R)$ becomes an Abelian group. If $A = (a_{ij})$ is an $m \times n$ matrix with $a_{ij} \in R$ and α is any real number then $\alpha A = \alpha(a_{ij}) = (b_{ij})$, where $b_{ij} = \alpha a_{ij}$ for all $i, j, 1 \le i \le m, 1 \le j \le n$ is again in $M_{m,n}(R)$ and the following properties hold:

$$(\alpha + \beta)A = \alpha A + \beta A, \quad \alpha(\beta A) = (\alpha\beta)A$$

$$\alpha(A + B) = \alpha A + \alpha B, \quad 1 \cdot A = A$$

for all $\alpha, \beta \in R$ and for all $A, B \in M_{m,n}(R)$.

Example 11.17. Let n be a positive integer, X be a variable and $R_n[X]$ be the set of all polynomials in X of degree $\le n$ over the field R of real numbers. The sum of two polynomials of degree $\leq n$ is again a polynomial of degree $\leq n$ and w.r.t. the addition of polynomials $R_n[X]$ is an Abelian group. Also for any $\alpha \in R$ and $a(X) = \sum_{i=0}^{n} a_i X^i$, $\alpha a(X) = \sum_{i=0}^{n} (\alpha a_i) X^i$ is again a polynomial of degree $\leq n$. Moreover, we have the following properties:

$$(\alpha + \beta)a(X) = \alpha a(X) + \beta a(X), \quad \alpha(\beta a(X)) = (\alpha \beta)a(X)$$

$$\alpha(a(X) + b(X)) = \alpha a(X) + \alpha b(X), \quad 1a(X) = a(X)$$

for all $\alpha, \beta \in R$ and all $a(X), b(X) \in R_n[X]$ with 1 the real number.

In the polynomial ring $R[X]$, two polynomials are equal if and only if the coefficients of various powers of X in the two polynomials are equal. Thus, every element of $R_n[X]$ can be uniquely written in the form $a_0 + a_1 X + \cdots + a_n X^n$, where $a_0, a_1, \ldots, a_n \in R$.

Example 11.18. Let F be a field and n be a positive integer. Let $V(n, F)$ denote the set of all ordered n-tuples $(a_1, a_2, \ldots, a_n)$, where $a_1, a_2, \ldots, a_n$ belong to F. Two ordered n-tuples $a = (a_1, a_2, \ldots, a_n)$ and $b = (b_1, b_2, \ldots, b_n)$ in $V(n, F)$ are equal if and only if $a_i = b_i$ for all $i, 1 \leq i \leq n$. For $a = (a_1, a_2, \ldots, a_n)$, $b = (b_1, b_2, \ldots, b_n)$ define $a + b = (a_1 + b_1, a_2 + b_2, \ldots, a_n + b_n)$. It is fairly easy to check that $V(n, F)$ w.r.t. the addition defined is an Abelian group. For $a = (a_1, a_2, \ldots, a_n) \in V(n, F)$ and $\alpha \in F$, define $\alpha a = \alpha(a_1, a_2, \ldots, a_n) = (\alpha a_1, \alpha a_2, \ldots, \alpha a_n)$. The following properties are then easy to check.

$$(\alpha + \beta)a = \alpha a + \beta a, \quad \alpha(\beta a) = (\alpha \beta)a,$$

$$\alpha(a + b) = \alpha a + \alpha b, \quad 1.a = a$$

for all $\alpha, \beta \in F$ and $a, b \in V(n, F)$. Making use of the definition of addition and of αa, we find that for any $a = (a_1, a_2, \ldots, a_n) \in V(n, F)$

$$a = (a_1, a_2, \ldots, a_n)$$

$$= (a_1, 0, \ldots, 0) + (0, a_2, 0, \ldots, 0) + \cdots + (0, 0, \ldots, 0, a_n)$$

$$= a_1(1, 0, \ldots, 0) + a_2(0, 1, 0, \ldots, 0) + \cdots + a_n(0, 0, \ldots, 0, 1)$$

$$= a_1 e_1 + a_2 e_2 + \cdots + a_n e_n,$$

where e_i is an n-tuple in which the ith component is 1 and every other component or entry is 0. Since every element of $V(n, F)$ can be uniquely written in the form $a = (a_1, a_2, \ldots, a_n)$, every $a \in V(n, F)$ can be uniquely written as $a_1 e_1 + a_2 e_2 + \cdots + a_n e_n$, where $a_i \in F$.

The four examples discussed above have the common properties, namely each one of the non-empty sets is an Abelian group and the product defined between the elements of the field R or F and the elements of the set V share the properties as in (11.1) below. There are several other objects which share these properties. Taking into consideration these common properties, we now define the following.

A non-empty set V with an additive composition is called a **vector space** over a field F if

(a) V is an Abelian group w.r.t. addition,
(b) to every ordered pair of elements (α, v) in the Cartesian product $F \times V$ there corresponds a unique element denoted by αv of V such that for all $\alpha, \beta \in F, u, v \in V$,

$$\text{(i) } (\alpha + \beta)v = \alpha v + \beta v; \quad \text{(ii) } \alpha(\beta v) = (\alpha\beta)v;$$

$$\text{(iii) } \alpha(u + v) = \alpha u + \alpha v; \quad \text{(iv) } 1v = v \qquad (11.1)$$

where 1 is the identity of the field F.

Elements of the vector space V are called vectors and those of the base field F are called scalars. Also the map $(\alpha, v) \to \alpha v$ from $F \times V$ to V is called the **scalar product** of the elements of V by the elements of the base field F.

We consider a couple of examples in addition to the Examples 11.15 to 11.18 discussed.

Example 11.19. Let K be a field and F be a subfield of K. If $\alpha \in F$ and $x \in K$, then both $\alpha, x \in K$ and, so, $\alpha x \in K$. Since K is a field, the properties $(b)(i) - (iv)$ of the scalar product are satisfied. Also K is an Abelian group w.r.t. addition. Hence K is a vector space over F. Thus every field is a vector space over any subfield of itself. As particular cases of this we have

(a) every field can be regarded as a vector space over itself.
(b) The field C of complex numbers is a vector space over the field R of real numbers. Since two complex numbers are equal if and only if their real and imaginary parts are equal, every complex number can be uniquely written in the from $a + ib$, where $a, b \in R$.

Example 11.20. Let $R[X]$ be the set of all polynomials in the variable X with real coefficients. $R[X]$ is an Abelian group w.r.t. the addition of polynomials. We can multiply any polynomial $a(X) = a_0 + a_1 X + a_2 X^2 +$

$\cdots + a_n X^n$ by a real number a in the usual way, i.e.

$$aa(X) = (aa_0) + (aa_1)X + (aa_2)X^2 + \cdots + (aa_n)X^n.$$

It is, then, fairly easy to check that $R[X]$ is a vector space over R.

Example 11.21. Let $L_3(R)$ denote the set of all lower triangular matrices over R. The set $L_3(R)$ is a group w.r.t. addition of matrices. We can also multiply any matrix

$$A = \begin{pmatrix} a & 0 & 0 \\ b & c & 0 \\ d & e & f \end{pmatrix}$$

by a real number α (the usual way) to again get a lower triangular matrix:

$$\alpha A = \begin{pmatrix} \alpha a & 0 & 0 \\ \alpha b & \alpha c & 0 \\ \alpha d & \alpha e & \alpha f \end{pmatrix}.$$

Then $L_3(R)$ becomes a vector space over R. Moreover, any element of $L_3(R)$ can be uniquely written in the form

$$a e_{11} + b e_{21} + c e_{22} + d e_{31} + e e_{32} + f e_{33},$$

where $a, b, c, d, e, f \in R$ and e_{ij} is the 3×3 matrix with the (i, j)th entry 1 and every other entry equal to 0.

11.2.1. *Basis of a vector space*

Definition. Let V be a vector space over a field F. A finite subset $S = \{v_1, \ldots, v_n\}$ of V is called **linearly independent** if whenever

$$\alpha_1 v_1 + \alpha_2 v_2 + \cdots + \alpha_n v_n = 0,$$

where $\alpha_i \in F$, then $\alpha_1 = \alpha_2 = \cdots = \alpha_n = 0$. A better way of saying this is: if

$$\alpha_1 v_1 + \alpha_2 v_2 + \cdots + \alpha_n v_n = 0,$$

for $\alpha_i \in F$, then there is no $i, 1 \le i \le n$, for which $\alpha_i \ne 0$. The subset S of V is said to be **linearly dependent set of vector** if S is not linearly independent. Thus, the subset $S = \{v_1, \ldots, v_n\}$ of V is linearly dependent if there exist scalars $\alpha_1, \alpha_2, \ldots, \alpha_n$ in F not all zero such that $\alpha_1 v_1 + \alpha_2 v_2 + \cdots + \alpha_n v_n = 0$.

An arbitrary subset S of V is said to be linearly independent if every finite subset of S is linearly independent and is said to be linearly dependent otherwise. Thus, a subset S of V is linearly dependent if there exists a finite subset S_1 of S which is linearly dependent.

Let $S = \{v_1, \ldots, v_n\}$ be a finite subset of a vector space V over a field F. Let $L(S) = \{\alpha_1 v_1 + \cdots + \alpha_n v_n \mid \alpha_i \in F\}$.

Then $L(S)$ is also a vector space over F and is called the **linear span** or just the **span** of S over F. We say that V is a **finite dimensional vector space** over F if there exists a finite subset S of V such that $V = L(S)$.

Remarks.

1. Observe that the vector spaces of Examples 11.15, 11.17, 11.18, 11.19 and 11.21 are all finite dimensional. The vector space of

   (a) Example 11.15 is spanned by $S = \{\hat{i}, \hat{j}, \hat{k}\}$;
   (b) Example 11.17 is spanned by $S = \{1, X, X^2, \ldots, X^n\}$;
   (c) Example 11.18 is spanned by $S = \{e_1, e_2, \ldots, e_n\}$;
   (d) Example 11.19 (a) is spanned by $S = \{1\}$, where 1 is the identity of the field;
   (e) Example 11.19 (b) is spanned by $\{1, i\}$, where $i = \sqrt{-1}$;
   (f) Example 11.21 is spanned by $S = \{e_{11}, e_{21}, e_{22}, e_{31}, e_{32}, e_{33}\}$.

2. Any subset containing the zero vector of a vector space is always linearly dependent. For, if $v_1, \ldots, v_m$ are elements of a vector space V over a field F with $v_m = 0$, we may take $\alpha_1 = \cdots = \alpha_{m-1} = 0$ and $\alpha_m = 1$ in F. Then $\alpha_1 v_1 + \alpha_2 v_2 + \cdots + \alpha_m v_m = 0$ and the α_is are not all zero.

3. If S_1, S_2 are subsets of a vector space V over F, S_1 is linearly dependent and $S_1 \subseteq S_2$, then S_2 is also linearly dependent.

4. Every subset of a linearly independent set of vectors of a vector space is linearly independent.

5. Every set consisting of a single non-zero vector of a vector space is linearly independent.

Theorem 11.5. Let $\{v_1, v_2, \ldots, v_m\}$ be a linearly dependent set of non-zero vectors of a vector space V over F. Then there exists a $k, 2 \leq k \leq m$ such that v_k is a linear combination of $v_1, v_2, \ldots, v_{k-1}$.

Proof. Since the elements $v_1, v_2, \ldots, v_m$ are linearly dependent, there exist scalars $\alpha_1, \alpha_2, \ldots, \alpha_m \in F$ not all zero such that

$$\alpha_1 v_1 + \alpha_2 v_2 + \cdots + \alpha_m v_m = 0. \tag{11.2}$$

Let k be the largest positive integer such that $\alpha_k \neq 0$. Then $\alpha_{k+1} = \cdots = \alpha_m = 0$. If $k = 1$, we get $\alpha_1 v_1 = 0$ which gives $\alpha_1^{-1}(\alpha_1 v_1) = 0$ or $v_1 = 0$. This is a contradiction and, hence, $k \geq 2$. The relation (11.2) becomes

$$\alpha_1 v_1 + \alpha_2 v_2 + \cdots + \alpha_k v_k = 0.$$

This relation then leads to

$$v_k = (-\alpha_k^{-1}\alpha_1)v_1 + \cdots + (-\alpha_k^{-1}\alpha_{k-1})v_{k-1}$$

and v_k is a linear combination of $v_1, \ldots, v_{k-1}$.

Definition. A set S of vectors in a vector space V over a field F is called a **basis** of V if (a) $V = L(S)$ and (b) S is linearly independent.

Theorem 11.6. A finite dimensional vector space over a field has a basis consisting of a finite number of elements.

Proof. Let V be a finite dimensional vector space over a field F. Then there exists a finite subset S of V such that $V = L(S)$. Let $S = \{v_1, \ldots, v_m\}$. We may assume that none of v_is are 0. If S is linearly independent, then S is a basis of V and we are through. Suppose that S is linearly dependent. Then there exists a $k, 2 \leq k \leq m$ such that v_k is a linear combination of $v_1, \ldots, v_{k-1}$. Let

$$v_k = \alpha_1 v_1 + \cdots + \alpha_{k-1} v_{k-1}, \tag{11.3}$$

where $\alpha_i \in F$. If $v \in V$, then

$$v = \beta_1 v_1 + \cdots + \beta_k v_k + \cdots + \beta_m v_m \tag{11.4}$$

where β_is are in F. Substituting for v_k from (11.3) in (11.4), we find that $v \in L(S_1)$, where $S_1 = S \backslash \{v_k\} = \{v_1, \ldots, v_{k-1}, v_{k+1}, \ldots, v_m\}$. This proves that $V = L(S_1)$.

If S_1 is linearly independent, then S_1 is a finite basis of V and we are through. If S_1 is linearly dependent, we repeat the argument used for obtaining S_1 from S, we can remove one element from S_1 to obtain S_2 such that $L(S_2) = V$. We can continue this process until we arrive at a subset S' of V such that $V = L(S')$ and S' is linearly independent. Since S has only a finite number of elements, we must arrive at S' after a finite number of steps. We have thus obtained a finite basis of V.

Theorem 11.7. If V is a finite dimensional vector space over a field F and $x_1, \ldots, x_m$ are linearly independent elements of V, then either $x_1, \ldots, x_m$ already form a basis or there exist elements $x_{m+1}, \ldots, x_{m+p}$ such that

$x_1, \ldots, x_m, x_{m+1}, \ldots, x_{m+p}$ form a basis of V over F. This is equivalent to saying that every set of linearly independent elements of a finite dimensional vector space can be extended to a basis of V.

Proof. If $x_1, x_2, \ldots, x_m$ already form a basis we have nothing to prove. So, suppose that $x_1, x_2, \ldots, x_m$ do not form a basis of V. Since V is finite dimensional, there exists a finite subset $\{y_1, y_2, \ldots, y_n\} = S$ of V such that $V = L(S)$. Now consider the set S_1 of elements $x_1, \ldots, x_m, y_1, y_2, \ldots, y_n$. Since every element of V can be expressed as a linear combination of $y_1, y_2, \ldots, y_n$, it is trivially a linear combination of the elements of S_1. Therefore, some element of S_1 is a linear combination of the preceding elements of S_1. This element cannot be an 'x' as $x_1, \ldots, x_m$ are linearly independent. Therefore, this element must be a y. By rearranging $y_1, \ldots, y_n$ if necessary, we may suppose that this element is y_n i.e. y_n is a linear combination of the subset $S_2 = \{x_1, \ldots, x_m, y_1, \ldots, y_{n-1}\}$. But then $L(S_2) = V$ (which can be proved as in the previous theorem). Either S_2 is linearly independent or is linearly dependent. If S_2 is linearly dependent, we repeat the above process and come to a subset $S_3 = \{x_1, \ldots, x_m, y_1, y_2, \ldots, y_{n-2}\}$ such that $V = L(S_3)$. We continue this process until we get a subset S' of S which is linearly independent and $V = L(S')$. At every step we have omitted a y from the set which spans V. Therefore, the subset S' we arrive at must contain $x_1, \ldots, x_m$. Hence $S' = \{x_1, x_2, \ldots, x_m, y_1, \ldots, y_p\}$. We put $y_i = x_{m+i}, 1 \leq i \leq p$ and we get a basis $S' = \{x_1, \ldots, x_m, x_{m+1}, \ldots, x_{m+p}\}$ of V which contains the linearly independent elements $x_1, \ldots, x_m$ we started with.

Theorem 11.8. Every basis of a finite dimensional vector space over a field has a finite number of elements.

Proof. Let V be a finite dimensional vector space over a field F. As already proved V has a finite basis. Let $S = \{v_1, v_2, \ldots, v_m\}$ be a finite basis of V. Then S is linearly independent and $V = L(S)$.

Let S_1 be any basis of V. Let $y_1 \in S_1$. Since y_1 is a linear combination of $v_1, v_2, \ldots, v_m$, the elements of $y_1, v_1, v_2, \ldots, v_m$ are linearly dependent. Therefore, one of these elements is a linear combination of the preceding elements. It must be one of the v_is. By renaming the elements, if necessary, we can assume that v_m is a linear combination of $y_1, v_1, \ldots, v_{m-1}$. As any element of V is a linear combination of the elements $v_1, \ldots, v_m$ and v_m is a linear combination of the elements $y_1, v_1, \ldots, v_{m-1}$, every element of V is a linear combination of the elements $y_1, v_1, \ldots, v_{m-1}$.

Next, consider another element y_2 of S_1. As y_2 is a linear combination of the elements $y_1, v_1, \ldots, v_{m-1}$, the elements $y_2, y_1, v_1, \ldots, v_{m-1}$ are linearly dependent. Then one of these elements is a linear combination of the preceding ones. This element cannot be a y as the ys are in S_1 and S_1 is linearly independent. By reordering the vs, if necessary, we may assume that v_{m-1} is a linear combination of $y_1, y_2, v_1, \ldots, v_{m-2}$. As $V = L(\{y_1, v_1, \ldots, v_{m-1}\})$ and v_{m-1} is a linear combination of $y_1, y_2, v_1, \ldots, v_{m-2}$, $V = L(\{y_1, y_2, v_1, \ldots, v_{m-2}\})$. We continue this process of inserting an element of S_1 and removing a v to get elements which span V. If $o(S_1) > m = o(S)$ we can remove all the elements of S and get elements $y_1, y_2, \ldots, y_m$ from S_1 which span V. Again $o(S_1) > m$ shows that there exists $y_{m+1} \in S_1$. As $V = L(\{y_1, \ldots, y_m\}), y_1, \ldots, y_m, y_{m+1}$ are linearly dependent which is a contradiction as S_1 is a basis of V and, so, S_1 is linearly independent. This proves that $o(S_1) \leq m = o(S)$.

Corollary. The number of elements in any basis of a finite dimensional vector space is the same as the number of elements in another basis.

Proof. Let V be a finite dimensional vector space over F and let S, S_1 be two bases of V. Then $o(S) < \infty$, $o(S_1) < \infty$. Let $o(S) = m$ and $o(S_1) = n$. As proved in the last theorem $m = 0(S) \leq 0(S_1) = n$ and, similarly, $n = o(S_1) \leq o(S) = m$. Hence $m = n$ or that $o(S) = o(S_1)$.

Corollary. Number of elements in any basis of a vector space is equal to the number of elements in any other basis.

Definition. The number of elements in any basis of a vector space V over F is called the **dimension** of V over F and is denoted by dim V.

Theorem 11.9. If V is a finite dimensional vector space of dimension n, then any $n + 1$ elements of V are linearly dependent.

Proof. Consider $n + 1$ elements $v_1, v_2, \ldots, v_{n+1}$ of V. If these elements are linearly independent, then either these elements form a basis of V or there exist elements $y_1, y_2, \ldots, y_p$ such that $v_1, v_2, \ldots, v_{n+1}, y_1, y_2, \ldots, y_p$ form a basis of V. Then $\dim_F V \geq n + 1 > n$ which is a contradiction. Hence the elements $v_1, v_2, \ldots, v_{n+1}$ are linearly dependent.

11.2.2. Subspaces and quotient spaces

Throughout, V is a vector space over a field F.

Definition. A non-empty subset W of V is called a **subspace** of V if for every pair of elements x, y in W and α, β in F, $\alpha x + \beta y \in W$.

Observe that if W is a subspace of V and $x, y \in W$, then taking $\alpha = \beta = 1$, we find that $x + y \in W$ i.e. the additive composition in V induces an additive composition in W. Also taking $y = x$ and $\alpha = 1, \beta = -1$, we get $x - x \in W$, i.e. $o \in W$. It is fairly easy to check that W is a subgroup of the additive Abelian group V. Also for $x \in W, \alpha \in F$, taking $\beta = 0$, we find that $\alpha x \in W$. Then W becomes a vector space over F.

Theorem 11.10. If U and W are subspaces of V, then $U \cap W$ is also a subspace of V.

Proof. Since $0 \in U$ and $0 \in W, 0 \in U \cap W$ and $U \cap W \neq \phi$. Let $x, y \in U \cap W$ and $\alpha, \beta \in F$. Then $x, y \in U$ and $x, y \in W$. As U and W are subspaces of V, $\alpha x + \beta y \in U$ and $\alpha x + \beta y \in W$. Thus $\alpha x + \beta y \in U \cap W$. As $x, y \in U \cap W$ and $\alpha, \beta \in F$ were chosen arbitrarily, it follows that $U \cap W$ is a subspace of V.

We can prove the following on similar lines:

Theorem 11.11. The intersection of any family of subspaces of V is a subspace of V.

Let S be a non-empty subset of V. Consider the family of all subspaces W of V such that $S \subseteq W$. This family of subspaces of V is non-empty as V is a subspace of V and $S \subseteq V$. Let W be the intersection of all the subspaces of V containing S. Then W also contains S and if W' is a subspaces of V containing S, then $W \subseteq W'$. Thus W is the smallest subspace of V containing S. We call W the **subspace** of V **spanned** by S.

Theorem 11.12. Let U and W be subspaces of V. Then the subset $U + W = \{x + y \,|\, x \in U, y \in W\}$ is a subspace of V and both U and W are subspaces of $U + W$.

Proof. Taking $x \in U, y = 0$ in $W, x = x + 0 \in U + W$ and then taking $x = 0$ in U and $y \in W$, $y = 0 + y \in U + W$. Thus, both $U \subseteq U + W$ and $W \subseteq U + W$. In particular, $U + W \neq \phi$. Let $x, x' \in U, y, y' \in W$ and $\alpha, \beta \in F$. Then

$$\alpha(x + y) + \beta(x' + y') = \alpha x + \alpha y + \beta x' + \beta y'$$

$$= (\alpha x + \beta x') + (\alpha y + \beta y')$$

which is again in $U + W$, as $x, x' \in U, \alpha, \beta \in F$ imply that $\alpha x + \beta x' \in U$ and $y, y' \in W, \alpha, \beta \in F$ imply $\alpha y + \beta y' \in W$. Hence $U + W$ is a subspace of V. That U and W are subspaces of $U + W$ is clear.

Theorem 11.13. If a vector space V is of finite dimension n, then any subspace W of V is of finite dimension and dimension of W is $\leq n$.

Proof. If the subspace $W = 0$, then W is spanned by 0 and W is zero dimensional and we are through. So, suppose that $W \neq 0$. Let $w_1 \in W$, $w_1 \neq 0$. If $W = W_1$ is the subspace of V spanned by w_1, then $\dim W = 1 < \infty$. If $W \neq W_1$, let $w_2 \in W, w_2 \notin W_1$. Let W_2 be the subspace of V spanned by $\{w_1, w_2\}$. Then $\dim W_2 = 2$. If $W_2 = W$, then W is finite dimensional. If not, we continue with the above procedure and suppose we have constructed subspaces $W_1 \subsetneqq W_2 \subsetneqq W_3 \subsetneqq \cdots \subsetneqq W_i \subsetneqq \cdots$ of W where W_i is spanned by $\{w_1, w_2, \ldots w_i\}$ and $w_i \notin W_{i-1} = L(\{w_i, \ldots, w_{i-1}\})$. As $w_i \notin W_{i-1}, w_i$ is not a linear combination of $w_1, \ldots, w_{i-1}$. As this is true for every $i \geq 2, w_1, w_2, \ldots, w_i$ are linearly independent. If W has not been reached in n steps i.e. if $W \neq W_n$, we can construct W_{n+1} which is spanned by $w_1, \ldots, w_{n+1}$ and $w_1, \ldots, w_{n+1}$ are linearly independent. But $w_1, \ldots, w_{n+1}$ are in V and $\dim V = n$ implies that any $n + 1$ elements in V are linearly dependent. In particular $w_1, \ldots, w_{n+1}$ are linearly dependent. This leads to a contradiction. Hence the process must stop at some $i \leq n$ and $W = W_i$ which leads to $\dim W = i \leq n$.

Corollary. Given any m-dimensional subspace W of V, there exists a basis $\{v_1, v_2, \ldots, v_m, \ldots, v_n\}$ of V such that $v_1, v_2, \ldots, v_m$ are in W and form a basis of W.

Example 11.22. Let V be the vector space of all square matrices of order 2 with real entries over the field R of real numbers. Let W be the subspace of V consisting of all lower (or upper) triangular matrices.

Let W_1 be the subspace of V consisting of all diagonal matrices and W_2 be the subspace of V consisting of matrices of the form $\begin{pmatrix} 0 & 0 \\ a & 0 \end{pmatrix}$ Now the vector space V is of dimension 4 with basis $\{e_{11}, e_{12}, e_{21}, e_{22}\}$ where e_{ij} is the matrix with (i, j)th entry 1 and every other entry 0. W is a subspace of V with $\dim W = 3$ and basis consisting of e_{11}, e_{21}, e_{22}; W_1 is a subspace of dimension 2 with basis e_{11}, e_{22}; and W_2 is a subspace of dimension 1 with basis consisting of the only element e_{21}.

Example 11.23. Let $V = R[x]$, the set of all polynomials in x with coefficients in the field R of real numbers and P_n or $R_n[x]$ be the set of all polynomials which are of degree $\leq n$. Then V, as seen before, is a vector space over R with $1, x, x^2, \ldots, x^k, \ldots$ as a basis, so that V is infinite dimensional. However, P_n is a subspace of $R[x]$ and dimension of P_n is $n+1$. A basis of $P_n(x)$ consists of the elements $1, x, \ldots, x^n$.

Example 11.24. Consider the vector space $V = R[x]$ as in Example 11.23 above. A polynomial $f(x) \in R[x]$ is called an **even polynomial** if $f(-x) = f(x)$ and is called an **odd polynomial** if $f(-x) = -f(x)$. The set $E(x)$ of all even polynomials in V is a subspace of V and so is the set $O(x)$ of all odd polynomials in x. Observe that $1 + x^2 + x^4 + x^8$ is an even polynomial while $x + x^3 + x^7$ is an odd polynomial. Every polynomial $f(x) \in V$ can be uniquely written in the form $f(x) = a(x) + b(x)$, where $a(x) \in E(x)$ and $b(x) \in O(x)$. It then follows that $V = E(x) + O(x)$. If $f(x) \in E(x) \cap O(x)$, then $f(-x) = f(x)$ and $f(-x) = -f(x)$ so that $f(x) = -f(x)$ which gives $f(x) = 0$. Hence $V = E(x) + O(x)$ and $E(x) \cap O(x) = 0$. If A, B are subspace of a vector space V with $V = A + B$ and $A \cap B = 0$, then A is called a **complement** of B and B is called a **complement** of A. Thus $E(x)$ and $O(x)$ are complements of each other in $V = R[x]$.

Example 11.25. Let A, B, C be subspaces of a vector space V over F with $C \subseteq A$. Let $x \in A \cap (B + C)$. Then $x \in A$ and $x = B + C$. As $x \in B + C$, $x = y + z$ with $y \in B, z \in C$. Now $z \in C \subseteq A$ implies that $z \in A$. Then $y = x - z \in A$, as $x \in A, z \in A$ or that $y \in A \cap B$. Therefore, $x \in A \cap B + C$. Hence

$$A \cap (B + C) \subseteq A \cap B + C.$$

Now $B \subseteq B + C$ implies $A \cap B \subseteq A \cap (B + C)$. Also $C \subseteq B + C$ and $C \subseteq A$. Therefore, $C \subseteq A \cap (B + C)$. Thus

$$A \cap B + C \subseteq A \cap (B + C).$$

Combining the pieces together, we get $A \cap (B + C) = A \cap B + C$.

Let V be a vector space over a field F (not necessarily finite dimensional) and W be a subspace of V. Then W is a subgroup of the additive Abelian group V. Let V/W be the quotient group of V modulo W. Then $V/W = \{v + W \mid v \in V\}$ with $(u + W) + (v + W) = (u + v) + W$.

For $v+W \in V/W$ and $\alpha \in F$, define $\alpha(v+W) = \alpha v + W$. If $v+W = v'+W$, then $v' = v + w$ for some $w \in W$. Therefore,

$$\alpha(v'+W) = \alpha v' + W = \alpha(v+w) + W = \alpha v + \alpha w + W = \alpha v + W = \alpha(v+W).$$

Thus the scalar product $\alpha(v + W) = \alpha v + W$ is independent of the choice of the representative v of $v + W$. It is then routine to check that

$$\alpha(u + W + v + W) = \alpha(u + W) + \alpha(v + W),$$

$$(\alpha + \beta)(v + W) = \alpha(v + W) + \beta(v + W),$$

$$\alpha(\beta(v + W)) = (\alpha\beta)(v + W)$$

and $1(v + W) = v + W$.

Thus V/W is a vector space over F and is called the **quotient space** of V modules the subspace W of V.

Theorem 11.14. Let V be a finite dimensional vector space over a field F and W be a subspace of V. Then $\dim V = \dim W + \dim V/W$.

Proof. Since W is a subspace of V which is finite dimensional, W is finite dimensional. Let $\dim W = m$ and $\dim V = n$ so that $m \le n$. Choose a basis $\{w_1, \ldots, w_m\}$ of W. By Theorem 11.7 we can find $k = n - m$ elements $v_1, \ldots, v_k$ in V such that $\{w_1, \ldots, w_m, v_1, \ldots, v_k\}$ is a basis of V. The result will be proved if we can prove that $\{v_1 + W, \ldots, v_k + W\}$ is a basis of V/W. Let $\alpha_1, \ldots, \alpha_k$ be elements of the base field F such that

$$\alpha_1(v_1 + W) + \cdots + \alpha_k(v_k + W) = 0$$

in V/W.

Then

$$\alpha_1 v_1 + \cdots + \alpha_k v_k + W = 0$$

which implies that $\alpha_1 v_1 + \cdots + \alpha_k v_k \in W$. Since $\{w_1, \ldots, w_m\}$ is a basis of W, there exist elements $\beta_1, \ldots, \beta_m$ in F such that

$$\alpha_1 v_1 + \alpha_2 v_2 + \cdots + \alpha_k v_k = \beta_1 w_1 + \beta_2 w_2 + \cdots + \beta_m w_m$$

or

$$(-\beta_1)w_1 + \cdots + (-\beta_m)w_m + \alpha_1 v_1 + \cdots + \alpha_k v_k = 0.$$

As the elements $w_1, \ldots, w_m, v_i, \ldots, v_k$ are linearly independent, it follows that $\beta_i = 0$ for $1 \le i \le m$ and $\alpha_j = 0$ for $1 \le j \le k$.

This proves that $v_1 + W, v_2 + W, \ldots, v_k + W$ are linearly independent.

Consider any element $v + W \in V/W$ so that $v \in V$. Then there exist elements $\alpha_1, \ldots, \alpha_k, \beta_1, \ldots, \beta_m$ in F such that

$$v = \alpha_1 v_1 + \cdots + \alpha_k v_k + \beta_1 w_1 + \cdots + \beta_m w_m.$$

Therefore,

$$v + W = \alpha_1 v_1 + \cdots + \alpha_k v_k + \beta_1 w_1 + \cdots + \beta_m w_m + W$$

$$= \alpha_1 v_1 + \cdots + \alpha_k v_k + W, \quad \text{as } \beta_1 w_1 + \cdots + \beta_m w_m \in W,$$

$$= \alpha_1 (v_1 + W) + \cdots + \alpha_k (v_k + W).$$

Thus the set $\{v_1 + W, \ldots, v_k + W\}$ spans V/W and, hence, it forms a basis of V/W. In particular $\dim V/W = k = n - m$ or that $n = \dim V = m + k = \dim W + \dim V/W$.

Example 11.26. Let V be the vector space and W, W_1, W_2 be the subspaces of V as in Example 11.22. For any matrix $\begin{pmatrix} a & b \\ c & d \end{pmatrix} \in V$,

$$\begin{pmatrix} a & b \\ c & d \end{pmatrix} = \begin{pmatrix} a & 0 \\ c & d \end{pmatrix} + b \begin{pmatrix} 0 & 1 \\ 0 & 0 \end{pmatrix}$$

$$= \begin{pmatrix} a & 0 \\ 0 & d \end{pmatrix} + c \begin{pmatrix} 0 & 0 \\ 1 & 0 \end{pmatrix} + b \begin{pmatrix} 0 & 1 \\ 0 & 0 \end{pmatrix}$$

$$= \begin{pmatrix} 0 & 0 \\ c & 0 \end{pmatrix} + a \begin{pmatrix} 1 & 0 \\ 0 & 0 \end{pmatrix} + b \begin{pmatrix} 0 & 1 \\ 0 & 0 \end{pmatrix} + d \begin{pmatrix} 0 & 0 \\ 0 & 1 \end{pmatrix}$$

from which it follows that $V/W, V/W_1$ and V/W_2 are spanned by

$$\left\{ \begin{pmatrix} 0 & 1 \\ 0 & 0 \end{pmatrix} + W \right\}; \left\{ \begin{pmatrix} 0 & 0 \\ 1 & 0 \end{pmatrix} + W_1, \begin{pmatrix} 0 & 1 \\ 0 & 0 \end{pmatrix} + W_1 \right\}$$

and

$$\left\{ \begin{pmatrix} 1 & 0 \\ 0 & 0 \end{pmatrix} + W_2, \begin{pmatrix} 0 & 1 \\ 0 & 0 \end{pmatrix} + W_2, \begin{pmatrix} 0 & 0 \\ 0 & 1 \end{pmatrix} + W_2 \right\}$$

respectively.

Since $W \neq V, V/W \neq 0$ and $\left\{ \begin{pmatrix} 0 & 1 \\ 0 & 0 \end{pmatrix} + W \right\}$ is a basis of V/W and $\dim V/W = 1$. Also $a \begin{pmatrix} 0 & 0 \\ 1 & 0 \end{pmatrix} + b \begin{pmatrix} 0 & 1 \\ 0 & 0 \end{pmatrix} \in W_1$ or that $\begin{pmatrix} 0 & b \\ a & 0 \end{pmatrix}$ is a diagonal matrix if and only if $a = 0 = b$. Therefore, the elements $\begin{pmatrix} 0 & 0 \\ 1 & 0 \end{pmatrix} + W_1$ and

$\begin{pmatrix} 0 & 1 \\ 0 & 0 \end{pmatrix} + W_1$ are linearly independent. Hence $\left\{ \begin{pmatrix} 0 & 0 \\ 1 & 0 \end{pmatrix} + W_1, \begin{pmatrix} 0 & 1 \\ 0 & 0 \end{pmatrix} + W_1 \right\}$ is a basis of V/W_1 and $\dim V/W_1 = 2$.

For $a, b, c \in R$,

$$a \begin{pmatrix} 1 & 0 \\ 0 & 0 \end{pmatrix} + b \begin{pmatrix} 0 & 1 \\ 0 & 0 \end{pmatrix} + c \begin{pmatrix} 0 & 0 \\ 0 & 1 \end{pmatrix} \in W_2$$

or that

$$\begin{pmatrix} a & b \\ 0 & c \end{pmatrix} = \begin{pmatrix} 0 & 0 \\ d & 0 \end{pmatrix}$$

for some $d \in R$ if and only if $a = b = c = 0$.

This proves that elements

$$\begin{pmatrix} 1 & 0 \\ 0 & 0 \end{pmatrix} + W_2, \begin{pmatrix} 0 & 1 \\ 0 & 0 \end{pmatrix} + W_2, \begin{pmatrix} 0 & 0 \\ 0 & 1 \end{pmatrix} + W_2$$

are linearly independent and, so, form a basis of V/W_2. Hence $\dim V/W_2 = 3$.

Observe that W_1, W_2 are subspaces of W while W_2 is not a subspace of W_1. It may be proved that $\left\{ \begin{pmatrix} 0 & 0 \\ 1 & 0 \end{pmatrix} + W_1 \right\}$ is a basis of W/W_1 and $\dim W/W_1 = 1$ while $\left\{ \begin{pmatrix} 1 & 0 \\ 0 & 0 \end{pmatrix} + W_2, \begin{pmatrix} 0 & 0 \\ 0 & 1 \end{pmatrix} + W_2 \right\}$ is a basis of W/W_2 and $\dim W/W_2 = 2$.

Let W_3 be the subspace of V consisting of all matrices of the form $\begin{pmatrix} 0 & a \\ 0 & 0 \end{pmatrix}, a \in R$.

Exercise. With V, W, W_1, W_2 and W_3 as above, prove that

(a) $V = W + W_3$, $W \cap W_3 = 0$
(b) $W = W_1 + W_2$, $W_1 \cap W_2 = 0$

11.2.3. Linear transformations

Definition. Let V, W be vector spaces over the same field F. A map $T : V \to W$ is called a **linear transformation** if

$$T(u + v) = T(u) + T(v), \quad u, v \in V,$$

$$T(\alpha v) = \alpha T(v), \quad \alpha \in F, v \in V.$$

If V, W are vector spaces over the same field F and $T: V \to W$ is a linear transformation, then T is, in particular, a homomorphism of Abelian groups. Therefore, $T(0) = 0$ and $T(-a) = -T(a), a \in V$. Let

$$\ker T = \{v \in V \mid T(v) = 0\}.$$

The subset $\ker T$ is non-empty and is called the **kernel** of the linear transformation T. $\ker T$ is a subgroup of the additive Abelian group V. If $\alpha \in F$ and $v \in \ker T$, then $T(\alpha v) = \alpha T(v) = \alpha 0 = 0$ and $\alpha v \in \ker T$. Hence $\ker T$ is a subspace of V.

A linear transformation $T: V \to W$ is called a **monomorphism**, an **epimorphism** or an **isomorphism** if the map T is respectively one–one, onto or one–one and onto. Two vector spaces V, W over a field F are said to be **isomorphic** if there exists a linear transformation $T: V \to W$ which is one–one and onto.

Theorem 11.15. Let V, W be vector spaces over the same field F and $T: V \to W$ be a linear transformation which is one–one. If $v_1, v_2, \ldots, v_n$ in V are linearly independent, then $\{T(v_1), \ldots, T(v_n)\}$ are linearly independent.

Proof. Let $\alpha_1, \alpha_2, \ldots, \alpha_n$ be elements of F such that

$$\alpha_1 T(v_1) + \alpha_2 T(v_2) + \cdots + \alpha_n T(v_n) = 0.$$

Then

$$T(\alpha_1 v_1 + \alpha_2 v_2 + \cdots + \alpha_n v_n) = 0 = T(0).$$

Since T is one–one, we get $\alpha_1 v_1 + \alpha_2 v_2 + \cdots + \alpha_n v_n = 0$. The elements $v_1, v_2, \ldots, v_n$ being linearly independent, we get

$$\alpha_1 = \alpha_2 = \cdots = \alpha_n = 0.$$

This completes the proof that the subset $\{T(v_1), T(v_2), \ldots, T(v_n)\}$ of W is linearly independent.

Theorem 11.16. Let V, W be finite dimensional vector spaces over a field F. The V and W are isomorphic if and only if $\dim_F V = \dim_F W$.

Proof. Let V and W be isomorphic. Then there exists a linear transformation $T: V \to W$ which is one–one and onto. Let $\dim_F V = n$ and $v_1, \ldots, v_n$ be a basis of V. Then, by Theorem 11.15, $T(v_1), T(v_2), \ldots, T(v_n)$ are linearly independent. If $\{T(v_1), \ldots, T(v_n)\}$ does not span W, there exists a $w \in W$ such that $w \notin L(\{T(v_1), T(v_2), \ldots, T(v_n)\})$. As T is onto, there exists a $v \in V$ such that $w = T(v)$. But $v = \alpha_1 v_1 + \cdots + \alpha_n v_n$, since $\{v_1, \ldots, v_n\}$ is a basis of V. Therefore,

$$w = T(v) = T(\alpha_1 v_1 + \cdots + \alpha_n v_n)$$
$$= \alpha_1 T(v_1) + \cdots + \alpha_n T(v_n)$$

which is in $L(\{T(v_1), \ldots, T(v_n)\})$. This is a contradiction. Hence $W = L(\{T(v_1), \ldots, T(v_n)\})$ and $\{T(v_i), \ldots, T(v_n)\}$ is a basis of W and $\dim W = n$.

Conversely, suppose that $\dim_F V = \dim_F W = n < \infty$. Let $\{v_i, \ldots, v_n\}$ be a basis of V and $\{w_1, \ldots, w_n\}$ be a basis of W. Define a map $T : V \to W$ by

$$T(\alpha_1 v_1 + \cdots + \alpha_n v_n) = \alpha_1 w_1 + \cdots + \alpha_n w_n, \quad \alpha_i \in F.$$

Since every element of V can be uniquely written as $\alpha_1 v_1 + \cdots + \alpha_n v_n$, with $\alpha_i \in F$, the map T is well defined. It is clear from the definition that T is a linear transformation and is onto. If $\alpha_i \in F$ such that

$$T(\alpha_1 v_1 + \cdots + \alpha_n v_n) = 0,$$

then

$$\alpha_1 w_1 + \cdots + \alpha_n w_n = 0$$

which in turn implies, $\{w_1, \ldots, w_n\}$ being a basis of W, that $\alpha_1 = \alpha_2 = \cdots = \alpha_n = 0$. Hence $\alpha_1 v_1 + \cdots + \alpha_n v_n = 0$ and T is one–one. Therefore, T is an isomorphism and the vector spaces V and W are isomorphic.

Example 11.27. Let F be a field, n be a positive integer and $V_n = F^{(n)}$ be the set of all ordered n-tuples $(\alpha_1, \alpha_2, \ldots, \alpha_n), \alpha_i \in F$. Defining addition component wise and scalar multiplication by multiplying every component by the scalar, we find that V_n is a vector space over F. Let e_i be the ordered n-tuples in which the ith component is 1 and every other components is 0. The elements $e_1, e_2, \ldots, e_n$ are linearly independent and span V_n. Thus V_n is a vector space of dimension n over F. If W is any vector space of dimension n over F, then W is isomorphic to V_n. Thus, up to isomorphism, there is only one vector space of finite dimension n over F, namely $V_n = F^{(n)}$.

Exercise 11.2.

1. In the polynomial ring $R[x]$ over the field R of real numbers, let V be the set of all polynomials of degree ≤ 2 and W be the set of all

polynomials of degree ≤ 3. Prove that both V, W are vector spaces over R. Also prove that dim $V = 3$ and dim $W = 4$.

2. Prove that in a vector space V over a field F, $\alpha \cdot 0 = 0$ for all $\alpha \in F$, $0 \in V$ and $0 \cdot v = 0$ for $0 \in F$ and all $v \in V$.

3. If V, W are vector spaces over the same field F and $T : V \to W$ is a linear transformation, prove that ker T is a subspace of V.

4. Let $R^{(2)}$ be the vector space of all ordered pairs $(x, y), x, y \in R$. Let $T : R^{(2)} \to R^{(2)}$ be the map defined by

$$T(x, y) = (\alpha x + \beta y, rx + sy),$$

where α, β, r, s are some fixed real numbers.

   (a) Prove that T is a linear transformation.
   (b) Find necessary and sufficient conditions in terms of α, β, r, s so that T is both one–one and onto.

5. Prove that the elements $(1, 1, 0), (0, 1, -1)$ of $R^{(3)}$ are linearly independent. Find an element of $R^{(3)}$ which together with the given two elements form a basis of $R^{(3)}$.

6. If x, y, z are linearly independent elements in a vector space V, prove that so are $x + y, y + z, z + x$.

7. Find a basis of $R^{(3)}$ which contains the elements $(1, 0, 0), (1, 0, 1)$.

8. Are the elements $(1, 1), (1, 2)$ of $R^{(2)}$ linearly independent? Justify.

9. If A, B, C are subspaces of a vector space V, is $A \cap (B + C) = A \cap B + A \cap C$? Justify.

10. Under what conditions on $\alpha \in C$ (the field of complex numbers) do the vectors $(\alpha, 0, 1), (1 + \alpha, 1, \alpha), (1, 1, 0)$ form a basis of $C^{(3)}$?

11. Find two different bases of $R^{(3)}$ which have $(1, 1, 0)$ in common.

12. Find conditions on the scalar α so that the vectors $(1 + \alpha, 1 - \alpha)$ and $(1 - \alpha, 1 + \alpha)$ in $R^{(2)}$ are

   (a) linearly dependent;
   (b) linearly independent.

13. Are the elements $(1, -1, 2), (2, 1, -1), (1, 1, 1)$ linearly dependent? Justify.

14. Let V be the vector space of all square matrices of order 2 over R, W, W_1 be respectively the subspace of lower triangular and upper triangular matrices. Find a basis of the subspace $W \cap W_1$.

Chapter 12

LATTICES AND BOOLEAN ALGEBRA

An important class of algebraic structures was introduced around 1854 by George Boole (1815–1864) (refer to Bell (1937), and Rosen, (2003/2005)) in connection with his study on mathematical logic. These algebraic structures are now called Boolean algebra. At more or less the same time, a similar and related concept of lattices was developed by R. Dedekind (1831–1916) (refer to Bell (1937) and Lidl and Pilz (1998)) in his study on groups and ideals. Lattices as introduced by Dedekind are what are now known as modular and distributive lattices. However, a major boost to the study of lattices was provided by the contribution made by G. Birkhoff. Our aim in the present chapter is to give the concepts and elementary results on lattices and Boolean algebra, which are important from the point of view of applications.

12.1. Lattices

Definition. A partially ordered set $(A, \leq)$ is called a **lattice**, if every two elements of A have a unique least upper bound (*l.u.b.*) and a unique greatest lower bound (*g.l.b.*).

If we accept the alternative definition of greatest lower bound and least upper bound of a pair of elements in a poset $(A, \leq)$, namely for elements a, b in a poset $(A, \leq)$,

(a) an element $x \in A$ is called a least upper bound of a, b if $a \leq x$, $b \leq x$ and if $y \in A$ is an element such that $a \leq y$, $b \leq y$, then $x \leq y$;

(b) an element $x \in A$ is called a greatest lower bound of a, b if $y \leq a$, $y \leq b$ and if $z \in A$ is an element such that $z \leq a$, $z \leq b$, then $z \leq y$; then it is immediate from the definition that if least upper bound of $a, b \in A$ exists, it is unique, and if greatest lower bound of $a, b \in A$ exists, then that is also unique. In view of this, we may define a lattice simply as follows.

A partially ordered set $(A, \leq)$ is called a lattice if every pair of elements of A has a greatest lower bound and a least upper bound. Observe that in a linearly ordered set the two definitions of greatest lower bound and of least upper bound of a pair of elements are identical.

Throughout the rest of this chapter and while discussing lattices, we shall adopt the above definition of greatest lower bound and least upper bound of a pair of elements. Then maximal element and greatest element of a lattice are identical and so are the minimal element and the least element.

Theorem 12.1. Every finite, non-empty subset S of a lattice $(L, \leq)$ has a least upper bound and a greatest lower bound.

Proof. Let $o(S) = n$. We prove by induction on n that S has a *g.l.b.* If $n = 1$ and $S = \{a\}$, then *g.l.b.* $S = a = $ *l.u.b.* S. If $n = 2$ and $S = \{a, b\}$, it follows from the definition of a lattice that *g.l.b.* and *l.u.b.* exit. Thus, the theorem holds for $n = 1, 2$. Suppose that $n \geq 2$ and that the theorem holds for all subsets of L having $\leq n$ elements. Suppose that $S = \{a_1, a_2, \ldots, a_{n+1}\}$. Consider $T = \{a_1, a_2, \ldots, a_n\}$. Then, by induction hypothesis, there exist $a = $ *g.l.b.* T and $b = $ *g.l.b.* $\{a, a_{n+1}\}$. Therefore $a \leq a_i$, $1 \leq i \leq n$, $b \leq a$ and $b \leq a_{n+1}$ so that $b \leq a_i$, $1 \leq i \leq n + 1$ showing that b is a lower bound of S. Let c be another lower bound of S. Then $c \leq a_i$, $1 \leq i \leq n + 1$. In particular, c is a lower bound of T. The element $a \in L$ being a *g.l.b.* T, $c \leq a$. Thus c is a lower bound of $\{a, a_{n+1}\}$, which has b as a *g.l.b.* Hence $c \leq b$. This completes the proof that $b = $ *g.l.b.* S. The induction is thus complete. A dual argument proves that every finite subset of L has an *l.u.b.*

Definition. A lattice $(L, \leq)$ is said to be **complete** if every subset (finite or infinite) S of L has a *g.l.b.* and an *l.u.b.*

As an immediate corollary of the above theorem, we have the following.

Corollary. Every finite lattice is complete.

We have already defined a zero element denoted by 0(say) and a unit element denoted by 1(say) in a poset $(P, \leq)$. A zero element in $(P, \leq)$, if it exists, is unique and a unit element in $(P, \leq)$, if it exists, is unique. A finite poset $(P, \leq)$ may have both, one or neither of these special elements (zero, one) (see Example 12.1) but in the case of lattices, we have the following.

Corollary. Every finite lattice has both a zero and a unit.

Example 12.1. Consider the sets $P = \{2, 3, 6, 12, 18\}$, $Q = \{1, 2, 3, 6, 12, 18\}$, $R = \{2, 3, 6, 12, 18, 36\}$ and $S = \{1, 2, 3, 6, 12, 18, 36\}$. In each of these four subsets of N, define a relation $\leq$ by $a \leq b$ if a divides b, for $a, b \in P, Q, R, S$ respectively. Then we have posets $(P, \leq)$, $(Q, \leq)$, $(R, \leq)$ and $(S, \leq)$, each one of which is finite. Observe that the poset $(P, \leq)$ has neither 0 nor 1, $(Q, \leq)$ has zero, namely, the number 1 but does not have a unit, $(R, \leq)$ does not have a zero but has a unit, which is the number 36 and $(S, \leq)$ has both 0, which is the number 1 and the unit, which is the number 36. The Hasse diagrams of these posets are, respectively:

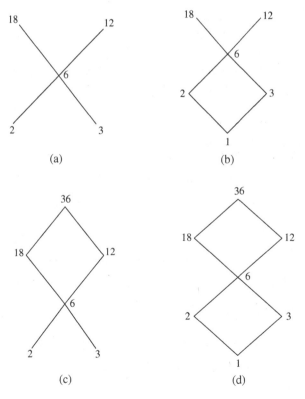

Fig. 12.1.

Example 12.2. The partially ordered set $(N, \leq)$ with '$\leq$' defined by $m \leq n$ if m divides n for $m, n \in N$, is a lattice. This lattice has infinitely many chains of infinite length and it also has antichains of infinite order. The integer 1 is the zero element of this lattice, but the lattice does not have a unit element.

Example 12.3. Let $(A, \leq)$ and $(X, \leq)$ be two lattices. Consider the Cartesian product $A \times X$ and define a relation '$\leq$' on $A \times X$ by: for (a, x), $(b, y) \in A \times X$, say that $(a, x) \leq (b, y)$ if and only if $a \leq b$ and $x \leq y$ under the given relation on A and X respectively. The relations on A and X being partial order relations $a \leq a$ for every $a \in A$ and $x \leq x$ for every $x \in X$. Therefore, $(a, x) \leq (a, x)$ for every $(a, x) \in A \times X$. Let $(a, x), (b, y) \in A \times X$ and suppose that $(a, x) \leq (b, y)$ and $(b, y) \leq (a, x)$. Then $a \leq b$, $x \leq y$ and $b \leq a$, $y \leq x$, which imply that $b = a$ and $y = x$. Thus $(a, x) = (b, y)$.

Next, consider (a, x), (b, y), $(c, z) \varepsilon A \times X$ and suppose that $(a, x) \leq (b, y)$, $(b, y) \leq (c, z)$. Then $a \leq b$, $x \leq y$, $b \leq c$ and $y \leq z$. The original relations on A, X being transitive, we get $a \leq c$, $x \leq z$ so that $(a, x) \leq (c, z)$. Thus the relation defined on $A \times X$ is reflexive, anti-symmetric and transitive and, hence, it is a partial order relation. Consider any two elements (a, x), $(b, y) \in A \times X$. Then $a, b \in A$, $x, y \in X$. Now $(A, \leq)$, $(X, \leq)$ being lattices, there exists a unique least upper bound $c \in A$ of a, b and a unique least upper bound z of x, y. In particular, $a \leq c$, $b \leq c$ and $x \leq z$, $y \leq z$. Therefore, $(a, x) \leq (c, z)$ and $(b, y) \leq (c, z)$. Thus (c, z) is an upper bound of (a, x), (b, y). Let $(e, u) \in A \times X$ be another upper bound of (a, x), (b, y). It follows from the definition of the relation on $A \times X$ that e is an upper bound of a, b and u is an upper bound of x, y. By definition of least upper bound, we cannot have $e \leq c$ and $u \leq z$ and so, we cannot have $(e, u) \leq (c, z)$. Thus (c, z) is a least upper bound of (a, x), (b, y). Hence (c, z) is the unique least upper bound of (a, x), (b, y). In a similar fashion it can be proved that every two elements of $A \times X$ have a unique greatest lower bound. Hence $A \times X$ is again a lattice.

Example 12.4. The set Z of integers with the usual order relation '$\leq$' is a partially ordered set. Given two integers a and b, one of these two elements is less than or equal to the other. Suppose that $a \leq b$. Then a is the greatest lower bound of a, b and b is the least upper bound. These bounds are unique. Hence $(Z, \leq)$ is a lattice. This lattice is not complete and has neither a zero nor a unit.

Example 12.5. Consider the set $D_{30} = \{1, 2, 3, 5, 6, 10, 15, 30\}$ and the relation $\leq$ as divides ($|$), i.e. $a \leq b$ if a divides b. This is a partially ordered set. The greatest lower bound of 1 and any $a \in D_{30}$ is clearly 1 while that of any $a \in D_{30}$ and 30 is a, and least upper bound of 30 and any $a \in D_{30}$ is 30 while that of 1 and any $a \in D_{30}$ is a. We then need to consider the pairs $(2, 3)$, $(2, 5)$, $(2, 6)$, $(2, 10)$, $(2, 15)$; $(3, 5)$, $(3, 6)$, $(3, 10)$, $(3, 15)$; $(5, 6)$, $(5, 10)$, $(5, 15)$; $(6, 10)$, $(6, 15)$; and $(10, 15)$ Observe that greatest lower bound (*g.l.b.*) of each of $(2, 3)$, $(2, 5)$, $(2, 15)$, $(3, 5)$, $(3, 10)$, $(5, 6)$ is 1 while that of $(2, 6)$, $(2, 10)$, $(6, 10)$ is 2, that of $(3, 6)$, $(3, 15)$, $(6, 15)$ is 3 and that of $(5, 10)$, $(5, 15)$, $(10, 15)$ is 5.

On the other hand, least upper bound (*l.u.b.*) of $(2, 3)$, $(2, 6)$, $(3, 6)$ is 6; that of $(2, 5)$, $(2, 10)$, $(5, 10)$ is 10; that of $(3, 5)$, $(3, 15)$, $(5, 15)$ is 15 and that of $(2, 15)$, $(3, 10)$, $(5, 6)$, $(6, 10)$, $(6, 15)$, $(10, 15)$ is 30.

Thus every pair of elements in D_{30} has *g.l.b.* and *l.u.b.* Hence D_{30} is a lattice. $1 \leq 2 \leq 6 \leq 30$, $1 \leq 3 \leq 6 \leq 30$; $1 \leq 2 \leq 10 \leq 30$, $1 \leq 3 \leq 15 \leq 30$, $1 \leq 5 \leq 10 \leq 30$ are all the possible chains of length 4 in the lattice. Observe that $D_{30} = \{1\} \cup \{2, 3, 5\} \cup \{6, 10, 15\} \cup \{30\}$ is the union of four antichains, as it should be in view of the corollary to Theorem 6.2. The integer 1 is the zero element of this lattice and 30 is the unit.

Example 12.6. Let n be a positive integer and D_n be the set of all positive divisors of n. If $a, b \in D_n$, then both the *g.c.d.* of a, b and *l.c.m.* of a, b are divisors of n. Now D_n w.r.t. the relation '$\leq$' with $a \leq b$ if a divides b is a partially ordered set. Also with respect to this relation, observe that for $a, b \in D_n$, *g.c.d.*(a, b) is the greatest lower bound of a, b and *l.c.m.* of a, b is the least upper bound of a, b. In view of the observation that *g.c.d.*(a, b) and *l.c.m.*(a, b) belong to D_n, $(D_n, |)$ becomes a lattice. The integer 1 is the least element of D_n and n is the greatest element.

Example 12.7. Let S be a non-empty set and $p(S)$ be the power set of S, i.e. the set of all subsets of S. For $A, B \in p(S)$, say that $A \leq B$ if A is a subset of B. Then $(p(S), \leq)$ is a partially ordered set. Observe that for $A, B \in p(S)$, $A \cap B \in p(S)$ and is the greatest lower bound of A, B while $A \cup B \in p(s)$ is the least upper bound of A, B. Hence $(p(S), \leq)$ is a lattice. The empty set ϕ is the least element of this lattice and S is the greatest or largest element. Also, the lattice is complete.

Example 12.8. We consider a particular case of the above example. Let $S = \{a, b, c\}$. Then $p(S) = \{\phi, \{a\}, \{b\}, \{c\}, \{a, b\}, \{b, c\}, \{a, c\}, \{a, b, c\}\}$.

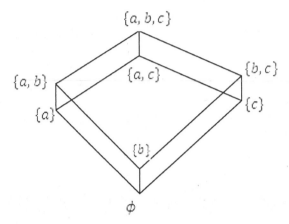

Fig. 12.2.

The Hasse diagram of this poset is as in Fig. 12.2.

Now $\{a\} \cap \{b\} = \{a\} \cap \{c\} = \{b\} \cap \{c\} = \{a\} \cap \{b, c\} = \{b\} \cap \{a, c\} = \{c\} \cap \{a, b\} = \phi$ so that ϕ is the greatest lower bound of (i) $\{a\}$, $\{b\}$; (ii) $\{a\}$, $\{c\}$; (iii) $\{b\}$, $\{c\}$; (iv) $\{a\}$, $\{b, c\}$; (v) $\{c\}$, $\{a, b\}$; (vi) $\{b\}$, $\{a, c\}$ while the least upper bounds of these pairs of elements are, respectively $\{a, b\}$, $\{a, c\}$, $\{b, c\}$, $\{a, b, c\}$, $\{a, b, c\}$, $\{a, b, c\}$. The greatest lower bound and the least upper bound of the pairs of element are as follows.

$\{a\}$, $\{a, b\}$ are $\{a\}$, $\{a, b\}$; $\{a\}$, $\{a, c\}$ are $\{a\}$, $\{a, c\}$;

$\{b\}$, $\{a, b\}$ are $\{b\}$, $\{a, b\}$; $\{b\}$, $\{b, c\}$ are $\{b\}$, $\{b, c\}$;

$\{c\}$, $\{a, c\}$ are $\{c\}$, $\{a, c\}$; $\{c\}$, $\{b, c\}$ are $\{c\}$, $\{b, c\}$;

$\{a\}$, $\{a, b, c\}$ are $\{a\}$, $\{a, b, c\}$; $\{b\}$, $\{a, b, c\}$ are $\{b\}$, $\{a, b, c\}$;

$\{c\}$, $\{a, b, c\}$ are $\{c\}$, $\{a, b, c\}$; $\{a, b\}$, $\{b, c\}$ are $\{b\}$, $\{a, b, c\}$;

$\{a, b\}$, $\{a, c\}$ are $\{a\}$, $\{a, b, c\}$; $\{a, b\}$, $\{a, b, c\}$ are $\{a, b\}$, $\{a, b, c\}$;

$\{b, c\}$, $\{a, c\}$ are $\{c\}$, $\{a, b, c\}$; $\{b, c\}$, $\{a, b, c\}$ are $\{b, c\}$, $\{a, b, c\}$;

$\{a, c\}$, $\{a, b, c\}$ are $\{a, c\}$, $\{a, b, c\}$.

Since $\phi \cap A = \phi$ and $\phi \cup A = A$ for any subset A of S, ϕ is the greatest lower bound and A is least upper bound of ϕ, A. Also A is the greatest lower bound and S is the least upper bound of A, S.

This completes the verification that $(p(S), \leq)$ is a lattice. Also ϕ is the least element and $\{a, b, c\}$ is the greatest element of this lattice. Observe that

$\phi \leq \{a\} \leq \{a, b\} \leq \{a, b, c\}$; $\phi \leq \{a\} \leq \{a, c\} \leq \{a, b, c\}$;

$\phi \leq \{b\} \leq \{a, b\} \leq \{a, b, c\}$; $\phi \leq \{b\} \leq \{b, c\} \leq \{a, b, c\}$;

$\phi \leq \{c\} \leq \{a, c\} \leq \{a, b, c\}$; $\phi \leq \{c\} \leq \{b, c\} \leq \{a, b, c\}$

are the only possible chains of length 4 and there is no chain of length 5 or more. Now $\{\phi\}$, $\{\{a\}, \{b, c\}\}$, $\{\{b\}, \{a, c\}\}$, $\{\{c\}, \{a, b\}\}$, $\{\{a\}, \{b\}, \{c\}\}$, $\{\{a, b\}\}$, $\{\{a, c\}\}$, $\{\{b, c\}\}$, $\{\{a, b, c\}\}$, $\{\{a\}\}$, $\{\{b\}\}$, $\{\{c\}\}$, $\{\{a, b\}, \{a, c\}\}$, $\{\{a, b\}, \{b, c\}\}$, $\{\{b, c\}, \{a, c\}\}$, $\{\{a, b\}, \{a, c\}, \{b, c\}\}$ are all the possible antichains. Observe that

$$p(S) = \{\phi\} \cup \{\{a\}, \{b\}, \{c\}\} \cup \{\{a, b\}, \{a, c\}, \{b, c\}\} \cup \{\{a, b, c\}\}$$

is a union of four antichains which is again in conformity with the result of corollary to Theorem 6.2.

Example 12.9. The lattices $(R, \leq)$ and $(Q, \leq)$ with the usual order relation $\leq$ of the number system are not complete. Neither of the two lattices has a zero or a unit.

Exercise. Is the partially ordered set with Hasse diagram Fig. 12.3 a lattice? Justify.

12.2. Lattices as Algebraic Systems

Let $(A, \leq)$ be a lattice. With every pair of elements a, b in A, there are associated uniquely determined elements namely $g.l.b.(a, b)$ and $l.u.b.(a, b)$ again in A. Thus we get two binary operations $\vee$ called **join** and $\wedge$ called **meet** on A which are defined by

$$a \vee b = l.u.b.(a, b), \quad a \wedge b = g.l.b.(a, b)$$

for $a, b \in A$.

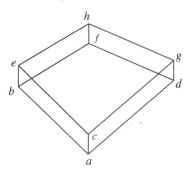

Fig. 12.3.

As the greatest lower bound and least upper bound of a, b are the same as the greatest lower bound and the least upper bound respectively of b, a, we have

$$a \vee b = b \vee a, \quad a \wedge b = b \wedge a \quad \text{for all } a, b \in A.$$

Thus, both the join and the meet operations on A are commutative. We prove some of the very basic properties of the two operations on lattices.

Theorem 12.2. For any a, b, c, d in a lattice $(A, \leq)$,

(a) $a \leq a \vee b$, $a \wedge b \leq a$;
(b) if $a \leq b$ and $c \leq d$, then $a \vee c \leq b \vee d$ and $a \wedge c \leq b \wedge d$.

Proof.

(a) By definition $a \vee b$ is *l.u.b.* of a, b. In particular, $a \vee b$ is an upper bound of a. Therefore, $a \leq a \vee b$. On the other hand, $a \wedge b$ is the *g.l.b.* of a, b and so, it is in particular, a lower bound of a. Therefore, $a \wedge b \leq a$.
(b) By (a) above $b \leq b \vee d$. Also $a \leq b$. Therefore, by transitivity of the relation $\leq$, we get $a \leq b \vee d$. Again $d \leq b \vee d$ and $c \leq d$. Therefore, $c \leq b \vee d$. Thus, we have proved that $a \leq b \vee d$ and $c \leq b \vee d$. Then $l.u.b.(a, c) \leq b \vee d$ i.e. $a \vee c \leq b \vee d$.
(c) Again by (a) above, $a \wedge c \leq a$. Also $a \leq b$. Therefore, $a \wedge c \leq b$. Also, $a \wedge c \leq c$ and $c \leq d$. Therefore, $a \wedge c \leq d$. Thus, $a \wedge c$ is a lower bound of b, d and, therefore, $a \wedge c \leq g.l.b.(b, d) = b \wedge d$.

Before proving some further properties of lattices, we describe a procedure called the **principle of duality**. This is a principle which helps us in proving many results more economically, in the sense that, for every result that we prove for a lattice, we can get another parallel result by just using this principle and by not going over the whole proof *ab initio*.

Let $(A, \leq)$ be a partially ordered set. Define another relation $\leq'$ on A by saying $a \leq' b$ if and only if $b \leq a$, $a, b \in A$. It is fairly easy to check that $\leq'$ is again a partial order relation on A. Now suppose that $(A, \leq)$ is a lattice. Let $a, b \in A$. Then $a \vee b = l.u.b.(a, b)$ and $a \wedge b = g.l.b.(a, b)$. Therefore, $a \leq a \vee b$, $b \leq a \vee b$ and $a \wedge b \leq a$, $a \wedge b \leq b$. These relations then imply that

$$a \vee b \leq' a, a \vee b \leq' b \quad \text{and} \quad a \leq' a \wedge b, b \leq' a \wedge b. \tag{12.1}$$

Let $x, y \in A$ be such that

$$x \leq' a, x \leq' b \quad \text{and} \quad a \leq' y, b \leq' y. \tag{12.2}$$

Then $a \leq x$, $b \leq x$ and $y \leq a$, $y \leq b$. i.e. x is an upper bound of a, b and y is a lower bound of a, b in $(A, \leq)$. Hence $a \vee b \leq x$ and $y \leq a \wedge b$. But then

$$x \leq' a \vee b \quad \text{and} \quad a \wedge b \leq' y. \tag{12.3}$$

The relations (12.1), (12.2), (12.3) then show that $a \vee b$ is the greatest lower bound of a, b and $a \wedge b$ is the least upper bound of a, b in $(A, \leq')$. This proves that $(A, \leq')$ is also a lattice. We can thus define two binary operations $\wedge'$ and $\vee'$ on A by $a \wedge' b = g.l.b.(a, b)$ in $(A, \leq')$, and $a \vee' b = l.u.b.(a, b)$ in $(A, \leq')$, $a, b \in A$. By what we have proved above, we get $a \wedge' b = a \vee b$, $a \vee' b = a \wedge b$, $a, b \in A$.

Let f, g be two propositions or statements involving the operations $\leq, \wedge$ and $\vee$. Then g is called the **dual** of f if it can be obtained from f by replacing $\wedge$ by $\vee$, $\vee$ by $\wedge$ and $\leq$ by $\geq$ or $\leq'$ (if $a \leq b$, we also write this as $b \geq a$). Now, if f is a proposition in the lattice $(A, \leq)$, then the dual proposition g is the same as the proposition f with $\leq$ replaced by $\leq', \wedge$ replaced by $\wedge'$ and $\vee$ replaced by $\vee'$. Thus, if f is a true proposition in $(A, \leq)$, g is true in $(A, \leq')$. This leads us to the **principle of duality**, which says that if a proposition is true, then so is its dual.

Thus, whenever we want to prove two propositions which are dual, we need to prove only one of the two and then say that the other follows from the principle of duality. We are now ready to prove some basic properties of lattices.

Theorem 12.3. Let $(A, \vee, \wedge)$ be an algebraic system defined by a lattice $(A, \leq)$. Then both the join and meet operations are (a) commutative; (b) associative.

Proof. The commutativity of the join and meet operations has already been proved.
(b) Let $a, b, c \in A$. Let $(a \wedge b) \wedge c = g$ and $a \wedge (b \wedge c) = h$.
Since g is the meet of $a \wedge b$ and c, $g \leq a \wedge b$ and $g \leq c$. As $a \wedge b \leq a$ and $a \wedge b \leq b$, we have, $g \leq a$, $g \leq b$ and $g \leq c$.

As $g \leq b$, $g \leq c$ and $b \wedge c$ is the greatest lower bound of b, c, we have $g \leq b \wedge c$. Thus $g \leq a$ and $g \leq b \wedge c$ and it follows from the definition of greatest lower bound that $g \leq a \wedge (b \wedge c) = h$. In a similar fashion, it follows that $h \leq g$. Thus $g \leq h$ and $h \leq g$. Therefore, $g = h$ or that

$(a \wedge b) \wedge c = a \wedge (b \wedge c)$. It then follows from the principle of duality that $(a \vee b) \vee c = a \vee (b \vee c)$, i.e. the join operation is associative.

Theorem 12.4. (The idempotent property of join and meet.)
For every a in A, $a \vee a = a$ and $a \wedge a = a$.

Proof. By what we have proved earlier, $a \wedge a \leq a$ (as $a \wedge a$ is the *g.l.b.* of a and a).

Again, $a \leq a$ and $a \leq a$, i.e. a is a lower bound of a, a. Therefore, $a \leq g.l.b.(a, a) = a \wedge a$. Thus $a \wedge a \leq a$, and $a \leq a \wedge a$. Therefore, $a \wedge a = a$. That $a \vee a = a$ then follows from the principle of duality.

Theorem 12.5. For any a, b in a lattice $(A, \leq)$, $a \vee (a \wedge b) = a$; $a \wedge (a \vee b) = a$.
The above relations are called the absorption laws.

Proof. Since $a \wedge c$ is the *g.l.b.* of a, c, $a \wedge c \leq a$ and $a \wedge c \leq c$. Thus

$$a \wedge (a \vee b) \leq a. \tag{12.4}$$

On the other hand $a \vee b$ is the *l.u.b.* of a, b. Therefore $a \leq a \vee b$. Also $a \leq a$. Therefore,

$$a \leq a \wedge (a \vee b). \tag{12.5}$$

Combining (12.4) and (12.5), we get $a \wedge (a \vee b) = a$. That $a \vee (a \wedge b) = a$ then follows by the principle of duality.

Before proceeding further, we will consider a connection between an algebraic system with two binary compositions and a lattice.

Theorem 12.6. Let A be a non-empty set with two binary compositions $\wedge$ and $\vee$ satisfying for $a, b, c \in A$,

L1 (i) $a \wedge b = b \wedge a$;　　　　　(i)$'$ $a \vee b = b \vee a$;
L2 (ii) $(a \wedge b) \wedge c = a \wedge (b \wedge c)$;　　(ii)$'$ $(a \vee b) \vee c = a \vee (b \vee c)$;
L3 (iii) $a \wedge a = a$;　　　　　　(iii)$'$ $a \vee a = a$;
L4 (iv) $a \vee (a \wedge b) = a$;　　　(iv)$'$ $a \wedge (a \vee b) = a$.

Then $(A, \leq)$ is a lattice where for $a, b \in A$, $a \leq b$ if and only if $a \wedge b = a$.

Proof. It follows immediately from (iii) that $a \leq a$ for all a in A. Thus the relation $\leq$ is reflexive.

Let $a, b \in A$ and suppose that $a \leq b$ and $b \leq a$. Then $a = a \wedge b$ and $b = b \wedge a$. By the commutative law L_1 satisfied by $\wedge$, we get $a = b$. Hence the relation '$\leq$' is anti-symmetric. Next suppose that $a, b, c \in A$ and $a \leq b$, $b \leq c$. Then $a = a \wedge b$; $b = b \wedge c$.

Now

$$a \wedge c = (a \wedge b) \wedge c$$
$$= a \wedge (b \wedge c) \quad \text{(by } L_2)$$
$$= a \wedge b = a.$$

Therefore, $a \leq c$. This completes the proof that $(A, \leq)$ is a partially ordered set.

Let $a, b \in A$. Then by (L_3) and (L_2),

$$a \wedge b = (a \wedge a) \wedge b = a \wedge (a \wedge b)$$

which implies that $a \wedge b \leq a$.

Also

$$a \wedge b = a \wedge (b \wedge b) = (a \wedge b) \wedge b$$

which implies that $a \wedge b \leq b$.

Thus $a \wedge b$ is a lower bound of a, b.

Let $c \in A$ and suppose that $c \leq a$, $c \leq b$. Then $c = a \wedge c$; $c = b \wedge c$;

Now (by L_2 and L_3)

$$c = a \wedge c = a \wedge (b \wedge c) = (a \wedge b) \wedge c.$$

Therefore, $c \leq a \wedge b$.

This completes the proof that $a \wedge b$ is the *g.l.b.* of a, b.

Again $a \wedge (a \vee b) = a$ (by (iv)$'$), which implies that $a \leq a \vee b$

Also by (i)$'$ and by (iv)$'$

$$b \wedge (a \vee b) = b \wedge (b \vee a) = b$$

which implies that $b \leq a \vee b$.

This proves that $a \vee b$ is an upper bound of a, b. Before proving that $a \vee b$ is the least upper bound of a, b, we prove that

(1) $a \wedge b = a$ if and only if $a \vee b = b$

Suppose that $a \wedge b = a$. Then $a \vee b = (a \wedge b) \vee b = b \vee (b \wedge a) = b$ (by (i), (i)$'$, (iv)) If $a \vee b = b$, then $a \wedge b = a \wedge (a \vee b) = a$ (by (iv)$'$). This proves the claims as at (1).

Let $c \in A$ be such that $a \leq c$, $b \leq c$. Then $a = a \wedge c$, $b = b \wedge c$. These imply (by (1)) that $a \vee c = c$ and $b \vee c = c$. Now $(a \vee b) \vee c = a \vee (b \vee c) = a \vee c = c$, which implies that $a \vee b \leq c$. Hence $a \vee b$ is the least upper bound of a, b. This completes the proof that $(A, \leq)$ is a lattice.

Remark. In view of this result, a lattice could equally well be defined as a non-empty set with two binary compositions $\wedge$ and $\vee$ satisfying the conditions (i)–(iv) and (i)'–(iv)' as in Theorem 12.6. The two definitions of a lattice are equivalent as follows from Theorem 12.6 and because a lattice $(A, \leq)$ gives rise to an algebraic system satisfying the conditions as in Theorem 12.6 having been proved already (Theorems 12.3 to 12.5).

Example 12.10. For all a, b, c in any lattice $(L, \leq)$,

$$[(a \wedge b) \vee (a \wedge c)] \wedge [(a \wedge b) \vee (b \wedge c)] = a \wedge b.$$

Solution. Let $(L, \leq)$ be a lattice. Since $x \leq x \vee y$ for all x, y in L,

$$a \wedge b \leq (a \wedge b) \vee (a \wedge c) \quad \text{and} \quad a \wedge b \leq (a \wedge b) \vee (b \wedge c). \tag{12.6}$$

Therefore,

$$a \wedge b \leq [(a \wedge b) \vee (a \wedge c)] \wedge [(a \wedge b) \vee (b \wedge c)].$$

On the other hand, $(a \wedge b) \leq a$, $a \wedge c \leq a$ together imply $[(a \wedge b) \vee (a \wedge c)] \leq a$. Similarly, $[(a \wedge b) \vee (b \wedge c)] \leq b$. Therefore,

$$[(a \wedge b) \vee (a \wedge c)] \wedge [(a \wedge b) \vee (b \wedge c)] \leq a \wedge b. \tag{12.7}$$

The relations (12.6) and (12.7) imply

$$[(a \wedge b) \vee (a \wedge c)] \wedge [(a \wedge b) \vee (b \wedge c)] = a \wedge b.$$

Example 12.11. Let L be the set of all integers (rational numbers or real numbers) which are ≤ 0. For $a, b \in L$, let $a \leq b$ be the usual less than or equal to relation for the real numbers. Then $(L, \leq)$ is a partially ordered set. For $a, b \in L$, a is the unique *g.l.b.* and b is the unique *l.u.b.* of a, b if $a \leq b$ (we can interchange the role of a, b if $b \leq a$). Thus $(L, \leq)$ is a lattice. This lattice has the number zero as unit but it does not have a 'zero'.

Example 12.12. Let $n = p_1 p_2 \ldots p_m$ where p_i are distinct primes and let L be the lattice of all positive divisors of n with $a \leq b$, for $a, b \in L$, if a divides b. Then for $a, b \in L$, a is a product of some of the primes

$p_1, p_2, \ldots, p_m$ without repetition and so is b. Now $a \wedge b = g.c.d.(a, b)$ and so $a \wedge b$ is the product of those primes out of $p_1, p_2, \ldots, p_m$, which occur in the factorization of both a as well as b. On the other hand, $a \vee b = l.c.m.(a, b)$. Therefore, $a \vee b$ is the product of primes which occur in the factorization of a together with primes which occur in the factorization of b but not in the factorization of a.

Example 12.13. Let $(L, \leq)$ be a lattice and a, b, c be elements of L such that $a \neq b$ and c covers both a and b. Then $a \leq c$, $b \leq c$ and, therefore, $a \vee b \leq c \vee c = c$. Also $a \leq a \vee b$ and $b \leq a \vee b$. Since c covers a, either $a \vee b = a$ or $a \vee b = c$. Similarly, as c covers b, either $a \vee b = b$ or $a \vee b = c$. If $a \vee b \neq c$, then $a = a \vee b = b$, which is a contradiction to $a \neq b$. Hence $c = a \vee b$.

Example 12.14. If a, b, c are in a lattice $(L, \leq)$, $a \neq b$ and c is covered by a as well as by b, then $c = a \wedge b$.

Example 12.15. Let a, b, c be elements in a lattice $(L, \leq)$. If $a \leq b$, then $a \vee (b \wedge c) \leq b \wedge (a \vee c)$.

Solution. Since $a \leq b$ and $b \wedge c \leq b$, we have

$$a \vee (b \wedge c) \leq b. \tag{12.8}$$

Also $a \leq a \vee c$ and $b \wedge c \leq c \leq a \vee c$.
 Therefore,

$$a \vee (b \wedge c) \leq a \vee c. \tag{12.9}$$

(12.8) and (12.9) together imply $a \vee (b \wedge c) \leq b \wedge (a \vee c)$.

Exercise. Prove that for all x, y, z in any lattice $(L, \leq)$,

$$x \wedge (y \vee z) \geq (x \wedge y) \vee (x \wedge z) \quad \text{and}$$
$$x \vee (y \wedge z) \leq (x \vee y) \wedge (x \vee z).$$

12.3. Sublattices and Homomorphisms

Let $(L, \leq)$ be a lattice and S be a non-empty subset of L. The order relation $\leq$ on L induces a relation '$\leq$' on S and $(S, \leq)$ is a partially ordered set. If $a, b \in S$, then $a, b \in L$ and $l.u.b.\{a, b\}, g.l.b.\{a, b\}$ exist in L. However, $g.l.b.\{a, b\}$ and $l.u.b.\{a, b\}$ may or may not belong to S. If both $g.l.b.\{a, b\}$

and $l.u.b.\{a, b\}$ belong to S for every pair of elements $a, b \in S$, then $(S, \leq)$ is called a **sublattice** of $(L, \leq)$.

Example 12.16. Consider the lattice $L = D_{30}$ of all positive integral divisors of 30 with the relation '|' i.e. for $a, b \in D_{30}, a|b$ if a divides b. Consider the subset $S = \{2, 3, 5, 6\}$ of D_{30}. Observe that in L, $g.l.b.\{2, 3\} = g.l.b.\{2, 5\} \neq g.l.b.(3, 5) = g.l.b.\{5, 6\} = 1$ which is in L (as it should be) but is not in S. Similarly, $l.u.b.\{5, 3\} = 15$, $l.u.b.\{5, 6\} = 30$, which are again not in S. Therefore, $(S, \leq)$ is not a sublattice of D_{30}. On the other hand, if we consider the subset $T = \{1, 2, 3, 6\}$ of D_{30}, then trivially $g.l.b.\{a, b\}$ and $l.u.b.\{a, b\}$ exist in T for every $a, b \in T$. Hence $(T, \leq)$ is a sublattice of D_{30}.

We may equally well define a sublattice of a lattice $(L, \leq)$ to be a non-empty subset of L which is a lattice under the induced partial order relation. Equivalently, if $(L, \wedge, \vee)$ is the algebraic system associated with the lattice $(L, \leq)$, then a non-empty subset S of L is a sublattice of L if and only if $a, b \in S$ imply $a \wedge b$ and $a \vee b$ are in S.

Example 12.17. There do exist situations where a non-empty subset S of a lattice $(L, \leq)$ is itself a lattice but is not a sublattice of L. For this, let G be a group, $p(G)$ the power set of the set G and $S(G)$ the set of all subgroups of G. Then $p(G)$ is a lattice under the relation '$\subseteq$', i.e. if $A, B \in p(G)$, then $A \leq B$ if A is a subset of B. Also $S(G)$ is a lattice under the relation $\leq$ defined by $H \leq K$ for $H, K \in S(G)$ if H is a subgroup of K. However, if $H, K \in S(G)$, then $H, K \in p(G)$ and $g.l.b.\{H, K\}$ in $p(G)$ is the set intersection $H \cap K$ which is also in $S(G)$, but $l.u.b.\{H, K\}$ in $p(G)$ is the set union $H \cup K$ which, in general, does not belong to $S(G)$. Hence $S(G)$ is not a sublattice of $p(G)$.

Let $(L, \leq)$ be a lattice and $a \leq b$ be elements of L. Let $I[b, a] = \{x \in L | a \leq x \leq b\}$. Since both $a, b \in I[b, a], I[b, a] \neq \phi$.

Theorem 12.7. $I[b, a]$ is a sublattice of $(L, \leq)$.

Proof. Let $x, y \in I[b, a]$. Then $a \leq x \leq b$, $a \leq y \leq b$. Then $a \leq g.l.b.\{x, y\}$ and $l.u.b.\{x, y\} \leq b$. Again $g.l.b.\{x, y\} \leq x \leq b$ and $x \leq l.u.b.\{x, y\} \leq b$ and so, $a \leq l.u.b.\{x, y\}$. Thus $a \leq g.l.b.\{x, y\} \leq b$ and $a \leq l.u.b.\{x, y\} \leq b$. Therefore, both $g.l.b.\{x, y\}$ and $l.u.b.\{x, y\} \in I[b, a]$. Hence $I[b, a]$ is a sublattice of $(L, \leq)$.

The sublattice $I[b, a]$ is called an **interval**.

Definition. Let L, M be two lattices. A map $f : L \to M$ is called a

(a) **join-homomorphism** if $f(a \vee b) = f(a) \vee f(b)$ for $a, b \in L$;

(b) **meet-homomorphism** if $f(a \wedge b) = f(a) \wedge f(b)$ for all $a, b \in L$;

(c) **homomorphism** if it is both a join homomorphism and a meet homomorphism;

(d) **order-homomorphism** if $a \leq b$ in L implies $f(a) \leq f(b)$.

Theorem 12.8. Let L, M be two lattices. If $f : L \to M$ is either a join-homomorphism or a meet-homomorphism, then f is an order-homomorphism.

Proof. Suppose that f is a join-homomorphism. Let $a \leq b$ in L. Then $a \vee b = b$ and, so, $f(b) = f(a \vee b) = f(a) \vee f(b)$, which implies that $f(a) \leq f(b)$. Thus f is an order-homomorphism.

That f is a meet-homomorphism implies f is an order homomorphism, which follows similarly.

Corollary. If $f : L \to M$ is a homomorphism, them f is an order-homomorphism.

Example 12.18. The converse of this result is not true in general. For example, consider the lattices L, M the Hasse diagrams of which are given in Fig. 12.4.

Let $f : L \to M$ be a map defined by

$$f(0) = 0 = f(a), \; f(b) = x, \; f(1) = 1.$$

Then f is clearly an order-homomorphism. However,

$$f(a \vee b) = f(1) = 1 \neq x = 0 \vee x = f(a) \vee f(b)$$

and, so, f is not a homomorphism.

However, in the case of one–one and onto maps, we have the following. (A homomorphism $f : L \to M$ which is one–one and onto, is called an **isomorphism**.)

Theorem 12.9. A map f from a lattice L onto a lattice M is an isomorphism if and only if $a \geq b$ in L implies and is implied by $f(a) \geq f(b)$ in M.

Proof. Let $f : L \to M$ be an isomorphism. In particular, f is a homomorphism. It follows from Theorem 12.8 that f is an order-homomorphism.

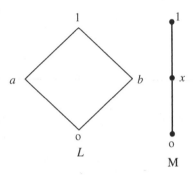

Fig. 12.4.

Let $g : M \to L$ be the inverse of the set map f which is one–one and onto. Then $gf(a) = a$ for every $a \in L$ and $fg(x) = x$ for every $x \in M$. Let $x, y \in M$ and suppose that $x = f(a), y = f(b), a, b \in L$. Then

$$g(x \vee y) = g(f(a) \vee f(b)) = g(f(a \vee b)) = a \vee b = g(x) \vee g(y)$$

and

$$g(x \wedge y) = g(f(a) \wedge f(b)) = g(f(a \wedge b)) = a \wedge b = g(x) \wedge g(y).$$

Hence g is a homomorphism and, so, g is an order homomorphism. Thus $f(a) \geq f(b)$ implies $a \geq b$.

Conversely, suppose that the one–one, onto map f is such that $a \leq b$ in L if and only if $f(a) \leq f(b)$ in M. Let $a, b \in L$ and suppose that $a \vee b = c$. Then $c \geq a$, $c \geq b$ and, therefore, $f(c) \geq f(a)$, $f(c) \geq f(b)$, which imply that $f(c) \geq f(a) \vee f(b)$. Let z be any element of M such that $z \geq f(a), f(b)$. Let $d \in L$ such that $z = f(d)$ so that $f(d) \geq f(a)$ and $f(d) \geq f(b)$. By the given condition $d \geq a$ and $d \geq b$ or that $d \geq a \vee b = c$ and, hence, $z = f(d) \geq f(c)$. This implies that $f(c) = f(a) \vee f(b)$ or that $f(a \vee b) = f(a) \vee f(b)$. In a similar fashion we can prove that $f(a \wedge b) = f(a) \wedge f(b)$. Hence f is a homomorphism and so, an isomorphism.

Definition. A subset A of a lattice L is called an **ideal** if (i) $a, b \in A$ implies $a \wedge b \in A$ and (ii) $a \in A$ and $x \in L$ imply $x \vee a \in A$. Also A is called a **principal ideal** generated by a and denoted by (a) if $A = \{x \in L | x \geq a\}$.

Observe that every ideal of L is a sublattice of L. However, every sublattice of L need not be an ideal of L. For example, we have seen that $T = \{1, 2, 3, 6\}$ is a sublattice of D_{30} but with $6 \in T$, $10 \in D_{30}$, $6 \vee 10 = l.u.b.\{6, 10\} = 30$ which is not in T. Therefore, T is not an ideal of D_{30}.

Example 12.19. For any set S, the power set $p(s)$ of S is a lattice. Also, it is bounded as $\phi \leq A \leq S$ for every subset A of S. In particular, if $S = N$, the set of all natural numbers, then $p(N)$ is a bounded lattice. Consider $S = \{A | A \text{ is a finite subset of } N\}$.

For $A, B \in S$, $A \wedge B = A \cap B$ is again a finite subset of N and so is $A \vee B = A \cup B$. Therefore, S is a sublattice of $p(N)$. If S were bounded above, there exists a finite subset B of N such that $A \leq B$ for every A in S. Let n be a natural number such that $n \notin B$. Such an n certainly exists as N has infinitely many elements, whereas B has only a finite number of elements. Let $C = B \cup \{n\}$. Then C is a finite subset of N so that $C \in S$ but $C \nleq B$. Hence S is not bounded. Thus, every sublattice of a bounded lattice need not be bounded.

If D is the set of all even natural numbers, D is an infinite subset of N and $D \notin S$ but $D \in p(N)$. For any finite subset A of N. $D \vee A = D \cup A$ is an infinite subset of N. Therefore, $D \vee A \notin S$ for any $A \in S$. Hence S is not an ideal in $p(N)$.

Definition. A subset A of a lattice $(L, \leq)$ is called a **dual ideal** if (i) for all $x, y \in A$, $x \vee y \in A$ and (ii) for all $x \in A$, $y \in L$, $x \wedge y \in A$. Also A is called a **dual principal ideal generated by** a if $A = \{x \in L | x \leq a\}$.

12.4. Distributive and Modular Lattices

In this section, we study lattices satisfying certain special properties.

Let $(L, \leq)$ be a lattice with $\wedge$ and $\vee$, the meet and the join operations respectively, induced by the relation $\leq$. If for all $a, b, c \in L$,

$$(\text{D1}) \qquad a \wedge (b \vee c) = (a \wedge b) \vee (a \wedge c)$$

then we say that the meet operation $(\wedge)$ distributes over the join operation $(\vee)$. On the other hand, if for all $a, b, c \in L$,

$$(\text{D2}) \qquad a \vee (b \wedge c) = (a \vee b) \wedge (a \vee c),$$

we say that the join operation distributes over the meet operation.

Theorem 12.10. In any lattice $(L, \leq)$ the conditions (D1) and (D2) are equivalent.

Proof. Suppose that for all a, b, c in L, the condition (D2) holds. Then, by (D2) above,

$$
\begin{aligned}
(a \wedge b) \vee (a \wedge c) &= ((a \wedge b) \vee a) \wedge ((a \wedge b) \vee c) \\
&= a \wedge (c \vee (a \wedge b)) \\
&= a \wedge ((c \vee a) \wedge (c \vee b)) \\
&= (a \wedge (c \vee a)) \wedge (c \vee b) \\
&= a \wedge (c \vee b) = a \wedge (b \vee c)
\end{aligned}
$$

which proves that (D1) holds.

Conversely, suppose that (D1) holds. Then, for any a, b, c in L,

$$
\begin{aligned}
(a \vee b) \wedge (a \vee c) &= ((a \vee b) \wedge a) \vee ((a \vee b) \wedge c) \\
&= a \vee ((a \vee b) \wedge c) = a \vee (c \wedge (a \vee b)) \\
&= a \vee ((c \wedge a) \vee (c \wedge b)) \\
&= (a \vee (c \wedge a)) \vee (c \wedge b) \\
&= a \vee (c \wedge b) \\
&= a \vee (b \wedge c).
\end{aligned}
$$

Thus (D2) holds.

Definition. A lattice $(L, \leq)$ is said to be **distributive** if either of the equivalent conditions (D1) and (D2) is satisfied.

Examples 12.20. There do exist several lattices that are distributive and several others that are not distributive. It follows from the distributive laws for sets that the lattice $p(S)$ of the power set of a non-empty set S with respect to the usual '$\leq$' relation is distributive.

To give another example of a distributive lattice, we first have the following.

Theorem 12.11. If a, b, c are positive integers, then

$$
g.c.d.(a, l.c.m.(b, c)) = l.c.m.(g.c.d.(a, b), g.c.d.(a, c)).
$$

Proof. Let $g.c.d.(a, b) = d$ and $g.c.d.(a, c) = e$. Then $d|a$ and $d|l.c.m.(b, c)$. Therefore, $d|g.c.d.(a, l.c.m.(b, c))$. Similarly, $e|g.c.d.(a, l.c.m.(b, c))$. Hence $l.c.m.(d, e)| \; g.c.d.(a, l.c.m.(b, c))$.

Let p be a prime and p^α be the largest power of p that divides $g.c.d.(a, l.c.m.(b, c))$. Then $p^\alpha|a$ and $p^\alpha|l.c.m.(b, c)$. It follows from the procedure for finding the $l.c.m.$ of two numbers by using prime factorizations of the two numbers that $p^\alpha|b$ and $p^\beta|c$ for some $\beta \leq \alpha$ or $p^\alpha|c$ and $p^\gamma|b$ for some $\gamma \leq \alpha$. Then, in the first case, $p^\alpha|d$ and $p^\beta|e$ while in

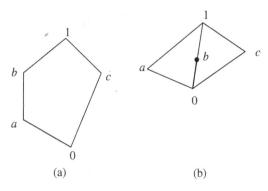

Fig. 12.5.

the second case $p^\alpha|e$ and $p^\gamma|d$. In either case $p^\alpha|l.c.m.(d,e)$. It follows from this that $g.c.d.(a,l.c.m.(b,c))|$ $l.c.m.(d,e)$. Hence $g.c.d.(a,l.c.m.(b,c)) = l.c.m.(g.c.d.(a,b), g.c.d.(a,c))$.

Corollary. The lattice $(N,|)$ of natural numbers with $a \leq b$ if $a|b$ i.e. if a divides b, is distributive.

Example 12.20. Consider the lattices with Hasse diagrams as given in Fig. 12.5.

In the case of lattice with Hasse diagram (a), we have $b \wedge (a \vee c) = b \wedge 1 = b$ and $(b \wedge a) \vee (b \wedge c) = a \vee 0 = a$. Thus $b \wedge (a \vee c) \neq (b \wedge a) \vee (b \wedge c)$ and this lattice is not distributive. In the case of the other lattice with Hasse diagram (b), we have $a \wedge (b \vee c) = a \wedge 1 = a$ and $(a \wedge b) \vee (a \wedge c) = 0 \vee 0 = 0$. This proves that the second lattice is also not distributive.

Example 12.21. Let G be a group and L be the set of all subgroups of G. If $\leq$ is the relation defined by $A \leq B$, for $A, B \in L$, if A is a subgroup of B, then $A \wedge B = A \cap B$ is the usual set intersection and $A \vee B$ is the subgroup of G generated by A and B. As already seen, L is a lattice. This lattice is not, in general, distributive. For this we take $G = S_3$ the symmetric group of degree 3 and $S_3 = \{1, (12), (13), (23), (123), (132)\}$.

Consider $A = \{1, (12)\}$, $B = \{1, (13)\}$, $C = \{1, (23)\}$. Then $A \wedge B = \{1\}$, $A \wedge C = \{1\}$, $B \vee C = S_3$ itself. Therefore, $A \wedge (B \vee C) = A \wedge S_3 = A$ but $(A \wedge B) \vee (A \wedge C) = \{1\} \vee \{1\} = \{1\}$.

This proves that the lattice of subgroups of S_3 is not distributive. Observe that the Hasse diagram for this lattice is as given in Fig. 12.6(a).

Example 12.22. Consider the lattice, the Hasse diagram of which is as in Fig. 12.6(b).

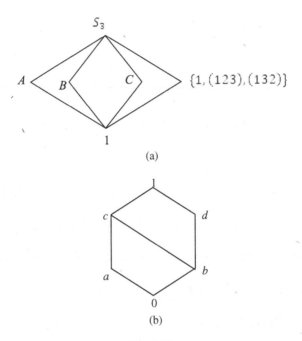

(a)

(b)

Fig. 12.6.

We claim that this lattice is distributive. We need to check the property (D1):

$$x \wedge (y \vee z) = (x \wedge y) \vee (x \wedge z).$$

Observe that if $x = 0$, then $x \wedge (y \vee z) = 0$, $x \wedge z = 0$ and $x \wedge y = 0$. When $y = 0$, $x \wedge (y \vee z) = x \wedge z$, $x \wedge y = 0$ and $(x \wedge y) \vee (x \wedge z) = x \wedge z$.
When $z = 0$, $x \wedge (y \vee z) = x \wedge y$ and $(x \wedge y) \vee (x \wedge z) = (x \wedge y) \vee 0 = x \wedge y$.
Thus (D1) holds if any one of x, y, z is 0. Similarly, it is easy to check that (D1) holds if any one of x, y, z is 1. Suppose $x = y$. Then $x \wedge (y \vee z) = x \wedge (x \vee z) = x$ and $(x \wedge y) \vee (x \wedge z) = (x \wedge x) \vee (x \wedge z) = x \vee (x \wedge z) = x$.
Thus (D1) holds in this case. It is equally easy to check that (D1) holds if $x = z$ or $y = z$. Thus, to check (D1) for any lattice, it is enough to check it when x, y, z are all distinct and each one of these is non-zero and non-unit.

Given any subset $\{x, y, z\}$ of a lattice L, in order to check the law (D1), we need to consider six possible permutations, but for every two permutations x, y, z and x, z, y, the laws (D1) are identical. Hence, for every such subset of L, we need to consider only three cases, namely: x, y, z; y, x, z and z, x, y.

Now, in the case of the lattice under consideration, we need to consider the subsets $\{a, b, c\}$, $\{a, b, d\}$, $\{a, c, d\}$ and $\{b, c, d\}$. We check the law (D1) for three possible permutations for each of the four sets. Now

(a) $\quad a \wedge (b \vee c) = a \wedge 1 = a; \quad (a \wedge b) \vee (a \wedge c) = 0 \vee a = a$

$\quad b \wedge (a \vee c) = b \wedge c = b; \quad (b \wedge a) \vee (b \wedge c) = 0 \vee b = b;$

$\quad c \wedge (a \vee b) = c \wedge c = c; \quad (c \wedge a) \vee (c \wedge b) = a \vee b = c;$

(b) $\quad a \wedge (b \vee d) = a \wedge d = 0; \quad (a \wedge b) \vee (a \wedge d) = 0 \vee 0 = 0;$

$\quad b \wedge (a \vee d) = b \wedge 1 = b; \quad (b \wedge a) \vee (b \wedge d) = 0 \vee b = b;$

$\quad d \wedge (a \vee b) = d \wedge c = b; \quad (d \wedge a) \vee (d \wedge b) = 0 \vee b = b;$

(c) $\quad a \wedge (c \vee d) = a \wedge 1 = a; \quad (a \wedge c) \vee (a \wedge d) = a \vee 0 = a;$

$\quad c \wedge (a \vee d) = c \wedge 1 = c; \quad (c \wedge a) \vee (c \wedge d) = a \vee b = c;$

$\quad d \wedge (a \vee c) = d \wedge c = b; \quad (d \wedge a) \vee (d \wedge c) = 0 \vee b = b;$

(d) $\quad b \wedge (c \vee d) = b \wedge 1 = b; \quad (b \wedge c) \vee (b \wedge d) = b \vee b = b;$

$\quad c \wedge (b \vee d) = c \wedge d = b; \quad (c \wedge b) \vee (c \wedge d) = b \vee b = b;$

$\quad d \wedge (b \vee c) = d \wedge c = b; \quad (d \wedge b) \vee (d \wedge c) = b \vee b = b.$

Thus, we find that the law (D1) holds for all x, y, z all distinct and each one non-zero and non-unit. Hence the lattice is distributive.

Example 12.23. Let L be the set of all ideals of the ring Z of integers. With $A \wedge B = A \cap B$, the usual set interaction and $A \vee B = A + B$, for $A, B \in L$, L becomes a lattice. Every ideal of Z is of the form aZ for some non-negative integer a. For positive integers a, b, c, let $A = aZ, B = bZ$ and $C = cZ$. Now $A \cap (B + C) = $ ideal generated by $l.c.m.(a, g.c.d.(b, c))$ and $A \cap B + A \cap C = $ ideal generated by $g.c.d.$ $(l.c.m.(a, b), l.c.m.(a, c))$. It follows from Theorem 12.11 that $A \cap (B + C) = A \cap B + A \cap C$. Hence the lattice of ideals of Z is distributive.

Definition. A lattice $(L, \leq)$ is said to be **modular** if for all x, y, z in L, the modular law

$$(M1) \quad x \wedge (y \vee (x \wedge z)) = (x \wedge y) \vee (x \wedge z)$$

holds.

Observe that if $x = 0$, then *l.h.s* of the modular law is 0 and so is the *r.h.s*. If $x = 1$, then *l.h.s* $= y \vee z = $ *r.h.s*. In a similar fashion, it is easy to see that the modular law holds if $y = 0$ or 1 or $z = 0$ or 1. Next, suppose that $x = y$. Then

$$x \wedge (y \vee (x \wedge z)) = x \wedge (x \vee (x \wedge z)) = x \wedge x = x \quad \text{and}$$

$$(x \wedge y) \vee (x \wedge z) = (x \wedge x) \vee (x \wedge z) = x \vee (x \wedge z) = x.$$

Thus the modular law holds in this case. It is equally easy to check that the modular law holds if $y = z$ or $z = x$. Hence the modular law is non-trivial and needs to be checked in the case when none of x, y, z is 0 or 1 and no two of these are equal.

The following result gives a more useful and convenient characterization of modular lattices.

Theorem 12.12. A lattice $(L, \leq)$ is modular if and only if

$$(M2) \qquad x \wedge (y \vee z) = (x \wedge y) \vee z$$

for $x, y, z \in L$ with $z \leq x$.

Proof. Let $(L, \leq)$ be modular and let $x, y, z \in L$ with $z \leq x$. Then $x \wedge (y \vee (x \wedge z)) = x \wedge (y \vee z)$ and $(x \wedge y) \vee (x \wedge z) = (x \wedge y) \vee z$. Therefore, by the modular law (M1), $x \wedge (y \vee z) = (x \wedge y) \vee z$ and (M2) holds. Conversely, suppose that (M2) holds for all $x, y, z \in L$ with $z \leq x$. Then, for any $x, y, z \in L$, we know that $x \wedge z \leq x$. Therefore, using (M2), we get (thinking of $x \wedge z = t$ with $t \leq x$) $x \wedge (y \vee (x \wedge z)) = (x \wedge y) \vee (x \wedge z)$ and (M1) holds.

Example 12.24. The lattices with Hasse diagrams (a) and (b) of Example 12.20 were shown to be non-distributive.

(a) For the lattice with diagram (a) $b \geq a$ and $b \wedge (a \vee c) = b \wedge 1 = b$ while $a \vee (b \wedge c) = a \vee 0 = a$. Thus $b \wedge (a \vee c) \neq a \vee (b \wedge c)$. Hence the lattice is not modular.

   We thus have an example of a lattice which is neither distributive nor modular.

(b) For the lattice with diagram (b), let $\{x, y, z\} = \{a, b, c\}$ as sets. Observe that $x \wedge y = x \wedge z = y \wedge z = 0$ and $x \vee y = x \vee z = y \vee z = 1$. Therefore,

$$x \wedge (y \vee (x \wedge z)) = x \wedge (y \vee 0) = x \wedge y = 0,$$

and $(x \wedge y) \vee (x \wedge z) = 0 \vee 0 = 0$.

Hence this lattice is modular. We thus have an example of a lattice which is modular but is not distributive.

Example 12.25. We have already proved (refer to Example 12.21) that the lattice of subgroups of S_3 is not distributive. For checking the modular law (M1) for this lattice we have to consider subsets of $\{A, B, C, D\}$, where $D = \{1, (123), (132)\}$, each having three elements. It is clear from the Hasse diagram of this lattice that for each subset of order 3, the modular law (M1) holds. Hence this lattice is modular. This gives another example of a lattice which is modular but is not distributive.

Example 12.26. Let G be a group and L be the lattice of all normal subgroups of G. Here, for subgroups A, B of G, $A \wedge B = A \cap B$, the usual set intersection and $A \vee B = AB = \{ab | a \in A, b \in B\}$, which is a normal subgroup of G and contains both A as well as B as subgroups. Consider normal subgroups A, B, C of G with $B \subseteq A$. Let $a \in A \cap BC$, so that $a \in A$ and $a \in BC$. Then $a = bc$ for some $b \in B, c \in C$. Now $b \in B \subseteq A$ and so, $c = b^{-1}a \in A$. Thus $c \in A \cap C$ and $a = bc \in B(A \cap C)$. This proves that $A \cap BC \subseteq B(A \cap C)$. The reverse inclusion being trivially true, we have $A \cap BC = B(A \cap C)$. Hence the lattice is modular.

Example 12.27. Let R be a ring and M a (left) R-module. For submodules A, B of M, define $A \wedge B = A \cap B$, the usual set intersection and $A \vee B = A + B$. The set of submodules of M then becomes a lattice. A simple argument, similar to the one used in Example 12.26 above shows that the lattice is modular.

Theorem 12.13. Every distributive lattice is modular.

Proof. Let $(L, \leq)$ be a distributive lattice. Then the distributive law (D1) holds for all $x, y, z \in L$. Let $x, y, z \in L$ and suppose that $x \geq z$. Then, by the law (D1), $x \wedge (y \vee z) = (x \wedge y) \vee (x \wedge z) = (x \wedge y) \vee z$ (as $z \leq x$) Thus the modular law (M2) holds and $(L, \leq)$ is a modular lattice.

Distributive law (D1) and, therefore, also (D2) is a sufficient condition for a lattice to be distributive and so, modular. The following theorem gives another sufficient condition for a lattice to be modular.

Theorem 12.14. If for all a, b, c in a lattice L,

(D3) $\qquad (a \vee b) \wedge (b \vee c) \wedge (c \vee a) = (a \wedge b) \vee (b \wedge c) \vee (c \wedge a)$

holds, then L is a modular lattice.

Proof. Let L be a lattice for every three elements of which, the law (D3) holds. Let $a, b, c \in L$ and suppose $a \geq c$. Then $a \vee c = a$ and $a \wedge c = c$. Then the law (D3)

$$(a \vee b) \wedge (b \vee c) \wedge (c \vee a) = (a \wedge b) \vee (b \wedge c) \vee (c \wedge a)$$

takes the form

$$(a \vee b) \wedge (b \vee c) \wedge a = (a \wedge b) \vee (b \wedge c) \vee c$$

or

$$[a \wedge (a \vee b)] \wedge (b \vee c) = (a \wedge b) \vee [(b \wedge c) \vee c]$$

or

$$a \wedge (b \vee c) = (a \wedge b) \vee (a \wedge c) \text{ (by } L4)$$

which proves the modular law (M2). Hence L is a modular lattice.

We have proved that for any lattice $(L, \leq)$, the distributive laws (D1) and (D2) and equivalent. We can also prove the following.

Theorem 12.15. For any lattice $(L, \leq)$, the laws (D1), (D2) and (D3) are equivalent.

Proof. Exercise.

In the set Z of integers, we know that for $a, b, c \in Z$, (i) $ab = ac, a \neq 0$ implies that $b = c$; (ii) $a + b = a + c$ implies that $b = c$. Thus cancellation law holds for multiplication as well as addition. However, in an arbitrary lattice $(L, \leq)$, we cannot conclude that for $a, b, c \in L$, (i) $a \wedge b = a \wedge c$ implies $b = c$; (ii) $a \vee b = a \vee c$ implies $b = c$. For example, consider the lattice of all subgroups of the symmetric groups S_3 of degree 3. Let $A = \{1, (12)\}$, $B = \{1, (13)\}$ and $C = \{1, (23)\}$. Then $A \wedge B = A \wedge C = \{1\}$ but $B \neq C$ and $A \vee B = S_3 = A \vee C$ but $B \neq C$. Thus, in a lattice in general neither the cancellation law for $\wedge$ nor for $\vee$ holds. In the case of a lattice, we can however, talk of the joint cancellation law. Namely, for a, b, c in a lattice L, $a \wedge b = a \wedge c$ and $a \vee b = a \vee c$ together imply $b = c$. As seen in the case of lattice of subgroups of S_3, even this (joint) cancellation law does not hold.

Theorem 12.16. If $(L, \leq)$ is a distributive lattice and $a, b, c \in L$ such that $a \vee b = a \vee c, a \wedge b = a \wedge c$, then $b = c$.

Proof. Since $a \wedge b \leq b$,

$$
\begin{aligned}
b &= b \vee (a \wedge b) = (b \vee a) \wedge (b \vee b) &&\text{(by (D2))} \\
&= (b \vee a) \wedge b = (a \vee b) \wedge b = (a \vee c) \wedge b = b \wedge (a \vee c) \\
&= (b \wedge a) \vee (b \wedge c) &&\text{(by (D1))} \\
&= (a \wedge b) \vee (b \wedge c) = (a \wedge c) \vee (b \wedge c) \\
&= ((a \wedge c) \vee b) \wedge ((a \wedge c) \vee c) &&\text{(by (D1))}
\end{aligned}
$$

$$= ((a \wedge c) \vee b) \wedge c$$

$$= ((a \vee b) \wedge (b \vee c)) \wedge c \qquad \qquad \text{(by (D2))}$$

$$= (a \vee b) \wedge ((b \vee c) \wedge c) = (a \vee b) \wedge c = (a \vee c) \wedge c = c$$

and the proof is complete.

The converse of this result is also true and is left as an exercise.

The following result gives another characterization of modular lattices.

Theorem 12.17. A lattice L is modular if and only if $a \geq b, a, b \in L$ and $a \wedge c = b \wedge c$, $a \vee c = b \vee c$ for any c in L imply that $a = b$.

Proof. Let L be modular and let a, b, c be elements of L such that $a \geq b$ and $a \wedge c = b \wedge c$, $a \vee c = b \vee c$. Then

$$a = a \wedge (a \vee c) = a \wedge (b \vee c) = b \vee (a \wedge c)$$

$$= b \vee (b \wedge c) = b.$$

Conversely, suppose that L is any lattice in which $a \geq b$ and $a \wedge c = b \wedge c$, $a \vee c = b \vee c$ imply $a = b$. Let $a \geq b$ in L and c be any element of L. Then

$$a \wedge (b \vee c) \geq b \wedge (b \vee c) = b = b \vee (b \wedge c) = b \vee (a \wedge c) \text{ or that}$$

$$a \wedge (b \vee c) \geq b \vee (a \wedge c). \qquad \qquad (12.10)$$

Also

$$(a \wedge (b \vee c)) \wedge c = a \wedge ((b \vee c) \wedge c) = a \wedge c \text{ and}$$

$$a \wedge c = (a \wedge c) \wedge c \leq (b \vee (a \wedge c)) \wedge c \leq (a \vee (a \wedge c)) \wedge c = a \wedge c.$$

Thus

$$(b \vee (a \wedge c)) \wedge c = a \wedge c = (a \wedge (b \vee c)) \wedge c.$$

By the duality principle, we have

$$(a \wedge (b \vee c)) \vee c = b \vee c = (b \vee (a \wedge c)) \vee c.$$

Setting $a \wedge (b \vee c) = y$ and $b \vee (a \wedge c) = z$, we find that

$$y \vee c = z \vee c, y \wedge c = z \wedge c \text{ and } y \geq z.$$

Therefore, $y = z$. Hence $a \wedge (b \vee c) = b \vee (a \wedge c)$ which proves that the lattice L is modular.

Theorem 12.18. If L is a modular lattice and a, b are any elements of L, then the intervals $I[a \vee b, a]$ and $I[b, a \wedge b]$ are isomorphic.

Proof. Let L be a modular lattice and $a, b \in L$. Let $x \in I[a \vee b, a]$. Then $a \leq x \leq a \vee b$. Therefore $a \wedge b \leq x \wedge b \leq (a \vee b) \wedge b = b$ so that $x \wedge b \in I[b, a \wedge b]$. Hence, we get a map $f : I[a \vee b, a] \to I[b, a \wedge b]$ given by

$$f(x) = x \wedge b, x \in I[a \vee b, a].$$

On the other hand, if $y \in I[b, a \wedge b]$, then $a \wedge b \leq y \leq b$ and, therefore, $a = (a \wedge b) \vee a \leq a \vee y \leq a \vee b$. Thus $a \vee y \in I[a \vee b, a]$. Define a map $g : I[b, a \wedge b] \to I[a \vee b, a]$ by

$$g(y) = a \vee y, \ y \in I[b, a \wedge b].$$

The maps f and g are clearly order homomorphisms. We now prove that the maps f, g are inverse of each other. For this we need to check that $g(f(x)) = x$ for every $x \in I[a \vee b, a]$ and $fg(y) = y$ for every $y \in I[b, a \wedge b]$. Let $x \in I[a \vee b, a]$ and $y \in I[b, a \wedge b]$. Then $a \leq x \leq a \vee b, a \wedge b \leq y \leq b$. Now

$$a \vee (x \wedge b) = x \wedge (a \vee b) = x \qquad \text{(by modular law (M2))}$$

and

$$(a \vee y) \wedge b = y \vee (a \wedge b) = y \qquad \text{(by modular law (M2))}.$$

Hence f and g are inverse maps of each other. It follows from Theorem 12.9 that f is an isomorphism with g as its inverse isomorphism.

Definition. Let L be a modular lattice with 0 and 1. A finite set of elements $a_1, a_2, \ldots, a_n$ of L is called **(join) independent** if for $i = 1, 2, \ldots, n$,

$$a_i \wedge (a_1 \vee \ldots \vee a_{i-1} \vee a_{i+1} \vee \ldots \vee a_n) = 0. \qquad (12.11)$$

Theorem 12.19. If the elements $a_1, \ldots, a_n$ in a modular lattice with 0 and 1 are independent, then

$$(a_1 \vee \ldots \vee a_r \vee a_{r+1} \vee \ldots \vee a_s) \wedge (a_1 \vee \ldots \vee a_r \vee a_{s+1} \vee \ldots \vee a_n)$$
$$= a_1 \vee \ldots \vee a_r. \qquad (12.12)$$

Proof. We first prove, by induction on s, that

$$(a_1 \vee \ldots \vee a_s) \wedge (a_{s+1} \vee \ldots \vee a_n) = 0. \qquad (12.13)$$

Observe that (12.13) is true for $s = 1$ by the very definition of independent elements. Assume that (12.13) is true for $s - 1$. Then

$$(a_1 \vee \ldots \vee a_s) \wedge (a_{s+1} \vee \ldots \vee a_n)$$
$$\leq (a_1 \vee \ldots \vee a_s) \wedge (a_s \vee a_{s+1} \vee \ldots \vee a_n)$$
$$= ((a_1 \vee \ldots \vee a_{s-1}) \wedge (a_s \vee a_{s+1} \vee \ldots \vee a_n)) \vee a_s$$
$$= 0 \vee a_s = a_s$$

first by modularity and then by induction hypothesis for $s - 1$. Then

$$(a_1 \vee \ldots \vee a_s) \wedge (a_{s+1} \vee \ldots \vee a_n)$$
$$= (a_1 \vee \ldots \vee a_s) \wedge (a_{s+1} \vee \ldots \vee a_n) \wedge a_s$$
$$= (a_1 \vee \ldots \vee a_s) \wedge ((a_{s+1} \vee \ldots \vee a_n) \wedge a_s)$$
$$\leq (a_1 \vee \ldots \vee a_s) \wedge ((a_1 \vee \ldots \vee a_{s-1} \vee a_{s+1} \ldots \vee a_n) \wedge a_s)$$
$$= (a_1 \vee \ldots \vee a_s) \wedge 0 = 0.$$

This completes induction and (12.13) holds for all $s \geq 1$. Now, let $a_1 \vee \ldots \vee a_r = b$, $c = a_{r+1} \vee \ldots \vee a_s$ and $d = a_{s+1} \vee \ldots \vee a_t$. Then, by (12.13) above $(b \vee c) \wedge d = 0$. Now, using modularity,

$$(a_1 \vee \ldots \vee a_r \vee a_{r+1} \vee \ldots \vee a_s) \wedge (a_1 \vee \ldots \vee a_r \vee a_{s+1} \vee \ldots \vee a_t)$$
$$= (b \vee c) \wedge (b \vee d)$$
$$= b \vee ((b \vee c) \wedge d) = b \vee 0 = b = a_1 \vee \ldots \vee a_r.$$

Theorem 12.20. Let $a_1, a_2, \ldots, a_n$ be a set of independent elements in a modular lattice with 0 and 1 and that $a_1 \vee a_2 \vee \ldots \vee a_n = 1$. Let $b_i = a_1 \vee \ldots \vee a_{i-1} \vee a_{i+1} \vee \ldots \vee a_n$, for $i = 1, 2, \ldots, n$. Then, for $i = 1, 2, \ldots, n$,

$$b_i \vee (b_1 \wedge \ldots \wedge b_{i-1} \wedge b_{i+1} \wedge \ldots b_n) = 1;$$
$$b_1 \wedge b_2 \wedge \ldots \wedge b_n = 0;$$
$$a_i = b_1 \wedge \ldots \wedge b_{i-1} \wedge b_{i+1} \wedge \ldots \wedge b_n.$$

Proof. It is clear from the definition of b_i that $b_i \vee b_j = a_1 \vee a_2 \vee \ldots \vee a_n = 1$ for $i \neq j$. Also $b_i \vee a_i = 1$, $a_i \wedge b_i = 0$ and $a_i \leq b_j$ for all $j \neq i$. As an application of Theorem 12.19, we find that for $i < j$,

$$b_i \wedge b_j = (a_1 \vee \ldots \vee a_{i-1} \vee a_{i+1} \vee \ldots \vee a_n) \wedge (a_1 \vee \ldots \vee a_{j-1}$$
$$\vee a_{j+1} \vee \ldots \vee a_n)$$
$$= a_1 \vee \ldots \vee a_{i-1} \vee a_{i+1} \vee \ldots \vee a_{j-1} \vee a_{j+1} \vee \ldots \vee a_n.$$

A repeated application of this then shows that $b_1 \wedge b_2 \ldots \wedge b_{i-1}$ is the join of $a_i, a_{i+1}, \ldots, a_n$ i.e.

$$b_1 \wedge b_2 \wedge \ldots \wedge b_{i-1} = a_i \vee a_{i+1} \vee \ldots \vee a_n.$$

Similarly

$$b_{i+1} \wedge b_{i+2} \wedge \ldots \wedge b_n = a_1 \vee a_2 \vee \ldots \vee a_i.$$

Therefore

$$b_1 \wedge b_2 \wedge \ldots \wedge b_{i-1} \wedge b_{i+1} \wedge \ldots \wedge b_n$$
$$= (a_i \vee a_{i+1} \vee \ldots \vee a_n) \wedge (a_1 \vee a_2 \vee \ldots \vee a_i) = a_i.$$

Then

$$b_i \wedge b_2 \wedge \ldots \wedge b_n = b_i \wedge (b_1 \wedge \ldots \wedge b_{i-1} \wedge b_{i+1} \wedge \ldots \wedge b_n) = b_i \wedge a_i = 0$$

and

$$b_i \vee (b_1 \wedge \ldots \wedge b_{i-1} \wedge b_{i+1} \wedge \ldots \wedge b_n) = b_i \vee a_i = 1.$$

Theorem 12.21. If the elements $a_1, \ldots, a_n$ of a modular lattice L with 0, 1 are independent and $(a_1 \vee a_2 \vee \ldots \vee a_n) \wedge a_{n+1} = 0$, then the elements $a_1, a_2, \ldots, a_{n+1}$ are independent.

Proof. For this we only need to prove that for any i, $1 \le i \le n$,

$$a_i \wedge (a_1 \vee \ldots \vee a_{i-1} \vee a_{i+1} \vee \ldots \vee a_n \vee a_{n+1}) = 0.$$

Let us write $a_1 \vee \ldots \vee a_{i-1} \vee a_{i+1} \vee \ldots \vee a_n = b_i$. Then, the elements $a_1, a_2, \ldots, a_n$ being independent and, by the given hypothesis, we have $a_i \wedge b_i = 0$, $(a_i \vee b_i) \wedge a_{n+1} = 0$ and $b_i \wedge a_{n+1} = 0$. Therefore $(a_i \vee b_i) \wedge a_{n+1} = b_i \wedge a_{n+1}$. In a modular lattice, $(x \vee y) \wedge z = y \wedge z$ implies $x \wedge (y \vee z) = x \wedge y$ (refer Example 12.33 below). Therefore $a_i \wedge (b_i \vee a_{n+1}) = a_i \wedge b_i = 0$.

Corollary. Prove that in a modular lattice with 0 and 1, the elements $a_1, a_2, \ldots, a_n$ are independent if and only if $(a_1 \vee \ldots \vee a_i) \wedge a_{i+1} = 0$, for $i = 1, 2, \ldots, n-1$.

Proof. Suppose that $a_1, a_2, \ldots, a_n$ are independent. Then, for any i, $1 \le i \le n-1$,

$$(a_1, \vee \ldots \vee a_i) \wedge a_{i+1} \le (a_1 \vee \ldots \vee a_i \vee a_{i+2} \vee \ldots \vee a_n) \wedge a_{i+1} = 0.$$

Conversely, suppose that for $i = 1, 2, \ldots, n - 1$

$$(a_1 \vee \ldots \vee a_i) \wedge a_{i+1} = 0.$$

The single element a_1 is always independent. For $n = 2$, $a_1 \wedge a_2 = 0$ and it follows from the theorem that a_1, a_2 are independent. Now suppose that $a_1, \ldots, a_i$ are independent. Then, in view of the given condition, it follows from the theorem that $a_1, \ldots, a_i, a_{i+1}$ are independent. This completes induction and, therefore, $a_1, a_2, \ldots, a_n$ are independent.

Definition. An element p of a lattice L with 0 is called a **point** or an **atom** if p is a cover of 0. In a lattice with 1, an element a is called a **co-atom** if 1 is a cover of a.

Let L be any lattice. We say that **descending chain condition** (d.c.c.) holds in L if there exists no infinite properly descending chain $a_1 > a_2 > a_3 > \cdots$ in L; and an **ascending chain condition (a.c.c.)** holds in L if there exists no infinite properly ascending chain $a_1 < a_2 < a_3 < \cdots$ in L.

Remark. It is immediate from the definitions that both ascending and descending chain conditions hold in any finite lattice.

Theorem 12.22. *Every lattice L with 0 satisfying descending chain condition contains atoms. Every lattice with 1 satisfying ascending chain condition contain co-atoms.*

Proof. Let L be a lattice with 0 satisfying d.c.c. Let $a \in L$. If a covers 0, then a is an atom. If a does not cover 0, there exists an a_1 in L with $a > a_1 > 0$. If a_1 covers 0, then a_1 is an atom. If a_1 does not cover 0, there exists an a_2 in L with $a > a_1 > a_2 > 0$. Since L satisfies descending chain condition, the above process must terminate in a finite number of steps. But the process terminates only when we arrive at an atom in L. The proof for existence of co-atoms in a lattice with 1 satisfying a.c.c. follows on similar lines.

Corollary. *Every finite lattice with 0 has at least one atom. Every finite lattice with 1 has at least one co-atom.*

Example 12.28. 1. If a covers $a \wedge b$ in a modular lattice L, prove that $a \vee b$ covers b.

Proof. Let L be a modular lattice, let $a, b \in L$ and suppose that a covers $a \wedge b$. Let $d \in L$ and suppose that $b \leq d \leq a \vee b$. Then

$$a \wedge b \leq a \wedge d \leq a \wedge (a \vee b) = a.$$

As a covers $a \wedge b$, we have either $a \wedge d = a \wedge b$ or $a \wedge d = a$. If $a \wedge d = a \wedge b$, then $b = (a \wedge b) \vee b = (a \wedge d) \vee b = (a \vee b) \wedge d$ (by modularity of L) $= d$. On the other hand, if $a \wedge d = a$, then $a \leq d$. Therefore, $a \vee b \leq b \vee d = d \leq a \vee b$ (by assumption). Hence $d = a \vee b$. This completes the proof that $a \vee b$ covers b.

Definition. A lattice in which for every pair of elements a, b, if a covers $a \wedge b$ implies $a \vee b$ covers b, then the lattice is called **semi-modular**.

The above example shows that every modular lattice is semi-modular. There do exist lattices which are semi-modular but are not modular.

Example 12.29. Show that the lattice with Hasse diagram given in Fig. 12.7 is semi-modular but is not modular.

Solution. Observe that $a \leq d$ and $d \wedge (a \vee e) = d \wedge (1) = d$ and $a \vee (d \wedge e) = a \vee 0 = a \neq d$. Thus $d \wedge (a \vee e) \neq a \vee (d \wedge e)$ and the lattice is not modular. For checking semi-modularity, there are 21 pairs of elements of the lattice which need to be considered. If x, y are elements of a lattice and x covers y, then it is trivially true that if x covers $x \wedge y$ then $x \vee y$ covers y. Hence we need consider only those pairs of elements neither of which covers the other. Also if x covers $x \wedge y$, then $x \vee y$ covers y is trivially true if one of x, y is 0 or one of x, y is 1. Therefore, we need to consider only the pairs a, b; a, e; b, d; c, d; c, e and d, e. We now take up these pairs one-by-one. For

a, b : a covers $a \wedge b = 0$ and $a \vee b = c$ covers b;
a, e : a covers $a \wedge e = 0$ and $a \vee e = 1$ covers e;
b, d : b covers $b \wedge d = 0$ and $b \vee d = 1$ covers d;

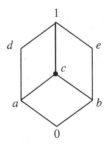

Fig. 12.7.

c, d : c covers $c \wedge d = a$ and $c \vee d = 1$ covers d;

c, e : c covers $c \wedge e = b$ and $c \vee e = 1$ covers e;

d, e : d does not cover $d \wedge e = 0$ and we need not check $d \vee e = 1$ covers e which of course is true.

Changing the order of the above pairs of elements, we find,

a, b : b covers $a \wedge b = 0$ and $a \vee b = c$ covers a;

a, e : since e does not cover $a \wedge e = 0$, we need not consider this case;

b, d : since d does not cover $b \wedge d = 0$, we are through in this case;

c, d : d covers $c \wedge d = a$ and $c \vee d = 1$ covers c;

c, e : e covers $c \wedge e = b$, $c \vee e = 1$ covers c;

d, e: e does not cover $d \wedge e = 0$, this case need not be considered.

Since all possible pairs of elements have been considered, it follows that the lattice is semi-modular.

Example 12.30. Show that a partially ordered set with an all element (a unit element) in which every non-empty set has a *g.l.b.* is a complete lattice.

Solution. Let L be a partially ordered set with a unit in which every non-empty set has a *g.l.b.* Then, in particular, every pair of elements has a *g.l.b.* In order to prove that L is a lattice and is complete, it is enough to prove that every non-empty subset S of L has a *l.u.b.* Consider the subset $T = \{x \in L | x \geq a, \text{for every } a \in S\}$ of L. Since $1 \in L$ and $1 \geq x$ for every $x \in L$ and, in particular, $1 \geq a$ for every $a \in S$, $1 \in T$. Thus T is non-empty. By the given hypothesis, T has a *g.l.b.* y (say). Then $y \leq x$ for every $x \in T$. Indeed, $y = \wedge_{x \in T} x$. Let $a \in S$. By definition of T, $a \leq x$ for every $x \in T$. Therefore $a \leq$ *g.l.b.* of $T = \wedge_{x \in T} x = y$. Hence $a \leq y$ for every $a \in S$ and y is an upper bound of S. If z is an upper bound of S, then $z \in T$ and y is *g.l.b.* of T, $y \leq z$. This proves that y is a *l.u.b.* of S. Hence L is a complete lattice.

Example 12.31. The lattice of all subgroups of the alternating group A_4 of degree 4 is not modular.

Proof. The alternating group A_4 is of order 12, has no subgroup of order 6, Klein's four-group V_4 is subgroup (indeed a normal subgroup of order 4), $B = \{1, (12)(34)\}$ is a subgroup of V_4 and $C = \{1, (123), (132)\}$ is a subgroup of A_4. Set $A = V_4$. Now, $B \cup C$, the subgroup of A_4 generated by B and C has a subset $\{1, (12)(34), (123), (132)\}$ of order 4 and it must also have the element $(12)(34)(123) = (243)$ and its inverse (234).

Since the group A_4 has no subgroup of order 6, $B \cup C = A_4$. Therefore $A \cap (B \cup C) = A \cap A_4 = A$ whereas $B \cup (A \cap C) = B \cup \{1\} = B \neq A$. Hence the lattice is not modular.

Example 12.32. For all x, y, z in a lattice L, prove that

$$((x \wedge y) \vee (x \wedge z)) \wedge ((x \wedge y) \vee (y \wedge z)) = x \wedge y.$$

Proof. Since $x \wedge y \leq (x \wedge y) \vee (x \wedge z)$ and $x \wedge y \leq (x \wedge y) \vee (y \vee z)$,

$$x \wedge y \leq ((x \wedge y) \vee (x \wedge z)) \wedge ((x \wedge y) \vee (y \wedge z)). \tag{12.14}$$

On the other hand, $x \wedge y \leq x, x \wedge z \leq x$ so that $(x \wedge y) \vee (x \wedge z) \leq x$. Similarly $(x \wedge y) \vee (y \wedge z) \leq y$. Therefore,

$$((x \wedge y) \vee (x \wedge z)) \wedge ((x \wedge y) \vee (y \wedge z)) \leq x \wedge y. \tag{12.15}$$

Combining (12.14) and (12.15), we get

$$((x \wedge y) \vee (x \wedge z)) \wedge ((x \wedge y) \vee (y \wedge z)) = x \wedge y.$$

Example 12.33. In a modular lattice, if $(x \vee y) \wedge z = y \wedge z$, then show that $x \wedge (y \vee z) = x \wedge y$.

Solution. Let $(L, \leq)$ be a modular lattice and suppose that for $x, y, z \in L$

$$(x \vee y) \wedge z = y \wedge z.$$

Then

$$x \wedge ((x \vee y) \wedge z) = x \wedge (y \wedge z)$$
$$\text{or} \quad (x \wedge (x \vee y)) \wedge z = x \wedge y \wedge z$$
$$\text{or} \quad x \wedge z = x \wedge y \wedge z$$

which implies that $x \wedge z \leq y$. But then $x \wedge z \leq x \wedge y$.

The principle of duality applied to the given condition $(x \vee y) \wedge z = y \wedge z$ shows that $(x \wedge y) \vee z = y \vee z$. Then

$$x \wedge (y \vee z) = x \wedge ((x \wedge y) \vee z) = (x \wedge y) \vee (x \wedge z) \quad (\text{by } M_2)$$
$$= x \wedge y.$$

Example 12.34. Show that a lattice L is distributive if and only if for any $x, y, z \in L$.

$$(x \vee y) \wedge z \leq x \vee (y \wedge z).$$

Proof. If L is distributive and x, y, z are in L, then

$$(x \vee y) \wedge z = (x \wedge z) \vee (y \wedge z) \leq x \vee (y \wedge z).$$

Conversely, suppose that whenever $x, y, z \in L$, then $(x \vee y) \wedge z \leq x \vee (y \wedge z)$. That $x \wedge z, y \wedge z \leq (x \vee y) \wedge z$ is clear and, so,

$$(x \wedge z) \vee (y \wedge z) \leq (x \vee y) \wedge z.$$

On the other hand,

$$(x \vee y) \wedge z = ((x \vee y) \wedge z) \wedge z \leq (x \vee (y \wedge z)) \wedge z$$
$$= ((y \wedge z) \vee x) \wedge z \leq (y \wedge z) \vee (x \wedge z) = (x \wedge z) \vee (y \wedge z).$$

Hence $(x \vee y) \wedge z = (x \wedge z) \vee (y \wedge z)$ and the lattice is distributive.

As an immediate consequence of this result (or the definition) it follows that every sublattice of a distributive lattice is distributive.

Example 12.35. Let $(L, \leq)$ be a distributive lattice and $a, b \in L$. Then the subset $A = \{x \in L | a \vee x = b \vee x\}$ is an ideal of L.

Proof. Let $x, y \in A$ and $z \in L$. Then $a \vee x = b \vee x$ and $a \vee y = b \vee y$. Now

$$a \vee (x \wedge y) = (a \vee x) \wedge (a \vee y) = (b \vee x) \wedge (b \vee y) = b \vee (x \wedge y).$$

Therefore $x \wedge y \in A$.

Next, $a \vee (x \vee z) = (a \vee x) \vee z = (b \vee x) \vee z = b \vee (x \vee z)$.

Thus $x \vee z \in A$. Hence A is an ideal of L.

12.5. Complemented Lattices

Let S be a non-empty set and $p(S)$ the power set of S which as seen earlier is a lattice with the usual set intersection and set union as the meet and join operations respectively. The set S itself is the unit element and the empty set ϕ is the zero element in this lattice. For every subset A of S, the complement A' of A has the property that $A \cup A' = S$ and $A \cap A' = \phi$. It is not hard to prove that A' is the unique element of $p(S)$ with the two properties listed. Guided by this, we can introduce the concept of complement in any lattice with a unit and zero element.

Let L be a lattice with 0 and 1 and $a \in L$. An element $a' \in L$, if it exists, is called a **complement** of a if $a \vee a' = 1$ and $a \wedge a' = 0$. Also if $a, b \in L$ and $b \leq a$, then an element $c \in L$ is called **complement** of b

relative to a if $b \vee c = a$ and $b \wedge c = 0$. It is clear from the definition of complement that

(a) complement of the complement a' of a is again a;
(b) complement relative to a of complement c of b relative to a is b again;
(c) complement of 1 is 0 and that of 0 is 1.

There do exist lattices with 0 and 1 in which (i) not every element has a complement; (ii) an element may have more than one complements; and (iii) every element has a unique complement (one such example is $p(S)$ considered above).

Example 12.36. Consider the lattice L consisting of all divisors of 42 in which the partial order is '|' i.e. is a divisor of. Then $L = \{1, 2, 3, 6, 7, 14, 21, 42\}$, the element 1 is zero and 42 is a unit in L. Since 42 is a square free number, if $a \in L$, then $g.c.d.\ (a, 42/a) = 1$ and $l.c.m.(a, 42/a) = 42$. Thus every element of L has a complement. Observe that complement of (a) 2 is 21 (b) 3 is 14; (c) 6 is 7. Thus every element of L has a unique complement.

In general, let n be a square free positive integer and L be the lattice of all positive divisors of n with the partial order '|' being 'is a divisor of'. Every element of this lattice has a unique complement.

Example 12.37. Let L be the lattice of all positive divisors of 18 with the partial order '|' being is a divisor of. Since $l.c.m.(2, 9) = 18$ and $g.c.d.(2, 9) = 1, 9$ is a complement of 2. Also 2 is the only integer other than 1 in L such that $g.c.d.(2, 3) = 1$ but $l.c.m.(2, 3) = 6 \neq 18$, the element 3 has no complement. Similarly the element 6 does not have a complement.

Example 12.38. In the lattice L of all positive divisors of 36, except for the elements 4 and 9, no non-trivial element has a complement. The elements 4 and 9 are complement of each other.

Example 12.39. In the lattice L of all positive divisors of 16, no element other than 0 and unit has a complement. The same is true in the lattice L of all positive divisors of n which is power of a prime number.

Example 12.40. In the lattices the Hasse diagrams of which are as in Fig. 12.8(i) and (ii), the element b has two complements a and c. In the case of lattice with Hasse diagram (ii) every one of a, b, c has two complements.

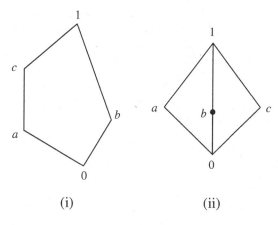

(i)   (ii)

Fig. 12.8.

In the case of lattice with Hasse diagram (i) both a and c have exactly one complement each. In fact b is the complement of a as well as of c in this case.

Definition. A lattice L with 0 and 1 is said to be **complemented** if every element of L has a complement. When we talk of a complemented lattice, it is understood that the lattice is with 0 and 1.

Theorem 12.23. In a distributive lattice with 0 and 1, if an element has a complement then this complement is unique.

Proof. Let $(L, \leq)$ be a distributive lattice with 0 and 1. Suppose that b, c in L are two complements of an element a in L. Then

$$a \vee b = 1, \quad a \wedge b = 0 \quad \text{and} \quad a \vee c = 1, \quad a \wedge c = 0.$$

Now

$$b = b \wedge 1 = b \wedge (a \vee c) = (b \wedge a) \vee (b \wedge c) \qquad \text{(byD1)}$$
$$= (a \wedge b) \vee (b \wedge c) = 0 \vee (b \wedge c) = (a \wedge c) \vee (b \wedge c) = (a \vee b) \wedge c$$
$$= 1 \wedge c = c.$$

Hence complement of a, if it exists, is unique.

Theorem 12.24. Let L be a complemented modular lattice and $a \in L$. Then the sublattice $L_a = \{x \in L | x \leq a\}$ of L is complemented.

Proof. The lattice L being complemented, L contains 0 and 1. As $0 \leq a$ and $a \leq a$, both $0, a \in L_a$. The element a is the unit in L_a. Let $b \in L_a$ i.e. $b \leq a$ and b' be complement of b in L so that $b \wedge b' = 0$ and $b \vee b' = 1$. Now

$$a = a \wedge 1 = a \wedge (b \vee b') = (a \wedge b') \vee b \qquad \text{(by modular law } M2\text{)}.$$

Also $(a \wedge b') \wedge b = a \wedge (b' \wedge b) = a \wedge 0 = 0$. Moreover $a \wedge b' \leq a$ and $a \wedge b' \in L_a$. Hence $a \wedge b'$ is complement of b in L_a.

The above result could alternatively be stated as follows.

Theorem 12.25. If L is a complemented modular lattice and $a \in L$, then every $b \leq a$ has a complement relative to a in L.

Lemma. Let $(L, \leq)$ be a distributive lattice with 0 and 1. If an element a in L has a complement a' (say) then for every b in L

$$a \vee (a' \wedge b) = a \vee b \quad \text{and} \quad a \wedge (a' \vee b) = a \wedge b.$$

Proof. Let a be an element of L having complement a' (say). Then

$$a \vee a' = 1, \ a \wedge a' = 0.$$

For any b in L, the lattice L being distributive,

$$a \vee (a' \wedge b) = (a \vee a') \wedge (a \vee b) = 1 \wedge (a \vee b) = a \vee b$$

and

$$a \wedge (a' \vee b) = (a \wedge a') \vee (a \wedge b) = 0 \vee (a \wedge b) = a \wedge b.$$

Theorem 12.26. Let $(L, \leq)$ be a distributive lattice with 0 and 1. Then the subset A of all elements of L that have complements is a sublattice of L.

Proof. Let $a, b \in A$ and let a', b' be complements of a, b respectively. Then

$$a \vee a' = 1, \quad a \wedge a' = 0, \quad b \vee b' = 1, b \wedge b' = 0.$$

Observe that $0' = 1$ and $1' = 0$ so that both 0 and 1 are in A. Now

$$
\begin{aligned}
(a \vee b) \vee (a' \wedge b') &= a \vee (b \vee (a' \wedge b')) \\
&= a \vee (a' \vee b) \qquad \text{(by the above lemma)} \\
&= (a \vee a') \vee b = 1 \vee b = 1
\end{aligned}
$$

and

$$(a \vee b) \wedge (a' \wedge b') = ((a \vee b) \wedge a') \wedge b' = (((a')' \vee b) \wedge a') \wedge b'$$
$$= (a' \wedge b) \wedge b' = a' \wedge (b \wedge b') = a' \wedge 0 = 0.$$

This proves that $a \vee b \in A$. In a similar fashion it can be proved that $(a \wedge b) \vee (a' \vee b') = 1$ and $(a \wedge b) \wedge (a' \vee b') = 0$. Hence $a \wedge b \in A$. This completes the proof that A is a sublattice of L.

12.6. Boolean Algebras

Definition. A complemented and distributive lattice is called a **Boolean algebra.**

Let L be a Boolean algebra. Then L being a complemented lattice, it has both 0 and 1. We have proved that in a distributive lattice with 0 and 1, complement of an element, if it exists, is unique. Thus every element a of L has a unique complement which we write as a'. Thus we obtain a map: $L \to L$ given by $a \to a'$. If $b \in L$ is complement of a', then $a' \vee b = 1$ and $a' \wedge b = 0$. Already $a \vee a' = 1$ and $a \wedge a' = 0$. By the uniqueness of the complement of a', we get $b = (a')'$ (or a'') $= a$. This proves that the complementation map $a \to a'$ is of period 2.

Theorem 12.27 (De Morgan's Laws) If L is a Boolean algebra, then for any a, b in L, $(a \vee b)' = a' \wedge b'$, $(a \wedge b)' = a' \vee b'$.

Proof. Let a, b be elements of L and let a', b' denote respectively the complement of a, b. Then, L being distributive

$$(a \vee b) \vee (a' \wedge b') = ((a \vee b) \vee a') \wedge ((a \vee b) \vee b')$$
$$= ((a \vee a') \vee b) \wedge (a \vee (b \vee b'))$$
$$= (1 \vee b) \wedge (a \vee 1) = 1 \wedge 1 = 1$$

and

$$(a \vee b) \wedge (a' \wedge b') = (a \wedge (a' \wedge b')) \vee (b \wedge (a' \wedge b'))$$
$$= ((a \wedge a') \wedge b') \vee ((b \wedge b') \wedge a')$$
$$= (0 \wedge b') \vee (0 \wedge a') = 0 \vee 0 = 0.$$

Hence $a' \wedge b'$ is the complement of $a \vee b$ or $(a \vee b)' = a' \wedge b'$.

That $a' \vee b'$ is the complement of $a \wedge b$ follows on similar lines and we also have $(a \wedge b)' = a' \vee b'$.

Our aim in the present section is twofold, namely: Every Boolean algebra determines a ring and every Boolean ring determines a Boolean algebra. Secondly, that every finite Boolean algebra is of order 2^n for some $n > 0$. Also that for every $n > 0$ there exists a unique Boolean algebra of order 2^n. First we have a couple of examples.

Example 12.41. Let S be a non-empty set and $p(S)$ be the power set of S. With $\wedge$ and $\vee$ as the usual set intersection and union, $p(S)$ becomes a lattice with ϕ as zero and S as unit (1). For every subset A of S, $A' = S - A$, the usual complement of A in S and $p(S)$ becomes a complemented lattice. Since, for subsets A, B, C of S,

$$A \cap (B \cup C) = (A \cap B) \cup (A \cap C) \quad \text{and} \quad A \cup (B \cap C) = (A \cup B) \cap (A \cup C)$$

and $p(S)$ is distributive. Hence $p(S)$ w.r.t. $\cap$, $\cup$ and $'$ becomes a Boolean algebra.

Example 12.42. Consider the lattice $(D_{30}, |)$, where for $a, b \in D_{30}$, $a|b$ means that a divides b. This is a lattice with 0 and 1 and is a complemented lattice. Since for positive integers a, b, c, $g.c.d.(a, l.c.m.(b, c)) = l.c.m.(g.c.d.(a, b), g.c.d.(a, c))$, this is in particular so if $a, b, c \in D_{30}$. Hence D_{30} is a distributive lattice as well and, so, $(D_{30}, |)$ is a Boolean algebra and its order is $8 = 2^3$.

Example 12.43. Next consider the lattice $(D_{16}, |)$ of all positive divisors of 16. It is a distributive lattice and has a zero and a unit. Since none of the elements 2, 4, 8 has a complement, the lattice D_{16} is not a Boolean algebra.

Example 12.44. In general, if n is a product of distinct primes $p_1, p_2, \ldots, p_m$, then $(D_n, |)$ is a Boolean algebra and the atoms in this Boolean algebra are $p_1, p_2, \ldots, p_m$. It is fairly easy to check that the order of this algebra is $1 + \binom{m}{1} + \binom{m}{2} + \cdots + \binom{m}{m} = 2^m$. Here $\binom{m}{i}$ are binomial coefficients.

Alternatively, we could say that if n is a square free positive integer, then $(D_n, |)$ is a Boolean algebra. On the other hand, if n is a positive integer which is not square free, then although the lattice $(D_n, |)$ is distributive and has 0 and 1, it is not complemented and, so, is not a Boolean algebra. For example, if $n = p^2 m, p$ a prime, then there is no element a in D_n other than n such that $l.c.m. (p, a) = n$ i.e. a unit. This is similar for pm. Therefore neither p nor pm has a complement. As a particular case of this (in addition to the case of $n = 16$ discussed above) we may consider the lattice $(D_{18}, '|')$ of all positive divisors of 18. The elements of D_{18} are 1, 2, 3, 6, 9, 18. The elements 2, 3 are the only atoms and 6, 9 are the only co-atoms in this lattice. The elements 2 and 9 are complements of each other but neither 3 nor 6 has a complement. The complement $3'$ of 3 if it exists has to satisfy $3 \wedge 3' = 1$ and $3 \vee 3' = 18$. But the only non-trivial element in D_{18} the *g.c.d.* of which with 3 is 1 is the element 2. But *l.c.m.*$(2, 3) = 6 \neq 18$. Hence 3 does not have a complement in D_{18}. Similarly for 6 there is no element in D_{18}, except 1 such that *g.c.d.* of this element with 6 is 1. Hence D_{18} is not a Boolean algebra.

Observe that $110 = 2 \times 5 \times 11$ is a square free number and the elements in the lattice D_{110} are 1, 2, 5, 10, 11, 22, 55, 110. We know that $(D_{110}, '|')$ is a distributive lattice. The elements 2, 55 are complements of each other and so are the element 5, 22 and 10, 11. Hence $(D_{110}, '|')$ is a Boolean algebra. The atoms in this lattice are 2, 5, 11, the only co-atom is 55 and its order is $8 = 2^3$.

Let $(B, \wedge, V, ')$ be a Boolean algebra. For $a, b \in B$, define

$$a + b = (a \wedge b') \vee (a' \wedge b) \quad \text{and} \quad a.b = ab = a \wedge b.$$

Thus we get two compositions '+' and '.' in the non empty set B.

Theorem 12.28. If $(B, \wedge, V, ')$ is a Boolean algebra, then B w.r.t. the compositions + and . introduced above is a commutative ring with identity.

Proof. Let $a, b, c \in B$ and $0, 1 \in B$ be respectively the zero and the unit. Then

$$a + b = (a \wedge b') \vee (a' \wedge b) = (b \wedge a') \vee (b' \wedge a) = b + a,$$
$$ab = a \wedge b = b \wedge a = ba.$$

So, both + and . are commutative. Also

$$a + 0 = (a \wedge 0') \vee (a' \wedge 0) = (a \wedge 1) \vee 0 = a \vee 0 = a$$

and $a.1 = a \wedge 1 = a$.

Thus, 0 is additive identity and 1 is multiplicative identity. Again,

$$a + a = (a \wedge a') \vee (a' \wedge a) = 0 \vee 0 = 0.$$

Therefore, a is its own inverse w.r.t. +.

Now

$$(a + b) + c = ((a \wedge b') \vee (a' \wedge b)) + c$$

$$= \{((a \wedge b') \vee (a' \wedge b)) \wedge c'\} \vee \{((a \wedge b') \vee (a' \wedge b))' \wedge c\}$$

$$= \{((a \wedge b') \wedge c') \vee ((a' \wedge b) \wedge c')\} \vee \{(a \wedge b')' \wedge (a' \wedge b)' \wedge c\}$$

$$= \{(a \wedge b' \wedge c') \vee (a' \wedge b \wedge c')\} \vee \{(a' \vee b) \wedge (a \vee b') \wedge c\}$$

$$= \{(a \wedge b' \wedge c') \vee (a' \wedge b \wedge c')\} \vee \{[(a' \wedge (a \vee b'))$$

$$\vee (b \wedge (a \vee b'))] \wedge c\}$$

$$= (a \wedge b' \wedge c') \vee (a' \wedge b \wedge c') \vee \{[((a' \wedge a) \vee (a' \wedge b'))$$

$$\vee ((b \wedge a) \vee (b \wedge b'))] \wedge c\}$$

$$= (a \wedge b' \wedge c') \vee (a' \wedge b \wedge c') \vee \{[(0 \vee (a' \wedge b'))$$

$$\vee ((b \wedge a) \vee 0)] \wedge c\}$$

$$= (a \wedge b' \wedge c') \vee (a' \wedge b \wedge c') \vee \{((a' \wedge b') \vee (a \wedge b)) \wedge c\}$$

$$= (a \wedge b \wedge c) \vee (a \wedge b' \wedge c') \vee (a' \wedge b \wedge c') \vee (a' \wedge b' \wedge c).$$

On giving a cyclic shift to a, b, c, the right-hand side of the above relation remains uncharged. Therefore,

$$(a + b) + c = (b + c) + a = a + (b + c).$$

Thus the associative law for $+$ holds. That the associative law for multiplication holds follows trivially from the definition. Now

$$a(b + c) = a \wedge [(b \wedge c') \vee (b' \wedge c)] = (a \wedge b \wedge c') \vee (a \wedge b' \wedge c).$$

On the other hand

$$ab + ac = \{(a \wedge b) \wedge (a \wedge c)'\} \vee \{(a \wedge b)' \wedge (a \wedge c)\}$$

$$= \{(a \wedge b) \wedge (a' \vee c')\} \vee \{(a' \vee b') \wedge (a \wedge c)\}$$

$$= (a \wedge b \wedge a') \vee (a \wedge b \wedge c') \vee (a' \wedge a \wedge c) \vee (b' \wedge a \wedge c)$$

$$= 0 \vee (a \wedge b \wedge c') \vee 0 \vee (a \wedge b' \wedge c)$$

$$= (a \wedge b \wedge c') \vee (a \wedge b' \wedge c).$$

Hence $a(b + c) = ab + bc$.

This completes the proof that $(B, +, .)$ is a commutative ring with identity.

For every $a \in B, a^2 = a\,a = a \wedge a = a$.

Corollary. $(B, +, .)$ is a Boolean ring.

Recall that a ring R in which $x^2 = x$ for every $x \in R$ is called a **Boolean ring**. Also an element $x \in R$ with $x^2 = x$ is called an **idempotent**.

Lemma. A Boolean ring R is always commutative and $2x = 0$ for every $x \in R$.

Proof. Let $x, y \in R$, where R is a Boolean ring. Then

$$x + y + xy + yx = x^2 + y^2 + xy + yx = (x + y)^2 = x + y.$$

Therefore $xy + yx = 0$.

Taking $x = y$ in this relation and using $x^2 = x$, we get $2x = 0$ and, so, $x = -x$. Using this in the relation $xy + yx = 0$, we get $xy = yx$. Hence the ring is commutative.

We now prove the converse of Theorem 12.28.

Theorem 12.29. A Boolean ring R with identity defines a Boolean algebra.

Proof. We need to define compositions $\wedge, \vee$ and $'$ in R. Let $a, b \in R$. Define

$$a \vee b = a + b - ab \quad \text{and} \quad a \wedge b = ab.$$

Because of the ring R being commutative, we have $a \vee b = b \vee a$ and $a \wedge b = b \wedge a$ for all $a, b \in R$. If 0 denotes the additive identity of R and 1 its identity element, then for every $a \in R$,

$$a \vee 0 = a, \ a \wedge 0 = 0, a \vee 1 = 1, \ a \wedge 1 = a, \quad \text{and} \quad a \vee a = a, a \wedge a = a$$

(a being an idempotent). For $a, b, c \in R$, that $(a \wedge b) \wedge c = a \wedge (b \wedge c)$ is clear. Also

$$(a \vee b) \vee c = (a + b - ab) + c - (a + b - ab)c$$
$$= a + b + c - ab - ac - bc + abc.$$

Giving a cyclic shift to a, b, c does not change the right-hand side of the above relation. Therefore,

$$(a \vee b) \vee c = (b \vee c) \vee a = a \vee (b \vee c).$$

Finally

$$(a \vee b) \wedge a = (a + b - ab)a = a^2 + ab - ab = a,$$
$$(a \wedge b) \vee a = ab + a - aba = a.$$

Hence $(R, \wedge, \vee)$ is a lattice with 0 and 1.

Again, for $a, b, c \in R$,

$$a \vee (b \wedge c) = a \vee (bc) = a + bc - a(bc) = a + bc - abc$$

and

$$(a \vee b) \wedge (a \vee c) = (a + b - ab)(a + c - ac)$$
$$= a^2 + ab - a^2b + ac + bc - abc - a^2c - abc + a^2bc$$
$$= a + bc - abc.$$

Hence

$$a \vee (b \wedge c) = (a \vee b) \wedge (a \vee c).$$

Also

$$a \wedge (b \vee c) = a(b + c - bc) = ab + ac - abc$$
$$= ab + ac - abac = (ab) \vee (ac) = (a \wedge b) \vee (a \wedge c).$$

This proves that the lattice is distributive.

For every $a \in R$,

$$a \vee (1 - a) = a + 1 - a - a(1 - a) = 1 - a + a^2 = 1$$

and

$$a \wedge (1 - a) = a(1 - a) = a - a^2 = a - a = 0.$$

Hence every element a of R has a complement $a' = 1 - a$. This completes the proof that $(R, \vee, \wedge)$ is a Boolean algebra.

Remark. In view of the lemma proved above we could have defined $a \vee b = a + b + ab$ instead of $a \vee b = a + b - ab$. In fact the two are the same.

To get a characterization of finite Boolean algebras, we need a few subsidiary results which we consider now.

Definition. Let B_1 and B_2 be two Boolean algebras. Then a map f from B_1 to B_2 is called a **(Boolean) homomorphism** if f is a lattice homomorphism (i.e. $f(x \vee y) = f(x) \vee f(y)$, $f(x \wedge y) = f(x) \wedge f(y)$ for all $x, y \in B_1$) and $f(x') = (f(x))'$ for all $x \in B_1$.

Theorem 12.30. Let B_1, B_2 be Boolean algebras and $f : B_1 \to B_2$ be a Boolean homomorphism. Then

(a) $f(0) = 0$ and $f(1) = 1$,
(b) for all $x, y \in B_1$ with $x \le y, f(x) \le f(y)$.

Proof. (b) Let $x, y \in B_1$ and $x \le y$. Then $x \vee y = y$. Therefore $f(y) = f(x \vee y) = f(x) \vee f(y)$ which implies that $f(x) \le f(y)$.

(a) Let $x \in B_1$. Then

$$f(0) = f(x \wedge x') = f(x) \wedge f(x') = f(x) \wedge (f(x))' = 0$$

and

$$f(1) = f(x \vee x') = f(x) \vee f(x') = f(x) \vee (f(x))' = 1.$$

Theorem 12.31. Every finite non-zero lattice with 0 has at least one atom.

Proof. Let $(L, \le)$ be finite lattice with 0. Let $a \in L, a \ne 0$. If a is an atom, we are through. If a is not an atom, there exists an element a_1 in L such that $a > a_1 > 0$. If a_1 is an atom, we are through. If a_1 is not an atom, there exists an element a_2 in L such that $a > a_1 > a_2 > 0$. We continue this process which comes to an end only when we arrive at an atom. Since L is finite, this process must terminate after a finite number of steps. Hence L has at least one atom.

Theorem 12.32. If L is a distributive lattice with 0, 1 and $a, b \in L$ and b' is complement of b with $a \wedge b' = 0$, then $a \le b$.

Proof. $a \wedge b' = 0$ together with the distribute law implies

$$b = 0 \vee b = (a \wedge b') \vee b = (a \vee b) \wedge (b' \vee b) = (a \vee b) \wedge 1$$

$$= a \vee b.$$

Therefore $a \le b$, as $a \vee b$ is the least upper bound of a, b.

Lemma. If a, b are distinct atoms in a lattice L with 0, 1, then $a \wedge b = 0$.

Proof. If $a \wedge b \ne 0$, then $a \ge a \wedge b > 0$ so that either $a > a \wedge b$ or $a = a \wedge b$. If $a > a \wedge b > 0$, we get a contradiction to a being an atom. If $a = a \wedge b$, then $b > a > 0$ (a, b being distinct) and we get a contradiction to b being an atom. Hence $a \wedge b = 0$.

Theorem 12.33. Let $(A, \vee, \wedge)$ be a finite Boolean algebra. Then every element of A can be uniquely represented as join of a finite number of atoms.

Proof. Let b be any element of A and $a_1, a_2, \ldots, a_m$ be all the atoms of A with $a_i \leq b, i = 1, 2, \ldots, m$

(a) Since $a_i \leq b$, for $i = 1, 2, \ldots, m$, $a_1 \vee a_2 \vee \ldots \vee a_m \leq b$.
For the sake of notational convenience, let $a_1 \vee a_2 \vee \ldots \vee a_m = c$. Then we have

$$c \leq b. \tag{12.16}$$

Suppose $b \wedge c' \neq 0$. Then there exists an atom a in A with $a \leq b \wedge c' \leq b$. Since $a_1, a_2, \ldots, a_m$ are all the atoms in A which are $\leq b$, a is one of $a_1, a_2, \ldots, a_m$. Let $a = a_i$. Again, $a \leq b \wedge c' \leq c'$. So that $a_i \leq c'$. Also $a_i \leq c$. Therefore, $a_i \leq c \wedge c' = 0$ or $a_i = 0$, which is a contradiction a_i being an atom. Hence it follows that $b \wedge c' = 0$. Theorem 12.32 then shows that

$$b \leq c. \tag{12.17}$$

Combining (12.16) and (12.17), we get $b = c$. Thus b is join of all the atoms in A which are $\leq b$.
(b) Now, suppose that we also have

$$b = b_1 \vee b_2 \vee \ldots \vee b_k \tag{12.18}$$

where $b_1, b_2, \ldots, b_k$ are certain atoms in A. It is then immediate from (12.18) that $b_j \leq b$ for every $j = 1, 2, \ldots, k$. Then $\{b_1, b_2, \ldots, b_k\}$ is a subset of $\{a_1, a, \ldots, a_m\}$. In particular $k \leq m$ and every b_i equals some a_j. For any $i, 1 \leq i \leq m, a_i \leq b$ so that

$$a_i = a_i \wedge b = a_i \wedge (b_1 \vee b_2 \vee \ldots b_k)$$
$$= (a_i \wedge b_1) \vee (a_i \wedge b_2) \vee \ldots \vee (a_i \wedge b_k). \tag{12.19}$$

If $b_1, b_2, \ldots, b_k$ are all different from a_i, then $a_i \wedge b_j = 0$ and the r.h.s. of (12.19) is the join of k elements every one of which is 0. Then, we get $a_i = 0$ which is a contradiction. Hence there exists a $j, 1 \leq j \leq k$ such that $a_i = b_j$. This proves that $\{a_1, a_2, \ldots, a_m\}$ is a subset of $\{b_1, b_2, \ldots, b_k\}$. Therefore $m \leq k$ and so $k = m$ and $\{b_1, b_2, \ldots, b_m\} = \{a_1, a_2, \ldots, a_m\}$. Hence the representation of b as a join of atoms is unique (except for the order of atoms).

Let $(B, \vee, \wedge, ')$ be a finite Boolean algebra and S be the set of all atoms in B. If $B \neq \{0\}$, there is at least one atom in B and $S \neq \emptyset$. Let $p(S)$ be the power set of S. We have already proved that $(p(S), \cap, \cup, ')$ is a finite Boolean algebra and cardinality of $p(S) = 2^m$ if m is the number of all the atoms in B. For a $b \in B$, let $A(b) = \{a \in S | a \leq b\}$ which is a subset of S. If $b \neq 0$, then $A(b) \neq \phi$ and $A(0) = \phi$. Define a map

$$f : B \to p(S) \text{ by } f(b) = A(b), b \in B.$$

Theorem 12.34. The map f is Boolean algebra isomorphism.

Proof. Let $b, c \in B$. For any atom a in B,

(a) $a \leq b \vee c$ if and only if $a \leq b$ or $a \leq c$ which is so if and only if $a \in A(b)$ or $a \in A(c)$ i.e. if and only if $a \in A(b) \cup A(c)$. Therefore, $A(b \vee c) = A(b) \cup A(c)$ and so $f(b \vee c) = f(b) \cup f(c)$.

(b) $a \leq b \wedge c$ if and only if $a \leq b$ and $a \leq c$ which is so if and only if $a \in A(b)$ and $a \in A(c)$ i.e. if and only if $a \in A(b) \cap A(c)$. Therefore, $A(b \wedge c) = A(b) \cap A(c)$ and so $f(b \wedge c) = f(b) \cap f(c)$.

(c) $a \leq b'$ if and only if $a \wedge (b')' = 0$ or $a \wedge b = 0$ which happens if and only if $a \not\leq b$ i.e. if and only if $a \in S - A(b)$. Therefore, $f(b') = S - A(b) = A(b)'$ the complement of $A(b)$ in S. This completes the proof f is a Boolean homomorphism. Let A be any subset of S and $A = \{a_1, a_2, \ldots, a_k\}$. Take $b = a_1 \vee a_2 \vee \ldots \vee a_k$. It follows from Theorem 12.33 that $a_1, a_2, \ldots, a_k$ are all atoms in B which are $\leq b$. Hence $A = A(b) = f(b)$. Therefore the homomorphism f is ontò. Let b, c be element B such that $f(b) = f(c)$. Then $A(b) = A(c)$. Let $A(b) = A(c) = \{a_1, a_2, \ldots, a_r\}$. Again, it follows from Theorem 12.33 that $b = a_1 \vee a_2 \vee \ldots \vee a_r$ and $c = a_1 \vee a_2 \vee \ldots a_r$. Therefore $b = c$ and the map f is one–one. Hence f is an isomorphism.

Corollary. The cardinality or order of the finite lattice B is 2^m where m is the number of all the atoms in B.

Let S and T be two finite sets of order m and let $\propto : S \to T$ be a one–one, onto map from S to T. Set $2^m = n$ and let $A_1, \ldots, A_n$ be all the possible subsets of S. For every i, $1 \leq i \leq n$, let $a(A_i) = B_i$. Then $B_1, B_2, \ldots, B_n$ are all the subsets of T and it is clear that for

$$1 \leq i \leq j \leq n, a(A_i \cup A_j) = B_i \cup B_j \text{ and } a(A_i \cap A_j) = B_i \cap B_j.$$

Also

$$a(A_i') = a(S - A_i) = T - a(A_i) = T - B_i = B_i'.$$

We have thus proved.

Theorem 12.35. α induces an isomorphism from the Boolean algebra $(p(S), \cap, \cup,')$ to the Boolean algebra $(p(T), \cap, \cup,')$.

Corollary. Up to isomorphism, there is only one finite Boolean algebra of order 2^m.

12.7. Boolean Polynomials and Boolean Functions

Definition. Let A and L be two lattices. In the Cartesian product $A \times L$ of A by L which is the set of all ordered pairs (a, x), where $a \in A$ and $x \in L$, define binary operations $\vee, \wedge$ by:

$$(a, x) \vee (b, y) = (a \vee b, x \vee y),$$

$$(a, x) \wedge (b, y) = (a \wedge b, x \wedge y).$$

It is fairly easy to check that $A \times L$ becomes a lattice. Moreover, if both A and L are distributive lattices, then so is $A \times L$. In case both A and L are complemented lattices, we may define a unary operation $'$ on $A \times L$ by $(a, x)' = (a', x'), a \in A, x \in L$ and $A \times L$ becomes a complemented lattice. Observe that $(0, 0)$ and $(1, 1)$ are respectively the zero element and the unit element of $A \times L$. The lattice $A \times L$ is called the direct sum of the lattices A and L. When A and L are Boolean algebras, then $A \times L$ is a Boolean algebra called the **direct sum** of the Boolean algebras A and L.

We can define direct sum of any finite number of lattices or Boolean algebras similarly. If $A_1, A_2, \ldots, A_n$ are lattices (Boolean algebras), instead of taking ordered pairs, we now take ordered n-tuples $(a_1, a_2, \ldots, a_n)$ of elements where $a_i \in A_i, 1 \leq i \leq n$ and define $\vee$ and $\wedge$ by

$$(a_1, a_2, \ldots, a_n) \vee (b_1, b_2, \ldots, b_n) = (a_1 \vee b_1, a_2 \vee b_2, \ldots, a_n \vee b_n),$$

$$(a_1, a_2, \ldots, a_n) \wedge (b_1, b_2, \ldots, b_n) = (a_1 \wedge b_1, a_2 \wedge b_2, \ldots, a_n \wedge b_n).$$

In case of Boolean algebras, we also define

$$(a_1, a_2, \ldots, a_n)' = (a_1', a_2', \ldots, a_n').$$

With these operations $A_1 \times A_2 \times \cdots \times A_n$ becomes a lattice (Boolean algebra) and is called the **direct sum of the lattices (Boolean algebras)** $A_1, A_2, \ldots, A_n$. In case $A_i = A$ for every i, $1 \leq i \leq n$, we write A^n for the direct sum $A \times A \times \cdots \times A$ of n of copies of A.

Let A be a Boolean algebra and $n \geq 2$ be an integer. Then any function (map) from A^n to A is called a **Boolean function** from A^n to A. If A is a finite Boolean algebra of order m (say), then as already seen for arbitrary finite sets, the number of Boolean functions from A^n to A is $o(A)^{o(A^n)} = m^{m^n}$.

Let us recall that if p and q are two propositions each of which takes two possible values namely true (T) or false (F), then $p \vee q$, $p \wedge q$ and p' (the negation of p) take values which are given by the truth tables

p	q	$p \vee q$	$p \wedge q$		p	p'
F	F	F	F		F	T
F	T	T	F		T	F
T	F	T	F			
T	T	T	T			

Instead of taking propositions and their join, meet and negation, we can take these operations in the set $\{F,T\}$. As is clear from the above tables, we have

$$F \vee F = F, \quad F \vee T = T \vee F = T \vee T = T, F \wedge F = F \wedge T = T \wedge F = F,$$

$$T \wedge T = T, \quad F' = T, \quad T' = F.$$

We may simplify our notations still further by replacing F by 0 and T by 1 and also taking $+$ for $\vee$ and for $\wedge$.

We then have a set $\{0, 1\}$ with operations

$$0 + 0 = 0, \quad 0 + 1 = 1 + 0 = 1 + 1 = 1, \quad 0.0 = 0.1 = 1.0 = 0,$$

$$1.1 = 1 \quad \text{and} \quad 0' = 1, \quad 1' = 0.$$

It is clear that $\{F, T\}$ with $\vee, \wedge$ and $'$ becomes a Boolean algebra. Hence $B = \{0, 1\}$ w.r.t. $+,$. and $'$ is a Boolean algebra. Observe that the elements $0, 1$ of B are not binary numbers as for binary numbers $1 + 1 = 0$ but here $1 + 1 = 1$.

Throughout this section we use B for the Boolean algebra $\{0,1\}$ with operations as defined above.

Let $X = \{x_1, x_2, \ldots, x_n\}$ be a set of n symbols called variables or indeterminates. We also call a symbol x a Boolean variables if x takes values only from B. Any function or map from B^n to B is called a **Boolean function** of degree n. The order of B^n being 2^n, there one exactly 2^{2^n} Boolean functions.

Boolean expressions or **Boolean polynomials** in the variables $x_1, x_2, \ldots, x_n$ are defined recursively by

1. $0, 1, x_1, \ldots, x_n$ are Boolean expressions.
2. If e, e_1, e_2 are Boolean expressions or polynomials, then $e_1 + e_2, e_1 e_2$ and e' are also Boolean expressions.

Let E_n or $E_n(x_1, x_2, \ldots, x_n)$ denote the set of all Boolean expressions. Observe that any Boolean expression in the variables $x_1, x_2, \ldots, x_m$ can also be regarded as a Boolean expression in the variables $x_1, x_2, \ldots, x_m, x_{m+1}$. Therefore we have $E_n \subseteq E_{n+1}$ for every $n \geq 1$.

Given a Boolean expression $e(x_1, x_2, \ldots, x_n)$, let us give the values $b_1, b_2, \ldots, b_n$ to $x_1, x_2, \ldots, x_n$ respectively, where $b_i \in B$, $1 \leq i \leq n$. There then results, on applications of Boolean algebra properties of B, an element $e(b_1, \ldots, b_n)$ of B. Thus every Boolean expression gives rise to a Boolean function. We shall soon prove the converse that every Boolean function arises from a Boolean expression.

Two Boolean expressions e_1 and e_2 in $x_1, x_2, \ldots, x_n$ are said to be equivalent if they give rise to the same Boolean function i.e. if and only if $e_1(b_1, b_2, \ldots, b_n) = e_2(b_1, b_2, \ldots, b_n)$ for all values of $b_1, b_2, \ldots, b_n$ in B.

Example 12.45. We determine the Boolean functions determined by the Boolean expressions $xy, xy', x'y, x'y'$. The functions are given by the table below

x	y	x'	y'	xy	xy'	$x'y$	$x'y'$
0	0	1	1	0	0	0	1
0	1	1	0	0	0	1	0
1	0	0	1	0	1	0	0
1	1	0	0	1	0	0	0

Example 12.46. There are 2^2 Boolean functions of degree 1 namely the functions from B to B. There is one function in which both the elements of B get mapped onto o, two functions under which one element gets mapped onto o and the other onto 1 and one function under which both the elements of B get mapped onto 1. The four functions are given by the table

x	f_1	f_2	f_3	f_4
0	0	0	1	1
1	0	1	0	1

The functions f_1 to f_4 are represented by the expressions xx', x, x', $x + x'$ respectively.

Example 12.47. The Boolean algebra B^2 has four elements namely $(0,0), (0,1), (1,0)$, and $(1,1)$. Since each one of these elements has two possible images, there are 2^4 possible functions from B^2 to B. There is one function in which every element of B^2 gets mapped onto 0, four functions in which exactly one element of B^2 gets mapped onto 1 and the other three get mapped onto 0, six functions in which two elements gets mapped onto 0 and the other two get mapped onto 1, three maps under which one element gets mapped onto 0 and other three get mapped onto 1 and finally one map under which every element of B^2 gets mapped onto 1. These 16 maps are given in tabular form by the table

x	y	f_1	f_2	f_3	f_4	f_5	f_6	f_7	f_8	f_9	f_{10}	f_{11}	f_{12}	f_{13}	f_{14}	f_{15}	f_{16}
0	0	0	0	0	0	1	0	0	1	0	1	1	0	1	1	1	1
0	1	0	0	0	1	0	0	1	0	1	0	1	1	0	1	1	1
1	0	0	0	1	0	0	1	0	0	1	1	0	1	1	0	1	1
1	1	0	1	0	0	0	1	1	1	0	0	0	1	1	1	0	1

Observe that the function f_1 is represented by the Boolean expression $xx' + yy'$ and f_2, f_3, f_4 and f_5 are represented respectively by the expressions $xy, xy', x'y, x'y'$. Then the functions of f_6 to f_{16} are represented by the expressions. $xy + xy', xy + x'y, xy + x'y', xy' + x'y, xy' + x'y', x'y + x'y', xy + xy' + x'y, xy + xy' + x'y', xy + x'y + x'y', x'y + xy' + x'y'$, and $xy + xy' + x'y + x'y'$ respectively.

Example 12.48. By simplifying a Boolean expression we mean to manipulate with them using the laws of Boolean algebra for the variables $x_1, x_2, \ldots, x_n$ and $x'_1, x'_2, \ldots, x'_n$. We consider, for example, some of the expressions obtained in the last example. We find that

$$xy + xy' = x(y + y') = x \cdot 1 = x,$$

$$xy + x'y = (x + x')y = 1 \cdot y = y,$$

$$xy' + x'y' = (x + x')y' = 1 \cdot y' = y',$$

$$x'y + x'y' = x'(y + y') = x' \cdot 1 = x',$$

$$xy + xy' + x'y = x(y + y') + x'y = x \cdot 1 + x'y = x + x'y,$$

$$xy + xy' + x'y' = x(y + y') + x'y' = x \cdot 1 + x'y' = x + x'y',$$

$$= x \cdot 1 + x'y' = x(y + y') + x'y'$$

$$= xy + xy' + x'y' = xy + (x + x')y' = y' + xy,$$

$$xy + x'y + x'y' = (x + x')y + x'y' = 1 \cdot y + x'y' = y + x'y',$$

$$xy' + x'y + x'y' = xy' + x'(y + y') = xy' + x' \cdot 1 = x' + xy',$$

$$xy + xy' + x'y + x'y' = x(y + y') + x'(y + y') = x \cdot 1 + x' \cdot 1$$

$$= x + x' = 1.$$

It is, then, clear from the above computations that the pairs of expressions

$xy + xy'$ and x;	$xy + x'y$ and y,
$xy' + x'y'$ and y';	$x'y + x'y'$ and x',
$xy + xy' + x'y$ and $x + x'y$;	$xy + xy' + x'y'$ and $x + x'y'$,
$xy + xy' + x'y'$ and $y' + xy$;	$xy + x'y + x'y'$ and $y + x'y'$,
$xy' + x'y + x'y'$ and $x' + xy'$;	$xy + xy' + x'y + x'y'$ and 1

are equivalent. The above also suggests that an expression may be simplified in more than one ways. In addition to the expression $xy' + xy + x'y'$ being equivalent to $x + x'y'$ as also to $y' + xy$, we find that $xy + xy' + x'y$ is equivalent to $x + x'y$ as also to $y + xy'$. Moreover, $xy + x'y + x'y'$ is equivalent to $y + x'y'$ as also to $x' + xy$ and $xy' + x'y + x'y'$ is equivalent to $x' + xy'$ as also to $y' + x'y$.

Example 12.49. Simplify the Boolean expressions

(a) $xy + x'yz' + yz$
(b) $(xy' + z)(x + y')z$
(c) $xy + xy'z + yz$

Solution. For (a) we have

$$xy + x'yz' + yz$$

$$= xy + (x'z')y + yz = (x + x'z')y + yz$$

$$= (x + x'z')y + zy$$

$$= (x + z + x'z')y = (x \cdot 1 + z + x'z')y = (x(z + z') + z + x'z')y$$

$$= (xz + xz' + z + x'z')y = (xz + (x + x')z' + z)y = (xz + z' + z)y$$
$$= (1 + xz)y = y + xyz.$$

For (b) we have

$$(xy' + z)(x + y')z$$
$$= (xy'x + xz + xy'y' + y'z)z$$
$$= (xy' + xz + xy' + y'z)z = (xy' + xz + y'z)z$$
$$= xy'z + xzz + y'zz$$
$$= xy'z + xz + y'z = xz(y' + 1) + y'z = xz(y' + y + y') + y'z$$
$$= xz(y + y') + y'z = xz \cdot 1 + y'z = xz + y'z.$$

For (c) we have

$$xy + xy'z + yz = xy + (xy' + y)z = xy + (xy' + (x + x')y)z$$
$$= xy + [x(y' + y) + x'y]z = xy + (x + x'y)z$$
$$= xy + xz + x'yz.$$

Example 12.50. The Boolean algebra B^3 has $2^3 = 8$ elements and then $2^8 = 256$ Boolean functions. Consider a Boolean function $f : B^3 \to B$ in which

$(0, 0, 0)$ is mapped onto 1, $(0, 0, 1)$ is mapped onto 0,
$(0, 1, 0)$ is mapped onto 1, $(1, 0, 0)$ is mapped onto 0,
$(0, 1, 1)$ is mapped onto 1, $(1, 0, 1)$ is mapped onto 0,
$(1, 1, 0)$ is mapped onto 1, $(1, 1, 1)$ is mapped onto 0.

Let us represent the above function along with two other functions in tabular form as under:

x	y	z	f	g	h
0	0	0	1	1	0
0	0	1	0	1	0
0	1	0	1	0	1
1	0	0	0	1	1
0	1	1	1	0	0
1	0	1	0	1	1
1	1	0	1	0	0
1	1	1	0	0	1

There is only one Boolean expression namely $x'y'z'$ which takes the value 1 at $(0,0,0)$ and at no other element of B^3, the only Boolean expression that takes the value 1 at $(0,1,0)$ and at no other point of B^3 is $x'yz'$; the one and only expression that takes the value 1 at $(0,1,1)$ and at no other point of B^3 is $x'yz$. The only expression that takes the value 1 at $(1,1,0)$ is xyz'. Then the expression

$$x'y'z' + x'yz' + x'yz + xyz' \qquad (12.20)$$

is such that it takes the value 1 at every one of the points $(0,0,0), (0,1,0), (0,1,1), (1,1,0)$ and takes the value 0 at the remaining four points. Hence the Boolean function f is represented by the Boolean expression (12.20). Similarly we find that the functions g and h are represented, respectively, by the Boolean expressions:

$$x'y'z' + x'y'z + xy'z' + xy'z \quad \text{and} \quad x'yz' + xyz' + xy'z + xyz.$$

Definition. Let $x_1, x_2, \ldots, x_n$ be n Boolean variables. A Boolean expression $y_1 y_2 \ldots y_n$, where for every $i, 1 \leq i \leq n, y_i = x_i$ or x'_i, is called a **minterm**. Also an expression which is the sum of certain minterms is called **a disjunctive normal form**. On the other hand an expression $y_1 + y_2 + \cdots + y_n$, where $y_1 = x_1$ or x'_1, $y_2 = x_2$ or x'_2, $\ldots, y_n = x_n$ or x'_n, is called a **maxterm** in the variables $x_1, x_2, \ldots, x_n$. An expression which is a product of some maxterms is called a **conjunctive normal form**.

We have discussed earlier minterm, disjunctive normal form, maxterm and conjunctive normal form for a given collection of subsets of a set. The discussion there is a particular case of the discussion here. Observe that there are 2^n minterms (maxterms) in n variables and $2^{2^n} - 1$ disjunctive (conjunctive) normal forms. A minterm $y_1 y_2 \ldots y_n$, takes the value 1 if and only if $y_i = 1$ for every i. This occurs if and only if $x_i = 1$ when $y_i = x_i$ and $x_i = 0$ when $y_i = x'_i$. Thus *a minterm gets the value 1 for one and only one combination of values of the variables.*

Theorem 12.36. Every Boolean function arises from a Boolean expression.

Proof. Let f be a Boolean function from B^n to B. Let $z_1, z_2, \ldots, z_t$ be elements of B^n such that $f(z_1) = f(z_2) = \cdots = f(z_t) = 1$ and $f(z) = 0$ for every other element of B^n. Choose an $i, 1 \leq i \leq t$. Now z_i being an element of B^n is an n-tuple over B. Suppose that the non-zero entries of z_i occur in the positions $k_1 < k_2 < \cdots < k_m$. The minterm $y_1 y_2 \ldots y_n$, where $y_i = x_i$ for $j = k_1, k_2, \ldots, k_m$ and $y_j = x'_j$ for other values of j, is such that it takes the value 1 for z_i. Do this for every $i, 1 \leq i \leq t$ and take the disjunctive normal form which is the sum of these t minterms. Then the function f arises from this disjunctive form.

Remark. Indeed the disjunctive normal form giving rise to the function f is called the **disjunctive normal form of the function** f.

Theorem 12.37. Every Boolean function from B^n to B arises from a conjunctive normal form.

Example 12.51. Find the sum-of-products expansion or the disjunctive normal form of the Boolean functions

(a) $f(x, y, z) = (x + z)y$; (b) $g(x, y, z) = xy'$.

Solution. There are two ways that a disjunctive normal form of a Boolean function may be obtained. We illustrate that for the two examples.

(a)
$$f(x, y, z) = (x + z)y = xy + zy = xy + yz = xy1 + 1yz$$
$$= xy(z + z') + (x + x')yz = xyz + xyz' + xyz + x'yz$$
$$= xyz + xyz + xyz' + x'yz = xyz + xyz' + x'yz.$$

(b) $g(x, y, z) = xy' = xy' \cdot 1 = xy'(z + z') = xy'z + xy'z'$.
Alternatively, we consider the table:

x	y	z	$x + z$	$(x + z)y$	y'	xy'
0	0	0	0	0	1	0
0	0	1	1	0	1	0
0	1	0	0	0	0	0
1	0	0	1	0	1	1
0	1	1	1	1	0	0
1	0	1	1	0	1	1
1	1	0	1	1	0	0
1	1	1	1	1	0	0

Therefore

$$f(x,y,z) = x'yz + xyz' + xyz \quad \text{and}$$
$$g(x,y,z) = xy'z' + xy'z.$$

Example 12.51. Find the disjunctive normal form or the sum-of-products expansion of the Boolean function $f(w,x,y,z)$ that has the value 1 if and only if an odd number of w, x, y, z have the value 1.

Solution. By the given condition $f(w,x,y,z)$ has the value 1 if and only if one or three of w, x, y, z have the value 1. The minterms which take the value 1 when

(a) one of w, x, y, z have the value 1 are

$$wx'y'z', w'xy'z', w'x'yz' \quad \text{and} \quad w'x'y'z;$$

(b) three of w, x, y, z have the value 1 are

$$w'xyz, wx'yz, wxy'z \quad \text{and} \quad wxyz'.$$

Hence the disjunctive normal form of $f(w,x,y,z)$ is

$$wx'y'z' + w'xy'z' + w'x'yz' + w'x'y'z + w'xyz + wx'yz + wxy'z + wxyz'.$$

Example 12.52. Let

(a) $E(x,y,z) = xy + xz + y'z$,
(b) $E(x,y,z) = \{(x+y)' + (x'z)\}'$.

Write $E(x,y,z)$ in both disjunctive and conjunctive normal form.

Solution.

(a)
$$E(x,y,z) = xy + xz + y'z = xy\cdot 1 + x\cdot 1z + 1\cdot y'z$$
$$= xy(z + z') + x(y + y')z + (x + x')y'z$$
$$= xyz + xyz' + xyz + xy'z + xy'z + x'y'z$$
$$= xyz + xyz' + xy'z + x'y'z$$

which is disjunctive normal form of $E(x,y,z)$

(b) $E(x,y,z) = \{(x+y)' + (x'z)\}' = \{x'y' + x'z\}' = (x'y')'(x'z)'$
$$= (x+y)(x+z') = xx + xy + xz' + yz'$$

$$= x + xy + xz' + yz'$$

$$= x \cdot 1 \cdot 1 + xy \cdot 1 + x \cdot 1 \cdot z' + 1 \cdot yz'$$

$$= x(y + y')(z + z') + xy(z + z') + x(y + y')z' + (x + x')yz'$$

$$= x(yz + yz' + y'z + y'z') + xyz + xyz' + xyz' + xy'z'$$
$$\quad + xyz' + x'yz'$$

$$= xyz + xyz' + xy'z + xy'z' + xyz + xyz' + xyz' + xy'z'$$
$$\quad + xyz' + x'yz'$$

$$= (xyz + xyz) + (xyz' + xyz' + xyz' + xyz') + xy'z$$
$$\quad + (xy'z' + xy'z') + x'yz'$$

$$= xyz + xyz' + xy'z + xy'z' + x'yz'.$$

Thus the disjunctive normal form of $E(x, y, z)$ is determined.

(b) We next find the conjunctive normal forms of the two $E(x, y, z)$. In this case the $E(x, y, z)$ have to be expressed as product of maxterms. Consider a maxterm $y = y_1 + y_2 + \cdots + y_n$ in the n variables $x_1, x_2, \ldots, x_n$ so that for any $i, 1 \leq i \leq n$, either $y_i = x_i$ or $y_i = x'_i$. Observe that maxterm $y = 0$ if and only if $y_i = 0$ for every $i = 1, 2, \ldots, n$. This is so if and only if $x_i = 0$ when $y_i = x_i$ and $x_j = 1$ when $y_j = x'_j$. Thus, there is exactly one n-tuple $(a_1, a_2, \ldots, a_n) \in B^n$ at which the maxterm y takes the value 0. On the other hand, given any n-tuple $a = (a_1, a_2, \ldots, a_n) \in B^n$, there exists exactly one maxterm $y = y_1 + y_2 + \cdots + y_n$ which takes the value 0 at a. In fact, if $a_{i_1} = a_{i_2} = \cdots = a_{i_k} = 1$ and $a_j = 0$ for every other j, then $y_{i_1} = x'_{i_1}, \ldots, y_{i_h} = x'_{i_k}$ and $y_j = x_j$ for every other j.

Given a Boolean expression $e(x_1, x_2, \ldots, x_n)$ find all n-tuples $a^{(1)}$, $a^{(2)}, \ldots, a^{(k)}$ at which e takes the value 0. Then find the maxterms $y^{(1)}, y^{(2)}, \ldots, y^{(k)}$ such that $y^{(i)}$ takes the value 0 at $a^{(i)}$. Then the product $y^{(1)}y^{(2)} \ldots y^{(k)}$ is a conjunctive normal form which takes the value 0 at exactly k n-tuples $a^{(1)}, \ldots, a^{(k)}$. Since $e(x_1, x_2, \ldots, x_n)$ and $y^{(1)}y^{(2)} \ldots y^{(k)}$ take identical values at every element of B^n, $y^{(1)}y^{(2)} \ldots y^{(k)}$ is the conjunctive normal form of the Boolean expression $e(x_1, x_2, \ldots, x_n)$.

Now we come back to the two Boolean expressions. First we find the values that these expressions take at the elements of B^3.

x y z	x' y' z'	xy xz $y'z$	$xy + xz + y'z$	$(x+y)'(x'z)$	$\{(x+y)' + x'z\}'$
0 0 0	1 1 1	0 0 0	0	1	0	0
0 0 1	1 1 0	0 0 1	1	1	1	0
0 1 0	1 0 1	0 0 0	0	0	0	1
1 0 0	0 1 1	0 0 0	0	0	0	1
0 1 1	1 0 0	0 0 0	0	0	1	0
1 0 1	0 1 0	0 1 1	1	0	0	1
1 1 0	0 0 1	1 0 0	1	0	0	1
1 1 1	0 0 0	1 1 0	1	0	0	1

The above table shows that the expression $e(x, y, z) = xy + xz + y'z$ takes the value 0 on the elements $(0, 0, 0)$, $(0, 1, 0)$, $(1, 0, 0)$ and $(0, 1, 1)$ of B^3. The maxterms corresponding to these elements are $x + y + z, x + y' + z, x' + y + z$ and $x + y' + z'$. Therefore the conjunctive normal form of the expression is

$$(x + y + z)(x + y' + z)(x' + y + z)(x + y' + z').$$

The zeros of the other Boolean expression are $(0, 0, 0)$, $(0, 0, 1)$ and $(0, 1, 1)$. The maxterms corresponding to these elements of B^3 are $x + y + z, x + y + z', x + y' + z'$ and, therefore, the conjunctive normal form of $((x + y)' + (x'z))'$ is

$$(x + y + z)(x + y + z')(x + y' + z').$$

Exercise. Verify by Boolean algebra computations that the conjunctive normal form of the two expressions in the above example are indeed the ones obtained.

Remark. The method of obtaining conjunctive normal form of a Boolean expression explained in the last example is a bit involved. It involves writing the truth table for the expression which can involve a lot of work if the number of Boolean variables increases. There is a simpler method of obtaining conjunctive normal form of an expression from its disjunctive normal form. Recall that we simplify or transform Boolean expression by using the Boolean algebra properties which are $(xy)' = x'+y', (x+y)' = x'y'$ and $(x')' = x$, where x, y are variables. In addition to these laws being applied to variables, they may also be applied to expressions. Given an expression $E(x_1, x_2, \ldots, x_n) = E$ (say), using the law of double negation, the expression E is equivalent to $E'' = (E')'$. Simplify E' by using the De Morgan's laws and the law of negation. Once this is done, apply the law of negation to E' to get an expression which is equivalent to E'' and hence to E.

We apply this procedure to the two expressions of Example 12.52.

Example 12.53. First consider the expression $E(x, y, x) = xy + xz + y'z$, the disjunctive normal form of which is $xyz + xyz' + xy'z + x'y'z$. We write $p \sim q$ to indicate that the expression p is equivalent to the expression q (This relation is clearly symmetric; indeed it is an equivalence relation). Now

(a) $E(x, y, z) \sim xyz + xyz' + xy'z + x'y'z$
$\sim ((xyz + xyz' + xy'z + x'y'z)')'$
$\sim ((x' + y' + z')(x' + y' + z)(x' + y + z')(x + y + z'))'$
$\sim \{(x' + x'y' + x'z + y' + y'z + x'z' + y'z')(x'y + x'z' + y + yz' + xy + xz' + yz' + z')\}'$
$\sim \{(x' + y' + x'y')(z' + y + z'y)\}'$
$\sim \{z'x' + x'y + x'yz' + y'z' + x'y'z'\}'$
$\sim \{x'(y + y')z' + x'y(z + z') + x'yz' + (x + x')y'z' + x'y'z'\}'$
$\sim \{x'yz' + x'y'z' + x'yz + xy'z'\}'$
$\sim (x + y' + z)(x' + y + z)(x + y' + z')(x + y + z)$

which is the required conjunctive normal form of $xy + xz + y'z$.

(b) $E(x, y, z) = \{(x + y)' + (x'z)\}' \sim (x'y' + x'z)'$
$\sim \{x'y'(z + z') + x'(y + y')z\}'$
$\sim \{x'y'z + x'y'z' + x'yz + x'y'z\}'$
$\sim (x + y + z')(x + y + z)(x + y' + z')(x + y + z')$
$\sim (x + y + z)(x + y + z')(x + y' + z')$

which is the required conjunctive normal form.

The simplified method of obtaining conjunctive normal form used for the second expression is self explanatory and could also be used for the first expression.

Example 12.54. Find a Boolean polynomial $E(x, y, z)$ that induces the function f given by

x	y	z	$f(x, y, z)$
0	0	0	1
0	0	1	0
0	1	0	1
0	1	1	0
1	0	0	1
1	0	1	0
1	1	0	0
1	1	1	0

Solution. The function f takes the value 1 at the points $(0, 0, 0), (0, 1, 0)$ and $(1, 0, 0)$ of B^3. The minterms that take the value 1 at these elements of B^3 are $x'y'z', x'yz'$ and $xy'z'$. Therefore, the disjunctive normal form of the polynomial representing the function f is

$$E(x, y, z) = xy'z' + x'yz' + x'y'z' \sim xy'z' + x'(y + y')z'$$
$$\sim xy'z' + x'z'.$$

Also

$$E(x, y, z) = xy'z' + x'yz' + x'y'z' \sim (x + x')y'z' + x'yz'$$
$$\sim y'z' + x'yz'.$$

The conjunctive normal form of this polynomial can also be obtained as below:

$$E(x, y, z) \sim xy'z' + x'z'$$

$$\sim \{(xy'z' + x'z')'\}'$$

$$\sim \{(x' + y + z)(x + z)\}'$$

$$\sim \{x'z + xy + yz + xz + z\}'$$

$$\sim \{x'(y + y')z + xy(z + z') + (x + x')yz + x(y + y')z + (x + x')(y + y')z\}'$$

$$\sim \{x'yz + x'y'z + xyz + xyz' + xy'z + (xy + xy' + x'y + x'y')z\}'$$

$$\sim \{x'yz + x'y'z + xyz + xyz' + xy'z\}'$$

$$\sim (x + y' + z')(x + y + z')(x' + y' + z)(x' + y + z')(x' + y' + z').$$

12.8. Switching (or Logical) Circuits

One of the most important and oldest applications of lattice theory is the use of Boolean algebra in modeling and simplifying switching or relay circuits. The main aspect of switching circuits is to describe electrical or electronic switching circuits as a mathematical model which is a diagram of a circuit with given properties.

These days, instead of electrical switches, it is certain types of electronic blocks called semiconductor elements that are predominantly used in the logical design of digital building components of electronic computers. Then the switches are represented as what are commonly called **gates or combination of gates** and the representation is called **symbolic representation**. Each circuit receives certain inputs and gives some output(s). Input/output requirements for models of many computer circuits are represented by functions from the Boolean algebra B^n to the Boolean algebra $B = \{0, 1\}$. Each set of values assigned to the Boolean variables $x_1, x_2, \ldots, x_n$ is called an input for the circuit described by f and the value taken by f for this given set of values of $x_1, \ldots, x_n$ is called the output. Recall that Boolean function can be represented by a Boolean polynomial and a Boolean polynomial involves three basic operations $\vee$ or $(+)$, $\wedge$ (or.) and $/$. Thus each circuit is obtained as a combination of some or all three basic circuits which we now describe.

If x and y are Boolean variables, then the basic polynomials $x + y$ (or $x \vee y$), $x \cdot y$ (or $x \wedge y$) are x' are described diagrammatically in the Fig. 12.9. Each diagram has lines for the variables x, y on the left and a single line on the right representing the polynomial (i.e. giving the outcome).

The symbol or diagram for $x + y$ is called an **or gate**, the one for xy is called an **and gate** and that for x' is called an **inverter**. The names **or gate** and **and gate** arise from the fact that the truth table showing

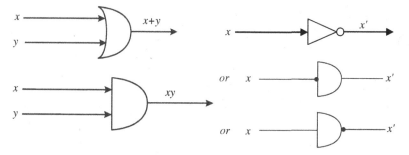

Fig. 12.9.

the functions represented by $x + y$ and xy are the exact analogs of the truth table for 'or' and 'and' respectively. The truth table for the inverter corresponds to the truth table for negation.

As already mentioned above, the functions from B^n to B can be used to describe the behavior of circuits with n $0 -$ or $- 1$ inputs and a single $0 -$ or $- 1$ output. This may be described diagrammatically by repeatedly using the diagrams for '$+$', '$.$' and "$'$".

We consider some examples for such description.

Example 12.55. Let $f : B^3 \to B$ be the function given by the polynomial

(a) $p(x, y, z) = xy + yz'$; (b) $p(x, y, z) = (x + y)x'$; (c) $x'(y + z')'$;
(d) $(x + y + z)x'y'z'$.

The truth tables for the functions f defined by the polynomials are

x	y	z	x'	z'	xy	yz'	$x+y$	$xy+yz'$	$(x+y)x'$
0	0	0	1	1	0	0	0	0	0
0	0	1	1	0	0	0	0	0	0
0	1	0	1	1	0	1	1	1	1
1	0	0	0	1	0	0	1	0	0
0	1	1	1	0	0	0	1	0	1
1	0	1	0	0	0	0	1	0	0
1	1	0	0	1	1	1	1	1	0
1	1	1	0	0	1	0	1	1	0

x	y	z	x'	y'	z'	$y+z'$	$(y+z')'$	$x'(y+z')'$	$x+y+z$	$x'y'z'$	$(x+y+z)$ $x'y'z'$
0	0	0	1	1	1	1	0	0	0	1	0
0	0	1	1	1	0	0	1	1	1	0	0
0	1	0	1	0	1	1	0	0	1	0	0
1	0	0	0	1	1	1	0	0	1	0	0
0	1	1	1	0	0	1	0	0	1	0	0
1	0	1	0	1	0	0	1	0	1	0	0
1	1	0	0	0	1	1	0	0	1	0	0
1	1	1	0	0	0	1	0	0	1	0	0

The circuit diagrams for these functions are respectively given in Fig. 12.10.

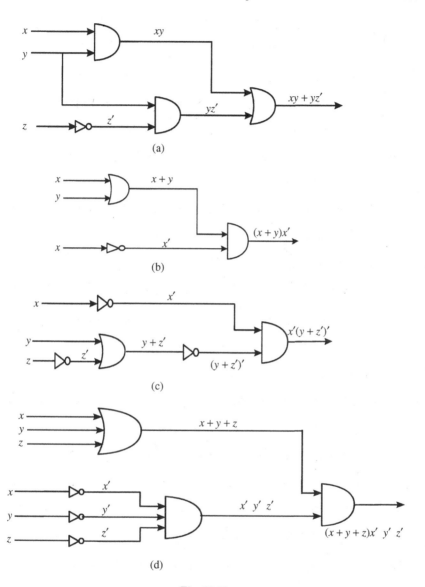

Fig. 12.10.

Example 12.56. Simplify the circuit given in Fig. 12.11.

Solution. The circuit in Fig. 12.11(a) is represented by the Boolean expression $((a' + b)(a' + d))((c' + b)(c' + d))$. Now using the laws of Boolean

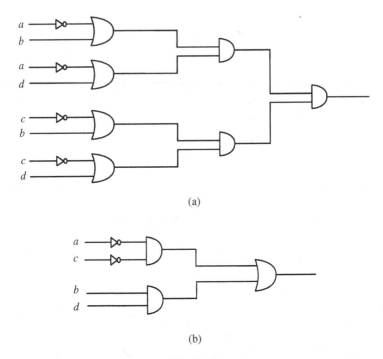

(a)

(b)

Fig. 12.11.

algebra — such as $(x + y)' = x'y', (x')' = x, (xy)' = x' + y'$, we get

$$(a' + b)(a' + d)(c' + b)(c' + d) = (ab')'(ad')'(b'c)'(cd')'$$

$$= (ab' + ad' + b'c + cd')' = (a(b' + d') + c(b' + d'))'$$

$$= ((a + c)(b' + d'))'$$

$$= (a + c)' + (b' + d')' = a'c' + bd.$$

Thus the Boolean expression representing the given circuit is equivalent to the expression $a'c' + bd$ and the circuit corresponding to the new expression is as given in Fig. 12.11(b) above which is the simplified form of the given circuit.

Example 12.57. Simplify the circuit given in Fig. 12.12(a).

Solution. The Boolean polynomial of the circuit is $(a_1 + a_2)'((a_2 + a_3)(a_2' + a_3') + a_1')$. This expression simplifies to

$$a_1'a_2'(a_2a_3' + a_2'a_3 + a_1') = a_1'a_2'a_3 + a_1'a_2'.$$

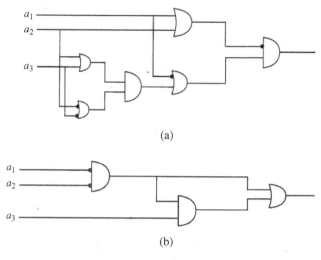

(a)

(b)

Fig. 12.12.

Therefore, the circuit given in Fig. 12.12(a) simplifies to the circuit given in Fig. 12.12(b).

Example 12.58. A large room has electrical switches next to the four doors to operate the central lighting. The four switches operate alternatively which means that each switch can switch on or switch off the lights. We like to determine symbolic representation of the switching circuit p and its contact diagram. Each switch has two positions which are either 'on' or 'off'. Let the switches be denoted by x_1, x_2, x_3, x_4 and let the two possible states of a switch x_i be $a_i \in \{0, 1\}$.

The light situation in the room is given by the value $p(a_1, a_2, a_3, a_4) = 0$ (or 1) if the lights are off or on respectively. Let us assume that $p(1, 1, 1, 1) = 1$.

(a) If we operate one or any three of the four switches, the light situation in the room changes from off to on or from on to off as the case may be.

(b) If we operate any two of the four switches or all the four switches, the light position in the room does not change.

We then get the following function values for the polynomial p which is the switching circuit we are looking for.

a_1	a_2	a_3	a_4	Minterm	$p(a_1\,a_2\,a_3\,a_4)$
1	1	1	1	$x_1x_2x_3x_4$	1
1	1	1	0	$x_1x_2x_3x_4'$	0
1	1	0	1	$x_1x_2x_3'x_4$	0
1	0	1	1	$x_1x_2'x_3x_4$	0
0	1	1	1	$x_1'x_2x_3x_4$	0
1	1	0	0	$x_1x_2x_3'x_4'$	1
1	0	1	0	$x_1x_2'x_3x_4'$	1
1	0	0	1	$x_1x_2'x_3'x_4$	1
0	1	1	0	$x_1'x_2x_3x_4'$	1
0	1	0	1	$x_1'x_2x_3'x_4$	1
0	0	1	1	$x_1'x_2'x_3x_4$	1
1	0	0	0	$x_1x_2'x_3'x_4'$	0
0	1	0	0	$x_1'x_2x_3'x_4'$	0
0	0	1	0	$x_1'x_2'x_3x_4'$	0
0	0	0	1	$x_1'x_2'x_3'x_4$	0
0	0	0	0	$x_1'x_2'x_3'x_4'$	1

Thus, the polynomial p takes value 1 in any one of the eight situations as given in the above table, we have

$$p(x_1, x_2, x_3, x_4) = x_1x_2x_3x_4 + x_1x_2x_3'x_4' + x_1x_2'x_3x_4' + x_1'x_2x_3x_4'$$

$$+ x_1x_2'x_3'x_4 + x_1'x_2x_3'x_4 + x_1'x_2'x_3x_4 + x_1'x_2'x_3'x_4'$$

The symbolic form of the circuit is as below (Fig. 12.13):

Example 12.59. Give the Boolean function described by the logic diagram in Fig. 12.14(a) and obtain a simplified equivalent logic diagram, if possible.

Solution. The given logic diagram gives the Boolean function $((x'y)' + yw) + (yw' + (yw + z'))$. The expression is equivalent to $xy' + yw + yw' + z' = xy' + y + z'$. The logic diagram corresponding to this expression is as given in Fig. 12.14(b).

Please note that an excellent source on the treatment of lattices and Boolean algebras is Birkhoff (1967). A beautiful treatment of the subject is also given in Dornhoff and Hohn (1978) and Lidl and Pilz (1998). Liu (2000) and Rosen (2003/2005) also contain some nice results. Mathematics majors need to refer to Jacobson (1964 and 1984) for a purely algebraic treatment binding it with study of groups and rings.

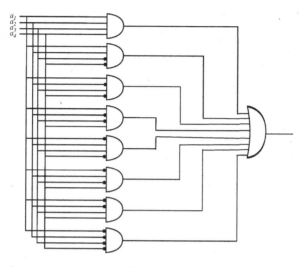

Fig. 12.13.

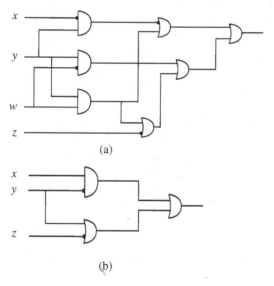

(a)

(b)

Fig. 12.14.

Exercises 12.1

1. Is the partially ordered set with Hasse diagram as in (i) Fig. 6.6, (ii) Fig. 6.7 of Chapter 6, a lattice? Justify.

2. Decide if the partially ordered sets $\{1, 2, 3, 4, 6, 8, 10\}$ and $\{1, 2, 3, 4, 6, 8, 12\}$ are lattices with the partial order relation '$\leq$' defined by $a \leq b$ if a divides b.

3. Find all possible chains and antichains in the partially ordered sets in Question 2 above.

4. Show that the set of all subgroups of a group G is a lattice.

5. Show that the set of normal subgroups of any group G is a lattice. If H, K are subgroups of G, say that $H \leq K$ if H is a subgroup of G. Although intersection $H \cap K$ of two subgroups of G is again a subgroup of G, the same is not true about the union of two subgroups. As a least upper bound of subgroups H, K, we take the subgroup generated by $H \cup K$ which is the smallest subgroups of G containing the subgroups H and K. In case H, K are normal subgroups of G, then $HK = \{x \in G \mid x = hk$ for some $h \in H\ k \in K\}$ is again a normal subgroup of G.

6. Prove that in any lattice $(L, \leq)$, the three distributive laws (D1), (D2) and (D3) are equivalent.

7. Prove that the set of all submodules of an R-module M with $A \wedge B = A \cap B$, the set intersection and $A \vee B = A + B$, for submodules A, B of M, is a lattice. Prove that this lattice is not distributive.

8. If $(L, \leq)$ is a lattice and for all $a, b, c \in L$, $a \wedge b = a \wedge c$ and $a \vee b = a \vee c$ together imply that $b = c$, then L is distributive.

9. Is the lattice of subgroups of the alternating group A_4 of degree 4 semi-modular? Justify.

10. Draw a Hasse diagram of the Klein's four-group.

11. Construct the Hasse diagram of the lattice of subgroups of the quaternion group $G = \{\pm 1, \pm i, \pm j, \pm k\}$ where

$$i^2 = j^2 = k^2 = -1, ij = -ji = k, jk = -kj = i, ki = -ik = j.$$

(Hint: Observe that G is a group of order 8 and its non-trivial proper subgroups have to be of order 2 and 4. The subgroup $\{1, -1\}$ is the only subgroup of order 2 and the only subgroups of order 4 are $\{\pm 1, \pm i\}, \{\pm 1, \pm j\}, \{\pm 1, \pm k\}$).

12. Let $(L, \leq)$ be a distributive lattice and $a, b \in L$. Prove that the subset $A = \{x \in L \mid a \wedge x = b \wedge x\}$ is a dual ideal of L.

13. Prove that in a modular lattice with 0 and 1, any subset of a linearly independent set of elements is independent.

14. If $(x \vee y) \wedge z = y \wedge z$ in a modular lattice, show that $x \wedge (y \vee z) = x \wedge y$.

15. Every finite lattice with unit 1 has at least one co-atom.

16. If L is a distributive lattice with 0, 1 and a, b are elements of L with $a \leq b$, then $a \wedge b' = 0$ where b' is complement of b.

17. If L is a distributive lattice with 0, 1 and a is an atom in L, then for any element b in L, $a \wedge b = 0$ if and if $a \nleq b$.

18. For all a, b in a Boolean algebra B,

    (a) $a \wedge b' = 0$ if and only if $a' \vee b = 1$.

    (b) $a' \wedge b = 1$ if and only if $a \wedge b = a$.

    (c) $a' \wedge b = 1$ if and only if $a \vee b = b$.

19. In a finite Boolean algebra, complement of an atom is a co-atom and complement of a co-atom is an atom.

20. If $a_1, a_2, \ldots, a_m$ are all the atoms in a finite Boolean algebra, then $a_1 \vee a_2 \vee \ldots \vee a_m = 1$.

21. The meet of all the co-atoms in a finite Boolean algebra is 0.

Chapter 13

MATRICES, SYSTEMS OF LINEAR EQUATIONS AND EIGEN VALUES

We have already encountered matrices while considering binary relations on sets and also as examples of groups and rings. Matrices play a vital role in the study of systems of linear equations. Consistency of a linear system of equations is determined through the rank of a matrix, which we will briefly discuss here. The Gaussian elimination method for solving a system of linear equations depends on elementary row operations, which are introduced and applied for that purpose. Also discussed are direct methods and methods of factorization for the solution of linear systems of equations. Eigen values and associated eigen vectors are discussed. Properties of eigen values of some special types of matrices are discussed.

13.1. Linear System of Equations

13.1.1. *Rank of a matrix*

Given a matrix A, the determinant of a square submatrix of order k of A obtained by omitting some rows and columns of A is called a minor of A of order k. The largest r such that every minor of A of order more than r is zero, but there is a minor of order r which is not zero, which is called the **rank** of A. The rank of A is usually denoted by $r(A)$ or $\rho(A)$. Observe that the rank of a zero matrix is 0 and the rank of an identity matrix of order n is n. The rank of a non-zero matrix is at least 1.

The rows of a matrix or its columns are vectors called 'row vectors' and 'column vectors'. Given a matrix A, the maximum number of linearly independent rows of A is called the **row rank** of A and the maximum number of linearly independent columns of A is called the **column rank** of A. We record without proof the following.

Theorem 13.1. For any matrix A, row rank of A = column rank of $A = r(A)$.

Example 13.1. Consider the matrix

$$A = \begin{pmatrix} 1 & 2 & 3 \\ 2 & 3 & 1 \\ 3 & 1 & 2 \end{pmatrix}. \text{ Then } \det(A) = \begin{vmatrix} 3 & 1 \\ 1 & 2 \end{vmatrix} - 2 \begin{vmatrix} 2 & 1 \\ 3 & 2 \end{vmatrix} + 3 \begin{vmatrix} 2 & 3 \\ 3 & 1 \end{vmatrix}$$

$$= 5 - 2 - 21 = -18 \neq 0.$$

Therefore, $r(A) = 3$.

Example 13.2. Consider the matrix $A = \begin{pmatrix} 1 & 2 & 3 \\ 2 & 3 & 1 \\ 2 & 4 & 6 \end{pmatrix}$. Since the third row is twice the first row, the three rows of A are linearly dependent. Therefore, $r(A) < 3$. The minor $\begin{vmatrix} 1 & 2 \\ 2 & 3 \end{vmatrix} = 3 - 4 = -1 \neq 0$ and, therefore, $r(A) = 2$.

Example 13.3. In the matrix $A = \begin{pmatrix} 1 & 2 & 3 \\ 3 & 6 & 9 \\ 2 & 4 & 6 \end{pmatrix}$, $-R_1 + R_2 - R_3 = 0$, where R_i denotes the ith row of A. So, the three rows are linearly dependent. Also $-2R_1 + R_3 = 0$, $-3R_1 + R_2 = 0$ and $2R_2 - 3R_3 = 0$, so that every two rows of A are linearly dependent. Hence the rank A is <2. Since every single non-zero vector is always linearly independent, $r(A) = 1$.

Example 13.4. Consider $A = \begin{pmatrix} 1 & 2 & 3 & 4 \\ 4 & 5 & 8 & 10 \\ 3 & 2 & 1 & 2 \end{pmatrix}$. Since there are three rows and four columns, $r(A) \leq 3$. Now the minor $B = \begin{vmatrix} 1 & 2 & 3 \\ 4 & 5 & 8 \\ 3 & 2 & 1 \end{vmatrix} = (5 - 16) - 2(4 - 24) + 3(8 - 15) = -11 + 40 - 21 = 8 \neq 0$. Hence $r(A) = 3$.

13.1.2. *Linear system of equations*

Consider the system

$$\begin{aligned} a_{11}x_1 + a_{12}x_2 + \cdots + a_{1n}x_n &= b_1, \\ a_{21}x_1 + a_{22}x_2 + \cdots + a_{2n}x_n &= b_2. \\ \vdots \qquad \vdots \qquad\quad \vdots \\ a_{m1}x_1 + a_{m2}x_2 + \cdots + a_{mn}x_n &= b_m. \end{aligned} \tag{13.1}$$

of m linear equations in n variables $x_1, x_2, \ldots, x_n$. The system of equations is said to be **consistent** if there exists a solution of the system, i.e. if we can find a set of values for the variables $x_1, x_2, \ldots, x_n$ which, when substituted in the given equations, satisfy all of these, i.e. the left-hand side of every equation becomes equal to the right-hand side of that equation. The matrix

$$A = \begin{pmatrix} a_{11} & a_{12} & \cdots & a_{1n} \\ a_{21} & a_{22} & \cdots & a_{2n} \\ \vdots & & & \\ a_{m1} & a_{m2} & \cdots & a_{mn} \end{pmatrix}$$

is called the **coefficient matrix** of the given system (13.1) of equations. Also the matrix

$$(A, b) = \begin{pmatrix} a_{11} & a_{12} & \cdots & a_{1n} & b_1 \\ a_{21} & a_{22} & \cdots & a_{2n} & b_2 \\ \vdots & & & & \\ a_{m1} & a_{m2} & \cdots & a_{mn} & b_m \end{pmatrix}$$

is called the **augmented matrix** of the given system. We again record without proof the following.

Theorem 13.2. The given system of equations (13.1) is consistent if and only if $r(A) = r(A, b)$ and there exists at least one solution for the given system.

There do exist situations when the given system (13.1) of linear equations may admit more than one solution. However, we do not consider this question here. The interested student may refer to Hoffman and Kunze (1972) or any other suitable book on linear algebra for details and proofs of Theorems 13.1 and 13.2.

Example 13.5. Check if the following systems of equations are consistent.

(a) $2x - y + z = 4$
$\quad 3x - y + z = 6$
$\quad 4x - y + 2z = 7$
$\quad -x + y - z = 9$

(b) $5x - 3y + 7z = 8$
$\quad 3x + 26y + 2z = 18$
$\quad 7x + 2y + 10z = 10$

(c) $x + 2y - z = 5$
$\quad x - y + 2z = 3$
$\quad 2x - 2y + 3z = 4$
$\quad x - y + z = 1$

(d) $5x - 2y + z = 2$
$\quad 7x + y - 5z = 3$
$\quad 3x + 7y + 4z = 5$

Solution. (a) The coefficient matrix and the associated augmented matrix are

$$A = \begin{pmatrix} 2 & -1 & 1 \\ 3 & -1 & 1 \\ 4 & -1 & 2 \\ -1 & 1 & -1 \end{pmatrix}, \quad (A, b) = \begin{pmatrix} 2 & -1 & 1 & 4 \\ 3 & -1 & 1 & 6 \\ 4 & -1 & 2 & 7 \\ -1 & 1 & -1 & 9 \end{pmatrix}$$

$$\det(A, b) = 2 \begin{vmatrix} -1 & 1 & 6 \\ -1 & 2 & 7 \\ 1 & -1 & 9 \end{vmatrix} + \begin{vmatrix} 3 & 1 & 6 \\ 4 & 2 & 7 \\ -1 & -1 & 9 \end{vmatrix} + \begin{vmatrix} 3 & -1 & 6 \\ 4 & -1 & 7 \\ -1 & 1 & 9 \end{vmatrix}$$

$$- 4 \begin{vmatrix} 3 & -1 & 1 \\ 4 & -1 & 2 \\ -1 & 1 & -1 \end{vmatrix}$$

$$= 2\{-(18 + 7) - (-9 - 7) + 6(1 - 2)\} + \{3(18 + 7) - (36 + 7)$$
$$+ 6(-4 + 2)\} + \{3(-9 - 7) + (36 + 7) + 6(4 - 1)\}$$
$$- 4\{3(1 - 2) + (-4 + 2) + (4 - 1)\}$$
$$= 2(-25 + 16 - 6) + (75 - 43 - 12) + (-48 + 43 + 18)$$
$$- 4(-3 - 2 + 3)$$
$$= -30 + 20 + 13 + 8 = 11 \neq 0.$$

Therefore, $r(A, b) = 4$. But the rank of A is ≤ 3. Therefore, $r(A) \neq r(A, b)$ and the given system of equations in inconsistent.

(b) In this case, the coefficient matrix and the augmented matrix are

$$A = \begin{pmatrix} 5 & -3 & 7 \\ 3 & 26 & 2 \\ 7 & 2 & 10 \end{pmatrix} \quad \text{and} \quad (A, b) = \begin{pmatrix} 5 & -3 & 7 & 8 \\ 3 & 26 & 2 & 18 \\ 7 & 2 & 10 & 10 \end{pmatrix}.$$

Now

$$\det(A) = 5(26 \times 10 - 4) + 3(30 - 14) + 7(6 - 182)$$
$$= 5 \times 256 + 48 - 1232$$
$$= 1280 + 48 - 1232 = 96 \neq 0.$$

Therefore, $r(A) = 3$. Since A is a submatrix of $(A, b), r(A, b) = 3$. Hence the given system of equations is consistent.

(c) The coefficient matrix and the associated augmented matrix are

$$A = \begin{pmatrix} 1 & 2 & -1 \\ 1 & -1 & 2 \\ 2 & -2 & 3 \\ 1 & -1 & 1 \end{pmatrix} \quad \text{and} \quad (A, b) = \begin{pmatrix} 1 & 2 & -1 & 5 \\ 1 & -1 & 2 & 3 \\ 2 & -2 & 3 & 4 \\ 1 & -1 & 1 & 1 \end{pmatrix}.$$

Now

$$det(A, b) = \begin{vmatrix} -1 & 2 & 3 \\ -2 & 3 & 4 \\ -1 & 1 & 1 \end{vmatrix} - 2 \begin{vmatrix} 1 & 2 & 3 \\ 2 & 3 & 4 \\ 1 & 1 & 1 \end{vmatrix} - \begin{vmatrix} 1 & -1 & 3 \\ 2 & -2 & 4 \\ 1 & -1 & 1 \end{vmatrix} - 5 \begin{vmatrix} 1 & -1 & 2 \\ 2 & -2 & 3 \\ 1 & -1 & 1 \end{vmatrix}.$$

In all the four determinants on the right-hand side of $det(A, b)$, the sum of the first and the third row equals the second row. Therefore, each one of the determinants is zero. Therefore, $det(A, b) = 0$. Hence $r(A, b) \leq 3$. The minor of A obtained by omitting the fourth row is

$$\begin{vmatrix} 1 & 2 & -1 \\ 1 & -1 & 2 \\ 2 & -2 & 3 \end{vmatrix} = \begin{vmatrix} 1 & 2 & -1 \\ 0 & -3 & 3 \\ 0 & -6 & 5 \end{vmatrix} = 3 \neq 0.$$

Therefore, $r(A) = 3$. Again, A being a submatrix of (A, b), $r(A, b) \geq r(A) = 3$. Hence $rank(A, b) = r(A) = 3$ and the system of equations is consistent.

(d) The coefficient and the associated augmented matrix are

$$A' = \begin{pmatrix} 5 & -2 & 1 \\ 7 & 1 & -5 \\ 3 & 7 & 4 \end{pmatrix} \quad \text{and} \quad (A, b) = \begin{pmatrix} 5 & -2 & 1 & 2 \\ 7 & 1 & -5 & 3 \\ 3 & 7 & 4 & 5 \end{pmatrix}.$$

Now

$$det(A) = 5(4 + 35) + 2(28 + 15) + (49 - 3)$$
$$= 195 + 86 + 46 = 327 \neq 0.$$

Therefore, $r(A) = 3$. Also, the number of rows of (A, b) being 3, $r(A, b) \leq 3$. Since A is a submatrix of (A, b), $r(A, b) \geq r(A) = 3$. Hence $r(A) = r(A, b)$ and the system of equations is consistent.

13.2. Elementary Row Operations, Gaussian Elimination

13.2.1. *Elementary row operations*

Consider an $m \times n$ matrix $A = (a_{ij})$, the rows of which are denoted by $R_1, R_2, \ldots, R_m$ respectively. The first non-zero entry in a row R_i is called the **leading non-zero element** of R_i. Observe that a zero row, i.e. a row in which every entry is zero, has no non-zero leading element.

The following operations performed on the rows of A are called **elementary row operations**.

(E_1) Interchange row R_i and row R_j indicated by interchange R_i and R_j or also written as R_{ij}.

(E_2) Multiply each element in a row R_i by a non-zero constant k. This operation is indicated by writing 'Multiply R_i by k' or 'kR_i'.

(E_3) Add a multiple of one row R_i to another row R_j, which is the same as saying replace the row R_j by $kR_i + R_j$. This operation is indicated by 'Add kR_i to R_j' or '$R_{ij}(k)$'.

To avoid fractions appearing as entries in the matrices, the operations E_2 and E_3 are sometimes applied simultaneously.

(E): Add a non-zero multiple of a row R_i to a non-zero multiple of row R_j which is same as saying, replace the row R_j by $kR_i + k'R_j$, and is written as 'Add kR_i to $k'R_j$'.

Definition. Two matrices, A and B are said to be **row equivalent** and are written as $A \sim B$, if the matrix B can be obtained from the matrix A by applying elementary row operations.

13.2.1.1. *Echlon matrices*

A matrix A is called an **echlon matrix** or is said to be in, **echlon form**, if the following conditions hold.

(a) All zero rows, if any, appear at the bottom of the matrix;

(b) Each non-zero leading entry in a row is to the right of the leading non-zero entry in the preceding row.

Definition. A row echlon matrix A is said to be in **row canonical form** or in **reduced row echlon form** if

(c) each leading non-zero entry in a row is 1;

(d) each leading non-zero entry is the only non-zero entry in its column.

The zero matrix O and the identity matrix I are clearly matrices in row canonical form. Each upper triangular matrix with non-zero diagonal entries is an echlon matrix, but not necessarily in row canonical form. We also say that a square matrix $A = (a_{ij})$ is in **triangular form** if $a_{11}, a_{22}, \ldots, a_{nn}$

are the leading non-zero entries. Every square matrix in triangular form is upper triangular, but not necessarily conversely. Observe that identity matrix is the only square matrix that is in triangular form and also in row canonical form. The following are echlon matrices.

$$
\begin{pmatrix}
2 & 1 & 2 & 0 & 3 & 5 & -6 \\
0 & 0 & 1 & -1 & -3 & 0 & 2 \\
0 & 0 & 0 & 0 & 5 & 1 & 3 \\
0 & 0 & 0 & 0 & 0 & 0 & 0
\end{pmatrix},
\begin{pmatrix}
1 & 2 & 3 \\
0 & 0 & 2 \\
0 & 0 & 0
\end{pmatrix},
\begin{pmatrix}
0 & 2 & 3 & 4 & 0 & 4 \\
0 & 0 & 0 & 1 & -1 & 3 \\
0 & 0 & 0 & 0 & 1 & 2
\end{pmatrix},
$$

$$
\begin{pmatrix}
2 & 3 & 6 \\
0 & 5 & 7 \\
0 & 0 & 8
\end{pmatrix}.
$$

None of the following is an echlon matrix

$$
\begin{pmatrix}
2 & 1 & 2 & 0 & 3 & 5 & -6 \\
0 & 0 & 0 & 1 & -1 & -3 & 2 \\
0 & 0 & 5 & 0 & 0 & 1 & 3 \\
0 & 0 & 0 & 0 & 0 & 0 & 0
\end{pmatrix},
\begin{pmatrix}
1 & 2 & 3 \\
0 & 0 & 0 \\
0 & 0 & 2
\end{pmatrix},
\begin{pmatrix}
0 & 0 & 2 & 3 & 4 & 5 \\
0 & 1 & -1 & 3 & 0 & 2 \\
0 & 0 & 1 & 4 & 5 & 6
\end{pmatrix}.
$$

13.2.2. *Gaussian elimination in matrix form*

Consider a matrix A. We give below two algorithms, first of which transforms the matrix A into echlon form by using elementary row operations and the second algorithm transforms the echlon matrix in row canonical form.

Algorithm 1. Forward elimination

Step 1. Find the first column with a non-zero entry. If no such column exists, then EXIT. (The matrix is in fact a zero matrix.) Otherwise, let j_1 be the number of this column, i.e. the j_1th column.

(a) Arrange so that $a_{1j_1} \neq 0$, i.e. if necessary, interchange rows so that a non-zero entry appears in the first row in j_1th column.

(b) Use a_{1j_1} as a **pivot** to obtain zeros below a_{1j_1}. That is, for $i > 1$,

 (i) Set $m = -\dfrac{a_{ij_1}}{a_{1j_1}}$.

 (ii) Add mR_1 to R_i.

 i.e. replace R_i by $mR_1 + R_i$.

Step 2. Repeat Step 1 with the submatrix formed by all the rows except the first row. Let j_2th be the first column in the submatrix with a non-zero entry. Arrange the rows so that this non-zero entry in column j_2 appears in the first row of the submatrix, i.e. $a_{2j_2} \neq 0$. Use a_{2j_2} entry as a pivot to make every entry in column j_2 below this entry zero.

Step 3. Continue this process until we arrive at a submatrix that has no non-zero entry.

Observe that at the end of the algorithm the pivots or the leading non-zero entries are $a_{1j_1}, a_{2j_2}, \ldots, a_{rj_r}$, where r is the number of non-zero rows in the matrix when expressed in the echlon form.

The number $m = \dfrac{-a_{ij_1}}{a_{1j_1}} = \dfrac{-\text{coefficient to be deleted}}{\text{pivot}}$ is called the **multiplier**. To avoid getting fractional entries in the matrix we could replace the operation in step 1(b) by adding $-a_{ij_1}R_1$ to $a_{1j_1}R_i$.

Algorithm 2. Backward elimination. The input is a matrix $A = (a_{ij})$, which is in echlon form with pivot entries $a_{1j_1}, a_{2j_2}, \ldots, a_{rj_r}$.

Step 1.

(a) Multiply the non-zero R_r by $\frac{1}{a_{rj_r}}$ so that the pivot entry becomes 1.
(b) Use $a_{rj_r} = 1$ to obtain zero above the pivot. For this do as below, for $i = r - 1, r - 2, \ldots, 1$;

   (i) Set $m = -a_{ij_i}$.
   (ii) Add mR_r to R_i.

This is equivalent to applying the elementary row transformation 'Add $-a_{ij_i}R_r$ to R_i' or replace R_i by $-a_{ij_i}R_r + R_i$.

Step 2. Repeat Step 1 for rows $R_{r-1}, R_{r-2}, \ldots, R_2$.

Step 3. Multiply R_1 by $\frac{1}{a_{1j_1}}$.

Example 13.6. Find the row canonical form of

$$A = \begin{pmatrix} 1 & 2 & -3 & 1 & 2 \\ 2 & 4 & -4 & 6 & 10 \\ 3 & 6 & -6 & 9 & 13 \end{pmatrix}.$$

Solution. We use the forward elimination method to first reduce A to echlon form. The first entry in the first row is 1. We use this as a pivot to make other entries in the first column zeros. For this we apply the row operations $-2R_1 + R_2$ and $-3R_1 + R_3$. Then

$$A \sim \begin{pmatrix} 1 & 2 & -3 & 1 & 2 \\ 0 & 0 & 2 & 4 & 6 \\ 0 & 0 & 3 & 6 & 7 \end{pmatrix} \sim \frac{-3}{2}R_2 + R_3 \begin{pmatrix} 1 & 2 & -3 & 1 & 2 \\ 0 & 0 & 2 & 4 & 6 \\ 0 & 0 & 0 & 0 & -2 \end{pmatrix}$$

which is the echlon form of the matrix A. We next apply the backward elimination to the echlon form of A. For this we apply the row transformations 'Replace R_3 by $-\frac{1}{2}R_3$' and 'Replace R_2 by $\frac{1}{2}R_2$'. We get

$$A \sim \begin{pmatrix} 1 & 2 & -3 & 1 & 2 \\ 0 & 0 & 1 & 2 & 3 \\ 0 & 0 & 0 & 0 & 1 \end{pmatrix}.$$

Next apply the row transformations 'Add $-3R_3$ to R_2' and 'Add $-2R_3$ to R_1'. We get

$$A \sim \begin{pmatrix} 1 & 2 & -3 & 1 & 0 \\ 0 & 0 & 1 & 2 & 0 \\ 0 & 0 & 0 & 0 & 1 \end{pmatrix}.$$

Now apply the row transformation 'Add $3R_2$ to R_1'. Then

$$A \sim \begin{pmatrix} 1 & 2 & 0 & 7 & 0 \\ 0 & 0 & 1 & 2 & 0 \\ 0 & 0 & 0 & 0 & 1 \end{pmatrix}$$

which is the row canonical form of the matrix A.

The forward and backward elimination algorithms show that every matrix can be reduced to a row canonical form. In fact every matrix can be reduced to a row canonical form uniquely.

Theorem 13.3. Any matrix A is row equivalent to a unique matrix in row canonical form.

13.2.3. *Gaussian elimination method*

We next use the above procedure of reducing a matrix to a row canonical form to solve a system of (non-homogeneous) linear equations. Consider a system $AX = B$ of linear equations. The augmented matrix of this system of equations is $M = [A, B]$. The system of equations is solved by applying the Gaussian elimination method to M:

Step 1. (Reduction). Reduce the augmented matrix M to echlon form. If a row of the form $(0, 0, \ldots, 0, b)$ appears with $b \neq 0$, then stop. The system does not have a solution.

Step 2. (Backward substitution). Reduce the echlon form of the matrix M to its row canonical form.

The row canonical form of M then shows if the system has a unique solution, or has a solution which is not unique or has a solution in the free variable form. We explain this procedure through some examples.

Example 13.7. Solve the system of equations

$$x + 2y + z = 3$$

$$2x + 5y - z = -4$$

$$3x - 2y - z = 5.$$

Solution. In matrix form the system is $AX = B$, where $A = \begin{pmatrix} 1 & 2 & 1 \\ 2 & 5 & -1 \\ 3 & -2 & -1 \end{pmatrix}$,
$X = \begin{pmatrix} x \\ y \\ z \end{pmatrix}$, $B = \begin{pmatrix} 3 \\ -4 \\ 5 \end{pmatrix}$ The augmented matrix of the given system of equations is

$$M = \begin{pmatrix} 1 & 2 & 1 & 3 \\ 2 & 5 & -1 & -4 \\ 3 & -2 & -1 & 5 \end{pmatrix}.$$

Apply the row transformations 'Add $-2R_1$ to R_2' and 'Add $-3R_1$ to R_3'. Then

$$M \sim \begin{pmatrix} 1 & 2 & 1 & 3 \\ 0 & 1 & -3 & -10 \\ 0 & -8 & -4 & -4 \end{pmatrix}.$$

Next apply 'Add $8R_2$ to R_3'. Then

$$M \sim \begin{pmatrix} 1 & 2 & 1 & 3 \\ 0 & 1 & -3 & -10 \\ 0 & 0 & -28 & -84 \end{pmatrix}$$

which is the echlon form of M. To obtain row canonical form of M, apply 'Multiply R_3 by $-\frac{1}{28}$'. We get

$$M \sim \begin{pmatrix} 1 & 2 & 1 & 3 \\ 0 & 1 & -3 & -10 \\ 0 & 0 & 1 & 3 \end{pmatrix} \begin{matrix} \text{'Add } 3R_3 \text{ to } R_2\text{'} \\ \underset{\sim}{} \\ \text{and 'Add } -R_3 \text{ to } R_1\text{'} \end{matrix} \begin{pmatrix} 1 & 2 & 0 & 0 \\ 0 & 1 & 0 & -1 \\ 0 & 0 & 1 & 3 \end{pmatrix}$$

$$\begin{matrix} \text{'Add } -2R_2 \text{ to } R_1\text{'} \\ \underset{\sim}{} \end{matrix} \begin{pmatrix} 1 & 0 & 0 & 2 \\ 0 & 1 & 0 & -1 \\ 0 & 0 & 1 & 3 \end{pmatrix}.$$

Therefore, the solution of the given system of equations is $x = 2$, $y = -1$ and $z = 3$.

Example 13.8. Solve the system of equations

$$x + 2y - 4z = -3$$

$$2x + 6y - 5z = 2$$

$$3x + 11y - 4z = 12.$$

Solution. The given system of equations is $AX = B$, where

$$A = \begin{pmatrix} 1 & 2 & -4 \\ 2 & 6 & -5 \\ 3 & 11 & -4 \end{pmatrix}, \quad X = \begin{pmatrix} x \\ y \\ z \end{pmatrix}, \quad B = \begin{pmatrix} -3 \\ 2 \\ 12 \end{pmatrix}.$$

The augmented matrix of this system of equations is

$$M = \begin{pmatrix} 1 & 2 & -4 & -3 \\ 2 & 6 & -5 & 2 \\ 3 & 11 & -4 & 12 \end{pmatrix}.$$

To reduce M to row echlon form, apply the transformations 'Add $-2R_1$ to R_2' and 'Add $-3R_1$ to R_3'.

$$M \sim \begin{pmatrix} 1 & 2 & -4 & -3 \\ 0 & 2 & 3 & 8 \\ 0 & 5 & 8 & 21 \end{pmatrix} \begin{matrix} \text{'Add } -5R_2 \text{ to } 2R_3\text{'} \\ \underset{\sim}{} \end{matrix} \begin{pmatrix} 1 & 2 & -4 & -3 \\ 0 & 2 & 3 & 8 \\ 0 & 0 & 1 & 2 \end{pmatrix}$$

$$\begin{matrix} \text{'Multiply } R_2 \text{ by } 1/2\text{'} \\ \underset{\sim}{} \end{matrix} \begin{pmatrix} 1 & 2 & -4 & -3 \\ 0 & 1 & 3/2 & 4 \\ 0 & 0 & 1 & 2 \end{pmatrix}$$

'Add $-3/2\ R_3$ to R_2'
and '$4R_3$ to R_1'
$\sim$
$\begin{pmatrix} 1 & 2 & 0 & 5 \\ 0 & 1 & 0 & 1 \\ 0 & 0 & 1 & 2 \end{pmatrix}$
'Add $-2R_2$ to R_1'
$\sim$
$\begin{pmatrix} 1 & 0 & 0 & 3 \\ 0 & 1 & 0 & 1 \\ 0 & 0 & 1 & 2 \end{pmatrix}$

which is the row canonical form of M and, therefore, the given system has a unique solution which is $x = 3$, $y = 1$, $z = 2$.

Example 13.9. Solve the system of equations

$$x - 3y + 2z - t = 2$$

$$3x - 9y + 7z - t = 7$$

$$2x - 6y + 7z + 4t = 7.$$

Solution. The given system of equations is $AX = B$, where

$$A = \begin{pmatrix} 1 & -3 & 2 & -1 \\ 3 & -9 & 7 & -1 \\ 2 & -6 & 7 & 4 \end{pmatrix}, \quad X = \begin{pmatrix} x \\ y \\ z \\ t \end{pmatrix}, \quad B = \begin{pmatrix} 2 \\ 7 \\ 7 \end{pmatrix}.$$

The augmented matrix of the given system of equations is

$$M = \begin{pmatrix} 1 & -3 & 2 & -1 & 2 \\ 3 & -9 & 7 & -1 & 7 \\ 2 & -6 & 7 & 4 & 7 \end{pmatrix}.$$

To reduce it to echlon form, apply 'Add $-3R_1$ to R_2' and 'Add $-2R_1$ to R_3'.

$$M \sim \begin{pmatrix} 1 & -3 & 2 & -1 & 2 \\ 0 & 0 & 1 & 2 & 1 \\ 0 & 0 & 3 & 6 & 3 \end{pmatrix}$$
'Add $-3R_2$ to R_3'
$\sim$
$$\begin{pmatrix} 1 & -3 & 2 & -1 & 2 \\ 0 & 0 & 1 & 2 & 1 \\ 0 & 0 & 0 & 0 & 0 \end{pmatrix}$$

'Add $-2R_2$ to R_1'
$\sim$
$$\begin{pmatrix} 1 & -3 & 0 & -5 & 0 \\ 0 & 0 & 1 & 2 & 1 \\ 0 & 0 & 0 & 0 & 0 \end{pmatrix}.$$

Thus the given system of equations reduces to:

$$x - 3y - 5t = 0$$

$$z + 2t = 1$$

which give $x = 3y + 5t$ and $z = -2t + 1$. This shows that the given system of equations has infinitely many solutions. For every arbitrary choice of y and t, we get a solution.

Example 13.10. Solve the system of equations

$$x + 2y + 3z = 7$$
$$x + 3y + z = 6$$
$$2x + 6y + 5z = 15$$
$$3x + 10y + 7z = 23.$$

Solution. The given system of equations is $AX = B$, where

$$A = \begin{pmatrix} 1 & 2 & 3 \\ 1 & 3 & 1 \\ 2 & 6 & 5 \\ 3 & 10 & 7 \end{pmatrix}, \quad X = \begin{pmatrix} x \\ y \\ z \end{pmatrix}, \quad B = \begin{pmatrix} 7 \\ 6 \\ 15 \\ 23 \end{pmatrix}.$$

The augmented matrix for this system of equations is

$$M = \begin{pmatrix} 1 & 2 & 3 & 7 \\ 1 & 3 & 1 & 6 \\ 2 & 6 & 5 & 15 \\ 3 & 10 & 7 & 23 \end{pmatrix}.$$

We first reduce M to row echlon form. We apply the row transformations 'Add $-R_1$ to R_2', 'Add $-2R_1$ to R_3' and 'Add $-3R_1$ to R_4'. Then

$$M \sim \begin{pmatrix} 1 & 2 & 3 & 7 \\ 0 & 1 & -2 & -1 \\ 0 & 2 & -1 & 1 \\ 0 & 4 & -2 & 2 \end{pmatrix} \quad \begin{array}{c} \text{'Add } -2R_2 \text{ to } R_3\text{'} \\ \text{and 'Add } -4R_2 \text{ to } R_4\text{'} \end{array} \quad \begin{pmatrix} 1 & 2 & 3 & 7 \\ 0 & 1 & -2 & -1 \\ 0 & 0 & 3 & 3 \\ 0 & 0 & 6 & 6 \end{pmatrix}$$

$$\begin{array}{c} \text{'Add } -2R_3 \text{ to } R_4\text{'} \end{array} \begin{pmatrix} 1 & 2 & 3 & 7 \\ 0 & 1 & -2 & -1 \\ 0 & 0 & 3 & 3 \\ 0 & 0 & 0 & 0 \end{pmatrix}$$

which is the row echlon form of M. To reduce M to row canonical form we apply the transformation 'Add $-2R_2$ to R_1'. Then

$$M \sim \begin{pmatrix} 1 & 0 & 7 & 9 \\ 0 & 1 & -2 & -1 \\ 0 & 0 & 3 & 3 \\ 0 & 0 & 0 & 0 \end{pmatrix} \begin{array}{c} \text{'Multiply } R_3 \text{ by } \frac{1}{3}\text{'} \end{array} \begin{pmatrix} 1 & 0 & 7 & 9 \\ 0 & 1 & -2 & -1 \\ 0 & 0 & 1 & 1 \\ 0 & 0 & 0 & 0 \end{pmatrix}$$

$$\begin{array}{c} \text{`Add } -7R_3 \text{ to } R_1\text{'} \\ \text{and `Add } 2R_3 \text{ to } R_2\text{'} \end{array} \begin{pmatrix} 1 & 0 & 0 & 2 \\ 0 & 1 & 0 & 1 \\ 0 & 0 & 1 & 1 \\ 0 & 0 & 0 & 0 \end{pmatrix}.$$

This then gives the solution $x = 2$, $y = 1$, $z = 1$.

Example 13.11. Solve the system of equations

$$x + 2y + z = 3, \quad 2x + 5y - z = 4; \quad 3x - 2y - z = 5.$$

Solution. In the matrix form the system of equations is $AX = B$, where

$$A = \begin{pmatrix} 1 & 2 & 1 \\ 2 & 5 & -1 \\ 3 & -2 & -1 \end{pmatrix}, \quad X = \begin{pmatrix} x \\ y \\ z \end{pmatrix} \quad \text{and} \quad B = \begin{pmatrix} 3 \\ 4 \\ 5 \end{pmatrix}.$$

The augmented matrix is

$$M = \begin{pmatrix} 1 & 2 & 1 & 3 \\ 2 & 5 & -1 & 4 \\ 3 & -2 & -1 & 5 \end{pmatrix}.$$

We apply elementary row transformations to M to reduce it to row canonical form:

$$M = \begin{pmatrix} 1 & 2 & 1 & 3 \\ 2 & 5 & -1 & 4 \\ 3 & -2 & -1 & 5 \end{pmatrix} \begin{array}{c} R_2 - 2R_1 \\ \sim \\ R_3 - 3R_1 \end{array} \begin{pmatrix} 1 & 2 & 1 & 3 \\ 0 & 1 & -3 & -2 \\ 0 & -8 & -4 & -4 \end{pmatrix}$$

$$\begin{array}{c} R_3 + 8R_2 \\ \sim \end{array} \begin{pmatrix} 1 & 2 & 1 & 3 \\ 0 & 1 & -3 & -2 \\ 0 & 0 & -28 & -20 \end{pmatrix} \begin{array}{c} -\frac{1}{28} R_3 \\ \sim \end{array} \begin{pmatrix} 1 & 2 & 1 & 3 \\ 0 & 1 & -3 & -2 \\ 0 & 0 & 1 & 5/7 \end{pmatrix}$$

$$\begin{array}{c} R_2 + 3R_3 \\ \sim \end{array} \begin{pmatrix} 1 & 2 & 1 & 3 \\ 0 & 1 & 0 & 1/7 \\ 0 & 0 & 1 & 5/7 \end{pmatrix} \begin{array}{c} R_1 - R_3 \\ \sim \end{array} \begin{pmatrix} 1 & 2 & 0 & 16/7 \\ 0 & 1 & 0 & 1/7 \\ 0 & 0 & 1 & 5/7 \end{pmatrix}$$

$$\begin{array}{c} R_1 - 2R_2 \\ \sim \end{array} \begin{pmatrix} 1 & 0 & 0 & 2 \\ 0 & 1 & 0 & 1/7 \\ 0 & 0 & 1 & 5/7 \end{pmatrix}.$$

Hence the solution to the given system of equations is

$$x = 2, \quad y = \frac{1}{7}, \quad z = \frac{5}{7}.$$

13.2.4. *Direct methods for the solution of linear system of equations*

Given a system of n non-homogeneous linear equations in n unknowns, the system can be solved by some direct methods. One of these is by the use of inverse of the coefficient matrix. The system will have a solution, provided the coefficient matrix is non-singular. Let the given system of equations be

$AX = B$, where $A = (a_{ij})$ is a square matrix of order n, $X = \begin{pmatrix} x_1 \\ x_2 \\ \vdots \\ x_n \end{pmatrix}$ and

$B = \begin{pmatrix} b_1 \\ b_2 \\ \vdots \\ b_n \end{pmatrix}$. If A is non-singular, we can find the inverse A^{-1} of A and

can pre-multiply the given equation in matrix form by A^{-1} and we get the solution $X = A^{-1}B$. If A is a singular matrix and $B \neq 0$, the given system has no solution.

Example 13.12. Solve the equations

$$3x + y + 2z = 3$$
$$2x - 3y - z = -3$$
$$x + 2y + z = 4.$$

Solution. The given system is equivalent to $AX = B$, where

$$A = \begin{pmatrix} 3 & 1 & 2 \\ 2 & -3 & -1 \\ 1 & 2 & 1 \end{pmatrix}, \quad X = \begin{pmatrix} x \\ y \\ z \end{pmatrix}, \quad B = \begin{pmatrix} 3 \\ -3 \\ 4 \end{pmatrix}.$$

Here $|A| = 3(-3+2) - 2(1-4) + 1(-1+6) = -3 + 6 + 5 = 8$ and

$$A^{-1} = \frac{1}{8} \begin{pmatrix} -1 & 3 & 5 \\ -3 & 1 & 7 \\ 7 & -5 & -11 \end{pmatrix}.$$

Therefore,

$$\begin{pmatrix} x \\ y \\ z \end{pmatrix} = \frac{1}{8} \begin{pmatrix} -1 & 3 & 5 \\ -3 & 1 & 7 \\ 7 & -5 & -11 \end{pmatrix} \begin{pmatrix} 3 \\ -3 \\ 4 \end{pmatrix} = \frac{1}{8} \begin{pmatrix} 8 \\ 16 \\ -8 \end{pmatrix} = \begin{pmatrix} 1 \\ 2 \\ -1 \end{pmatrix}$$

i.e.

$$x = 1, \quad y = 2, \quad z = -1.$$

The inverse of a square matrix A, which is non-singular can be computed by using a modification of the Gaussian elimination method.

Example 13.13. Find the inverse of the square matrix

$$A = \begin{pmatrix} 2 & 1 & 1 \\ 3 & 2 & 3 \\ 1 & 4 & 9 \end{pmatrix},$$

if it exists.

Solution. Observe that

$$|A| = 2(18 - 12) - 3(9 - 4) + (3 - 2) = 12 - 15 + 1 = -2 \neq 0.$$

Therefore, A is invertible. In order to find A^{-1}, we need to find out a matrix $\begin{pmatrix} x_{11} & x_{12} & x_{13} \\ x_{21} & x_{22} & x_{23} \\ x_{31} & x_{32} & x_{33} \end{pmatrix}$ such that

$$\begin{pmatrix} 2 & 1 & 1 \\ 3 & 2 & 3 \\ 1 & 4 & 9 \end{pmatrix} \begin{pmatrix} x_{11} & x_{12} & x_{13} \\ x_{21} & x_{22} & x_{23} \\ x_{31} & x_{32} & x_{33} \end{pmatrix} = \begin{pmatrix} 1 & 0 & 0 \\ 0 & 1 & 0 \\ 0 & 0 & 1 \end{pmatrix}.$$

This is equivalent to solving the three systems of linear equations

$$AX = \begin{pmatrix} 1 \\ 0 \\ 0 \end{pmatrix}, \quad AX = \begin{pmatrix} 0 \\ 1 \\ 0 \end{pmatrix}, \quad AX = \begin{pmatrix} 0 \\ 0 \\ 1 \end{pmatrix}.$$

Solving these we will get, respectively, the 1st, 2nd and the 3rd columns of A^{-1}. We can do this by considering the augmented matrix

$$\begin{pmatrix} 2 & 1 & 1 & \vdots & 1 & 0 & 0 \\ 3 & 2 & 3 & \vdots & 0 & 1 & 0 \\ 1 & 4 & 9 & \vdots & 0 & 0 & 1 \end{pmatrix}$$

and reducing it to row canonical form. We apply transformations 'Add$-\frac{3}{2}R_1$ to R_2' and 'Add $-\frac{1}{2}R_1$ to R_3'

$$\begin{pmatrix} 2 & 1 & 1 & \vdots & 1 & 0 & 0 \\ 0 & 1/2 & 3/2 & \vdots & -3/2 & 1 & 0 \\ 0 & 7/2 & 17/2 & \vdots & -1/2 & 0 & 1 \end{pmatrix}$$

'Add $-7R_2$ to R_3' $\underset{\sim}{}$ $\begin{pmatrix} 2 & 1 & 1 & \vdots & 1 & 0 & 0 \\ 0 & 1/2 & 3/2 & \vdots & -3/2 & 1 & 0 \\ 0 & 0 & -2 & \vdots & 10 & -7 & 1 \end{pmatrix}$.

Multiply R_3 by $-1/2$

$$\begin{pmatrix} 2 & 1 & 1 & \vdots & 1 & 0 & 0 \\ 0 & 1/2 & 3/2 & \vdots & -3/2 & 1 & 0 \\ 0 & 0 & 1 & \vdots & -5 & 7/2 & -1/2 \end{pmatrix}$$

'Add $-3/2R_3$ to R_2' $\underset{\sim}{}$
'Add $-R_3$ to R_1' $\begin{pmatrix} 2 & 1 & 0 & \vdots & 6 & -7/2 & 1/2 \\ 0 & 1/2 & 0 & \vdots & 6 & -17/4 & 3/4 \\ 0 & 0 & 1 & \vdots & -5 & 7/2 & -1/2 \end{pmatrix}$

$$\sim \begin{pmatrix} 2 & 0 & 0 & \vdots & -6 & 5 & -1 \\ 0 & 1 & 0 & \vdots & 12 & -17/2 & 3/2 \\ 0 & 0 & 1 & \vdots & -5 & 7/2 & -1/2 \end{pmatrix}$$

$$\sim \begin{pmatrix} 1 & 0 & 0 & \vdots & -3 & 5/2 & -1/2 \\ 0 & 1 & 0 & \vdots & 12 & -17/2 & 3/2 \\ 0 & 0 & 1 & \vdots & -5 & 7/2 & -1/2 \end{pmatrix} .$$

Hence

$$A^{-1} = \begin{pmatrix} -3 & 5/2 & -1/2 \\ 12 & -17/2 & 3/2 \\ -5 & 7/2 & -1/2 \end{pmatrix} .$$

13.2.5. *Method of factorization*

A given system of n non-homogeneous linear equations in n unknowns may also be solved by using a method called the factorization method. This method involves expressing the coefficient matrix as a product LU, where L is lower triangular matrix with every diagonal entry 1 and U is an upper triangular matrix. Given a system of equation $AX = B$, we write $A = LU$, where

$$L = \begin{pmatrix} 1 & 0 & 0 \\ l_{21} & 1 & 0 \\ l_{31} & l_{32} & 1 \end{pmatrix} \quad \text{and} \quad U = \begin{pmatrix} u_{11} & u_{12} & u_{13} \\ 0 & u_{22} & u_{23} \\ 0 & 0 & u_{33} \end{pmatrix}.$$

Then the given system becomes $LUX = B$. Now UX is a 3×1 matrix and suppose $UX = Y$. Then the given system becomes $LY = B$, which can be easily solved for Y. This system is

$$y_1 = b_1$$

$$l_{21}y_1 + y_2 = b_2$$

$$l_{31}y_1 + l_{32}y_2 + y_3 = b_3,$$

which give the values of y_1, y_2, y_3 quite easily. Once the y_1, y_2, y_3 are known, we solve the system $UX = Y$ by the Gaussian elimination method. So, the main problem reduces to writing A as LU. We explain this factorization with the help of an example.

Example 13.14. Factorize the matrix $A = \begin{pmatrix} 5 & -2 & 1 \\ 7 & 1 & -5 \\ 3 & 7 & 4 \end{pmatrix}$ in the form LU, where L is unit lower triangular and U is upper triangular and hence solve the system of equations

$$5x - 2y + z = 4$$

$$7x + y - 5z = 8$$

$$3x + 7y + 4z = 10.$$

Determine also L^{-1}, U^{-1} and hence find A^{-1}.

Solution. Let

$$L = \begin{pmatrix} 1 & 0 & 0 \\ l_{21} & 1 & 0 \\ l_{31} & l_{32} & 1 \end{pmatrix} \quad \text{and} \quad U = \begin{pmatrix} u_{11} & u_{12} & u_{13} \\ 0 & u_{22} & u_{23} \\ 0 & 0 & u_{33} \end{pmatrix}$$

so that

$$\begin{pmatrix} 1 & 0 & 0 \\ l_{21} & 1 & 0 \\ l_{31} & l_{32} & 1 \end{pmatrix} \begin{pmatrix} u_{11} & u_{12} & u_{13} \\ 0 & u_{22} & u_{23} \\ 0 & 0 & u_{33} \end{pmatrix} = \begin{pmatrix} 5 & -2 & 1 \\ 7 & 1 & -5 \\ 3 & 7 & 4 \end{pmatrix}.$$

Then

$$u_{11} = 5, \quad l_{21}u_{11} = 7, \quad l_{31}u_{11} = 3,$$
$$u_{12} = -2, \quad l_{21}u_{12} + u_{22} = 1, \quad l_{31}u_{12} + l_{32}u_{22} = 7,$$
$$u_{13} = 1, \quad l_{21}u_{13} + u_{23} = -5, \quad l_{31}u_{13} + l_{32}u_{23} + u_{33} = 4.$$

Thus u_{11}, u_{12} and u_{13} are determined. Using values of these, we get

$$l_{21} = \frac{7}{5}, \quad l_{31} = \frac{3}{5}.$$

Using the values of u_{11}, u_{12}, u_{13} and those of l_{21}, l_{31}, we get

$$u_{22} = 1 - \frac{7}{5} \cdot (-2) = \frac{19}{5};$$

$$u_{23} = -5 - \frac{7}{5} \cdot 1 = -\frac{32}{5}, 7 = l_{31}u_{12} + l_{32}u_{22} = \frac{3}{5} \cdot (-2) + l_{32} \cdot \frac{19}{5}$$

which gives $l_{32} = \frac{41}{19}$

$$u_{33} = -l_{31}u_{13} - l_{32}u_{23} + 4 = 4 - \frac{3}{5} \cdot 1 - \frac{41}{19} \cdot \left(-\frac{32}{5} \right)$$

$$= \frac{17}{5} + \frac{1312}{95} = \frac{19 \times 17 + 1312}{95} = \frac{1635}{95} = \frac{327}{19}.$$

Therefore,

$$U = \begin{pmatrix} 5 & -2 & 1 \\ 0 & 19/5 & -32/5 \\ 0 & 0 & 327/19 \end{pmatrix}, \quad L = \begin{pmatrix} 1 & 0 & 0 \\ 7/5 & 1 & 0 \\ 3/5 & 41/19 & 1 \end{pmatrix}.$$

Let $UX = Y = \begin{pmatrix} y_1 \\ y_2 \\ y_3 \end{pmatrix}$ so that

$$\begin{pmatrix} 1 & 0 & 0 \\ 7/5 & 1 & 0 \\ 3/5 & 41/19 & 1 \end{pmatrix} \begin{pmatrix} y_1 \\ y_2 \\ y_3 \end{pmatrix} = \begin{pmatrix} 4 \\ 8 \\ 10 \end{pmatrix}$$

which gives

$$y_1 = 4, \quad y_2 = 8 - \frac{7}{5}y_1 = \frac{12}{5},$$

$$y_3 = 10 - \frac{12}{5} - \frac{41}{19} \times \frac{12}{5} = 10 - \frac{12}{5}\left(1 + \frac{41}{19}\right) = 10 - \frac{144}{19} = \frac{46}{19}.$$

Then

$$\begin{pmatrix} 5 & -2 & 1 \\ 0 & 19/5 & -32/5 \\ 0 & 0 & 327/19 \end{pmatrix} \begin{pmatrix} x \\ y \\ z \end{pmatrix} = \begin{pmatrix} 4 \\ 12/5 \\ 46/19 \end{pmatrix}$$

which gives $z = \frac{46}{327}$,

$$\frac{19}{5}y - \frac{32}{5}z = \frac{12}{5} \quad \text{or} \quad 19y - 32 \times \frac{46}{327} = 12$$

or

$$y = \frac{4(3 \times 327 + 8 \times 46)}{19 \times 327} = \frac{4 \times 1349}{19 \times 327} = \frac{4 \times 71}{327} = \frac{284}{327}$$

and

$$5x - 2y + z = 4 \quad \text{or} \quad 5x - \frac{568}{327} + \frac{46}{327} = 4 \quad \text{or}$$

$$5x = 4 + \frac{522}{327} = \frac{1308 + 522}{327} = \frac{1830}{327} = \frac{610}{109}$$

which gives $x = \frac{122}{109}$.

Hence $x = \frac{122}{109}$, $y = \frac{284}{327}$ and $z = \frac{46}{327}$ is the solution.

Also

$$U^{-1} = \frac{1}{327}\begin{pmatrix} \dfrac{327}{5} & \dfrac{654}{19} & 9 \\ 0 & \dfrac{1635}{19} & 32 \\ 0 & 0 & 19 \end{pmatrix}, \quad L^{-1} = \begin{pmatrix} 1 & 0 & 0 \\ -\dfrac{7}{5} & 1 & 0 \\ \dfrac{46}{19} & -\dfrac{41}{19} & 1 \end{pmatrix}.$$

Then

$$A^{-1} = \frac{1}{327}\begin{pmatrix} 39 & 15 & 9 \\ -43 & 17 & 32 \\ 46 & -41 & 19 \end{pmatrix}.$$

Example 13.15. Factorize the matrix $A = \begin{pmatrix} 1 & -1 & 1 \\ 1 & -2 & 4 \\ 1 & 2 & 2 \end{pmatrix}$ into the form LU, where L is unit lower triangular and U is upper triangular matrix.

Solution. Let

$$L = \begin{pmatrix} 1 & 0 & 0 \\ l_{21} & 1 & 0 \\ l_{31} & l_{32} & 1 \end{pmatrix}, \quad U = \begin{pmatrix} u_{11} & u_{12} & u_{13} \\ 0 & u_{22} & u_{23} \\ 0 & 0 & u_{33} \end{pmatrix}$$

so that

$$\begin{pmatrix} 1 & 0 & 0 \\ l_{21} & 1 & 0 \\ l_{31} & l_{32} & 1 \end{pmatrix} \begin{pmatrix} u_{11} & u_{12} & u_{13} \\ 0 & u_{22} & u_{23} \\ 0 & 0 & u_{33} \end{pmatrix} = \begin{pmatrix} 1 & -1 & 1 \\ 1 & -2 & 4 \\ 1 & 2 & 2 \end{pmatrix}.$$

Comparing corresponding entries for the matrices on the two sides, we get $u_{11} = 1$, $u_{12} = -1$, $u_{13} = 1$; $l_{21}u_{11} = 1$, $l_{31}u_{11} = 1$ which give $l_{21} = 1$, $l_{31} = 1$, $l_{21}u_{12} + u_{22} = -2$, $l_{31}u_{12} + l_{32}u_{22} = 2$ which give $u_{22} = -2 - l_{21}u_{12} = -1$; $1 \cdot (-1) + l_{32}(-1) = 2$ or $l_{32} = -3$, $l_{21}u_{13} + u_{23} = 4$, $l_{31}u_{13} + l_{32}u_{23} + u_{33} = 2$ which give $u_{23} = 4 - 1 = 3$; $u_{33} = 2 - 1 \cdot 1 - (-3) \cdot 3 = 10$.

Hence

$$L = \begin{pmatrix} 1 & 0 & 0 \\ 1 & 1 & 0 \\ 1 & -3 & 1 \end{pmatrix}, \quad U = \begin{pmatrix} 1 & -1 & 1 \\ 0 & -1 & 3 \\ 0 & 0 & 10 \end{pmatrix}.$$

Observe that both L and U are non-singular and, therefore, A is non-singular. Also

$$L^{-1} = \begin{pmatrix} 1 & 0 & 0 \\ -1 & 1 & 0 \\ -4 & 3 & 1 \end{pmatrix} \quad \text{and} \quad U^{-1} = -\frac{1}{10}\begin{pmatrix} -10 & 10 & -2 \\ 0 & 10 & -3 \\ 0 & 0 & -1 \end{pmatrix}$$

and so

$$A^{-1} = U^{-1}L^{-1} = -\frac{1}{10}\begin{pmatrix} -10 & 10 & -2 \\ 0 & 10 & -3 \\ 0 & 0 & -1 \end{pmatrix}\begin{pmatrix} 1 & 0 & 0 \\ -1 & 1 & 0 \\ -4 & 3 & 1 \end{pmatrix}$$

$$= -\frac{1}{10}\begin{pmatrix} -12 & 4 & -2 \\ 2 & 1 & -3 \\ 4 & -3 & -1 \end{pmatrix}.$$

13.2.6. *Some additional examples*

Example 13.16. Test for solvability and find a solution, if it exists, to the following equations.

$$x + 2y - 3z + t = 1, \quad x + y + z + t = 0.$$

Solution. The given system of equations is $AX = B$, where

$$A = \begin{pmatrix} 1 & 2 & -3 & 1 \\ 1 & 1 & 1 & 1 \end{pmatrix}, \quad X = \begin{pmatrix} x \\ y \\ z \\ t \end{pmatrix}, \quad B = \begin{pmatrix} 1 \\ 0 \end{pmatrix}.$$

We reduce the augmented matrix $M = (A, B)$ to row echlon form.

$$M = \begin{pmatrix} 1 & 2 & -3 & 1 & 1 \\ 1 & 1 & 1 & 1 & 0 \end{pmatrix} \overset{R_2 - R_1}{\sim} \begin{pmatrix} 1 & 2 & -3 & 1 & 1 \\ 0 & -1 & 4 & 0 & -1 \end{pmatrix}$$

$$\sim \begin{pmatrix} 1 & 2 & -3 & 1 & 1 \\ 0 & 1 & -4 & 0 & 1 \end{pmatrix} \overset{R_1 - 2R_2}{\sim} \begin{pmatrix} 1 & 0 & 5 & 1 & -1 \\ 0 & 1 & -4 & 0 & 1 \end{pmatrix}$$

which is row canonical form of M. Since there is no row of the form $(0\ 0\ 0\ 0\ a)$, $a \neq 0$, the system has a solution. The given system then reduces to $x + 5z + t = -1$, $y - 4z = 1$. For every value of z, we get a value of y and also a value of $x + t$. For any z, $y = 4z + 1$ and $x + t = -1 - 5z$. Then giving any value to t, we get a corresponding value of x. Thus, for every choice of values of z and t, we get a solution. For example, if we take $z = t = 1$, then $y = 5$, $x = -7$ while if $z = t = 0$, $y = 1$, $x = -1$. As a general solution of the system, we get

$$x = -1 - 5z - t, \quad y = 4z + 1.$$

Example 13.17. Test the following equations for solvability and find a solution, if it exists

$$-3x + y + 4z = 1$$
$$x + y + z = 0$$
$$-x + z = -1$$
$$x + y - 2z = 0.$$

Solution. In matrix form, the given system of equations is $AX = B$, where

$$A = \begin{pmatrix} -3 & 1 & 4 \\ 1 & 1 & 1 \\ -1 & 0 & 1 \\ 1 & 1 & -2 \end{pmatrix}, \quad X = \begin{pmatrix} x \\ y \\ z \end{pmatrix}, \quad B = \begin{pmatrix} 1 \\ 0 \\ -1 \\ 0 \end{pmatrix}.$$

We reduce the augmented matrix $M = (A, B)$ to row canonical form.

$$M = \begin{pmatrix} -3 & 1 & 4 & 1 \\ 1 & 1 & 1 & 0 \\ -1 & 0 & 1 & -1 \\ 1 & 1 & -2 & 0 \end{pmatrix} \underset{\sim}{R_1 \leftrightarrow R_2} \begin{pmatrix} 1 & 1 & 1 & 0 \\ -3 & 1 & 4 & 1 \\ -1 & 0 & 1 & -1 \\ 1 & 1 & -2 & 0 \end{pmatrix}$$

$$\underset{\underset{\sim}{R_4 - R_1}}{\overset{R_2 + 3R_1}{R_3 + R_1}} \begin{pmatrix} 1 & 1 & 1 & 0 \\ 0 & 4 & 7 & 1 \\ 0 & 1 & 2 & -1 \\ 0 & 0 & -3 & 0 \end{pmatrix} \underset{\sim}{R_2 \leftrightarrow R_3} \begin{pmatrix} 1 & 1 & 1 & 0 \\ 0 & 1 & 2 & -1 \\ 0 & 4 & 7 & 1 \\ 0 & 0 & -3 & 0 \end{pmatrix}$$

$$\underset{\sim}{R_3 - 4R_1} \begin{pmatrix} 1 & 1 & 1 & 0 \\ 0 & 1 & 2 & -1 \\ 0 & 0 & -1 & 5 \\ 0 & 0 & -3 & 0 \end{pmatrix} \underset{\sim}{R_4 - 3R_3} \begin{pmatrix} 1 & 1 & 1 & 0 \\ 0 & 1 & 2 & -1 \\ 0 & 0 & -1 & 5 \\ 0 & 0 & 0 & -15 \end{pmatrix}$$

which is the row echlon form of M. Since this matrix contains a row $(0\ 0\ 0 -15)$, the system has no solution or the system of equations is inconsistent.

Example 13.18. For what values of α does the following system of equations have a solution?

$$3x - y + \alpha z = 1$$
$$3x - y + z = 5.$$

Solution. The given system of equations in the matrix form is $AX = B$, where

$$A = \begin{pmatrix} 3 & -1 & \alpha \\ 3 & -1 & 1 \end{pmatrix}, \quad B = \begin{pmatrix} 1 \\ 5 \end{pmatrix}, \quad X = \begin{pmatrix} x \\ y \\ z \end{pmatrix}.$$

We reduce the augmented matrix $M = (A, B)$ to row echlon form.

$$M = \begin{pmatrix} 3 & -1 & \alpha & 1 \\ 3 & -1 & 1 & 5 \end{pmatrix} \underset{\sim}{R_2 - R_1} \begin{pmatrix} 3 & -1 & \alpha & 1 \\ 0 & 0 & 1 - \alpha & 4 \end{pmatrix}.$$

If $1 - \alpha = 0$, then $M \sim \begin{pmatrix} 3 & -1 & \alpha & 1 \\ 0 & 0 & 0 & 4 \end{pmatrix}$ and the system has no solution. If $1 - \alpha \neq 0$, then

$$M \sim \begin{pmatrix} 3 & -1 & \alpha & 1 \\ 0 & 0 & 1-\alpha & 4 \end{pmatrix} \overset{R_2/1-\alpha}{\sim} \begin{pmatrix} 3 & -1 & \alpha & 1 \\ 0 & 0 & 1 & 4/1-\alpha \end{pmatrix}$$

$$\sim \begin{pmatrix} 3 & -1 & 0 & (1-5\alpha)/1-\alpha \\ 0 & 0 & 1 & 4/1-\alpha \end{pmatrix}$$

and the system reduces to $3x - y = \frac{1-5\alpha}{1-\alpha}$, $z = \frac{4}{1-\alpha}$ and for every value of $\alpha \neq 1$, the system is consistent.

Example 13.19. Test for solvability the following system of equations, and if solvable, find a solution

$$x + 4y + 3z = 1$$
$$3x + z = 1$$
$$4x + y + 2z = 1.$$

Solution. The given system in matrix form is $AX = B$, where

$$A = \begin{pmatrix} 1 & 4 & 3 \\ 3 & 0 & 1 \\ 4 & 1 & 2 \end{pmatrix}, \quad X = \begin{pmatrix} x \\ y \\ z \end{pmatrix} \quad \text{and} \quad B = \begin{pmatrix} 1 \\ 1 \\ 1 \end{pmatrix}.$$

We reduce the augmented matrix $M = (A, B)$ to row echlon form.

$$M = \begin{pmatrix} 1 & 4 & 3 & 1 \\ 3 & 0 & 1 & 1 \\ 4 & 1 & 2 & 1 \end{pmatrix} \overset{R_2 - 3R_1}{\underset{R_3 - 4R_1}{\sim}} \begin{pmatrix} 1 & 4 & 3 & 1 \\ 0 & -12 & -8 & -2 \\ 0 & -15 & -10 & -3 \end{pmatrix}$$

$$\overset{-\frac{1}{2}R_2}{\sim} \begin{pmatrix} 1 & 4 & 3 & 1 \\ 0 & 6 & 4 & 1 \\ 0 & -15 & -10 & -3 \end{pmatrix} \overset{R_3 + \frac{5}{2}R_2}{\sim} \begin{pmatrix} 1 & 4 & 3 & 1 \\ 0 & 6 & 4 & 1 \\ 0 & 0 & 0 & -1/2 \end{pmatrix}.$$

Since there is a row $(0\ 0\ 0\ -1/2)$ in this matrix, the given system of equations has no solution.

13.3. Eigen Values

All matrices considered in this section will be square matrices. Let A be a square matrix of order n over the field R of real numbers and I be the identity matrix of order n. Then $\det(A - xI)$ is a polynomial of degree n in the variable x. This polynomial is called the **characteristic polynomial** of A. The equation $\det(A - xI) = 0$ is called the **characteristic equation** of A. This equation being of degree n, it has exactly n roots in the field C of complex numbers. Each one of these roots is called a **characteristic root** or **eigen value**, or **latent root** of the matrix. The matrix A thus has n characteristic roots or eigen values not necessarily all distinct.

Example 13.20.

(a) If $A = \begin{pmatrix} 2 & 3 \\ -1 & 4 \end{pmatrix}$, then its characteristic polynomial is

$$\det\left(\begin{pmatrix} 2 & 3 \\ -1 & 4 \end{pmatrix} - x\begin{pmatrix} 1 & 0 \\ 0 & 1 \end{pmatrix}\right)$$

$$= \begin{vmatrix} 2 - x & 3 \\ -1 & 4 - x \end{vmatrix}$$

$$= (2 - x)(4 - x) + 3 = 8 - 6x + x^2 + 3 = x^2 - 6x + 11.$$

The characteristic equation is $x^2 - 6x + 11 = 0$ the root of which are $3 \pm \sqrt{2}i$. Hence the eigen values of A are $3 \pm \sqrt{2}i$.

(b) If $A = \begin{pmatrix} 1 & 0 & 0 \\ 2 & 3 & 0 \\ -1 & 3 & 4 \end{pmatrix}$, its characteristic equation is

$$\det\begin{pmatrix} 1 - x & 0 & 0 \\ 2 & 3 - x & 0 \\ -1 & 3 & 4 - x \end{pmatrix} = 0$$

or $(1 - x)(3 - x)(4 - x) = 0$. Hence the eigen values of A are 1, 3, 4, which are the diagonal entries of the matrix A.

Recall that a matrix in which all the entries above (below) the principle diagonal are zero is called a lower (upper) triangular matrix. A matrix is called a triangular matrix if it is either upper triangular or lower triangular. Observe that the determinant of a triangular matrix is the product of the entries on the main diagonal. In view of this we have:

The eigen values of a triangular matrix are the diagonal entries of the matrix.

There is an alternative way of defining eigen values. A number λ is called an eigen value of a square matrix A of order n if there exists a non-zero vector X of order n such that $AX = \lambda X$. In order for this equation to make sense, observe that X is a column vector. We can rewrite the above matrix in the form $(A - \lambda I)X = 0$, which is a set of n homogeneous linear equations in the variables $x_1, x_2, \ldots, x_n$, where $X = (x_1, x_2, \ldots, x_n,)^t$. We know that the above system of homogeneous linear equations has a non-zero solution if and only if $\det(A - \lambda I) = 0$, i.e. if and only if λ is an eigen value of A in the sense defined earlier. The non-zero vector X satisfying $AX = \lambda X$ is called the **characteristic vector** corresponding to the characteristic root λ. Observe that corresponding to an eigen value λ there is exactly one eigen vector, up to a non-zero constant multiple, X satisfying $AX = \lambda X$.

Before studying characteristic roots and vectors in detail, we prove a celebrated theorem known as Cayley–Hamilton theorem.

Theorem 13.4. Every square matrix satisfies its own characteristic equation.

Proof. Let A be a square matrix of order n and let $\det(A - xI) = a_0 + a_1 x + \cdots + a_n x^n = 0$ be its characteristic equation. Then we have to prove that $a_0 I + a_1 A + \cdots + a_n A^n = 0$. Observe that the co-factor of every entry of the matrix $A - xI$ is a polynomial of degree $\leq n - 1$ in the variable x. Since the adjoint of $A - xI$ is the matrix obtained from it by replacing the (i, j)th entry by the cofactor of the (j, i)th entry, we can write

$$\operatorname{adj}(A - xI) = B_0 + B_1 x + \cdots + B_{n-1} x^{n-1}$$

where every B_i is a square matrix of order n. Since

$$(A - xI)\operatorname{adj}(A - xI) = \det(A - xI)I,$$

we get

$$(A - xI)(B_0 + B_1 x + \cdots + B_{n-1} x^{n-1}) = (a_0 + a_1 x + \cdots + a_n x^n)I.$$

Comparing the coefficients of like powers of x on the two sides of this relation, we get

$$AB_0 = a_0 I$$
$$AB_1 - B_0 = a_1 I$$
$$AB_2 - B_1 = a_2 I$$
$$\vdots$$
$$AB_{n-1} - B_{n-2} = a_{n-1} I$$
$$-B_{n-1} = a_n I.$$

Pre-multiplying these equations respectively by $I, A, A^2, \ldots, A^n$ and adding, we get $a_0 I + a_1 A + a_2 A^2 + \cdots + a_n A^n = 0$, which is what we were to prove.

One of the advantages of the Cayley–Hamilton theorem is that we obtain the inverse of a non-singular matrix without going into the finding of adj A. As above, let

$$\det(A - xI) = a_0 + a_1 x + a_2 x^2 + \cdots + a_n x^n. \tag{13.2}$$

Taking $x = 0$ on both sides of this relation, we get $\det(A) = a_0$. Therefore, if A is non-singular, then $a_0 \neq 0$. The relation $a_0 I + a_1 A + a_2 A^2 + \cdots + a_n A^n = 0$ shows that $-A(a_1 I + a_2 A + \cdots + a_n A^{n-1}) = a_0 I$ or that

$$-A\left(\frac{a_1}{a_0}I + \frac{a_2}{a_0}A + \cdots + \frac{a_n}{a_0}A^{n-1}\right) = I$$

which gives

$$A^{-1} = -\left(\frac{a_1}{a_0}I + \frac{a_2}{a_0}A + \cdots + \frac{a_n}{a_0}A^{n-1}\right).$$

It is clear that the coefficient a_n of x^n in the characteristic polynomial (13.2) of A is $(-1)^n$. Also the product of all the roots of the polynomial (13.2) is $(-1)^n \frac{a_0}{a_n} = a_0 = \det(A)$. Hence the product of all the characteristic roots of A equals $\det(A)$.

13.3.1. *Eigen values and eigen vectors*

Example 13.21. We considered the matrix

$$A = \begin{pmatrix} 1 & 0 & 0 \\ 2 & 3 & 0 \\ -1 & 3 & 4 \end{pmatrix}$$

and found that its eigen values are 1, 3, 4. We now find eigen vectors corresponding to these eigen values. Let $\begin{pmatrix} x \\ y \\ z \end{pmatrix}$ be an eigen vector corresponding to 1. Then

$$\begin{pmatrix} 1 & 0 & 0 \\ 2 & 3 & 0 \\ -1 & 3 & 4 \end{pmatrix} \begin{pmatrix} x \\ y \\ z \end{pmatrix} = \begin{pmatrix} x \\ y \\ z \end{pmatrix}$$

which gives

$$\left\{ \begin{pmatrix} 1 & 0 & 0 \\ 2 & 3 & 0 \\ -1 & 3 & 4 \end{pmatrix} - I \right\} \begin{pmatrix} x \\ y \\ z \end{pmatrix} = 0 \text{ or } \begin{pmatrix} 0 & 0 & 0 \\ 2 & 2 & 0 \\ -1 & 3 & 3 \end{pmatrix} \begin{pmatrix} x \\ y \\ z \end{pmatrix} = 0.$$

This system is equivalent to $2y+2x = 0$, $-x+3y+3z = 0$ or that $y = -x$, $z = \frac{4}{3}x$. Therefore, eigen vector corresponding to the eigen value 1 is $a\begin{pmatrix} 1 \\ -1 \\ 4/3 \end{pmatrix}$, where a is an arbitrary non-zero constant.

For the eigen value 3, we have $(A - 3I)(x, y, z)' = 0$ or

$$\begin{pmatrix} -2 & 0 & 0 \\ 2 & 0 & 0 \\ -1 & 3 & 1 \end{pmatrix} \begin{pmatrix} x \\ y \\ z \end{pmatrix} = 0$$

which gives $x = 0$ and $-x + 3y + z = 0$ or that $z = -3y$.

Hence the corresponding eigen vector is $b\begin{pmatrix} 0 \\ 1 \\ -3 \end{pmatrix}$, where b is an arbitrary non-zero constant. Next, we consider the eigen value 4. For this, we have

$$\left\{ \begin{pmatrix} 1 & 0 & 0 \\ 2 & 3 & 0 \\ -1 & 3 & 4 \end{pmatrix} - 4I \right\} \begin{pmatrix} x \\ y \\ z \end{pmatrix} = 0 \quad \text{or} \quad \begin{pmatrix} -3 & 0 & 0 \\ 2 & -1 & 0 \\ -1 & 3 & 0 \end{pmatrix} \begin{pmatrix} x \\ y \\ z \end{pmatrix} = 0$$

which gives $x = 0$, $y = 0$. Therefore, a corresponding eigen vector is $c\begin{pmatrix} 0 \\ 0 \\ 1 \end{pmatrix}$, where c is an arbitrary non-zero constant.

Remarks

1. Every eigen value of a non-singular matrix is non-zero.

   Let A be a non-singular matrix and λ be a characteristic root of A. There exists a non-zero vector X such that $AX = \lambda X$. If $\lambda = 0$, we get $AX = 0$, which is a system of homogeneous linear equations and has a non-zero solution if and only if $\det(A) = 0$, which is not the case here. Hence $\lambda \neq 0$.

2. If λ is an eigen value of a non-singular matrix A, it follows trivially from the definition that $\frac{1}{\lambda}$ is a characteristic root of A^{-1}.

3. If P is a non-singular square matrix of order n and A is a square matrix of order n, then

$$\det(P^{-1}AP - xI) = \det(P^{-1}(A - xI)P) = \det(P^{-1})\det(A - xI)\det(P)$$
$$= \det(A - xI).$$

   Thus A and $P^{-1}AP$ have the same characteristic equation and, therefore, the same characteristic roots.

4. If follows immediately from the above observation that if A, B are square matrices of equal orders and at least one of these, say A, is non-singular,

then both AB, BA have the same characteristic roots. This is so because $BA = (A^{-1}A)BA = A^{-1}(AB)A$.

This result is, in fact, true even when both A and B are singular, but the proof follows on slightly different lines.

5. We have seen that the product of all the characteristic roots of a matrix A equals $\det(A)$. Therefore, if A is a singular matrix, then $\det(A) = 0$ and the characteristic equation has at least one root zero. Thus, at least one characteristic root of a singular matrix is zero.

6. Every square matrix A of odd order with real entries has at least one real characteristic root. This happens because the characteristic equation of A is then of odd degree (which equals the order of A) with real coefficients and every equation of odd degree with real coefficients has at least one real root. We now consider some examples.

Example 13.22. Find the eigen values and the corresponding eigen vectors of the following matrices:

$$\text{(a)} \begin{pmatrix} 3 & -5 & -4 \\ -5 & -6 & -5 \\ -4 & -5 & 3 \end{pmatrix} ; \quad \text{(b)} \begin{pmatrix} 5 & 4 & -4 \\ 4 & 5 & -4 \\ -1 & -1 & 2 \end{pmatrix}.$$

Solution. (a) The characteristic equation is

$$0 = \begin{vmatrix} 3-x & -5 & -4 \\ -5 & -6-x & -5 \\ -4 & -5 & 3-x \end{vmatrix}$$

$$= (3-x)\{-(6+x)(3-x) - 25\} + 5\{-5(3-x) - 20\}$$

$$\quad - 4\{25 - 4(6+x)\}$$

$$= (3-x)\{-18 + 3x + x^2 - 25\} + 5(5x - 35) - 4(1 - 4x)$$

$$= -129 + 52x - x^3 + 25x - 175 - 4 + 16x$$

$$= -x^3 + 93x - 308$$

$$= -x^3 + 16x + 77x - 308$$

$$= -x(x^2 - 16) + 77(x - 4)$$

$$= (x - 4)\{-x^2 - 4x + 77\}.$$

Therefore, the characteristic roots are: $4, \dfrac{-4 \pm \sqrt{16 + 4 \times 77}}{2}$ or $4, -11, 7$.

For the eigen value 4,

$$\begin{pmatrix} -1 & -5 & -4 \\ -5 & -10 & -5 \\ -4 & -5 & -1 \end{pmatrix} \begin{pmatrix} x \\ y \\ z \end{pmatrix} = 0 \quad \text{or} \quad \begin{pmatrix} 1 & 5 & 4 \\ 1 & 2 & 1 \\ 4 & 5 & 1 \end{pmatrix} \begin{pmatrix} x \\ y \\ z \end{pmatrix} = 0$$

or

$$\begin{pmatrix} 1 & 5 & 4 \\ 0 & -3 & -3 \\ 0 & -15 & -15 \end{pmatrix} \begin{pmatrix} x \\ y \\ z \end{pmatrix} = 0$$

which gives $y + z = 0$ and $x + 5y + 4z = 0$.

Therefore, the corresponding eigen vector is $a\begin{pmatrix} -1 \\ 1 \\ -1 \end{pmatrix}$, where a is a non-zero constant.

For the eigen value -11,

$$\begin{pmatrix} 14 & -5 & -4 \\ -5 & 5 & -5 \\ -4 & -5 & 14 \end{pmatrix} \begin{pmatrix} x \\ y \\ z \end{pmatrix} = 0 \quad \text{or} \quad \begin{pmatrix} 9 & 0 & -9 \\ -1 & 1 & -1 \\ -9 & 0 & 9 \end{pmatrix} \begin{pmatrix} x \\ y \\ z \end{pmatrix} = 0$$

which gives $z = x$ and $y = 2x$. Thus the corresponding eigen vector is $b\begin{pmatrix} 1 \\ 2 \\ 1 \end{pmatrix}$, where b is a non-zero constant.

Finally, for the eigen value 7, we have

$$\begin{pmatrix} -4 & -5 & -4 \\ -5 & -13 & -5 \\ -4 & -5 & -4 \end{pmatrix} \begin{pmatrix} x \\ y \\ z \end{pmatrix} = 0 \quad \text{or} \quad \begin{pmatrix} 4 & 5 & 4 \\ 5 & 13 & 5 \\ 0 & 0 & 0 \end{pmatrix} \begin{pmatrix} x \\ y \\ z \end{pmatrix} = 0$$

or

$$\begin{pmatrix} 4 & 5 & 4 \\ 0 & 27/4 & 0 \\ 0 & 0 & 0 \end{pmatrix} \begin{pmatrix} x \\ y \\ z \end{pmatrix} = 0$$

which gives $y = 0$ and $x + z = 0$. Hence a corresponding characteristic vector is $c\begin{pmatrix} 1 \\ 0 \\ -1 \end{pmatrix}$.

(b) The characteristic equation is

$$0 = \begin{vmatrix} 5 - x & 4 & -4 \\ 4 & 5 - x & -4 \\ -1 & -1 & 2 - x \end{vmatrix}$$

$$= (5 - x)\{(5 - x)(2 - x) - 4\} - 4\{4(2 - x) - 4\} - 4\{-4 + 5 - x\}$$

$$= (5-x)(6-7x+x^2) - 4(4-4x) - 4(1-x)$$

$$= 30 - 41x + 12x^2 - x^3 - 20 + 20x$$

$$= -x^3 + 12x^2 - 21x + 10$$

$$= -x^3 + x^2 + 11x^2 - 11x - 10x + 10$$

$$= -x^2(x-1) + 11x(x-1) - 10(x-1)$$

$$= -(x-1)(x^2 - 11x + 10) = -(x-1)^2(x-10).$$

Thus the characteristic roots are 1, 1, 10. For the characteristic root 1,

$$\begin{pmatrix} 4 & 4 & -4 \\ 4 & 4 & -4 \\ -1 & -1 & 1 \end{pmatrix} \begin{pmatrix} x \\ y \\ z \end{pmatrix} = 0 \quad \text{or} \quad \begin{pmatrix} 1 & 1 & -1 \\ 0 & 0 & 0 \\ 1 & 1 & -1 \end{pmatrix} \begin{pmatrix} x \\ y \\ z \end{pmatrix} = 0$$

which gives $x + y - z = 0$. Hence a corresponding characteristic vector is $\begin{pmatrix} a \\ b \\ a+b \end{pmatrix}$. where a, b are constants.

For the characteristic root 10,

$$\begin{pmatrix} -5 & 4 & -4 \\ 4 & -5 & -4 \\ -1 & -1 & -8 \end{pmatrix} \begin{pmatrix} x \\ y \\ z \end{pmatrix} = 0 \quad \text{or} \quad \begin{pmatrix} 0 & 9 & 36 \\ 0 & -9 & -36 \\ 1 & 1 & 8 \end{pmatrix} \begin{pmatrix} x \\ y \\ z \end{pmatrix} = 0$$

which gives $y = -4z$, $x + y + 8z = 0$ or $x = -4z$.

Hence a corresponding characteristic vector is $b\begin{pmatrix} 4 \\ 4 \\ -1 \end{pmatrix}$.

We next consider some typical results.

Example 13.23. Let $A = (a_{ij})$ be a square matrix in which $\sum_i a_{ij} = 1$ for every j. Then A has 1 as a characteristic root.

Proof. The characteristic equation of A is

$$0 = |A - xI| = \begin{vmatrix} a_{11} - x & a_{12} & \cdots & a_{1n} \\ a_{21} & a_{22} - x & \cdots & a_{2n} \\ \vdots & \vdots & \vdots & \vdots \\ a_{n1} & a_{n2} & \cdots & a_{nn} - x \end{vmatrix}$$

$$= \begin{vmatrix} \sum_i a_{i1} - x & \sum_i a_{i2} - x & \cdots & \sum_i a_{in} - x \\ a_{21} & a_{22} - x & \cdots & a_{2n} \\ \vdots & \vdots & \vdots & \vdots \\ a_{n1} & a_{n2} & \cdots & a_{nn} - x \end{vmatrix}$$

$$= \begin{vmatrix} 1 - x & 1 - x & \cdots & 1 - x \\ a_{21} & a_{22} - x & \cdots & a_{2n} \\ \vdots & \vdots & \vdots & \vdots \\ a_{n1} & a_{n2} & \cdots & a_{nn} - x \end{vmatrix}$$

$$= (1 - x) \begin{vmatrix} 1 & 1 & \cdots & 1 \\ a_{21} & a_{22} - x & \cdots & a_{2n} \\ \vdots & \vdots & \vdots & \vdots \\ a_{n1} & a_{n2} & \cdots & a_{nn} - x \end{vmatrix}$$

and $x - 1$ is a factor of the right-hand side and, therefore, 1 is a characteristic root of A.

The following result follows on similar lines.

Example 13.24. If $A = (a_{ij})$ is a square matrix in which for every i, $\sum_j a_{ij} = 1$, then 1 is a characteristic root of A.

Recall that a matrix A is called symmetric if $A' = A$ (A' is the transpose of A, which is the matrix obtained from A by interchanging its rows and columns) and is called **skew-symmetric** if $A' = -A$. A matrix A with complex coefficients is called **Hermitian** if $A^\Theta = (\overline{A})^t = \overline{A'} = A$ and it is called **skew-Hermitian** if $A^\Theta = -A$.

Theorem 13.5. The characteristic roots of a Hermitian matrix are all real.

Proof. Let A be a Hermitian matrix and λ be a characteristic root of A. Recall that A is Hermitian, which means that $A^\Theta = A$. There exists a column vector $X \neq 0$ such that $AX = \lambda X$. Pre-multiplying both sides of this relation with X^Θ, we get

$$X^\Theta A X = X^\Theta \lambda X = \lambda X^\Theta X.$$

If $X = \begin{pmatrix} x_1 \\ x_2 \\ \vdots \\ x_n \end{pmatrix}$, then $X^\Theta = (\overline{x_1}, \overline{x_2}, \ldots, \overline{x_n})$ where $\overline{x_i}$ means the complex conjugate of x_i. Then $X^\Theta X = \sum_{i=1}^{n} \overline{x_i} x_i \neq 0$ as the x_i are not all zero. Also, this is a real number. Since A is Hermitian and $(X^\Theta)^\Theta = X$, we have $(X^\Theta A X)^\Theta = X^\Theta A^\Theta (X^\Theta)^\Theta = X^\Theta A^\Theta X$. Moreover, $X^\Theta A X$ is a 1×1 matrix, i.e. it is a scalar. Therefore, $(X^\Theta A X)^\Theta = \overline{X^\Theta A X}$. Therefore, $\overline{X^\Theta A X} = X^\Theta A X$ so that $X^\Theta A X$ is a real number and we get $\lambda = \frac{X^\Theta A X}{X^\Theta X}$ which is a real number.

As a real symmetric matrix is Hermitian, it follows that every characteristic root of a real symmetric matrix is real.

Next, suppose that A is a skew-Hermitian matrix. Let λ be a characteristic root of A. Let X be a non-zero vector such that $AX = \lambda X$. Then $(iA)X = (i\lambda)X$, which shows that $i\lambda$ is a characteristic root of iA. But A being skew-Hermitian, iA is Hermitian and, therefore, $i\lambda$ is real. Hence λ is pure imaginary or $\lambda = 0$. As a consequence of this we also get:

A characteristic root of a real skew-symmetric matrix is either 0 or pure imaginary.

Definition. A square matrix A is **unitary** if $A^\Theta A = I$. Also, a matrix A is called **orthogonal** if A is real and $A'A = I$. Thus every orthogonal matrix is unitary.

Theorem 13.6. The modulus of each characteristic root of a unitary matrix is 1.

Proof. Let A be a unitary matrix and λ be a characteristic root of A. Then there exists a vector $X \neq 0$ such that $AX = \lambda X$. Taking Hermitian transpose of both sides, we get $X^\Theta A^\Theta = \overline{\lambda} X^\Theta$. Then, we get

$$X^\Theta A^\Theta A X = \lambda \overline{\lambda} X^\Theta X \quad \text{or} \quad X^\Theta I X = \lambda \overline{\lambda} X^\Theta X \quad \text{or} \quad X^\Theta X = \lambda \overline{\lambda} X^\Theta X$$

which implies $(1 - \lambda \overline{\lambda}) X^\Theta X = 0$.

As already seen in one of the results, $X^\Theta X \neq 0$ is a real number and, therefore, $1 - \lambda \overline{\lambda} = 0$ or $\lambda \overline{\lambda} = 1$ which implies that $|\lambda| = 1$.

Corollary. The modulus of each characteristic root of an orthogonal matrix is 1. Thus every eigen value of an orthogonal or a unitary matrix is non-zero.

Exercises 13.1.

1. Find the rank of each one of the following matrices.

(a) $\begin{pmatrix} 1 & 2 & -3 \\ 4 & -5 & 8 \\ 3 & 2 & 1 \end{pmatrix}$, (b) $\begin{pmatrix} 1 & -4 & 5 \\ 3 & 7 & -1 \\ 1 & -15 & -11 \end{pmatrix}$, (c) $\begin{pmatrix} 3 & -1 & 2 \\ 2 & 3 & -1 \\ 1 & -2 & 1 \end{pmatrix}$,

(d) $\begin{pmatrix} 1 & 2 & 3 \\ 2 & 4 & 5 \end{pmatrix}$, (e) $\begin{pmatrix} 1 & -2 & 3 \\ 2 & -4 & 6 \end{pmatrix}$, (f) $\begin{pmatrix} 2 & 1 & 10 & 1 \\ 3 & 3 & 18 & 2 \\ 1 & 9 & 16 & 4 \end{pmatrix}$,

(g) $\begin{pmatrix} 5 & 0 & 1 \\ 1 & -2 & 0 \\ 0 & 5 & -1 \end{pmatrix}$, (h) $\begin{pmatrix} -1 & 0 & 1 \\ 0 & -8 & 0 \\ 1 & 0 & -1 \end{pmatrix}$, (i) $\begin{pmatrix} 2 & -1 & 0 \\ -1 & 2 & -1 \\ 0 & -1 & 2 \\ 0 & 0 & -1 \end{pmatrix}$.

2. If A is a non-singular matrix and λ is a characteristic root of A, prove that $\frac{1}{\lambda}$ is a characteristic root of A^{-1}.

3. If λ is a characteristic root of an orthogonal matrix A, prove that $\frac{1}{\lambda}$ is also a characteristic root of A.

4. A matrix A is called an idempotent matrix if $A^2 = A$. If λ is a characteristic root of an idempotent matrix A, prove that $\lambda = 0$ or 1.

5. If λ is a characteristic root of a matrix A and $A^3 = I$, prove $\lambda = 0, 1$ or -1.

6. Show that the eigen values of a symmetric matrix are all real.

7. Two vectors X and Y of the same order are said to be orthogonal if $X'Y = 0$. Prove that the eigen vectors associated with two distinct eigen values of a symmetric matrix are orthogonal.

8. Using only 0s and 1s,

   (a) list all possible square matrices of order 2 in echlon form;
   (b) find the number of all possible square matrices of order 3 in row canonical form.

9. Find the characteristic roots and the corresponding characteristic vectors of the following matrices.

(a) $\begin{pmatrix} 0 & 1 & 0 \\ 1 & 0 & 1 \\ 0 & 1 & 0 \end{pmatrix}$, (b) $\begin{pmatrix} 0 & 1 & 0 \\ 0 & 0 & 1 \\ 0 & 0 & 0 \end{pmatrix}$, (c) $\begin{pmatrix} 0 & 1 & 0 & 0 \\ 0 & 0 & 1 & 0 \\ 0 & 0 & 0 & 1 \\ 0 & 1 & 2 & 0 \end{pmatrix}$,

(d) $\begin{pmatrix} 2 & 1 & 0 & 0 \\ 0 & 2 & 1 & 0 \\ 0 & 0 & 2 & 1 \\ 0 & 0 & 0 & 3 \end{pmatrix}$, (e) $\begin{pmatrix} 8 & -6 & 2 \\ -6 & 7 & -4 \\ -2 & 4 & -3 \end{pmatrix}$, (f) $\begin{pmatrix} 6 & 2 & -2 \\ -2 & 1 & 3 \\ 2 & 3 & -1 \end{pmatrix}$,

(g) $\begin{pmatrix} 6 & 20 & 10 \\ -4 & -6 & -8 \\ 6 & 10 & 14 \end{pmatrix}$, (h) $\begin{pmatrix} 2 & 1 & 0 \\ 0 & 2 & 1 \\ 0 & 0 & 2 \end{pmatrix}$, (i) $\begin{pmatrix} 2 & 0 & 1 \\ 1 & 2 & 0 \\ 0 & 1 & 1 \end{pmatrix}$.

10. Solve the following systems of linear equations.

(a) $x + y - 2z + 4t = 5$
$\quad 2x + 2y - 3z + t = 4$
$\quad 3x + 3y - 4z - 4t = 3.$

(b) $x - 2y + 4z = 2$
$\quad 2x - 3y + 5z = 3$
$\quad 3x - 4y + 6z = 7.$

11. Find the inverse of each of the following matrices by using the Gaussian elimination method.

$$\begin{pmatrix} 2 & 4 & 3 \\ 0 & 1 & 1 \\ 2 & 2 & -1 \end{pmatrix}, \begin{pmatrix} 1 & 6 & 4 \\ 0 & 2 & 3 \\ 0 & 1 & 2 \end{pmatrix}, \begin{pmatrix} 5 & -2 & 1 \\ 7 & 1 & -5 \\ 3 & 7 & 4 \end{pmatrix}, \begin{pmatrix} 1 & -1 & 1 \\ 1 & -2 & 4 \\ 1 & 2 & 2 \end{pmatrix}.$$

12. Solve the equations

$$2x + 3y + z = 9$$
$$x + 2y + 3z = 6$$
$$3x + y + 2z = 8$$

by the factorization method.

13. Compute the inverse of the matrix $\begin{pmatrix} 3 & 2 & 4 \\ 2 & 1 & 1 \\ 1 & 3 & 5 \end{pmatrix}$ and use the result to solve the system of equations $AX = B$, where $B = \begin{pmatrix} 7 \\ 7 \\ 2 \end{pmatrix}$.

14. Show that the equations

$$x + 2y - z = 3$$
$$x - y + 2z = 1$$
$$2x - 2y + 3z = 2$$
$$x - y + z = -1$$

are consistent and solve them.

15. Solve the system of equations

$$2x - y = 0$$
$$-x + 2y - z = 0$$
$$-y + 2z - u = 0$$
$$-z + 2u = 1.$$

16. Test for solvability the following systems of equations and find a solution whenever it is possible.

(a) $x + y + z = 8$
$x + y + t = 1$
$x + z + t = 14$
$y + z + t = 14.$

(b) $x + y - z = 3$
$x - 3y + 2z = 1$
$2x - 2y + z = 4.$

(c) $-3x + y + 4z = -5$
$x + y + z = 2$
$-2x + z = -3$
$x + y - 2z = 5.$

(d) $2x + y - z = 0$
$-x + z = 0$
$-x - y - z = 1.$

(e) $2x + y + 3z - t = 1$
$2x + y - 2z + t = 0$
$2x + y - z + 2t = -1.$

(f) $-x + y + t = 0$
$y + z = 1.$

(g) $x + 2y + 4z = 1$
$2x + y + 5z = 0$
$3x - y + 5z = 0.$

(h) $3x + 4y = -1$
$x + y = -1$
$-x + 2y = 0$
$2x + 3y = 0.$

BIBLIOGRAPHY

Abbott, J. C. (1969). *Sets, Lattices and Boolean Algebras*, Allyn and Bacon, Boston, MA. .

Ainger, M. (1997). *Combinatorial Theory*, Springer, New York, NY.

Apostol, T. M. (1989). *Introduction to Analytic Number Theory*, Springer International Student Edition, Narosa Publishing House, New Delhi.

Bell, E. T. (1937). *Men of Mathematics 2*, Penguin Books Australia, Victoria.

Biggs, N. L. (1993). *Algebraic Graph Theory*, 2nd edn., Cambridge University Press, Cambridge.

Birkhoff, G. (1967). *Lattice Theory*, 3rd edn., *Amer. Math. Soc. Collo. Publ.*, Providence, RI.

Birkhoff, G. and Bartee, T. C. (1975). *Modern Applied Algebra*, 2nd edn., Academic Press, New York, NY.

Bollobas, B. (1990). *Graph Theory, An Introductory Course*, 3rd edn. Springer-Verlag, New York, NY.

Bondy, J. A. and Murty, U. S. R. (1976). *Graph Theory with Applications*, North Holland, Amsterdam.

Cameron, P. J. and Vanlint, J. H. (1991). *Designs, Graphs, Codes and their Links*, Cambridge University Press, Cambridge.

Dilworth, R. P. (1950). A Decomposition Theorem for Partially Ordered Sets, *Annals Math.*, **51**, pp. 161–166.

Deo, N. (2007). *Graph Theory*, Prentice Hall of India, New Delhi.

Doerr, A. and Levasseur, K. (2001). *Applied Discrete Structures for Computer Science*, Galgotia Publications, New Delhi.

Dornhoff, L. L. and Hohn, F. E. (1978). *Applied Modern Algebra*, Macmillan, New York, NY.

Fleishuer, H. (1991). *Eulerian Graphs and Related Topics*, Part 1, Vol. 2, North Holland, Amsterdam.

Gill, A. (1976). *Applied Algebra for Computer Sciences*, Prentice Hall, Englewood Cliffs, NJ.

Godsil, L. D. (1993). *Algebraic Combinatorics*, Chapman & Hall, New York, NY.

Hall, M., Jr. (1986). *Combinatorial Theory*, 2nd edn., Wiley, New York, NY.

Halmos, P. R. (1965). *Finite-Dimensional Vector Spaces*, East-West Student Edition, East-Press, New Delhi.

Halmos, P. R. (1972). *Naive Set Theory*, Affiliated East West Press, New Delhi.

Harary, F. (1969). *Graph Theory*, Addison Wesley, Reading, MA.

Herstein, I. N. (1964). *Topics in Algebra*, Blaisdell Publishing, New York, NY.

Hoffman, K. and Kunze, R. (1972). *Linear Algebra*, Prentice Hall India, New Delhi.

Jacobson, N. (1964). *Lectures in Abstract Algebra*, Vol. 1. D. Van Nostrand, Affiliated East West Press, New Delhi.

Jacobson, N. (1984). *Basic Algebra*, Vol. 1, Hindustan Publishing Corporation, New Delhi.

Jensen, T. R. and Toft, B. (1995). *Graph Coloring Problems*, Wiley, New York, NY.

Jungnickel, D. (1999). *Graphs, Networks and Algorithms*, Springer-Verlag, New York, NY.

Kolman, B., Busby, R. C. and Ross, S. C. (2005). *Discrete Mathematical Structures*, Pearson Education, New Delhi.

Levy, H. and Lessman, F. (1961). *Finite Difference Equations*, Macmillan, New York, NY.

Lidl, R. and Niederreiter, H. (1986). *Introduction to Finite Fields and their Applications*, Cambridge University Press, Cambridge.

Lidl, R. and Pilz, G. (1998). *Applied Abstract Algebra*, Springer-Verlag, New York, NY.

Lipschutz, S. and Lipson, M. (1999/2002). *Discrete Mathematics*, Tata McGraw-Hill, New Delhi.

Liu, C. L. (2000). *Elements of Discrete Mathematics*, Tata McGraw-Hill, New Delhi.

Mirsky, L. (1971). A Dual of Dilworth's Decomposition Theorem, *Amer. Math. Monthly*, **78**, pp. 876–877.

Niven, I. and Zuckerman, H. S. (1976). *An Introduction to the Theory of Numbers*, Wiley Eastern Ltd., New Delhi.

Preparata, F. P. and Yeh, R. T. (1973). *Introduction to Discrete Structures*, Addison Wesley, Reading, MA.

Rosen, K. H. (2003/2005). *Discrete Mathematics and its Applications*, Tata McGraw-Hill, New Delhi.

Rutherford, D. E. (1965). *Introduction to Lattice Theory*, Oliver & Boyd, Edinburgh.

Sastry, S. S. (1994). *Introductory Methods of Numerical Analysis*, Prentice Hall, New Delhi.

Stoll, R. R. (1963). *Set Theory and Logic*, W. H. Freedman and Co., San Francisco, CA.

Tremblay, J. P. and Manohar, R. (1997). *Discrete Mathematical Structures with Applications to Computer Science*, Tata McGraw-Hill, New Delhi.

Veerarajan, T. (2007). *Discrete Mathematics with Graph Theory and Combinations*, Tata McGraw-Hill, New Delhi.

Vermani, L. R. (1996). *Elements of Algebraic Coding Theory*, Chapman & Hall, London.

INDEX

615